Table of Contents

CHAPTER 1

Solving Simple Equations **2**

1.1 Relating Words and Mathematical Symbols 4

1.2 Numerical Expressions and Algebraic Expressions 8

1.3 Equations and Logical Reasoning 14

1.4 Solving Simple Equations Using Addition 20

1.5 Solving Simple Equations Using Subtraction 22

1.6 Solving Simple Equations by Choosing Addition or Subtraction 24

1.7 Solving Simple Equations Using Multiplication 28

1.8 Solving Simple Equations Using Division 30

1.9 Solving Simple Equations by Choosing Multiplication or Division 32

Chapter 1 Assessment 36

CHAPTER 2

Operations With Real Numbers **38**

2.1 Real Numbers 40

2.2 Adding Real Numbers 46

2.3 Subtracting Real Numbers 50

2.4 Multiplying and Dividing Real Numbers 56

2.5 Solving Equations Involving Real Numbers 62

2.6 The Distributive Property and Combining Like Terms 66

2.7 Solving Two-Step Equations 70

2.8 Solving Multistep Equations 76

2.9 Deductive and Indirect Reasoning in Algebra 82

Chapter 2 Assessment 88

CHAPTER 3

Linear Equations in Two Variables 90

3.1 The Coordinate Plane 92

3.2 Functions and Relations 96

3.3 Ratio, Rate, and Direct Variation 102

3.4 Slope and Rate of Change 106

3.5 Graphing Linear Equations in Two Variables 112

3.6 The Point-Slope Form of an Equation 118

3.7 Finding the Equation of a Line Given Two Points 122

3.8 Relating Two Lines in the Plane 126

3.9 Linear Patterns and Inductive Reasoning 132

Chapter 3 Assessment 138

CHAPTER 4

Solving Inequalities in One Variable 140

4.1 Introduction to Inequalities 142

4.2 Solving Inequalities Using Addition or Subtraction 148

4.3 Solving Inequalities Using Multiplication or Division 154

4.4 Solving Simple One-Step Inequalities 160

4.5 Solving Multistep Inequalities 164

4.6 Solving Compound Inequalities 170

4.7 Solving Absolute-Value Equations and Inequalities 176

4.8 Deductive Reasoning With Inequalities 182

Chapter 4 Assessment 184

Solving Systems of Equations and Inequalities **186**

5.1 Solving Systems of Equations by Graphing 188

5.2 Solving Systems of Equations by Substitution........................ 190

5.3 Solving a System of Equations by Elimination 196

5.4 Classifying Systems of Equations.................................. 202

5.5 Graphing Linear Inequalities in Two Variables 206

5.6 Graphing Systems of Linear Inequalities in Two Variables 210

Chapter 5 Assessment.. 216

Operations With Polynomials **218**

6.1 Integer Exponents 220

6.2 The Power Functions $y = kx$, $y = kx^2$, and $y = kx^3$.................. 226

6.3 Polynomials.. 230

6.4 Adding and Subtracting Polynomials............................... 234

6.5 Multiplying and Dividing Monomials 240

6.6 Multiplying Polynomials.. 246

6.7 Dividing Polynomials... 252

Chapter 6 Assessment.. 256

Factoring Polynomials **258**

7.1 Using the Greatest Common Factor to Factor an Expression 260

7.2 Factoring Special Polynomials 266

7.3 Factoring $x^2 + bx + c$ 270

7.4 Factoring $ax^2 + bx + c$ 274

7.5 Solving Polynomial Equations by Factoring 280

7.6 Quadratic Functions and Their Graphs 286

7.7 Analyzing the Graph of a Quadratic Function 290

Chapter 7 Assessment.. 296

CHAPTER 8

Quadratic Functions and Equations **298**

8.1 Square Roots and the Equation $x^2 = k$ 300

8.2 Solving Equations of the Form $ax^2 + c = 0$ 306

8.3 Solving Quadratic Equations by Completing the Square 312

8.4 Solving Quadratic Equations by Using the Quadratic Formula 318

8.5 Solving Quadratic Equations and Logical Reasoning 324

8.6 Quadratic Patterns 328

8.7 Quadratic Functions and Acceleration 334

Chapter 8 Assessment 338

CHAPTER 9

Rational Expressions, Equations, and Functions **340**

9.1 Rational Expressions and Functions 342

9.2 Simplifying Rational Expressions 344

9.3 Multiplying and Dividing Rational Expressions 350

9.4 Adding Rational Expressions 356

9.5 Subtracting Rational Expressions 362

9.6 Simplifying and Using Complex Fractions 368

9.7 Solving Rational Equations 372

9.8 Inverse Variation 378

9.9 Proportions and Deductive Reasoning 384

Chapter 9 Assessment 386

Radical Expressions, Equations, and Functions 388

10.1 Square-Root Expressions and Functions 390

10.2 Solving Square-Root Equations 392

10.3 Numbers of the Form 398

10.4 Operations on Square-Root Expressions 404

10.5 Powers, Roots, and Rational Exponents 410

10.6 Assessing and Justifying Statements About Square Roots 416

Chapter 10 Test.. 422

Glossary 424

Selected Answers 435

Index 497

CHAPTER 1

Solving Simple Equations

▶ What You Already Know

For some years now, you have studied how to add, subtract, multiply, and divide many different kinds of numbers. For example, you have worked with whole numbers, fractions, decimals, and percents.

▶ What You Will Learn

In Chapter 1, you will learn to solve equations by using what you already know about addition, subtraction, multiplication and division.

In Chapter 1, you first learn how to translate verbal phrases into mathematical symbols. You will then reinforce your understanding of the order of operations by simplifying and evaluating various expressions.

Finally, you will examine different kinds of equations, some that are always true and others that are sometimes or never true.

VOCABULARY

Basic Properties of Equality
   reflexive   symmetric
   transitive
additive identity
algebraic expression
counterexample
equation
evaluate
expression
formula
identity
inverse operations
mathematical expression
mathematical proof
multiplicative identity
numerical expression

open sentence
opposite
order of operations
Properties of Real Numbers
   closure   commutative
   associative   identity
   inverse   distributive
Properties of Equality
   addition   subtraction
   multiplication division
reciprocal
replacement set
simplify
solution
Substitution Principle
variable

The diagram below shows how mathematical skills and mathematical reasoning are interrelated with the skills and concepts in Chapter 1. Notice that learning how to solve simple equations is a major focus of Chapter 1.

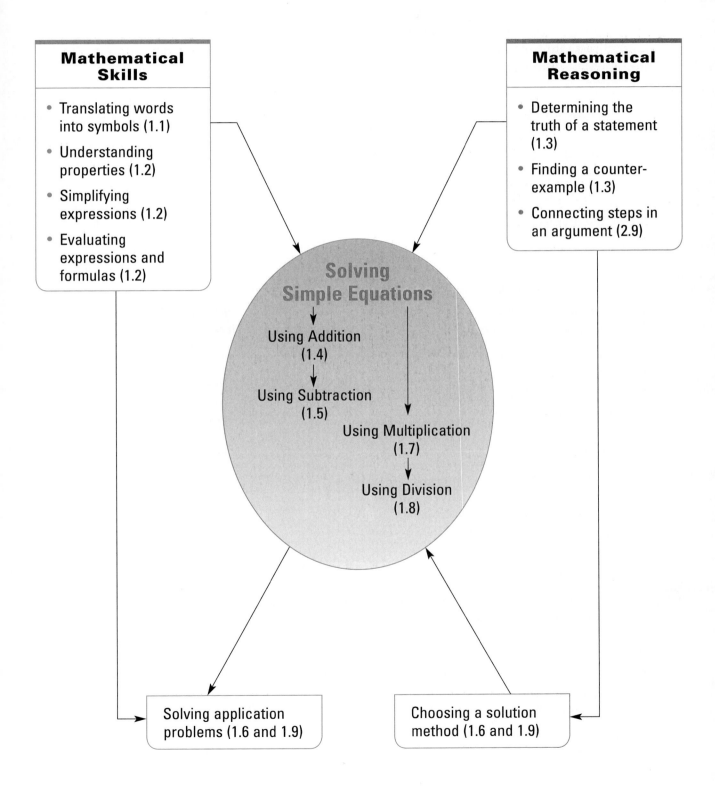

Mathematical Skills

- Translating words into symbols (1.1)
- Understanding properties (1.2)
- Simplifying expressions (1.2)
- Evaluating expressions and formulas (1.2)

Mathematical Reasoning

- Determining the truth of a statement (1.3)
- Finding a counter-example (1.3)
- Connecting steps in an argument (2.9)

Solving Simple Equations

Using Addition (1.4)

Using Subtraction (1.5)

Using Multiplication (1.7)

Using Division (1.8)

Solving application problems (1.6 and 1.9)

Choosing a solution method (1.6 and 1.9)

1.1 Relating Words and Mathematical Symbols

LESSON

You can use operation symbols, such as $+$, $-$, $\times$, or $\div$, to translate words and phrases into *mathematical expressions*. In algebra, letters can be used to represent numbers, such as the a and b in the table below.

Verbal Phrases	Expressions
a plus b, a increased by b, the sum of a and b, b added to a	$a + b$
b subtracted from a, b taken from a, a decreased by b, a less b, difference of a minus b	$a - b$
a multiplied by b, the product of a and b, a times b	$a \times b \quad a \cdot b \quad (a)(b) \quad ab$
a divided by b, the quotient of a and b, a per b	$a \div b \quad \dfrac{a}{b} \quad a/b$

EXAMPLE 1 **Write each phrase as a mathematical expression.**
  a. *the sum of 5 and 4*
  b. *the difference of 5 minus 4*
  c. *the product of 5 and 4*
  d. *the quotient of 5 divided by 4*

▶ **Solution**

Look for key words.

a. the sum of 5 and 4
  $5 + 4$
b. the difference of 5 minus 4
  $5 - 4$
c. the product of 5 and 4
  5×4
d. the quotient of 5 divided by 4
  $5 \div 4$

TRY THIS Write each phrase as a mathematical expression.
  a. the sum of 12 and 3
  b. 12 less 3
  c. 12 multiplied by 3
  d. the quotient of 12 divided by 3

To indicate grouping in mathematics, you can use parentheses ( ).

EXAMPLE 2 **Write the following phrase as a mathematical expression:**
  the difference of the product of 9 and 3 minus the sum of 4 and 2.

▶ **Solution**

Look for key words.

1. *Difference* indicates subtraction. (product of 9 and 3) $-$ (sum of 4 and 2)
2. Write the product and the sum with operation symbols. $9 \times 3 \quad\quad - \quad\quad 4 + 2$
3. Use parentheses to indicate grouping. $(9 \times 3) \quad\quad - \quad\quad (4 + 2)$

TRY THIS Write the following phrase as a mathematical expression:
  the difference of the sum of 4 and 3 minus the product of 10 and 2.

EXERCISES

Write each mathematical expression in words.

1. 6.5×4.5
2. $6.5 - 4.5$
3. $6.5 \div 4.5$
4. $6.5 + 4.5$
5. $12 - (3.5 + 1)$
6. $(3 + 4) - (3 + 1)$

PRACTICE

Write each phrase as a mathematical expression. Do not calculate.

7. the difference of 9 minus 9
8. the quotient of 9 divided by 9
9. the sum of 3 and 3
10. the product of 5 and 6
11. 12 less 4
12. 6 increased by 5
13. 18 divided by 6
14. the quotient of 6 divided by 18
15. 20 less than the sum of 6 and 2
16. 7 less than the sum of 5 and 10
17. the sum of 3 and 3, less the sum of 0 and 7
18. the sum of 7 and 12, decreased by the sum of 3 and 5
19. the sum of the product of 5 and 4 and the product of 3 and 6
20. the difference of 7 times 6 minus 5 times 4
21. the product of 3 and 4, plus the quotient of 20 divided by 4
22. the sum of the quotient of 18 divided by 9 and the quotient of 20 divided by 4

Critical Thinking   Write each phrase in as many ways as you can.

23. $\frac{1}{2}$ of y
24. a divided by $\frac{1}{2}$
25. b added a times
26. b subtracted a times

MIXED REVIEW

Find each sum, difference, product, or quotient. Round your answers to two decimal places. (previous courses)

27. $100 + 1$
28. $100 + 1.1$
29. $100.1 + 1.1$
30. $100.1 + 1.01$
31. $100.01 + 1.01$
32. $100 - 1$
33. $100 - 1.1$
34. $100.1 - 1.1$
35. $100.1 - 1.01$
36. $100.01 - 1.01$
37. 100×1
38. 100×1.1
39. 100.1×1.1
40. 100.1×1.01
41. 100.01×1.01
42. $100 \div 1$
43. $100 \div 1.1$
44. $100.1 \div 1.1$
45. $100.1 \div 1.01$
46. $100.01 \div 1.01$
47. 32×10
48. $32 \times 100,000$
49. $32 \div 10$
50. $32 \div 100,000$

You can begin to solve a real-world problem by writing a statement that contains both words and mathematical symbols. The symbol "=" indicates a statement.

Verbal Statement	Mathematical Statement
a equals b, a is b, a is the same as b	$a = b$

EXAMPLE 1

The price for one CD is $14.95 and the price for one tape is $4.95. Write a statement that gives the total cost of three CDs and five tapes. Do not calculate.

▶ **Solution**

Look for key words.

The total cost equals the cost of three CDs plus the cost of five tapes.
cost of three CDs: $3 \times \$14.95$ cost of five tapes: $5 \times \$4.95$
total cost $= (3 \times \$14.95) + (5 \times \$4.95)$

TRY THIS

The prices for one shirt and one pair of pants are $22.95 and $44.50, respectively. Write a statement that gives the total cost of two shirts and two pairs of pants.

EXAMPLE 2

The perimeter of a rectangle is found by multiplying the sum of its length and width by 2. The length of a rectangle is 5.5 feet and its width is 4.5 feet. Write a statement for the perimeter.

▶ **Solution**

The perimeter equals 2 times the sum of the length and width.
perimeter $= 2 \times (5.5 + 4.5)$
perimeter $= 2(5.5 + 4.5)$

TRY THIS

Write a statement for the perimeter of a rectangle which has a length of 0.02 in. and a width of 0.01 in.

EXAMPLE 3

One way of finding the average of the test scores 87, 90, and 78 is to add the scores and divide this sum by 3. Another way is to divide each score by 3 and then add these quotients. Write a mathematical statement to show that the results of these two methods are the same.

▶ **Solution**

Adding, then dividing   Dividing, then adding
$(87 + 90 + 78) \div 3 = (87 \div 3) + (90 \div 3) + (78 \div 3)$

$$\frac{87 + 90 + 78}{3} = \frac{87}{3} + \frac{90}{3} + \frac{78}{3}$$

TRY THIS

Rework Example 3 for the scores 80, 97, and 75.

EXERCISES

KEY SKILLS

**Write each statement using both words and mathematical symbols.
Do not calculate.**

Sample: the total number of miles a car traveled after 5 trips of 15 miles each
Solution: total miles traveled = 5 trips × 15 miles = 5 × 15

1. the total cost of 12 tickets if each ticket costs $5

2. the total amount of water in 12 glasses that each holds 8 ounces

3. the amount of money in the bank after $35 is withdrawn from $1200

4. the number of cookies each of four students receives when 24 cookies are shared equally

PRACTICE

**Write each statement using words and mathematical symbols as in
Examples 1 and 2. Do not calculate.**

5. the total cost of 7 CDs at $14.95 each plus 3 tapes at $4.50 each

6. the weight of 7 books each weighing 1.3 pounds added to the weight of 5 notebooks each weighing 0.3 pounds

7. the total cost of 7 cans of paint at $20.14 each, less 3 cans of paint costing $14.25 each

8. the perimeter of a rectangle whose length is 126 feet and whose width is 44.5 feet

9. the perimeter of a rectangle whose length and width are both 12.5 feet

Write each sentence as a mathematical statement, as in Example 3.

10. When 5 and 18 are added and the sum is then divided by 2, the result is the same as one half of 5 plus one half of 18.

11. When you add 3 to 5 and then add 6, you get the same result as adding 6 to 3 and then adding 5.

MIXED REVIEW

Solve each problem. (previous courses)

Sample: $60 \times \frac{3}{5}$ Solution: $\frac{\overset{12}{\cancel{60}}}{1} \times \frac{3}{\cancel{5}_1} = 36$

12. $42 \times \frac{1}{3}$

13. $42 \times \frac{1}{6}$

14. $42 \times \frac{2}{3}$

15. $42 \times \frac{5}{6}$

Sample: $33 \div \frac{3}{4}$ Solution: $\frac{\overset{11}{\cancel{33}}}{1} \times \frac{4}{\cancel{3}_1} = 44$

16. $42 \div \frac{1}{3}$

17. $42 \div \frac{1}{6}$

18. $42 \div \frac{2}{3}$

19. $42 \div \frac{6}{7}$

Numerical Expressions and Algebraic Expressions

SKILL A *Identifying and using basic properties of numbers*

In arithmetic, you learned how to add and multiply numbers. The basic properties of real numbers that govern these operations are shown below.

Let a, b, and c represent real numbers.

Property	Addition	Multiplication
Closure	$a + b$ is a real number.	ab is a real number.
Commutative	$a + b = b + a$	$a \cdot b = b \cdot a$
Associative	$a + b + c = a + (b + c)$ $= (a + b) + c$	$a \cdot b \cdot c = a \cdot (b \cdot c)$ $= (a \cdot b) \cdot c$
Identity	Because $a + 0 = 0 + a = a$, 0 is called the **additive identity.**	Because $a \cdot 1 = 1 \cdot a = a$, 1 is called the **multiplicative identity.**
Inverse	For every real number a, there is an **opposite** real number, $-a$, and $a + (-a) = -a + a = 0$.	For every nonzero real number a, there is a **reciprocal**, $\frac{1}{a}$, and $a \cdot \frac{1}{a} = \frac{1}{a} \cdot a = 1$.

The Distributive Property links addition and multiplication.

Distributive Property

If a, b, and c are real numbers, then $a(b + c) = ab + ac$.

EXAMPLE 1 **Which property is illustrated by this statement?**
Adding two numbers and then adding a third number gives the same answer as adding the sum of the second two numbers to the first.

▶ **Solution**

The statement illustrates the Associative Property of Addition, $(a + b) + c = a + (b + c)$.

TRY THIS   In Example 1, which property is illustrated if *sum* is replaced by *product* and *adding* is replaced by *multiplying*?

EXAMPLE 2 **Use properties to simplify** $(1)(10) + 3\left(\dfrac{1}{3}\right)$.

▶ **Solution**

$$(1)(10) + 3\left(\frac{1}{3}\right)$$

Identity Property of Multiplication $\quad 10 \quad + \quad 1 \quad$ *Inverse Property of Multiplication*

$$11$$

TRY THIS   Use properties to simplify $10(0.1) + 100(0.01)$.

EXERCISES

Which property is illustrated by each statement?

1. $2 + 0 = 2$

2. $(1)2.5 = 2.5$

3. $2.4 + 7.1 = 7.1 + 2.4$

4. $1 + (2 + 5) = (1 + 2) + 5$

5. $5\left(\dfrac{1}{5}\right) = 1$

6. $4 \times 10 = 10 \times 4$

PRACTICE

Which property is illustrated by each statement?

7. Multiplying 5 by 9 gives the same answer as multiplying 9 by 5.

8. Adding 5 to 9 gives the same number as adding 9 to 5.

9. Add 0 to any number and you get the same number.

10. Multiply any number by 1 and you get the same number.

11. Multiply any nonzero number by its reciprocal and the result is 1.

Simplify each expression. Use mental math where possible. Identify the property or properties used.

12. $5 \times \dfrac{1}{5}$

13. $\dfrac{5}{3} \times \dfrac{3}{5}$

14. $1 \times \dfrac{1}{7}$

15. $\dfrac{1}{3} \times 1$

16. $0 + \dfrac{5}{2}$

17. $3 + 0 + \dfrac{5}{2}$

18. $3 + \dfrac{5}{2} + 0$

19. $1 \times 3 \times 5$

20. $4 \times 1 \times 7$

21. $\left(\dfrac{5}{2} \times \dfrac{2}{5}\right) + \left(3 \times \dfrac{1}{3}\right)$

22. $\left(1 \times \dfrac{2}{5}\right) + \left(1 \times \dfrac{3}{5}\right)$

23. $(1 \times 6) + \left(\dfrac{5}{3} \times \dfrac{3}{5}\right)$

24. $(1 \times 7) + \left(\dfrac{2}{9} \times \dfrac{9}{2}\right)$

25. $2(4.5 + 0)$

26. $3(12 + 0)$

27. **Critical Thinking** Show a convenient way to calculate a 15% tip on a meal that costs $12.00 by using the Distributive Property. Do not calculate.

MIXED REVIEW

Find the value of each expression. (previous courses)

(Recall that a^2 means $a \times a$.)

28. 2^2

29. 3^2

30. 7^2

31. 5^2

32. 10^2

33. 100^2

34. 1^2

35. 0^2

36. $\left(\dfrac{1}{2}\right)^2$

37. $\left(\dfrac{3}{4}\right)^2$

38. $\left(\dfrac{5}{2}\right)^2$

39. $\left(\dfrac{7}{4}\right)^2$

40. $(0.5)^2$

41. $(0.2)^2$

42. $(0.05)^2$

43. $(0.002)^2$

A **numerical expression** contains operation symbols and numbers. To **simplify** a numerical expression, perform all the indicated operations using the following order of operations.

> ### Order of Operations
> 1. Simplify expressions within grouping symbols.
> 2. Simplify expressions involving *exponents*.
> 3. Multiply and divide from left to right.
> 4. Add and subtract from left to right.

In addition to parentheses ( ), grouping symbols include square brackets [ ] and braces { }. The fraction bar is also a grouping symbol.

EXAMPLE 1 **Simplify** $\dfrac{(12 - 3)^2}{2(4 + 1) - 2}$.

▶ **Solution**

$$\frac{(12 - 3)^2}{2(4 + 1) - 2} = \frac{9^2}{2 \times 5 - 2} \qquad \longleftarrow \text{ Work within parentheses first.}$$
$$12 - 3 = 9 \qquad 4 + 1 = 5$$
$$= \frac{81}{2 \times 5 - 2}$$
$$= \frac{81}{10 - 2} \qquad \longleftarrow \text{ Multiply before subtracting.}$$
$$= \frac{81}{8}, \text{ or } 10\frac{1}{8}$$

TRY THIS Simplify $3 + \dfrac{(2 + 3)^2}{2(2 + 1) + 2}$.

In some expressions, you may need to work with groups within groups. Begin with the innermost grouping symbols and work outward.

EXAMPLE 2 **Simplify** $\left[2(5 - 1)^2 - 3(10 - 8)^2\right]^2$.

▶ **Solution**

$$[2(5 - 1)^2 - 3(10 - 8)^2]^2$$
$$[2 \times (4)^2 - 3 \times (2)^2]^2 \qquad \longleftarrow \text{ Work in the innermost grouping symbols first.}$$
$$(2 \times 16 - 3 \times 4)^2 \qquad \longleftarrow 4^2 = 16 \text{ and } 2^2 = 4$$
$$(32 - 12)^2 \qquad \longleftarrow \text{ Multiply.}$$
$$20^2 \qquad \longleftarrow \text{ Subtract.}$$
$$400 \qquad \longleftarrow 20^2 = 20 \times 20$$

TRY THIS Simplify $4[3(10 - 8)^2 + 4(10 - 9)^2]$.

EXERCISES

Write the expression that results when you perform the operation(s) in parentheses. Do not continue after that step.

1. $4 + 3(10 - 3)$

2. $3(12 + 3) - 5(10 - 1)$

3. $(3 + 1)^2 + 5 \times 4$

4. $3 \times 7 - (5 - 1)^2$

5. $\dfrac{(7 + 2)^2}{3 \times 2}$

6. $\dfrac{(3 + 2)^2}{(10 - 8)^2}$

PRACTICE

Use the order of operations to simplify each expression.

7. $2(3 + 4) + 3 \times 5$

8. $3 \times 5 - 2(3 + 4)$

9. $(3 - 1)(10 + 4)$

10. $(13 - 3)(13 + 3)$

11. $\dfrac{(3 + 2)^2}{2(10 - 8) + 1}$

12. $\dfrac{(7 + 1)^2}{3(7 - 1) + 18}$

13. $\dfrac{3 + (5 + 2)^2}{7(7 - 1) - 16}$

14. $\dfrac{(5 + 2)^2 - 24}{6 - (2 - 1)}$

15. $\dfrac{(5 + 2)^2 - (5 + 1)^2}{16 - 2(5 + 2)}$

16. $\dfrac{(3 - 2)^2 + (1 + 1)^2}{2(5 + 3) - 9}$

17. $\dfrac{(3 - 2)^2 + (1 + 1)^2}{(6 - 2)^2 - (2 + 1)^2}$

18. $\dfrac{(6 - 3)^2 + (3 + 1)^2}{(6 + 4)^2 - (4 + 1)^2}$

19. $\dfrac{4 + 4}{2(1 + 1)} + \dfrac{(3 + 1)^2}{4(3 - 1)}$

20. $\dfrac{20 + 4}{3(3 - 1)} - \dfrac{(3 + 1)^2}{4(3 - 1)}$

21. $[20(2 + 3) - 3(5 + 5)]^2$

22. $[2(2 + 1) + 3(2 + 2)]^2$

23. $3[2(2 + 1)]^2$

24. $7[2(6 - 1)]^2$

25. $[2(6 - 1)]^2 + 3(10 - 7)^2$

26. $30(9 - 6)^2 - [2(6 - 1)]^2$

27. $\dfrac{[(3.2 + 0.8)^2 + (0.5 + 0.5)^2]}{[(6 - 1)^2 - (5 - 1)^2]^2}$

28. $\dfrac{[(7.3 + 2.7)^2 - (5.5 + 0.5)^2]}{[(3 - 1)^2 + (3 - 1)]^2}$

MIXED REVIEW

Find each sum. (previous courses)

29. $\dfrac{1}{2} + \dfrac{1}{2}$

30. $\dfrac{1}{3} + \dfrac{1}{3}$

31. $\dfrac{2}{3} + \dfrac{2}{3}$

32. $\dfrac{3}{4} + \dfrac{3}{4}$

Sample: $\dfrac{1}{4} + \dfrac{1}{2}$ Solution: $\left(\dfrac{2}{2}\right)\dfrac{1}{2} = \dfrac{2}{4} \rightarrow \dfrac{1}{4} + \dfrac{2}{4} = \dfrac{3}{4}$

33. $\dfrac{3}{4} + \dfrac{1}{2}$

34. $\dfrac{1}{6} + \dfrac{1}{3}$

35. $\dfrac{3}{10} + \dfrac{2}{5}$

36. $\dfrac{1}{2} + \dfrac{1}{6}$

Sample: $\dfrac{1}{5} + \dfrac{1}{6}$ Solution: $\left(\dfrac{6}{6}\right)\left(\dfrac{1}{5}\right) + \left(\dfrac{5}{5}\right)\left(\dfrac{1}{6}\right) = \dfrac{6}{30} + \dfrac{5}{30} = \dfrac{11}{30}$

37. $\dfrac{1}{2} + \dfrac{1}{3}$

38. $\dfrac{1}{2} + \dfrac{2}{3}$

39. $\dfrac{1}{3} + \dfrac{3}{4}$

40. $\dfrac{1}{3} + \dfrac{2}{5}$

41. $\dfrac{4}{5} + \dfrac{2}{3}$

42. $\dfrac{2}{5} + \dfrac{3}{4}$

43. $\dfrac{3}{5} + \dfrac{1}{6}$

44. $\dfrac{3}{5} + \dfrac{5}{6}$

A **variable** is a letter that is used to represent numbers. An **algebraic expression** is an expression that contains numbers, variables and operation symbols. To *evaluate* an algebraic expression, substitute numbers for the variables in the expression and then calculate. When you evaluate, you are using the Substitution Principle.

The Substitution Principle

For any numbers a and b, if $a = b$, then a and b may be substituted for each other.

EXAMPLE 1 Evaluate $2(x - 3) + 5$ given $x = 7$.

▶ **Solution**

$2(\boxed{x} - 3) + 5$
$2(\boxed{7} - 3) + 5$ ←——— *Replace x with 7, then simplify.*
$2(4) + 5$
13

TRY THIS Evaluate $3(a + 6) - 9$ given $a = 7$.

An algebraic expression may contain more than one variable.

EXAMPLE 2 Evaluate $\dfrac{3a + 5b + 1}{4a - 11b}$ given $a = 7$ and $b = 2$.

▶ **Solution**

$$\frac{3\,\boxed{a} + 5\,\boxed{b} + 1}{4\,\boxed{a} - 11\,\boxed{b}} \longrightarrow \frac{3\,\boxed{(7)} + 5\,\boxed{(2)} + 1}{4\,\boxed{(7)} - 11\,\boxed{(2)}} \quad \longleftarrow \text{\textit{Replace a with 7 and b with 2.}}$$

$$= \frac{21 + 10 + 1}{28 - 22} \quad \longleftarrow \text{\textit{Multiply.}}$$

$$= \frac{32}{6}, \text{ or } 5\frac{1}{3}$$

TRY THIS Evaluate $\dfrac{4r - 5s + 5}{3r + s}$ given $r = 10$ and $s = 5$.

A **formula** is an equation that shows a mathematical relationship between two or more quantities.

For example, the surface area, S, of a rectangular box is given by the formula $S = 2lw + 2wh + 2lh$. You can find the surface area of the box at right by evaluating the formula.

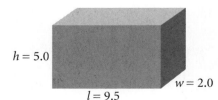

$h = 5.0$
$w = 2.0$
$l = 9.5$

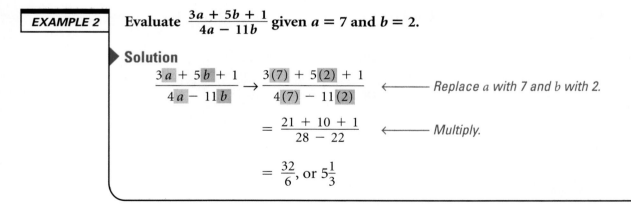

$$\overset{l \quad\; w}{} \quad \overset{w \quad\; h}{} \quad \overset{l \quad\; h}{}$$
$$S = 2(9.5)(2.0) + 2(2.0)(5.0) + 2(9.5)(5.0) = 153 \text{ units}^2$$

EXERCISES

KEY SKILLS

Write the numerical expression that results when you apply the
Substitution Principle. Do not evaluate.

Sample: $x + 3$, given $x = 2$ Solution: $2 + 3$

1. $3x^2 - 5x + 4$, given $x = 5$

2. $3(a + 3) - 2(a - 1)$, given $a = 9$

3. $\frac{3m + 3n}{4n}$, given $m = 3$ and $n = 2$

4. $\frac{7k - 3p}{4p + 1} + \frac{k}{p}$, given $k = 3$ and $p = 2$

PRACTICE

Evaluate each expression.

5. $3d + 4(d - 1)$, given $d = 7$

6. $8s - 4(s + 1)$, given $s = 3$

7. $8b^2 - 4b$, given $b = 5$

8. $2k^2 + 7k$, given $k = 10$

9. $0.5h^2 - 0.25h$, given $h = 4$

10. $0.5h^2 + 0.6h$, given $h = 10$

11. $\frac{1}{2}(a - 3) + \frac{1}{3}$, given $a = 9$

12. $\frac{m + 4}{2} + \frac{m - 2}{5}$, given $m = 6$

13. $\frac{5(n + 3) - 1}{4} - \frac{3(n + 1)}{3}$, given $n = 1$

14. $\frac{5(z - 3) + 1}{4} + \frac{2(z + 1)}{3}$, given $z = 5$

15. $a + 2(1 + b) - 2b$, given $a = 3$ and $b = 1$

16. $7x - 2(1 + y) + 1$, given $x = 5$ and $y = 2$

17. $\frac{7(m - n)}{3(m + n)}$, given $m = 7$ and $n = 1$

18. $\frac{5(2m - 3n)}{4(m - n)}$, given $m = 10$ and $n = 6$

19. $\frac{7(m - n)}{3(m + n)}$, given $m = \frac{3}{4}$ and $n = \frac{1}{4}$

20. $\frac{5(2m - 3m)}{4(m - n)}$, given $m = \frac{1}{2}$ and $n = \frac{1}{3}$

21. The amount in a savings account is calculated using the formula
$A = p(1 + r)t$, where A is the amount, p is the *principal*, r is the annual
interest rate, and t is the time in years. What is the amount in an account
after 1 year if the principal is $1200 and the interest rate is 0.05?

MIXED REVIEW

Write each fraction as a decimal and each decimal as a fraction.
(previous courses)

Sample: $\frac{1}{8}$ Solution: $8\overline{)1.000}$ (0.125)   Sample: 0.125 Solution: $\frac{125}{1000} = \frac{1}{8}$

22. $\frac{1}{2}$

23. $\frac{1}{4}$

24. $\frac{1}{8}$

25. $\frac{1}{20}$

26. $\frac{1}{40}$

27. $\frac{3}{4}$

28. $\frac{4}{5}$

29. $\frac{3}{8}$

30. $\frac{7}{8}$

31. 0.3

32. 0.33

33. 3.3

34. 3.03

35. $\frac{1}{6}$

36. $\frac{5}{6}$

37. $\frac{2}{3}$

38. $0.\overline{6}$

39. 0.66

1.3 LESSON

Equations and Logical Reasoning

An **equation** is a mathematical statement that two expressions are equal (=). An equation with at least one variable is called an **open sentence.** Examples of equations that are open sentences follow.

$$x + 5 = 8 \qquad x = x + 9 \qquad x + 4 = 4 + x$$

A **solution** to an open sentence is any value of the variable that makes the equation true. A solution is said to *satisfy* the equation.

- There is exactly one value of x, 3, that satisfies $x + 5 = 8$.
- No value of x satisfies $x = x + 9$. This equation has no solution.
- The equation $x + 4 = 4 + x$ is true for all values of x because of the Commutative Property of Addition. An equation that is true for all values of the variable is called an **identity.**

A set of possible replacements for a variable is called a **replacement set,** and the set is enclosed with braces { }.

EXAMPLE

Given each replacement set, find the solution to the equation $2x + 1 = 11$.

a. $\{0, 2, 4, 6, 8\}$ **b.** $\{1, 3, 5, 7, 9\}$

▶ **Solution**

a. $\{0, 2, 4, 6, 8\}$ **b.** $\{1, 3, 5, 7, 9\}$

Make a table.

x	$2x + 1$	Solution?
0	$2(0) + 1 = 1$	No, $1 \neq 11$.
2	$2(2) + 1 = 5$	No, $5 \neq 11$.
4	$2(4) + 1 = 9$	No, $9 \neq 11$.
6	$2(6) + 1 = 13$	No, $13 \neq 11$.
8	$2(8) + 1 = 17$	No, $17 \neq 11$.

x	$2x + 1$	Solution?
1	$2(1) + 1 = 3$	No, $3 \neq 11$.
3	$2(3) + 1 = 7$	No, $7 \neq 11$.
5	$2(5) + 1 = 11$	Yes, $11 = 11$. ✔
7	$2(7) + 1 = 15$	No, $15 \neq 11$.
9	$2(9) + 1 = 19$	No, $19 \neq 11$.

$2x + 1 = 11$ does not have a solution in $\{0, 2, 4, 6, 8\}$. $2x + 1 = 11$ has one solution, 5, in $\{1, 3, 5, 7, 9\}$.

TRY THIS

Given each replacement set, find the solution to the equation $3x - 5 = 13$.

a. $\{0, 2, 4, 6, 8\}$ **b.** $\{1, 3, 5, 7, 9\}$

The Example shows that an equation can have a solution in one replacement set but no solution in another replacement set. When no replacement set is specified, assume that the replacement set is all real numbers.

EXERCISES

KEY SKILLS

State whether each item is an equation. Write *yes* or *no*.

1. $2(x + 5) - 3(x + 1)$

2. $3(a + 5) = 4(a - 1) + 2$

3. $4x + 2y = 5$

State whether each item is an open sentence. Write *yes* or *no*.

4. $5x = 75$

5. $5x = 7y$

6. $4(5) = 20$

PRACTICE

Find the values in the replacement set {0, 1, 2, 3, 4, 5, 6} that are solutions
to each equation. If the equation has no solution, write *none*.

7. $x - 1 = 2$

8. $a - 1 = 4$

9. $w + 1 = 8$

10. $t + 1 = 0$

11. $2y = 8$

12. $5c = 10$

13. $2h + 1 = 7$

14. $2z - 1 = 11$

15. $3x - 2 = 7$

16. $4d + 1 = 8$

17. $3(r + 3) = 12$

18. $5(k - 2) = 20$

19. $2(n + 1) + 1 = 10$

20. $3(v - 1) + 2 = 11$

21. $5(u + 1) + 1 = 11$

Is each equation an identity? If not, explain.

22. $x - 1 = x - 1$

23. $2(p - 1) = (p - 1)(2)$

24. $2g + 3 = 9$

25. $2z - 1 = 5$

26. $2s + 8 = 8 + 2s$

27. $3(2w) = 2(3w)$

28. $(2q)(3) = (2)(3)q$

29. $3x - 5 = 13$

30. $7 = 4y - 2$

31. **Critical Thinking** Does the equation $2x + 1 = 6$ have a solution in
the replacement set of:
   a. all even numbers?
   b. all odd numbers?
   c. all mixed numbers?

MIXED REVIEW

Find each product or quotient. (previous courses)

Sample: $\frac{5}{12} \times \frac{3}{4}$ Solution: $\frac{5}{\underset{4}{12}} \times \frac{\overset{1}{3}}{4} = \frac{5}{16}$ Sample: $\frac{3}{5} \div \frac{8}{10}$ Solution: $\frac{3}{5} \times \frac{\overset{2}{10}}{\underset{4}{8}} = \frac{6}{7}$

32. $\frac{1}{3} \times \frac{1}{2}$

33. $\frac{1}{3} \times \frac{1}{5}$

34. $\frac{2}{3} \times \frac{1}{5}$

35. $\frac{2}{3} \times \frac{1}{2}$

36. $\frac{2}{3} \times \frac{3}{5}$

37. $\frac{2}{3} \times \frac{4}{5}$

38. $\frac{3}{7} \times \frac{7}{3}$

39. $\frac{4}{7} \times \frac{7}{4}$

40. $\frac{3}{7} \times \frac{14}{3}$

41. $\frac{3}{7} \times \frac{21}{9}$

42. $\frac{1}{3} \div \frac{1}{2}$

43. $\frac{1}{3} \div \frac{1}{5}$

44. $\frac{2}{3} \div \frac{1}{5}$

45. $\frac{2}{3} \div \frac{1}{2}$

46. $\frac{2}{3} \div \frac{3}{5}$

47. $\frac{2}{3} \div \frac{4}{5}$

48. $\frac{3}{7} \div \frac{7}{3}$

49. $\frac{4}{7} \div \frac{7}{4}$

50. $\frac{3}{7} \div \frac{14}{3}$

51. $\frac{3}{7} \div \frac{21}{9}$

An important part of studying mathematics is learning how to prove that a statement is true or false. If you know that a statement is true, you can use it to prove other statements. If a statement is not *always* true, it is considered to be false and it cannot be used to justify a statement in a proof.

If you can find a case where a statement is not true, that case is called a **counterexample.** One counterexample is all you need to prove that a statement is false.

EXAMPLE 1   **Give a counterexample to show that the statement is false.**

> **The sum of two even numbers is always an odd number.**

▶ **Solution**

Guess and check.

Choose two even numbers and find their sum.
$$2 + 4 = 6 \qquad \longleftarrow \text{6 is not an odd number.}$$
Because $2 + 4 = 6$ is a counterexample, the statement is false.

TRY THIS   Give a counterexample to show that the following statement is false.

> *The sum of two odd numbers is always an odd number.*

When you are searching for counterexamples, you may find some cases where the statement is true. But in mathematics, true examples are not enough to prove that a statement is always true unless you test every possible case.

Knowing that a statement is true in some cases and false in others can be useful. You can look for patterns among the examples and counterexamples to see if a statement is true for certain sets of numbers. You can then make a guess, or **conjecture,** about which numbers make the statement true.

EXAMPLE 2   **Is the following statement always, sometimes, or never true?**

> **The product of two numbers a and b is greater than both a and b.**

Justify your response.

▶ **Solution**

Make an organized list.

Search for examples and counterexamples:

$4 \times 8 = 32$	True ✔	The product, 32, is greater than both 4 and 8.
$4 \times 1 = 4$	False ✘	The product, 4, is not greater than 4.
$4 \times \frac{1}{2} = 2$	False ✘	The product, 2, is not greater than 4.
$4 \times 0 = 0$	False ✘	The product, 0, is not greater than both 4 and 0.

These examples show that the statement is *sometimes true.* It appears that the statement may be true only when both factors are greater than 1.

TRY THIS   Rework Example 2 using the following statement.

> *The quotient when a is divided by b is greater than both a and b.*

EXERCISES

Is each statement true for the value(s) of the variable(s) given?

1. The quantity $\frac{n}{10}$ is a whole number when $n = 120$.

2. The quantity $\frac{n}{10}$ is a whole number when $n = 125$.

3. The sum $a + b + c$ is even when $a = 2$, $b = 4$, and $c = 5$.

PRACTICE

Give a counterexample to show that each statement is false.

4. The quotient of two even numbers is an even number.

5. The product of an even number and an odd number is odd.

6. The quotient of two odd numbers is odd.

For what values of n is each statement true?

7. n divided by 4 is a whole number.

8. $n + 5$ is an odd number.

9. $3n$ is an even number.

10. $3n$ is an odd number.

Determine whether each statement is always, sometimes, or never true. Justify your response.

11. The sum of three whole numbers is odd.

12. The product of two whole numbers is even.

13. If the digits of a number add up to a multiple of 3, the number is divisible by 6.

14. If the digits of a number add up to a multiple of 2, then the number is even.

MIXED REVIEW

Write each fraction as a decimal and then as a percent. Write each percent as a decimal and then as a fraction. (previous courses)

Sample: $\frac{1}{5}$ Solution: $5)\overline{1.00}^{\,0.20}$ $0.20 \rightarrow 20\%$

15. $\frac{3}{5}$ 16. $\frac{3}{10}$ 17. $\frac{3}{8}$ 18. $\frac{4}{25}$ 19. $\frac{9}{4}$ 20. $\frac{12}{5}$

Sample: 4.5%   Solution: $4.5\% = 0.045 = \frac{45}{1000} = \frac{9}{250}$

21. 16%   22. 62%   23. 125%   24. 300%   25. 16.5%   26. 0.1%

In arithmetic, you may use some concepts that seem very obvious. But in algebra, it is important to understand, define, and name these concepts. These include the Properties of Numbers (1.2 Skill A) and the Basic Properties of Equality defined below.

Basic Properties of Equality

Reflexive Property: For all real numbers a, $a = a$.

Symmetric Property: For all real numbers a and b, if $a = b$, then $b = a$.

Transitive Property: For all real numbers a, b, and c, if $a = b$ and $b = c$, then $a = c$.

Properties of Numbers, Properties of Equality, and definitions are often used to write a **mathematical proof**, which is a convincing argument using *logic* to show that a statement is true.

EXAMPLE 1 **Prove that $\frac{1}{2} = 50\%$.**

▶ **Solution**

Statements	Reasons
1. $\frac{1}{2} = \frac{1}{2} \times \frac{50}{50}$	**1.** Identity Property of Multiplication $\left(1 = \frac{50}{50}\right)$
2. $\frac{1}{2} \times \frac{50}{50} = \frac{1 \times 50}{2 \times 50} = \frac{50}{100}$	**2.** Definition of multiplication of fractions
3. $\frac{50}{100} = 50\%$	**3.** Definition of percent
4. Thus, $\frac{1}{2} = 50\%$.	**4.** Transitive Property of Equality

TRY THIS Prove that $\frac{3}{4} = 75\%$.

You can use Properties of Numbers to show that a statement containing a variable is true.

EXAMPLE 2 **Prove that $3a = 2a + a$.**

▶ **Solution**

Statements	Reasons
1. $3a = a + a + a$	**1.** Definition of multiplication by 3
2. $\quad = (a + a) + a$	**2.** Associative Property of Addition
3. $\quad = 2a + a$	**3.** Definition of multiplication by 2

TRY THIS Prove that $4a = 2a + 2a$.

EXERCISES

Justify each step below.

1. $2n + (2n + 1) = (2n + 2n) + 1$
   $= (n + n + n + n) + 1$
   $= 4n + 1$

2. $2n + 1 + 2n + 3 = 2n + 2n + 1 + 3$
   $= (2n + 2n) + (1 + 3)$
   $= (n + n + n + n) + 1 + 3$
   $= 4n + 4$

PRACTICE

Prove the following statements.

3. $\frac{3}{5} = 60\%$

4. $\frac{1}{10} = 10\%$

5. $\frac{4}{10} = 40\%$

6. $\frac{3}{12} = 25\%$

7. $3(x + 3) + 4 = 3x + 13$

8. $2(x + 1) + 5 = 2x + 7$

9. $5a = 3a + 2a$

10. $6a = 5a + a$

11. $3 + 4(y + 2) = 4y + 11$

12. $6 + 4(r + 3) = 4r + 18$

Critical Thinking If n is a whole number, then $2n$ is an even number and $2n + 1$ is an odd number. Does the given expression represent an even number or an odd number? Justify your answer.

13. $2n + 2n$

14. $2n + (2n + 1)$

15. $(2n + 1) + (2n + 1)$

16. $(2n + 1) + (2n + 3)$

MID-CHAPTER REVIEW

Write each phrase as a mathematical expression. Do not evaluate. (1.1 Skill A)

17. 5 more than 3

18. the sum of 2.5 and 8

Write each sentence as an equation. (1.1 Skill B)

19. One plus twice ten is twenty-one.

20. The sum of 3 and 4 less 2 equals 5.

Use the order of operations to simplify each expression. (1.2 Skill B)

21. $\frac{(3 + 2)^2}{15 - 4 \times 3}$

22. $3[4(6 - 5)]^2$

Evaluate each expression. (1.2 Skill C)

23. $3.5(x + 3) + 4.5(x + 7)$, given $x = 5$

24. $4(rs) + 3(r + s)$, given $r = 3$ and $s = 9$

Find the values in the replacement set $\{0, 1, 2, 3, 4, 5, 6\}$ that are solutions to each equation. If the equation has no solution, write *none*. (1.3 Skill A)

25. $2x + 5 = 11$

26. $3x + 5 = 13$

27. $2(a - 1) - 5 = 11$

28. $y + (y + 5) = 13$

1.4 Solving Simple Equations Using Addition

LESSON

SKILL A   *Using addition to solve an equation in one step*

The *Addition Property of Equality* states the following:

If the same quantity is added to each side of a true equation, then the equation that results is also true. If *a* and *b* balance on the scale, then $a + c$ and $b + c$ will balance also.

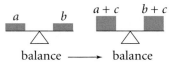

> **Addition Property of Equality**
> If *a*, *b*, and *c* are numbers and $a = b$, then $a + c = b + c$.

Use the Addition Property of Equality to solve equations involving subtraction.

EXAMPLE 1   Solve $x - 10 = 25$. **Check your solution.**

▶ **Solution**

$$x - 10 = 25$$
$$x - 10 \boxed{+ 10} = 25 \boxed{+ 10} \quad \longleftarrow \text{ Apply the Addition Property of Equality.}$$
$$x + 0 = 35 \quad \longleftarrow \text{ Apply the Additive Identity Property.}$$
$$x = 35$$

Check: $35 - 10 = 25$ ✓

TRY THIS   Solve $d - 5 = 4$. Check your solution.

The Symmetric Property of Equality allows you to switch the left and right sides of an equation.

EXAMPLE 2   Solve $12.5 = x - 4.6$. **Check your solution.**

▶ **Solution**

$$12.5 = x - 4.6$$
$$x - 4.6 = 12.5 \quad \longleftarrow \text{ Apply the Symmetric Property of Equality.}$$
$$x - 4.6 \boxed{+ 4.6} = 12.5 \boxed{+ 4.6} \quad \longleftarrow \text{ Apply the Addition Property of Equality.}$$
$$x + 0 = 17.1 \quad \longleftarrow \text{ Apply the Additive Identity Property.}$$
$$x = 17.1$$

Check: $17.1 - 4.6 = 12.5$ ✓

TRY THIS   Solve $100 = t - 14.5$. Check your solution.

EXERCISES

What number would you add to each side of the given equation to solve it? Do not solve.

1. $x - 2.3 = 12$ **2.** $17 = t - 5$ **3.** $22.5 = x - 3$ **4.** $a - 4 = 19$

PRACTICE

Solve each equation. Check your solutions.

5. $x - 2 = 4$ **6.** $x - 1 = 5$ **7.** $6 = q - 3$ **8.** $7 = q - 2$

9. $x - 2.5 = 5.0$ **10.** $x - 3.5 = 12$ **11.** $15 = d - 6.6$ **12.** $20 = r - 6.9$

13. $x - 7.5 = 3.5$ **14.** $x - 2.4 = 1.5$ **15.** $1.75 = k - 0.75$ **16.** $1.6 = k - 1.4$

17. $x - \frac{11}{3} = \frac{19}{3}$ **18.** $x - \frac{10}{3} = \frac{7}{3}$ **19.** $5\frac{1}{3} = z - 5\frac{1}{3}$ **20.** $2\frac{3}{4} = z - 1\frac{3}{4}$

21. $x - 0 = 10\frac{1}{2}$ **22.** $1\frac{3}{5} = k - 0$ **23.** $x - 2.75 = 0$ **24.** $0 = k - 7.2$

25. $x - 3 = 10 + 5$ **26.** $x - 3.5 = 10.5 + 5.5$ **27.** $3 + 5 = x - 5$

28. $13 + 13 = x - 7$ **29.** $x - 2\frac{3}{4} = 2\frac{1}{4} + 3\frac{1}{4}$ **30.** $x - 3\frac{3}{5} = 10\frac{1}{5} - 3\frac{1}{5}$

31. $9\frac{4}{5} - 4\frac{1}{5} = c - 2\frac{3}{5}$ **32.** $12\frac{6}{7} - 3\frac{1}{7} = w - 8\frac{3}{5}$ **33.** $12\frac{1}{4} - 3\frac{1}{4} = a - 6.6$

34. Critical Thinking Find b such that $x - b = 1$ has a solution of $4\frac{1}{2}$.

35. Critical Thinking Find c such that $x - 4\frac{1}{2} = c$ has a solution of $4\frac{1}{2}$.

MIXED REVIEW

Solve each problem. Round answers to two decimal places if necessary. (previous courses)

Sample: Find 45% of 20.   Solution: $0.45 \times 20 = 9$

36. Find 25% of 110.   **37.** Find 60% of 200.

38. What is 50% of 36?   **39.** What is 60% of 50?

40. 80% of 250   **41.** 120% of 18

Sample: What percent of 20 is 9?   Solution: $\frac{9}{20} = 0.45 = 45\%$

42. What percent of 100 is 55?   **43.** What percent of 24 is 16?

44. What percent of 75 is 90?   **45.** What percent of 15 is 20?

1.5 Solving Simple Equations Using Subtraction

LESSON

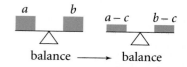

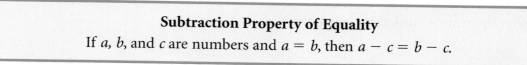

SKILL A   *Using subtraction to solve an equation in one step*

The *Subtraction Property of Equality* states the following:

If the same quantity is subtracted from each side of a true equation, then the equation that results is also true.

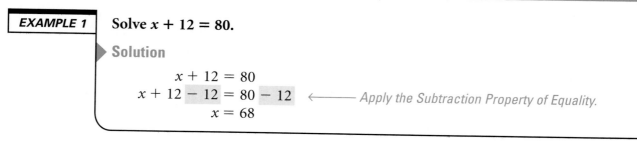

Subtraction Property of Equality
If a, b, and c are numbers and $a = b$, then $a - c = b - c$.

EXAMPLE 1   **Solve $x + 12 = 80$.**

▶ **Solution**

$$x + 12 = 80$$
$$x + 12 - 12 = 80 - 12 \quad \longleftarrow \text{Apply the Subtraction Property of Equality.}$$
$$x = 68$$

TRY THIS   Solve $a + 15.5 = 40$.

The Subtraction Property of Equality applies whether the variable is on the left side or the right side of the equation.

EXAMPLE 2   **Solve $80 = t + 24$.**

▶ **Solution**

$$80 = t + 24$$
$$80 - 24 = t + 24 - 24 \quad \longleftarrow \text{Apply the Subtraction Property of Equality.}$$
$$56 = t$$

TRY THIS   Solve $20 = c + 15.5$.

The Commutative Property of Addition allows you to rearrange equations.

EXAMPLE 3   **Solve $14.95 + z = 20$.**

▶ **Solution**

$$14.95 + z = 20$$
$$z + 14.95 = 20 \quad \longleftarrow \text{Apply the Commutative Property of Addition.}$$
$$z + 14.95 - 14.95 = 20 - 14.95 \quad \longleftarrow \text{Apply the Subtraction Property of Equality.}$$
$$z = 5.05$$

TRY THIS   Solve $13.95 + c = 16$.

EXERCISES

What number would you subtract from each side of the equation to solve it? Do not solve.

1. $p + 6.4 = 12$

2. $7.2 + t = 12.3$

3. $4 = 3.1 + y$

Solve each equation. Check your solution.

4. $t + 4 = 19$

5. $t + 6 = 16$

6. $19 + a = 19$

7. $1.2 + a = 1.2$

8. $1.7 = 1.4 + z$

9. $100 = 99.95 + z$

10. $t + \dfrac{10}{3} = 6$

11. $t + \dfrac{16}{7} = 7$

12. $\dfrac{2}{3} + w = 2$

13. $3\dfrac{2}{3} + w = 4$

14. $12\dfrac{5}{8} = 11\dfrac{5}{8} + x$

15. $1\dfrac{2}{5} = t + \dfrac{3}{5}$

16. $11\dfrac{2}{5} = s + \dfrac{1}{5}$

17. $t + 2.4 = 6 - 0.4$

18. $z + 3.4 = 6 - 0.6$

19. $z + 1.5 = 6 - 3.5$

20. $t + 2\dfrac{2}{5} = 6 - 0.4$

21. $a + 2\dfrac{2}{5} = 6 - 2\dfrac{2}{5}$

22. Solve $z + \left(\dfrac{1}{8} + \dfrac{1}{4} + \dfrac{1}{2} + 1\right) = 1 + 2 + 4 + 8.$

23. Solve $z + \left(\dfrac{7}{4} + \dfrac{5}{4} + \dfrac{3}{4} + \dfrac{1}{4} + 1\right) = 7 + 5 + 3 + 1.$

24. Critical Thinking Find a such that $x + a = 3\dfrac{2}{3}$ has a solution of $\dfrac{1}{2}$.

Find each sum. (previous courses)

Sample: $\dfrac{3}{8} + \dfrac{5}{6}$ Solution: $\dfrac{3}{2 \times 4} + \dfrac{5}{2 \times 3} = \left(\dfrac{3}{3}\right)\left(\dfrac{3}{2 \times 4}\right) + \left(\dfrac{4}{4}\right)\left(\dfrac{5}{2 \times 3}\right) = \dfrac{9}{24} + \dfrac{20}{24} = \dfrac{29}{24}$

25. $\dfrac{1}{3} + \dfrac{1}{6}$

26. $\dfrac{1}{4} + \dfrac{1}{12}$

27. $\dfrac{2}{3} + \dfrac{5}{6}$

28. $\dfrac{3}{4} + \dfrac{5}{12}$

29. $\dfrac{1}{6} + \dfrac{1}{9}$

30. $\dfrac{5}{6} + \dfrac{2}{9}$

31. $\dfrac{3}{8} + \dfrac{7}{12}$

32. $\dfrac{5}{8} + \dfrac{11}{12}$

33. $\dfrac{1}{12} + \dfrac{1}{18}$

34. $\dfrac{5}{12} + \dfrac{5}{18}$

35. $\dfrac{1}{8} + \dfrac{1}{20}$

36. $\dfrac{1}{12} + \dfrac{1}{20}$

Sample: $5\dfrac{2}{3} + 2\dfrac{2}{3}$ Solution: $5 + 2 = 7$ and $\dfrac{2}{3} + \dfrac{2}{3} = \dfrac{4}{3}$, or $1\dfrac{1}{3}$, so $7 + 1\dfrac{1}{3} = 8\dfrac{1}{3}$

37. $3\dfrac{3}{5} + 1\dfrac{1}{5}$

38. $4\dfrac{3}{8} + 3\dfrac{3}{8}$

39. $7\dfrac{1}{5} + 6\dfrac{4}{5}$

40. $5\dfrac{7}{8} + 3\dfrac{3}{8}$

Solve each equation. Check your solution. (1.4 Skill A)

41. $x - 2.5 = 19$

42. $0.1 = t - 15.9$

43. $100 = w - 100$

Solving Simple Equations by Choosing Addition or Subtraction

LESSON

SKILL A *Choosing addition or subtraction as the solution method*

When you solve an equation, you need to choose a solution method that will leave the variable by itself on one side of the equation. This is sometimes called isolating the variable.

Example 1 part **a** shows an equation involving subtraction that is solved by addition. Example 1 part **b** shows an equation involving addition that is solved by subtraction. Addition and subtraction are *inverse operations*. That is, addition will undo subtraction and subtraction will undo addition.

EXAMPLE 1

Solve each equation. Check your solution.

a. $x - 12 = 18$ b. $x + 12 = 18$

▶ **Solution**

a.
$$x - 12 = 18$$
$$x - 12 \; + \; 12 = 18 \; + \; 12$$
$$x = 30$$

⟵ *Because 12 is subtracted from x, add 12 to each side.*

Check: $30 - 12 = 18$ ✔

b.
$$x + 12 = 18$$
$$x + 12 \; - \; 12 = 18 \; - \; 12$$
$$x = 6$$

⟵ *Because 12 is added to x, subtract 12 from each side.*

Check: $6 + 12 = 18$ ✔

TRY THIS   Solve each equation. Check your solution.   a. $x - 9 = 9$ b. $x + 9 = 9$

EXAMPLE 2

Solve $10 = n + 3\frac{4}{5} - 2\frac{1}{5}$. Check your solution.

▶ **Solution**

$$10 = n + 3\frac{4}{5} - 2\frac{1}{5}$$

$$n + 3\frac{4}{5} - 2\frac{1}{5} = 10$$

⟵ *Use the Symmetric Property of Equality to rewrite the equation with the variable on the left.*

$$n + 1\frac{3}{5} = 10$$

⟵ *Simplify: $3\frac{4}{5} - 2\frac{1}{5} = 1\frac{3}{5}$*

$$n + 1\frac{3}{5} \; - \; 1\frac{3}{5} = 10 \; - \; 1\frac{3}{5}$$

⟵ *Apply the Subtraction Property of Equality.*

$$n = 8\frac{2}{5}$$

Check: $8\frac{2}{5} + 3\frac{4}{5} - 2\frac{1}{5} = \left(8\frac{2}{5} + 3\frac{4}{5}\right) - 2\frac{1}{5} = 12\frac{1}{5} - 2\frac{1}{5} = 10$ ✔

TRY THIS   Solve $4 = m + 2.3 - 1.6$. Check your solution.

EXERCISES

KEY SKILLS

Perform the given operation on each side of $x + 5 = 11$. Write the equation that results.

1. Subtract 3.

2. Subtract 4.

3. Subtract 5.

PRACTICE

Solve each equation. Check your solution.

4. $x + 2.5 = 12$

5. $2.4 + c = 11$

6. $13 = z - 5.2$

7. $y - 2.7 = 2.9$

8. $3.6 + y = 19$

9. $13 = 9 + w$

10. $v - 6.2 = 19$

11. $22 = a - 2.2$

12. $b + 13.5 = 20$

13. $d - 13.5 = 20$

14. $n - 3 = 4.5$

15. $m + \dfrac{10}{3} = \dfrac{14}{3}$

16. $p + \dfrac{10}{3} = \dfrac{17}{3}$

17. $z - 5 = 3 + 2$

18. $12 - 3 = w + 5$

19. $10 - 3 = 2 + h$

20. $\left(3\dfrac{1}{3} + 2\dfrac{1}{3}\right) = 3\dfrac{4}{5} + h$

21. $n - \left(7\dfrac{1}{3} + 5\dfrac{2}{3}\right) = 3\dfrac{4}{5}$

22. $6 - 2\dfrac{1}{4} = 2\dfrac{1}{2} + z$

23. $7 = z + 4\dfrac{1}{2} - 3\dfrac{3}{4}$

24. $n - \left(3\dfrac{1}{3} + 2\dfrac{2}{3} - 1\dfrac{1}{3}\right) = 7\dfrac{2}{3}$

25. **Critical Thinking** For what value of a will the solution to $x + a = b$ be the same as the solution to $x - a = b$?

26. **Critical Thinking** Find the difference of the solution to $x + a = b$ less the solution to $x - a = b$.

MIXED REVIEW

Which fraction is greater? (previous courses)

Sample: $\dfrac{2}{5}$ or $\dfrac{3}{7}$ Solution: $\left(\dfrac{7}{7}\right)\dfrac{2}{5} = \dfrac{14}{35}$ $\left(\dfrac{5}{5}\right)\dfrac{3}{7} = \dfrac{15}{35}$ $\dfrac{15}{35} > \dfrac{14}{35} \rightarrow \dfrac{3}{7}$

27. $\dfrac{1}{4}$ or $\dfrac{1}{3}$

28. $\dfrac{1}{4}$ or $\dfrac{1}{5}$

29. $\dfrac{3}{4}$ or $\dfrac{2}{3}$

30. $\dfrac{3}{4}$ or $\dfrac{4}{5}$

31. $\dfrac{6}{7}$ or $\dfrac{7}{8}$

32. $\dfrac{6}{8}$ or $\dfrac{7}{9}$

Order each list of numbers from least to greatest. (previous courses)

33. $\dfrac{3}{5}, \dfrac{1}{2}, \dfrac{5}{7}, \dfrac{2}{3}$

34. $0.145, 0.15, 0.155, 0.1$

35. $\dfrac{1}{21}, 0.05, 0.45, 0.14, \dfrac{11}{25}, \dfrac{8}{50}$

36. $\dfrac{1}{12}, 0.12, \dfrac{1}{20}, 0.20, \dfrac{1}{6}, 0.6, 0.06, 0.16$

When you read a real-world problem, look for key words that suggest how to write the equation. Key words usually express an action indicating some kind of change. Key words may also describe a relationship between two quantities.

EXAMPLE 1

At 11:00 P.M., a weather announcer reported that the air temperature had dropped 12 degrees from the temperature at 8:00 P.M. If the temperature at 11:00 P.M. was 42°F, what was the temperature at 8:00 P.M.?

▷ **Solution**

Look for key words.

Let t represent the air temperature in degrees Fahrenheit at 8:00 P.M. A key word in the weather report is *dropped*.

$$t - 12 = 42 \qquad \longleftarrow \text{A drop means a decrease.}$$
$$t - 12 + 12 = 42 + 12 \qquad \longleftarrow \text{Apply the Addition Property}$$
$$t = 54 \qquad\qquad\qquad\qquad \text{of Equality.}$$

The temperature at 8:00 P.M. was 54°.

Check: Because the temperature had dropped, the 8 P.M. temperature should be higher than the 11:00 P.M. temperature.

54° is higher than 42° ✔

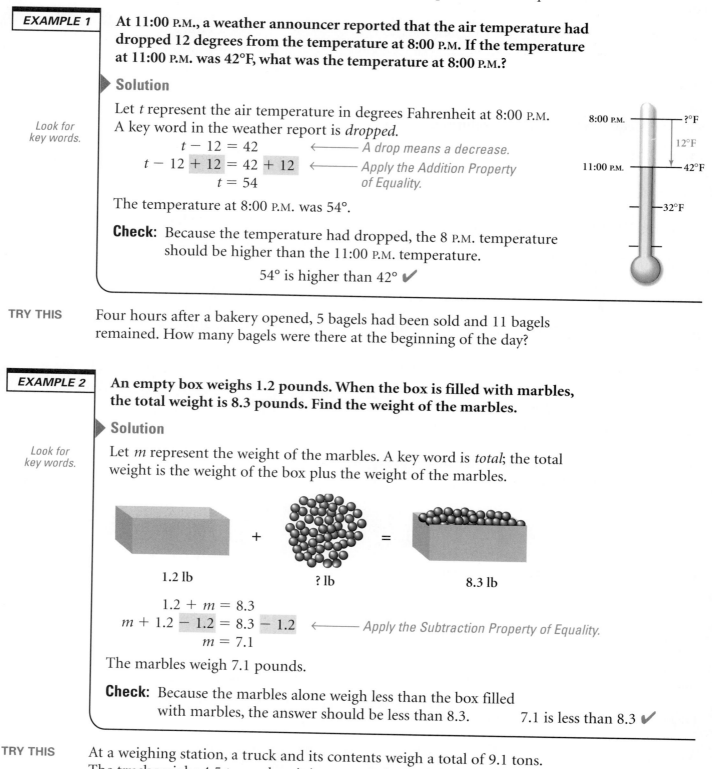

8:00 P.M. — ?°F

12°F

11:00 P.M. — 42°F

32°F

TRY THIS

Four hours after a bakery opened, 5 bagels had been sold and 11 bagels remained. How many bagels were there at the beginning of the day?

EXAMPLE 2

An empty box weighs 1.2 pounds. When the box is filled with marbles, the total weight is 8.3 pounds. Find the weight of the marbles.

▷ **Solution**

Look for key words.

Let m represent the weight of the marbles. A key word is *total*; the total weight is the weight of the box plus the weight of the marbles.

1.2 lb   + ? lb   = 8.3 lb

$$1.2 + m = 8.3$$
$$m + 1.2 - 1.2 = 8.3 - 1.2 \qquad \longleftarrow \text{Apply the Subtraction Property of Equality.}$$
$$m = 7.1$$

The marbles weigh 7.1 pounds.

Check: Because the marbles alone weigh less than the box filled with marbles, the answer should be less than 8.3. 7.1 is less than 8.3 ✔

TRY THIS

At a weighing station, a truck and its contents weigh a total of 9.1 tons. The truck weighs 4.5 tons when it is empty. What is the weight of the contents?

EXERCISES

KEY SKILLS

For each situation, write an equation and identify the key words. Do not solve.

1. Seven students left a room and 14 students remained. How many students were in the room before any students left?

2. After 7 students entered a room, there were 22 students in the room. How many students were in the room before any students entered?

PRACTICE

Write an equation to represent each situation. Solve the equation and then answer the question.

3. In 6 hours, the temperature has dropped 6 degrees Fahrenheit. If the current temperature is 72°F, what was the temperature 6 hours ago?

4. A customer gave a clerk 2 five-dollar bills and 2 one-dollar bills for a purchase of $11.52. How much change should the customer receive?

5. After 12 marbles were removed from a bag, 18 marbles remained. How many marbles were in the bag before any were removed?

6. A scale shows 1.2 pounds of flour. How much more flour should be added to make 1.5 pounds of flour on the scale?

7. A balance of $1245.35 remains in a bank account after a withdrawal of $320.39. How much was in the account before the withdrawal?

8. A motorist noticed that the odometer in her car read 12,345.2 miles. A few hours later, it read 12,511.1 miles. How far did she drive?

9. An empty container weighs 1.3 pounds. When filled with sand, the total weight is 10 pounds. How much does the sand weigh?

10. The following numbers represent the hourly changes in a stock market index during one trading day. If the index was 5420 at the beginning of the day, what was the final index?

$$+120, -80, +10, +35, +60, -5, -15, +10$$

MIXED REVIEW APPLICATIONS

Write each phrase as a mathematical expression. (1.1 Skill A)

11. the cost of a CD and 2 tapes at $19.95 and $7.95 each, respectively

12. total student attendance: three buses each containing 32 students and 7 cars each having 4 students

1.7 Solving Simple Equations Using Multiplication

LESSON

> **SKILL A** ▶ *Using multiplication to solve an equation in one step*

The Multiplication Property of Equality states the following:

Multiplication Property of Equality
If a, b, and c are numbers and $a = b$, then $ac = bc$.

Because multiplication and division are inverse operations, the Multiplication Property of Equality is useful when you solve an equation in which the variable is divided by a number.

EXAMPLE 1 | Solve $\frac{x}{5} = 2.5$. **Check your solution.**

▶ **Solution**

$$\frac{x}{5} = 2.5$$

$$(5)\frac{x}{5} = (5)2.5 \quad \longleftarrow \text{ Apply the Multiplication Property of Equality.}$$

$$x = 12.5 \qquad \textbf{Check: } \frac{12.5}{5} = 2.5 \checkmark$$

TRY THIS | Solve $10.2 = \frac{t}{3}$. Check your solution.

When a variable is isolated, as in $x = 5$, its coefficient is 1. You know from the Identity Property of Multiplication that you can obtain a coefficient of 1 by multiplying a fraction by its reciprocal. So, you can write $\frac{2n}{3}$ as $\frac{2}{3} \times n$ to identify its reciprocal, $\frac{3}{2}$, that can help you isolate the variable n.

EXAMPLE 2 | Solve $\frac{2n}{3} = 6.2$. **Check your solution.**

▶ **Solution**

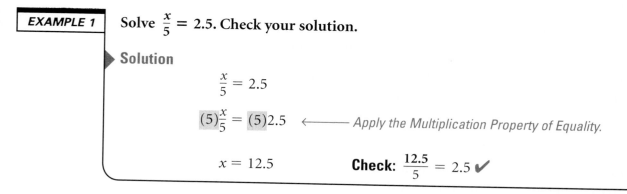

$$\frac{2}{3} \times n = 6.2 \quad \longleftarrow \text{ Write } \frac{2n}{3} \text{ as } \frac{2}{3} \times n.$$

$$\left(\frac{3}{2}\right)\frac{2}{3} \times n = \left(\frac{3}{2}\right)6.2 \quad \longleftarrow \text{ Apply the Multiplication Property of Equality.}$$

$$n = 9.3 \qquad \textbf{Check: } \frac{2(9.3)}{3} = \frac{18.6}{3} = 6.2 \checkmark$$

TRY THIS | Solve $\frac{5v}{3} = 20$. Check your solution.

EXERCISES

KEY SKILLS

What number would you use to multiply each side of the equation by to isolate the variable? Do not solve.

1. $\dfrac{a}{3} = 12$

2. $\dfrac{b}{4} = 1$

3. $\dfrac{1}{2}x = 9$

4. $\dfrac{2x}{5} = 6$

5. $\dfrac{4t}{9} = 8$

6. $\dfrac{9y}{10} = 12$

PRACTICE

Solve each equation. Check your solution.

7. $\dfrac{t}{2} = 2$

8. $\dfrac{1}{4}t = 4$

9. $5 = \dfrac{a}{6}$

10. $6 = \dfrac{q}{8}$

11. $\dfrac{v}{8} = 1.3$

12. $\dfrac{w}{5} = 5.5$

13. $7.0 = \dfrac{x}{7}$

14. $2.0 = \dfrac{y}{3}$

15. $\dfrac{w}{6} = \dfrac{3}{4}$

16. $\dfrac{r}{10} = \dfrac{3}{5}$

17. $\dfrac{3}{7} = \dfrac{z}{3}$

18. $\dfrac{3}{16} = \dfrac{s}{8}$

19. $\dfrac{3r}{7} = \dfrac{3}{2}$

20. $\dfrac{4r}{11} = \dfrac{11}{8}$

21. $\dfrac{14}{8} = \dfrac{7s}{2}$

22. $\dfrac{14}{10} = \dfrac{7x}{10}$

23. $\dfrac{3c}{10} = 1.5$

24. $\dfrac{8c}{7} = 5.6$

25. $5.6 = \dfrac{7z}{10}$

26. $12.1 = \dfrac{11m}{10}$

27. $\dfrac{7c}{2} = 12.1 + 8.9$

28. $\dfrac{10z}{2} = 1.3 + 8.7$

29. $1.3 + 7.1 = \dfrac{7n}{10}$

30. $2.5 + 5.1 = \dfrac{2d}{6}$

31. $\left(\dfrac{7}{4}\right)\left(\dfrac{5}{2}\right)\dfrac{2x}{3} = 1$

32. $\left(\dfrac{3}{2}\right)\left(\dfrac{5}{9}\right)\dfrac{9x}{3} = 14$

33. $\left(\dfrac{7}{2}\right)\left(\dfrac{5}{9}\right)\dfrac{2t}{7} = 5$

34. $\dfrac{x}{4} + 5 = 6$

35. $\dfrac{x}{3} - 2 = 4$

36. $\dfrac{2y}{9} + 6 = 10$

37. **Critical Thinking** Suppose that a and b are nonzero numbers and c is any number. Solve $\dfrac{ax}{b} = c$ for x.

MIXED REVIEW

Solve each equation. Check your solution. (1.4 Skill A, 1.5 Skill A)

38. $c - 2 = 6$

39. $g - 10 = 19$

40. $18.1 = z - 10.1$

41. $0.3 = x - 0.8$

42. $y + 0.8 = 1.2$

43. $a + 11.3 = 11.4$

44. $x + \dfrac{2}{3} = \dfrac{11}{3}$

45. $x + \dfrac{4}{5} = 10$

46. $s - \dfrac{9}{2} = 5$

1.8

LESSON

Solving Simple Equations Using Division

SKILL A *Using division to solve an equation in one step*

> ### Division Property of Equality
>
> If a, b, and c are numbers, c is nonzero, and $a = b$, then $\dfrac{a}{c} = \dfrac{b}{c}$.

Use the Division Property of Equality to solve an equation in which the variable is multiplied by a number.

EXAMPLE 1   Solve $\dfrac{21}{5} = 7d$. **Check your solution.**

▶ **Solution**

$$\frac{21}{5} = 7d$$

$$\frac{21}{5} \div 7 = 7d \div 7 \quad \longleftarrow \text{ Apply the Division Property of Equality.}$$

$$\frac{21}{5} \times \frac{1}{7} = d \quad \longleftarrow \begin{array}{l}\text{Recall that dividing by 7 is equivalent}\\ \text{to multiplying by } \frac{1}{7}.\end{array}$$

$$\frac{3}{5} = d \qquad \textbf{Check: } 7\left(\frac{3}{5}\right) = \frac{21}{5} ✔$$

TRY THIS   Solve $24 = 4c$. Check your solution.

EXAMPLE 2   **A carpenter wants to cut a 12 foot board into 4 shorter boards of equal length. How long will each board be?**

▶ **Solution**

Look for key words.

Let v represent the length of each short board.

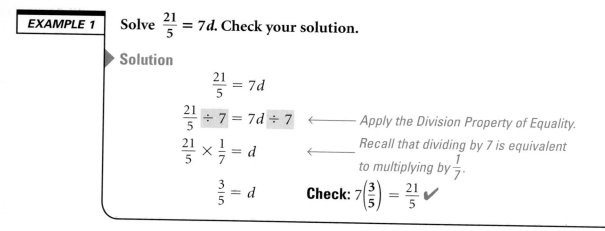

number of boards	times	length of each board	is	original board length
↓	↓	↓	↓	↓
4	×	v	=	12

Write an equation.

Write and solve an equation.

$$4v = 12$$

$$\frac{4v}{4} = \frac{12}{4} \quad \longleftarrow \text{ Apply the Division Property of Equality.}$$

$$v = 3$$

Each new board will be 3 feet long.

TRY THIS   Rework Example 2 if the carpenter wants 5 boards of equal length cut from a board 12.5 ft long.

EXERCISES

Which of the given numbers is the best divisor to use to isolate each variable?

1. $3x = 15$; 3 or 15

2. $2.5z = 13$; 2.5 or 13

3. $\frac{9}{2} = 5n$; $\frac{1}{2}$, $\frac{9}{2}$, or 5

4. $4t = 3.5$; 4, 3, or 3.5

Solve each equation. Check your solution.

5. $3x = 24$

6. $5x = 25$

7. $81 = 9b$

8. $121 = 11a$

9. $2s = 7$

10. $3s = 10$

11. $6.5 = 1.5d$

12. $28.7 = 0.7g$

13. $4r = \frac{9}{2}$

14. $7x = \frac{10}{3}$

15. $\frac{38}{5} = 8p$

16. $\frac{10}{7} = 2z$

17. $7x = 1\frac{3}{7} + 4\frac{4}{7}$

18. $3h = 10\frac{2}{5} - 4\frac{2}{5}$

19. $2\frac{9}{11} + 4\frac{2}{11} = 2c$

20. $12\frac{8}{13} + 4\frac{5}{13} = 5k$

21. $(3 + 1)x = 12$

22. $(15 - 12)x = 12$

23. $36 = (15 - 9)k$

24. $3 = (5 - 2)w$

25. $3n - 9 = 21$

26. $2x + 5 = 13$

27. $3 + 6x = 27$

28. $35 = 1.5x - 1$

Find the length of each board or the amount of each liquid.

29. a 12-foot board cut into 8 equal lengths

30. 18 gallons of water poured into 6 containers of equal capacity

31. a 12.6-foot board cut into 6 equal lengths

32. 20 gallons of water poured into 8 containers of equal capacity

33. Critical Thinking   Is it possible to cut a 17-foot board into 4 pieces of equal length and have the length of each board be a whole number? Justify your response.

Evaluate each expression using the given value for each variable.
(1.2 Skill C)

34. $3.5(x - 2) + 3$, $x = 4$

35. $3x + 4(x - 1)$, $x = 2.5$

36. $2.5(x + 3) - 10$, $x = 7$

37. $\frac{3}{4}(2c + 7) - 5$, $c = 0$

38. $\frac{3}{5}(2d + 7) - 4d$, $d = 1$

39. $0.6(3w + 1) - 0.6w$, $w = 3$

1.9 LESSON

Solving Simple Equations by Choosing Multiplication or Division

Choosing multiplication or division as the solution method

Multiplication and division are *inverse operations*. That is, multiplication will undo division and division will undo multiplication.

EXAMPLE 1 **Choose a method for solving $5h = 120$. Then solve and check.**

▶ **Solution**

In words: 5 multiplied by h is 120. Choose division.

$$\frac{5h}{5} = \frac{120}{5} \longleftarrow \text{Apply the Division Property of Equality.}$$
$$h = 24$$

Check: $5(24) = 120$ ✔

TRY THIS   Choose a method for solving $4z = 36$. Then solve and check.

EXAMPLE 2 **Choose a method for solving $\frac{n}{7} = 2.5$. Then solve and check.**

▶ **Solution**

In words: n divided by 7 is 2.5. Choose multiplication.

$$(7)\frac{n}{7} = (7)2.5 \longleftarrow \text{Apply the Multiplication Property of Equality.}$$
$$n = 17.5, \text{ or } 17\frac{1}{2}$$

Check: $\left(17\frac{1}{2}\right) \div 7 = \frac{35}{2} \times \frac{1}{7} = \frac{5}{2} = 2.5$ ✔

TRY THIS   Choose a method for solving $\frac{a}{4} = \frac{3}{2}$. Then solve and check.

Some equations can be solved by either multiplication or division.

EXAMPLE 3 **Choose a method for solving $2x = 4.6$. Then solve and check.**

▶ **Solution**

Method 1: Choose division.

$$\frac{2x}{2} = \frac{4.6}{2}$$
$$x = 2.3$$

Method 2: Choose multiplication.

$$\left(\frac{1}{2}\right)2x = \left(\frac{1}{2}\right)4.6$$
$$x = 2.3$$

Check: $2(2.3) = 4.6$ ✔

TRY THIS   Choose a method for solving $12 = \frac{4x}{5}$. Then solve and check.

EXERCISES

Write each equation in words.

Sample: $\frac{x}{2} = 10$ Solution: x divided by 2 is 10.

1. $3x = 22$ **2.** $\frac{c}{4} = 15$ **3.** $\frac{3c}{4} = 4$ **4.** $0.5d = 10$

PRACTICE

Write the operations you would choose to solve each equation. Then solve and check.

5. $3.5x = 7$ **6.** $\frac{1}{4}d = 1$ **7.** $4.3k = 8.6$

8. $\frac{1}{3}n = 13$ **9.** $\frac{4m}{3} = 10$ **10.** $\frac{3m}{4} = 10$

11. $\frac{n}{3.5} = 1$ **12.** $2.5p = 2.5$ **13.** $\frac{p}{2.5} = 2.5$

14. $(2.5 + 1.5)n = 8$ **15.** $\frac{n}{2.5 + 1.5} = 8$ **16.** $\frac{2n}{3} = \frac{2}{3}$

17. $\frac{3}{2}n = \frac{2}{3}$ **18.** $\frac{n}{\left(1\frac{1}{2} + 6\frac{1}{2}\right)} = 2$ **19.** $\left(1\frac{1}{2} + 6\frac{1}{2}\right)n = 32$

Solve each equation.

20. $5x + 6 = 11$ **21.** $3x - 4 = 8$ **22.** $9x + 1 = 10$

23. $\frac{x}{2} - 4 = 4$ **24.** $3 = \frac{x}{5} - 7$ **25.** $12 = \frac{3x}{4} + 8$

26. Critical Thinking Suppose that a and b are nonzero numbers. Show that $\frac{ax}{b} = 10$ and $\frac{x}{\left(\frac{b}{a}\right)} = 10$ have the same solution.

MIXED REVIEW

Solve each equation. Check your solution. (1.4 Skill A, 1.5 Skill A, 1.7 Skill A, 1.8 Skill A)

27. $4.5x = 27$ **28.** $z - 3.2 = 7.2$ **29.** $\frac{n}{7} = 3$

30. $z - \frac{5}{4} = \frac{31}{4}$ **31.** $c + 2.5 = 7.9$ **32.** $d + (4 - 3.8) = 10$

33. $2x = 7.8$ **34.** $\frac{3n}{2} = 3$ **35.** $t + 4.25 = 5$

When you read a real-world problem, look for key words or phrases that suggest how to write an equation.

EXAMPLE 1

An investor normally deposits a fixed amount into a bank account each month. One month, the investor deposits four times the normal monthly amount. If the deposit was $6000 that month, what is the normal monthly deposit?

▶ **Solution**

Look for key words.

1. Let n represent the normal monthly deposit in dollars. The key phrase is *four times*.

2. Write and solve an equation.

Write an equation.

current deposit = 4 times the normal monthly deposit

$$6000 = 4n$$
$$4n = 6000 \quad \longleftarrow \text{Apply the Symmetric Property of Equality.}$$
$$\frac{4n}{4} = \frac{6000}{4} \quad \longleftarrow \text{Apply the Division Property of Equality.}$$
$$n = 1500$$

The normal monthly deposit is $1500.

Check: $4(1500) = 6000$ ✓

TRY THIS   This week, Martha received $27, which is 3 times her normal allowance. How much is her normal allowance?

EXAMPLE 2

A box of crayons is evenly divided among 8 children. Each child receives 3 crayons. How many crayons were in the box?

▶ **Solution**

Look for key words.

1. Let c represent the total number of crayons in the box. The key word is *divided*.

2. Write and solve an equation.

Write an equation.

$$\frac{\text{total number of crayons}}{\text{number of children}} = \text{number of crayons per child}$$

$$\frac{c}{8} = 3$$

$$(8)\frac{c}{8} = (8)3 \quad \longleftarrow \text{Apply the Multiplication Property of Equality.}$$

$$c = 24$$

There were 24 crayons in the box.

Check: $\frac{24}{8} = 3$ ✓

TRY THIS   Mrs. Gomez withdrew some money from her bank to divide among her 6 children. Each child received $35.50. How much money did she withdraw?

EXERCISES

For each situation, identify the key word(s) and write an equation. Do not solve.

1. A father wants to divide a certain amount of money among 5 children and give each child $24. How much should he withdraw?

2. Jamal wants to divide 3 hours of study time so that he spends the same amount of time on each of 6 school subjects. How much time will he devote to each subject?

PRACTICE

For each situation, write and solve an equation.

3. A contractor wants to pay each of 6 workers an equal amount of money. How much should each worker get if the contractor distributes $1140?

4. A carpenter needs four boards of equal length cut from a board 16-foot long. How long will each board be?

5. A teenager wants to save $290 to buy a stereo. How many weeks will it take if he can save $25 every week? Round your answer to the nearest whole number.

6. How much time should Gina set aside so that she can spend 45 minutes studying each of 5 subjects? Give your answer in hours.

7. One bus can seat 32 students. How many buses will be needed to transport 380 students?

8. **Critical Thinking** Jack is twice the age of Alice. Sasha is twice the age of Jack. If Alice is 5, how old is Sasha?

9. **Critical Thinking** Deshon has finished 5 more homework exercises than Yuko. Yuko has finished 3 more than Andres. Andres has finished twice as many as Mary. If Mary has finished 7 exercises, how many has Deshon finished?

MIXED REVIEW

Solve each problem. (1.6 Skill A)

10. Rob's purchases total $11.53. How much change should he receive if he gives the cashier a ten-dollar bill and 2 one-dollar bills?

11. After a 12.5°F drop in temperature, a thermometer showed 63.4°F. What was the temperature before the drop?

12. A scale shows 1.13 pounds of shrimp on the tray. How much should be added so that the shopper will receive 1.25 pounds of shrimp?

Write each phrase as a mathematical expression.

1. the difference of 4 minus 17 less the sum of 6 and 1

2. the cost of 4 bicycles at $165 each plus the cost of 3 tricycles at $54 each

Simplify each expression. Use mental math where possible. Identify the property or properties used.

3. $\dfrac{2}{3} \cdot \dfrac{3}{2} + \dfrac{5}{7} \cdot \dfrac{7}{5}$

4. $1 \cdot 6 + 1 \cdot 7$

5. $5(0 + 3)$

Simplify each expression by following the order of operations.

6. $2(10 - 5)^2 + \dfrac{12 - 9}{3}$

7. $\dfrac{30 + 6}{12 - 6} + \dfrac{(3 + 1)^2}{(3 - 1)^2}$

Evaluate each expression.

8. $\dfrac{n + 18}{n - 6} + \dfrac{2n + 1}{n - 7}$; $n = 12$

9. $\dfrac{3a + b}{a - 6} + \dfrac{2b + 6}{a - b}$; $a = 12, b = 6$

Find the values in the replacement set {0, 1, 2, 3, 4, 5, 6} that are solutions to each equation. If the equation has no solution, write *none*.

10. $3(x - 1) = 21$

11. $4(n - 1) + 3 = 23$

12. $7z - 5 = 9$

13. Give a counterexample to show that the statement below is false.
   The sum of an even number and an odd number is always even.

14. Use Properties of Equality and properties of operations to show that the equation below is true.
$$3(a + 3) + 6 = 15 + 3a$$

Solve each equation. Check your solution.

15. $x - 2 = 6$

16. $4.6 = a - 6.2$

17. $t - 5 = 12.8$

18. $y + 19.95 = 20$

19. $18.65 = w + 12.3$

20. $x + 0.02 = 0.6$

Solve each equation. Check your solution.

21. $z + 75 = 138$

22. $19 = p + 7.5$

23. $12.5 = z - 7.4$

24. $2\frac{1}{2} = z - 5\frac{1}{3}$

25. $n - 4\frac{3}{5} = 4\frac{1}{5}$

26. $z + 13 = 7\frac{5}{8} + 9\frac{3}{8}$

27. Write an equation to represent the situation below. Solve the equation and answer the question.

After withdrawing $90 from a bank account, a customer had a balance of $1200.45. How much money did the customer have before the withdrawal?

Solve each equation. Check your solution.

28. $\frac{x}{4} = 13$

29. $2.4 = \frac{t}{12}$

30. $\frac{3a}{5} = 2.7$

31. $10 = \frac{10c}{3}$

32. $\frac{5}{2} = \frac{15n}{4}$

33. $\frac{3z}{7} = \frac{1}{2}$

34. $4k = 25$

35. $96 = 12h$

36. $4a = \frac{17}{2}$

37. $\frac{10}{3} = 9z$

38. $\frac{10}{3} - \frac{2}{3} = 8z$

39. $(6 + 5)b = 44$

40. $25 = \frac{2}{5}r$

41. $18b = 37$

42. $\frac{g}{2} = \frac{2}{5} + \frac{2}{5}$

43. $64 = (2.6 + 5.4)z$

44. $2s = 5\frac{2}{5} - 2\frac{1}{3}$

45. $\frac{x}{12.5 - 7} = 2$

Write an equation to represent each situation. Solve the equation and answer the question.

46. How much money should each of 7 workers receive if the boss distributes $1477 equally among them?

47. For how many weeks will a teenager need to save money if she can save $30 per week and she wants to save a total of $360?

Operations With Real Numbers

▶ **What You Already Know**

In Chapter 1, you learned how to choose a single property of equality and apply it to solve a simple equation. When you applied the property, you performed addition, subtraction, multiplication, or division of positive numbers and found positive solutions.

▶ **What You Will Learn**

In Chapter 2, you will have the opportunity to apply the same properties of equality to solve equations. Now, however, you will need to add, subtract, multiply, and divide both positive and negative numbers to arrive at the solutions. To help you, the chapter reviews the skills needed to perform operations on the integers.

Once you have extended your equation-solving skills to one-step equations involving both positive and negative numbers, you will then learn to solve equations that require two or more solution steps. You will also solve equations that involve the Distributive Property and the variable on both sides of the equation.

VOCABULARY	
absolute value	Multiplicative Property of −1
additive inverse	natural numbers
coefficient	negation
conclusion	Opposite of a Difference
conditional statement	Opposite of a Sum
deductive reasoning	opposites
equilateral triangle	origin
equivalent equations	rational numbers
hypothesis	real numbers
indirect reasoning	repeating decimal
infinite	scalene triangle
integers	subset
irrational numbers	terminating decimal
isosceles triangle	Transitive Property of Deductive
least common multiple (LCM)	Reasoning
like terms	whole numbers
monomial	

The diagram below shows how mathematical skills and mathematical reasoning are interrelated with the skills and concepts in Chapter 2. Notice that operations with real numbers are presented before you learn how to solve equations with real numbers.

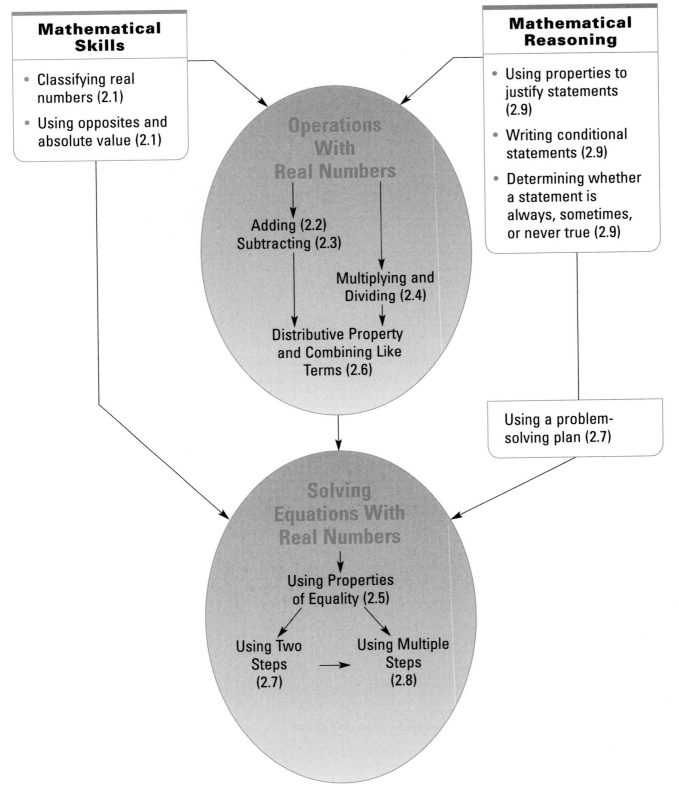

CHAPTER 2

Mathematical Skills

- Classifying real numbers (2.1)
- Using opposites and absolute value (2.1)

Mathematical Reasoning

- Using properties to justify statements (2.9)
- Writing conditional statements (2.9)
- Determining whether a statement is always, sometimes, or never true (2.9)

Operations With Real Numbers

Adding (2.2)
Subtracting (2.3)

Multiplying and Dividing (2.4)

Distributive Property and Combining Like Terms (2.6)

Using a problem-solving plan (2.7)

Solving Equations With Real Numbers

Using Properties of Equality (2.5)

Using Two Steps (2.7)

Using Multiple Steps (2.8)

2.1 Real Numbers

LESSON

SKILL A *Representing rational numbers as fractions and as decimals*

The numbers that you most commonly use are classified in these sets:

- **natural numbers**: the numbers that are used in counting
- **whole numbers**: the set of natural numbers plus 0
- **integers**: the set of whole numbers and their opposites
- **rational numbers**: the set of numbers that can be written as a fraction

  of the form $\frac{p}{q}$, where p and q are integers and q does not equal 0

Notice in the diagram that all integers are also rational numbers. This is because any integer can be written as a fraction. For example, 5 is also $\frac{5}{1}$.

Because a fraction can be written as a decimal, any rational number $\frac{p}{q}$ can be written in decimal form.

When you divide the denominator of a fraction into the numerator, the result is either a *terminating* or *repeating* decimal.

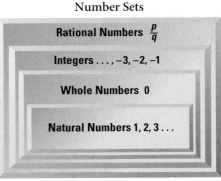

Number Sets

Rational Numbers $\frac{p}{q}$

Integers . . . , −3, −2, −1

Whole Numbers 0

Natural Numbers 1, 2, 3 . . .

A **terminating decimal** has a *finite* number of nonzero digits to the right of the decimal point, such as 0.25 or 0.10. A **repeating decimal** has a string of one or more digits that repeat *infinitely*, such as 1.555555 . . ., or 0.345634563456 . . . These repeating decimals can be indicated by a bar over the repeating digits, $1.\overline{5}$ or $0.\overline{3456}$.

EXAMPLE 1 Write 0.375 as a fraction in simplest form.

> *A fraction in simplest form is the same as a fraction in lowest terms.*

▶ **Solution**

$$0.375 \rightarrow \frac{375}{1000} = \frac{5 \cdot 5 \cdot 5 \cdot 3}{5 \cdot 5 \cdot 5 \cdot 8} = \frac{3}{8} \qquad \text{Thus, } 0.375 = \frac{3}{8}.$$

TRY THIS Write 0.625 as a fraction in simplest form.

EXAMPLE 2 Write $\frac{1}{3}$ as a decimal.

▶ **Solution**

Divide 1 by 3, as shown at right. You can see that this division process will repeat infinitely. Thus, $\frac{1}{3} = 0.\overline{3}$.

$$\begin{array}{r} 0.33 \\ 3\overline{)1.00} \\ \underline{9} \\ 10 \\ \underline{9} \\ 1 \end{array}$$

TRY THIS Write $\frac{5}{18}$ as a repeating decimal.

---

40 Chapter 2 Operations With Real Numbers

EXERCISES

KEY SKILLS

Write each decimal as a fraction. Do not simplify.

Sample: 1.02   Solution: $\frac{102}{100}$

1. 0.35   **2.** 1.5   **3.** 13.755   **4.** 0.0054

PRACTICE

Write each decimal as a fraction in simplest form.

5. 0.2   **6.** 0.7   **7.** 1.4   **8.** 3.9

9. 0.25   **10.** 0.75   **11.** 1.25   **12.** 1.75

13. 0.155   **14.** 0.666   **15.** 0.111   **16.** 1.555

Write each fraction or mixed number as a decimal. Identify whether the decimal that results is terminating or repeating.

17. $\frac{1}{2}$ **18.** $\frac{13}{100}$ **19.** $4\frac{3}{5}$ **20.** $5\frac{1}{4}$

21. $\frac{1}{9}$ **22.** $\frac{2}{11}$ **23.** $4\frac{2}{3}$ **24.** $2\frac{1}{3}$

25. $\frac{1}{6}$ **26.** $\frac{7}{18}$ **27.** $10\frac{3}{11}$ **28.** $10\frac{5}{6}$

29. Critical Thinking Show that the result of $\left(4\frac{1}{2}\right) \div \left(2\frac{3}{4}\right)$ is a rational number by
 a. writing it as a quotient of two integers.   **b.** writing it as a decimal.

MIXED REVIEW

Order each list of numbers from least to greatest. (previous courses)

30. $\frac{1}{30}, \frac{1}{33}, 0.30, \frac{1}{3}, 0.33, 0.03$ **31.** $\frac{1}{2}, 0.2, \frac{1}{20}, 0.22, \frac{1}{200}$

Solve each problem. Round to two decimal places if necessary.
(previous courses) (samples available on page 21)

32. Find 5% of 50.   **33.** Find 16% of 75.   **34.** Find 0.1% of 20.

35. Find 0.01% of 200.   **36.** Find 150% of 20.   **37.** Find 500% of 20.

38. What percent of 10 is 3?   **39.** What percent of 20 is 3?   **40.** What percent of 40 is 3?

41. What percent of 50 is 3?   **42.** What percent of 1 is 3?   **43.** What percent of $\frac{1}{2}$ is 3?

Solve each equation. Check your solution.
(1.4 Skill A, 1.5 Skill A, 1.7 Skill A, 1.8 Skill A)

44. $a - 3 = 8$ **45.** $t + 3 = 18$ **46.** $z + 1.2 = 10$ **47.** $y - 18.2 = 105$

48. $18y = 54$ **49.** $1.9y = 38$ **50.** $\frac{3x}{5} = 2$ **51.** $\frac{7x}{3} = 84$

There are two more sets of numbers that are important to know.

The set of **irrational numbers** consists of all decimals that do not terminate and do not repeat. The number 0.101001000100001. . . is an example of an irrational number. The number pattern does not repeat exactly, and the *ellipsis* (. . .) indicates that the numbers do not terminate. Irrational numbers *cannot be written in form* $\frac{p}{q}$, *where p and q are integers and q ≠ 0.*

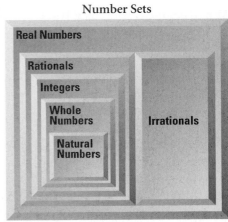

Number Sets

The set of **real numbers** consists of all rational and irrational numbers.

Notice that the sets of natural numbers, whole numbers, integers, rational numbers, and irrational numbers are all contained within the larger set of real numbers. When one set is contained in another set, we say that the smaller set is a **subset** of the larger set.

EXAMPLE 1

Classify each number in as many ways as possible.

a. 4   b. −4   c. $0.\overline{3}$

▶ **Solution**

a. 4 is a natural number, a whole number, an integer, a rational number and a real number.

b. −4 is an integer, a rational number, and a real number.

c. $0.\overline{3}$ is a rational number and a real number.

TRY THIS   Classify 10,000 in as many ways as possible.

EXAMPLE 2

Is each statement below true or false? Justify your answer.

a. *All natural numbers are rational numbers.*
b. *All integers are irrational numbers.*

▶ **Solution**

a. The statement is true. If n is a natural number, then $n = \frac{n}{1}$, a rational number.

b. The statement is false. For example, 2 is an integer and 2 can be written as $\frac{2}{1}$. Therefore, 2 is not irrational.

TRY THIS   Is the following statement true or false?
      All integers are natural numbers.
If the statement is false, give a counterexample.

EXERCISES

1. Identify integers that are not whole numbers: $5, 2, -3, 7, 0, -12, -4, 17$

2. Identify rational numbers that are not whole numbers: $-4, 3.7, 0, 5, 12.67, -1.5$

3. Identify rational numbers between 1 and 1.5: $1.55, 1.05, \frac{4}{5}, 1.\overline{5}, 1.\overline{41}, \frac{7}{4}$

Classify each number in as many ways as possible.

4. $\frac{1}{2}$

5. $-\frac{1}{2}$

6. 4.3

7. -6.125

8. -5

9. -105

10. $0.456456456\ldots$

11. $2.343434\ldots$

12. $0.151151115\ldots$

13. $3.334443334444\ldots$

14. $10{,}000{,}000$

15. 1

Is each statement true or false? If the statement is true, give reasons for your response. If the statement is false, give a counterexample.

16. There is no number that is both rational and irrational.

17. The counting numbers include the integers.

18. Every mixed number is a rational number.

19. Whenever you divide a counting number by 3, the result is another counting number.

20. There are rational numbers between $\frac{1}{4}$ and $\frac{1}{3}$.

21. **Critical Thinking** A well-known irrational number is π, or *pi*. Pi represents the ratio $\frac{circumference}{diameter}$ of a circle. How can you explain that an irrational number can be represented by a ratio?

Evaluate each expression. (previous courses)

22. $8\frac{1}{11} + \frac{10}{11}$

23. $1\frac{2}{9} - \frac{2}{9}$

24. $4\frac{1}{3} + 3\frac{1}{2}$

25. $8\frac{1}{3} - 5\frac{1}{2}$

Use the order of operations to simplify each expression. (1.2 Skill B)

26. $\dfrac{(2 \times 3 + 1)^2}{2 \times 7}$

27. $\dfrac{(2 \times 3 - 1)^2}{4 \times 5 - 5}$

28. $\dfrac{(3 \times 2)^2 - 4^2}{5^2 - 4}$

29. $\dfrac{(3 \times 2)^2 - (2 + 3)^2}{(1 + 1)^2 - 3}$

Real Numbers and the Number Line

Every real number can be represented as a point on a number line. Every point on a number line represents a real number.

The number 0 is called the **origin** of the number line. Numbers to the left of 0 are negative, and numbers to the right of 0 are positive. Zero is neither positive nor negative.

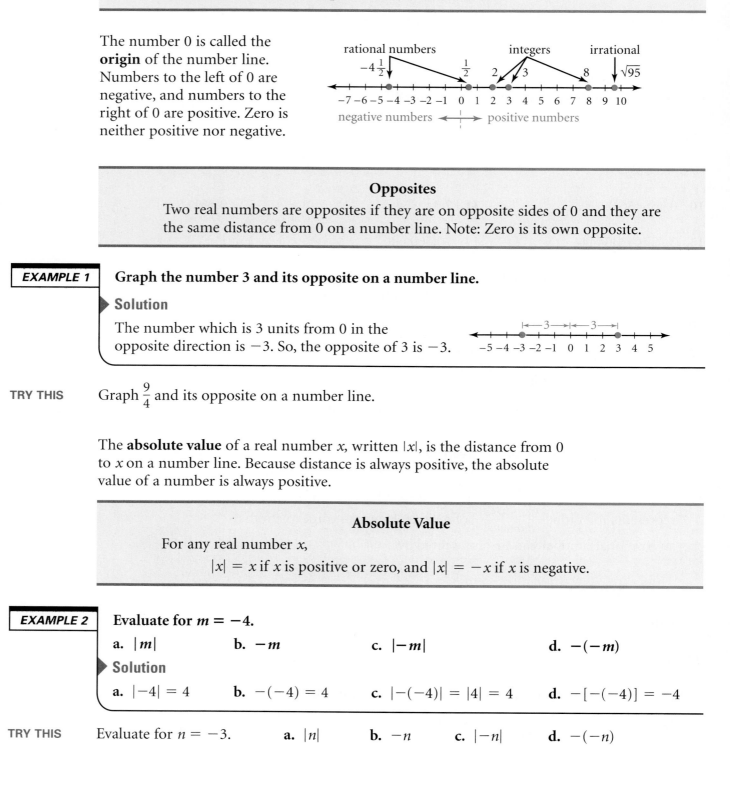

Opposites

Two real numbers are opposites if they are on opposite sides of 0 and they are the same distance from 0 on a number line. Note: Zero is its own opposite.

EXAMPLE 1 **Graph the number 3 and its opposite on a number line.**

▶ **Solution**

The number which is 3 units from 0 in the opposite direction is -3. So, the opposite of 3 is -3.

TRY THIS Graph $\frac{9}{4}$ and its opposite on a number line.

The **absolute value** of a real number x, written $|x|$, is the distance from 0 to x on a number line. Because distance is always positive, the absolute value of a number is always positive.

Absolute Value

For any real number x,

$$|x| = x \text{ if } x \text{ is positive or zero, and } |x| = -x \text{ if } x \text{ is negative.}$$

EXAMPLE 2 **Evaluate for $m = -4$.**

   a. $|m|$ b. $-m$ c. $|-m|$ d. $-(-m)$

▶ **Solution**

   a. $|-4| = 4$ b. $-(-4) = 4$ c. $|-(-4)| = |4| = 4$ d. $-[-(-4)] = -4$

TRY THIS Evaluate for $n = -3$.   a. $|n|$ b. $-n$ c. $|-n|$ d. $-(-n)$

EXERCISES

Graph all of the following numbers on the same number line.

1. 0

2. $3\frac{1}{2}$

3. 4.5

4. -2

5. -3.5

6. -4.25

Graph each number and its opposite on a number line.

7. 4

8. 1

9. 0

10. -5

11. $-2\frac{1}{2}$

12. $-4\frac{1}{7}$

13. 1.5

14. -1.5

15. -3.25

Evaluate each expression for the given value of the variable.

16. $-r;\ r = 0$

17. $-|v|;\ v = 5$

18. $-|w|;\ w = -6$

19. $-(-n);\ n = 5$

20. $-|(-p)|;\ p = 5$

21. $-d;\ d = 7$

22. $|-(-q)|;\ q = 0$

23. $-(-h);\ h = -1.3$

24. $-|(-z)|;\ z = -3.7$

25. $-|-(-t)|;\ t = 3$

26. $-|(-t)|;\ t = -0.8$

27. $-[-(-y)];\ y = -6$

Simplify each expression.

28. $|-2| + |2|$

29. $|0| + |-3| + |5|$

30. $(|-10| + |-9|) - (|-3| + |-1|)$

31. $(|-3| - |-1|) + (|-7| - |-1|)$

32. $(|3| + |-5|)^2$

33. $(|-8| - |-5|)^2$

34. $\left|\frac{3}{4} + \frac{1}{4}\right|^2$

35. $\left|2\frac{3}{4} - \frac{1}{4}\right|^2$

36. $\left|\frac{1}{2} + \frac{1}{4} + \frac{1}{8}\right|^2$

Evaluate each expression. (1.2 Skill C)

37. $3x^2 - 2x + 7;\ x = 2$

38. $2.5(x - 5) + 2.5;\ x = 5.5$

39. $3(x - 5)^2 + 3;\ x = 9$

Classify each number in as many ways as possible. (2.1 Skill B)

40. -3.9

41. $14.525252\ldots$

42. -120

2.2 Adding Real Numbers

LESSON

You have already learned how to add positive numbers. How do you add two real numbers when one or both numbers are negative?

You can show addition of real numbers as a series of moves on the number line.

- Represent a positive number as a move to the right.
- Represent a negative number as a move to the left.

Note: If a number is not preceded by $+$ or $-$, it is positive. Negative numbers are always preceded by $-$.

EXAMPLE 1   **Use a number line to find the sum $5 + (-7)$.**

▶ **Solution**

Start at the origin. Move 5 units to the right.
Then move 7 units to the left.

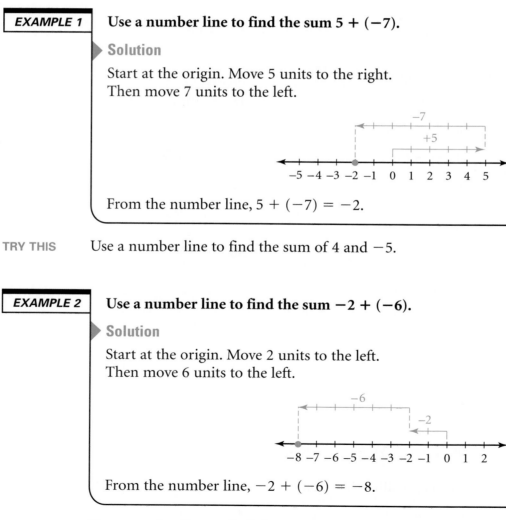

From the number line, $5 + (-7) = -2$.

TRY THIS   Use a number line to find the sum of 4 and -5.

EXAMPLE 2   **Use a number line to find the sum $-2 + (-6)$.**

▶ **Solution**

Start at the origin. Move 2 units to the left.
Then move 6 units to the left.

From the number line, $-2 + (-6) = -8$.

TRY THIS   Use a number line to find the sum $-1 + (-4)$.

EXERCISES

Use a number line such as the one below to find the number that results when you start at the origin and follow the instructions.

1. Move 3 units to the right. Move 8 units to the left.

2. Move 4 units to the left. Move 8 units to the right.

3. Move 2 units to the left. Move 2 units to the right.

4. Move 5 units to the right. Move 5 units to the left.

$$\longleftrightarrow$$
$$-5\ -4\ -3\ -2\ -1\quad 0\quad 1\quad 2\quad 3\quad 4\quad 5$$

Use a number line to find each sum.

5. $3 + (-4)$ 6. $-4 + 5$ 7. $-5 + 6$

8. $6 + (-9)$ 9. $-3 + 3$ 10. $-4 + 4$

11. $-3 + (-1)$ 12. $-5 + (-1)$ 13. $-6 + (-2)$

14. $-1 + (-1)$ 15. $-3 + (-3)$ 16. $-2 + (-2)$

Write the number that results when you follow the instructions.

17. Start at -3. Move 8 units to the right. Then move 2 units to the left.

18. Start at 5. Move 6 units to the right. Then move 2 units to the left.

19. Start at -2. Move 6 units to the left. Then move 2 units to the left.

Use a number line to find each sum.

20. $(-3 + 2) + 5$ 21. $[-3 + (-2)] + 6$ 22. $[4 + (-6)] + 3$

23. $[(-4) + 2] + 4$ 24. $[(-4) + (-2)] + 3$ 25. $[(-5) + (-1)] + (-4)$

26. **Critical Thinking**   Use a number line to show that when you add $\frac{1}{2} + \frac{1}{4} + \frac{1}{8} + \frac{1}{16}\cdots$, the sum is less than 1.

Simplify each expression. Use mental math where possible. Identify the properties used. (1.2 Skill A)

27. $1 \times \frac{3}{4} + 1 \times \frac{1}{4}$ 28. $\frac{4}{3} \times \frac{3}{4} + 1 \times 2$ 29. $\frac{5}{3} \times \frac{3}{5} + \frac{7}{4} \times \frac{4}{7}$ 30. $1 \times \frac{5}{3}(2 + 1)$

Use properties of equality and properties of numbers to show that each equation is true. (1.3 Skill C)

31. $\frac{9}{10} = 90\%$ 32. $6z = 4z + 2z$

You can add two real numbers without using a number line. Simply follow the rules below.

Rules for Addition of Two Real Numbers

Same signs: If both numbers have the same sign, add the two numbers as if both numbers are positive. The result will have the same sign as both of the given numbers.

Opposite signs: If the numbers have opposite signs, subtract the smaller from the larger as if both numbers were positive. The result will have the sign of the number with the greater absolute value.

EXAMPLE 1 | **Find the sum $5 + (-7)$.**

▶ **Solution**

1. 5 and -7 have opposite signs. Treat -7 as 7 and subtract 5.

$$7 - 5 = 2 \quad \longleftarrow \text{Subtract the smaller number from the larger.}$$

2. The result will be negative because $|-7|$ is greater than $|5|$.

Thus, $5 + (-7) = -2$.

TRY THIS   Find the sum $4 + (-8)$.

EXAMPLE 2 | **Find the sum $-2 + (-6)$.**

▶ **Solution**

1. -2 and -6 have the same sign. Add them as if they were positive.

$$2 + 6 = 8$$

2. The result will be negative because both numbers are negative.

Thus, $-2 + (-6) = -8$.

TRY THIS   Find the sum $-5 + (-3)$.

EXAMPLE 3 | **Find the sum $-2 + (-6) + 10$.**

▶ **Solution**

$$
\begin{aligned}
-2 + (-6) + 10 &= [-2 + (-6)] + 10 &&\longleftarrow \text{Apply the Associative Property of Addition.}\\
&= -8 + 10 &&\longleftarrow \text{Apply the rule for same signs.}\\
&= 2 &&\longleftarrow \text{Apply the rule for opposite signs.}
\end{aligned}
$$

TRY THIS   Find the sum $-1 + 7 + (-2)$.

EXERCISES

KEY SKILLS

Should you add or subtract the absolute values of the given numbers?

1. $-3 + 7$

2. $-4 + (-5)$

3. $-1.4 + (-2.4)$

4. $4.2 + (-7.1)$

PRACTICE

Find each sum. Write fractions in simplest form.

5. $5 + 6$

6. $-8 + 5$

7. $-1 + 9$

8. $-6 + (-9)$

9. $-7 + (-5)$

10. $-3 + (-24)$

11. $-100 + (-23)$

12. $35 + (-42)$

13. $-55 + 18$

14. $-12 + 12$

15. $13 + (-13)$

16. $-37 + (-38)$

17. $-68 + (-69)$

18. $6.5 + (-5.5)$

19. $-4.2 + 5.7$

20. $-12.1 + (-1.5)$

21. $-1.23 + 5.15$

22. $\frac{4}{7} + \left(-\frac{2}{7}\right)$

23. $-\frac{7}{9} + \frac{1}{9}$

24. $-\frac{2}{5} + \left(-\frac{1}{5}\right)$

25. $-\frac{5}{12} + \left(-\frac{1}{12}\right)$

26. $-3 + (-4) + 6$

27. $-10 + (-3) + (-5)$

28. $30 + (-40) + (-20)$

29. $-18 + (-2) + (-5)$

30. $-10 + (-10) + 10$

31. $15 + (-15) + (-10)$

32. $-27 + 18 + 12$

33. $-32 + (-79) + (-19)$

34. $2.5 + (-3.5) + (-1)$

35. $4.03 + 1.7 + (-3)$

36. $1.1 + (-1.45) + 2.1$

37. $0.2 + (-0.25) + 0.2$

Evaluate each expression for the given value of the variable.

38. $|3a + 4| + (-10); a = 2$

39. $|4z - 3| + [-3 + (-5)]; z = 1$

40. $|9b - 8| + [4 + (-7)]; b = 2$

MIXED REVIEW

Is each equation an identity? If not, explain. (1.3 Skill A)

41. $(1.4x)(1.3x) = (1.3x)(1.4x)$

42. $3(x + 5) + 3 = 4$

43. $2(x + 3) + 5 = 5 + 2(x + 3)$

44. $\frac{4}{3}(3x) = \frac{4}{3} \times 12$

45. $d + 7 - 7 = 13 - 7$

46. $s + 7 - 7 = s + 7 - 7$

Classify each number in as many ways as possible. (2.1 Skill B)

47. $0.112233\ldots$

48. $3.313113111\ldots$

49. -13

50. 82.5

51. -82.5

52. $\frac{1}{3}$

53. $\frac{1}{101010}$

54. $\frac{1}{1010101}$

2.3 Subtracting Real Numbers

LESSON

Subtracting real numbers

In Lesson 1.2, you learned various properties of numbers for addition and multiplication. In particular, you learned that every number *a* has an opposite, −*a*, which is also called an *additive inverse*.

> For every real number *a*, there is a real number −*a*, called its opposite, such that $a + (-a) = 0$.

Because you can add any two real numbers, and every real number has an additive inverse, you can define subtraction by addition of the additive inverse.

Subtraction of Real Numbers

If *a* and *b* represent real numbers, then $a - b = a + (-b)$.

The following examples show how you can rewrite subtraction of a number as addition of its opposite.

EXAMPLE 1 **Find the difference 13 − 25.**

▶ **Solution**

$$13 - 25 = 13 + (-25)$$ ⟵——— Rewrite subtraction as addition of the opposite.
$$= -12$$ ⟵——— Use the rule for addition of opposite signs.

TRY THIS Find the difference 31 − 52.

EXAMPLE 2 **Find the difference 15 − (−21).**

▶ **Solution**

$$15 - (-21) = 15 + 21$$ ⟵——— Add 21, the opposite of −21.
$$= 36$$

TRY THIS Find the difference 150.4 − (−97.5).

EXAMPLE 3 **Find the difference −13 − (−27).**

▶ **Solution**

$$-13 - (-27) = -13 + 27$$ ⟵——— Add 27, the opposite of −27.
$$= 14$$ ⟵——— Use the rule for addition of opposite signs.

TRY THIS Find the difference −30 − (−19).

EXERCISES

Rewrite each difference as a sum.

1. $12 - 15$

2. $10 - (-15)$

3. $1.2 - 0$

4. $-5 - (-7)$

5. $14 - 3$

6. $18 - 18$

Without performing the subtraction, tell whether the difference will be greater than 0, equal to 0, or less than 0.

7. $18 - 18$

8. $102 - 103$

9. $15 - 14$

10. $-18 - 4$

11. $-16 - (-12)$

12. $2.3 - (-2.3)$

13. $-33.5 - (-33.5)$

14. $7\frac{1}{3} - 7\frac{3}{5}$

15. $7\frac{3}{5} - 7\frac{1}{5}$

PRACTICE

Find each difference.

16. $10 - 4$

17. $15 - 10$

18. $-5 - 3$

19. $-9 - 6$

20. $4 - 10$

21. $12 - 18$

22. $-12 - 18$

23. $-15 - 10$

24. $6 - 12$

25. $15 - (-10)$

26. $-12 - (-18)$

27. $-15 - (-10)$

28. $-12 - (-7)$

29. $1.5 - 2.5$

30. $2.5 - 1.5$

31. $-2.45 - 1.5$

32. $-2.5 - (-1.15)$

33. $\frac{9}{2} - \frac{11}{2}$

34. $-\frac{9}{2} - \frac{10}{2}$

35. $-4\frac{1}{2} - \left(-5\frac{1}{2}\right)$

36. $4\frac{1}{2} - \left(-5\frac{1}{2}\right)$

Evaluate each expression for the given value of the variable.

37. $3x - |x - 4|; x = 2$

38. $2c - |4c - 5|; c = 3$

39. $|x - 3| - |x - 4|; x = 2$

40. $|t - 6| - |t - 9|; t = 0$

41. $3z - |6 + 5z|; z = 6$

42. $w - |6 - w|; w = -7$

MIXED REVIEW

Solve each equation. (1.4 Skill A, 1.5 Skill A)

43. $x - 1.5 = 7$

44. $x - 6 = 6$

45. $t + 7.3 = 8.9$

Solve each equation. (1.7 Skill A, 1.8 Skill A)

46. $\frac{n}{3} = 5$

47. $\frac{t}{4} = 6$

48. $3g = 27$

When there is a negative sign in front of an expression in parentheses, such as $-(3 + 4)$, there are two methods that you can use to simplify the expression.

One method is to first simplify within the parentheses and then apply the rule for the addition of opposite signs, as shown in Example 1.

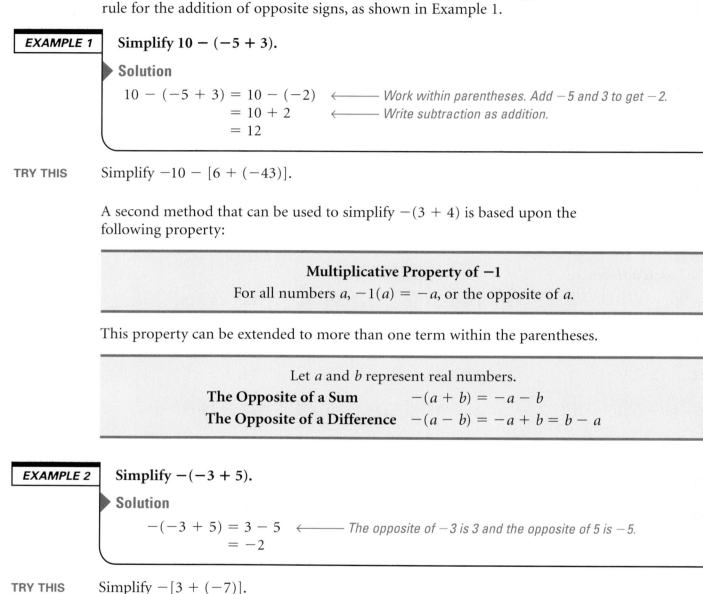

EXAMPLE 1

Simplify $10 - (-5 + 3)$.

▶ **Solution**

$$
\begin{aligned}
10 - (-5 + 3) &= 10 - (-2) &&\longleftarrow \textit{Work within parentheses. Add } -5 \textit{ and } 3 \textit{ to get } -2.\\
&= 10 + 2 &&\longleftarrow \textit{Write subtraction as addition.}\\
&= 12
\end{aligned}
$$

TRY THIS   Simplify $-10 - [6 + (-43)]$.

A second method that can be used to simplify $-(3 + 4)$ is based upon the following property:

Multiplicative Property of -1

For all numbers a, $-1(a) = -a$, or the opposite of a.

This property can be extended to more than one term within the parentheses.

Let a and b represent real numbers.

The Opposite of a Sum $-(a + b) = -a - b$

The Opposite of a Difference $-(a - b) = -a + b = b - a$

EXAMPLE 2

Simplify $-(-3 + 5)$.

▶ **Solution**

$$
\begin{aligned}
-(-3 + 5) &= 3 - 5 &&\longleftarrow \textit{The opposite of } -3 \textit{ is } 3 \textit{ and the opposite of } 5 \textit{ is } -5.\\
&= -2
\end{aligned}
$$

TRY THIS   Simplify $-[3 + (-7)]$.

EXAMPLE 3

Simplify $-[-10 + (-12)] - [-7 - (-15)]$.

▶ **Solution**

$$
\begin{aligned}
&-[-10 + (-12)] - [-7 - (-15)]\\
&[10 - (-12)] + [7 + (-15)]\\
&(10 + 12) + [7 + (-15)]\\
&22 + (-8)\\
&14
\end{aligned}
$$

TRY THIS   Simplify $-[-15 + (-15)] - [(-17) + (-13)]$.

EXERCISES

Use the Multiplicative Property of -1 as in the first step in Example 2.
Do not evaluate.

1. $-(2 + 1)$

2. $-(3 - 5)$

3. $-[-2 + (-3)]$

4. $-[3 + (-5)]$

5. $-[-4 - (-7)]$

6. $-[-1 - (-1)]$

PRACTICE

Simplify each expression.

7. $10 - (3 + 5)$

8. $10 - (-3 + 6)$

9. $-12 - (4 + 9)$

10. $-3 - (10 + 9)$

11. $-12 - (3 + 1)$

12. $100 - (82 + 18)$

13. $-(3 + 4) - (5 + 6)$

14. $-(3 - 7) - (5 - 6)$

15. $-(-3 - 7) - (-5 + 7)$

16. $12 - (3 + 4) - (5 + 6)$

17. $-20 - (5 + 5) - (-5 + 6)$

18. $-2 - (5 + 1) - (-5 - 8)$

19. $-(2x + 3)$

20. $-(3 - 4a)$

21. $-(2d - 5)$

22. $-(7z + 6)$

23. $-(x + 7)$

24. $-(-5w + 11)$

25. $-(3 + 4) - (5 + 6) + (2 - 5)$

26. $-(3 - 7) - (5 - 6) + (-2 + 4)$

27. $-(-3 - 7) - [(-5 + 7) - (1 - 9)]$

28. $-20 - (5 + 5) - [(-5 + 6) - (-2 - 3)]$

29. $-[(-2 - 8) - (-5 + 11)] - [(1 - 10) - (-3 - 3)]$

30. $-[7 - (4 + 4)] - (-5 + 7) - [(-2 - 3) + (-8 + 8)]$

Evaluate each expression for the given values of the variables.

31. $|3a| + |2b| - |2a + 3b|$; $a = -1$ and $b = -3$

32. $|-2r| + |5s| - |4r + 6s|$; $r = -2$ and $s = -5$

33. $(|x| - |y|)^2 - (|x| + |y|)^2$; $x = 1$ and $y = 1$

34. $2w - 3|w| - |x - w|$; $x = -2$ and $w = -3$

MIXED REVIEW

Find each sum. (2.2 Skill B)

35. $5.5 + (-3)$

36. $-5 + (-12)$

37. $-12 + (-20)$

38. $-18 + 18$

39. $-6 + (-9)$

40. $-6.4 + 6.4$

41. $-11 + 18 + (-6)$

42. $-7 + (-6) + (-2)$

43. $-6.1 + 6.2 + (-3.2)$

EXAMPLE 1

The thermostat in Rachel's freezer was set at 5°F. She lowered the thermostat by 7°F. Later, she raised the thermostat by 3°F. At what temperature was the thermostat finally set?

▶ **Solution**

Look for key words.

Translate the phrases into a numerical expression. The key words are *lowered* and *raised. Lowered* indicates subtraction and *raised* indicates addition.

$$5 - 7 + 3$$
$$= 5 + 3 - 7 \quad \longleftarrow \text{Commutative Property}$$
$$= (5 + 3) - 7 \quad \longleftarrow \text{Associative Property}$$
$$= 8 - 7$$
$$= 1$$

Notice that you can use the Commutative and Associative Properties to make your calculations easier. Adding all the positive numbers in an expression before subtracting can aid mental computation.

The thermostat was finally set at 1°F.

TRY THIS

One winter afternoon, the temperature was 8°F. By evening, the temperature had dropped 5°F, and by 11:00 P.M., the temperature had dropped another 6°F. What was the temperature at 11:00 P.M.?

EXAMPLE 2

A bank customer makes three withdrawals and two deposits. Find the resulting, or *net*, change in the customer's balance.

deposits: $120.50 and $76.45
withdrawals: $70.32, $18.59 and $112.75

▶ **Solution**

Find the sum of the deposits less the sum of the withdrawals.

$$\text{deposits} - \text{withdrawals}$$
$$= (120.50 + 76.45) - (70.32 + 18.59 + 112.75)$$
$$= 196.95 - 201.66$$
$$= -4.71$$

The net change is a decrease of $4.71 in the bank balance.

TRY THIS

A bank customer makes three withdrawals and two deposits. Find the net change in the customer's balance.

deposits: $100.50 and $86.55
withdrawals: $80.32, $20.55 and $112.45

You can also solve Example 2 by using the Multiplicative Property of -1.
$$(120.50 + 76.45) - (70.32 + 18.59 + 112.75)$$
$$= 120.50 + 76.45 - 70.32 - 18.59 - 112.75$$
$$= -4.71$$

EXERCISES

KEY SKILLS

Translate each phrase into numerical expressions.

1. the total restaurant bill with the following items: the price of the meal, $6.75, a discount coupon for $1.50 off, tax of $0.42, and a tip of $1.00

2. the total bill at the clothing store after the following: a return of a shirt, $25.50, purchase of two shirts for $17.25 and $21.75, and $0.68 tax

3. the net change in altitude of an airplane after climbs of 450 ft and 275 ft and descents of 45 ft, 120 ft, and 230 ft

PRACTICE

Solve each problem.

4. The temperature was 16°F, and then it fell 19°F. What was the final temperature?

5. The temperature was −5°F at midnight. It rose 7°F by 3:00 A.M. Find the temperature at that time.

6. On Tuesday afternoon, the temperature was −3°F. That night it dropped 8°F, and the next morning it rose 15°F. What was the temperature on Wednesday morning?

7. The net change in stock market averages after daily changes of: +36, −107, −56, +76, and +45.

Find the net change in each bank balance.

8. deposits: $110.33, $202.56, and $53.70
   withdrawals: $33.55, $300.56, and $45.00

9. deposits: $300.55, $250.00, and $79.90
   withdrawals: $313.55, $256.56, and $65.00

10. **Critical Thinking**   Write and evaluate an expression for the net value of the following transactions at the clothing store: the return of a shirt for $15, the purchase of two pairs of shoes at $30 a pair, a 30% discount on the shoes, and a 5% sales tax on the net purchase price.

MIXED REVIEW APPLICATIONS

Solve each problem. (1.9 Skill B)

11. Sam wants to deposit an equal amount of money into his new bank account each month. At the end of one year, he wants to have a total of $2700 in his account. How much money should he deposit each month?

12. **a.** How many people can be fed 6-ounce servings using 64 ounces of hamburger?
    b. How large is each portion of salad if there are 56 ounces of salad and 14 people to feed?

 **Multiplying and Dividing Real Numbers**

LESSON

SKILL A *Multiplying real numbers*

When you multiply two real numbers, use the following rules to find the sign of the product.

Rules for Multiplying Two Real Numbers

Same signs: If both numbers have the same sign, multiply as if the numbers were positive. The product will always be positive.

Opposite signs: If the numbers have opposite signs, multiply as if the numbers were positive. The product will always be negative.

EXAMPLE 1 **Find each product.**   **a.** $(-26)(2)$ **b.** $(-26)(-2)$

▶ **Solution**

a. Multiply as if the numbers were positive. $26 \times 2 = 52$
   Because -26 and 2 have opposite signs, use $-$. -52

b. Multiply as if the numbers were positive. $26 \times 2 = 52$
   Because -26 and -2 have the same sign, no
   change is necessary. 52

TRY THIS Find each product.   **a.** $(-120) \times (-4)$ **b.** $(1.5) \times (-4)$

EXAMPLE 2 **Find $\left(-4\frac{1}{2}\right)^2$.**

▶ **Solution**

Recall that $x^2 = x \cdot x$.

$$\left(-4\frac{1}{2}\right)^2 = \left(-4\frac{1}{2}\right)\left(-4\frac{1}{2}\right) = \left(-\frac{9}{2}\right)\left(-\frac{9}{2}\right) = \frac{81}{4}, \text{ or } 20\frac{1}{4}$$

TRY THIS Find $\left(-3\frac{1}{3}\right)^2$.

You can write a proof for the rules of multiplying two real numbers shown above. Let a and b represent positive real numbers. Then $-a$ is negative and $(-a)b$ is the product of a negative number and a positive number.

$$\begin{aligned} (-a)b + ab &= (-a + a)b \quad \longleftarrow \text{ Distributive Property} \\ &= 0 \cdot b \quad \longleftarrow \text{ Inverse Property of Addition} \\ &= 0 \end{aligned}$$

This means ab is the additive inverse of $(-a)b$. So, $(-a)b = -(ab)$.

EXERCISES

Write the sign of each product.

1. $(-4) \times (5)$ **2.** $(-4)(-5)$ **3.** $\left(4\frac{3}{5}\right)^2$ **4.** $\left(-\frac{3}{5}\right)\left(-2\frac{2}{3}\right)$

Find each product. Write fractions in simplest form.

5. $6(-2)$ **6.** $9(-9)$ **7.** $(-5)(-5)$ **8.** $(-3)(-4)$

9. $4(-11)$ **10.** $(12)(-13)$ **11.** $(-36) \times (-3)$ **12.** $(-5) \times (-11)$

13. $2.5 \times (-2)$ **14.** $(-1.2) \times (-1.2)$ **15.** $\frac{-4}{7} \times \frac{7}{3}$ **16.** $\frac{-2}{7} \times \frac{3}{-5}$

17. $3 \times \left(-3\frac{1}{3}\right)$ **18.** $(-2) \times \left(-5\frac{1}{2}\right)$ **19.** $\left(1\frac{3}{5}\right)^2$ **20.** $\left(-\frac{3}{5}\right)\left(-\frac{2}{3}\right)$

21. $(-3)(-4)(-2)$ **22.** $(-10)(5)(-2)$ **23.** $(-10)(-6)(-3)$

24. $(-12)(-6)(-5)$ **25.** $[(-6) \times (-10)] \times 4$ **26.** $[(-7) \times (-2)] \times (-4)$

Determine whether each product is positive or negative.

27. $-4a$; a is positive   **28.** $-4a$; a is negative   **29.** $4b \times -\frac{1}{2}$; b is positive

30. $\frac{-5}{-3} \times c$; c is negative   **31.** $\frac{3}{-4} \times n$; n is positive   **32.** $\frac{-4}{9} \times m$; m is negative

33. Critical Thinking Use mental math to simplify the following expression and then describe your strategy. $\left(\frac{2}{3}\right)\left(-\frac{5}{4}\right)\left(-\frac{5}{6}\right)\left(\frac{3}{2}\right)\left(\frac{6}{5}\right)\left(-\frac{4}{5}\right)$

Find each sum. (2.2 Skill B)

34. $-14.1 + (-3.8)$ **35.** $2.5 + (-18)$ **36.** $5 + (-3)$ **37.** $-7 + (-4)$

38. $-5 + (-12.1)$ **39.** $-5.6 + (-9.3)$ **40.** $-13 + (-13)$ **41.** $-13 + 13$

Find each difference. (2.3 Skill A)

42. $2 - (-2)$ **43.** $-18 - 18$ **44.** $27.2 - 27.2$ **45.** $100 - 19.90$

46. $2.2 - (-2)$ **47.** $-1.9 - 19$ **48.** $27.2 - 7.2$ **49.** $10 - 9.8$

In Lesson 1.2, you learned that every nonzero real number a has a reciprocal, also called the multiplicative inverse.

Because you can multiply any two real numbers, and every nonzero real number has a multiplicative inverse, you can define division using multiplication and multiplicative inverses.

Division of Real Numbers

Let a and b represent real numbers, where b is nonzero.

$$a \div b = a \times \frac{1}{b}$$

Same signs: If a and b have the same sign, divide as if the numbers were positive. The quotient will always be positive.

Opposite signs: If a and b have opposite signs, divide as if the numbers were positive. The quotient will always be negative.

EXAMPLE 1 Find $(-26) \div 2$.

▶ **Solution**

1. Divide as if the numbers were positive. $26 \div 2 = 13$

2. Make the quotient negative. -13

TRY THIS Find each quotient. **a.** $(-120) \div (4)$ **b.** $(-120) \div (-4)$

EXAMPLE 2 Find $\frac{3}{4} \div \left(-\frac{15}{8}\right)$.

▶ **Solution**

$$\frac{3}{4} \div \left(-\frac{15}{8}\right) = \frac{3}{4} \times \left(-\frac{8}{15}\right)$$ ⟵ *Multiply by $-\frac{8}{15}$, the reciprocal of $-\frac{15}{8}$.*

$$= -\left(\frac{3}{4} \times \frac{\overset{2}{8}}{\underset{5}{15}}\right) = -\frac{2}{5}$$

TRY THIS Find $\left(-3\frac{1}{3}\right) \div \left(1\frac{2}{3}\right)$.

The following is a proof of the rule for dividing a negative number by a positive number. Let a and b represent positive real numbers. Then $-a$ is negative.

$$\frac{-a}{b} = (-a) \times \frac{1}{b}$$ ⟵ *Definition of division*

$$= -\left(a \times \frac{1}{b}\right)$$ ⟵ *Multiplication with opposite signs*

$$= -\left(\frac{a}{b}\right)$$ ⟵ *Definition of division*

EXERCISES

KEY SKILLS

Write the sign of each quotient.

1. $(-45) \div (5)$

2. $(-15) \div (5)$

3. $\left(4\frac{3}{5}\right) \div 2$

4. $\left(-\frac{3}{5}\right) \div \left(-\frac{2}{3}\right)$

PRACTICE

Find each quotient. Write fractions in simplest form.

5. $(-60) \div 10$

6. $(-32) \div 8$

7. $(-72) \div 8$

8. $(-32) \div (-4)$

9. $(39) \div (-13)$

10. $(27) \div (-3)$

11. $\frac{-81}{9}$

12. $\frac{-54}{-27}$

13. $\frac{35}{-7}$

14. $(-55) \div (-11)$

15. $2.5 \div (-2)$

16. $(-1.2) \div (-1.2)$

17. $10 \div \left(-3\frac{1}{3}\right)$

18. $(-11) \div \left(-5\frac{1}{2}\right)$

19. $\frac{-4}{7} \div \frac{3}{7}$

20. $\frac{-2}{7} \div \frac{3}{-5}$

21. $\left(4\frac{3}{5}\right) \div \left(-4\frac{3}{5}\right)$

22. $\left(-\frac{3}{5}\right) \div \left(-\frac{3}{2}\right)$

23. $(-13) \div (-4)$

24. $(-17) \div (5)$

25. $(-25) \div (-5)$

26. $(22) \div (-5)$

27. $(-6) \div (-10)$

28. $(-12) \div (-15)$

Determine whether each quotient is positive or negative.

29. $\frac{a}{-2}$; a is positive

30. $\frac{-2}{a}$; a is negative

31. $\frac{4b}{-2}$; b is positive

32. $\frac{-5c}{-3}$; c is negative

33. $\frac{-3n-3}{-4}$; n is positive

34. $\frac{m+4}{-9}$; m is positive

MIXED REVIEW

Simplify. (previous courses)

35. 2^2

36. 3^2

37. 4^2

38. 5^2

39. 6^2

40. 7^2

41. 8^2

42. 9^2

43. $5^2 - 4^2$

44. $6^2 - 5^2$

45. $10^2 - 9^2$

46. $100^2 - 99^2$

Solve each equation. Check your solution. (1.6 Skill A)

47. $3.45 + g = 13.5$

48. $t - 2 = 18.5$

49. $100 = 13.8 + r$

50. $12.56 = s - 5$

Solve each equation. Check your solution. (1.9 Skill A)

51. $\frac{2w}{7} = 10$

52. $13 = \frac{13z}{6}$

53. $44 = \frac{4a}{11}$

54. $\frac{2n}{2} = 1$

You have learned the order of operations and the rules for adding, subtracting, multiplying, and dividing real numbers. You can now simplify many expressions that involve positive and negative numbers.

EXAMPLE 1   **Simplify** $(-8 + 5)^2 - \left(\dfrac{-24}{6}\right).$

▶ **Solution**

$(-8 + 5)^2 - \left(\dfrac{-24}{6}\right)$

$\quad = (-3)^2 - (-4)$ ←——— *Perform the addition and division in parentheses.*

$\quad = 9 - (-4)$ ←——— *Calculate $(-3)^2 = (-3)(-3) = 9$.*

$\quad = 13$ ←——— *Subtract.*

TRY THIS   Simplify $\left(\dfrac{14 \times (-2)}{-7}\right) + (-5)^2 - 3.$

Notice in Example 1 that $(-3)^2$ equals 9, not -9. This can be generalized for any real number a: $(-a)^2 = a^2$, not $-a^2$.

When you simplify a product of fractions, look for common factors of the numerators and denominators. Divide both the numerator and a denominator by the common factors; this process is sometimes called *canceling*.

EXAMPLE 2   **Simplify** $\dfrac{(1 - 4)^2}{(-2)(5)} \times \dfrac{(-2)(-6)}{(6)(6)}.$

▶ **Solution**

$\dfrac{(1 - 4)^2}{(-2)(5)} \times \dfrac{(-2)(-6)}{(6)(6)}$

$= \dfrac{(-3)(-3)}{(-2)(5)} \times \dfrac{(-2)(-6)}{3 \times 2 \times 3 \times 2}$

$= \dfrac{(-3)(-3)}{(-2)(5)} \times \dfrac{(-2)(-6)}{3 \times 2 \times 3 \times 2}$

$= \dfrac{(-1)(-1)(-3)}{(5)(2)}$

$= -\dfrac{3}{10}$

TRY THIS   Simplify $\dfrac{(1 - 6)^2}{(-2)(5)} \times \dfrac{(2)(-5)}{(-5)(-5)}.$

EXERCISES

KEY SKILLS

Simplify each power.

1. $(-5)^2$

2. $-(-5)^2$

3. $-(5)^2$

4. 5^2

PRACTICE

Simplify each expression.

5. $3 - 5(1 - 8)$

6. $2 \times 8 \div [4 - (-4)]$

7. $2 \div [4 - (-4)] \times 8$

8. $\dfrac{2[-3 + (-2)]}{5}$

9. $\dfrac{-3[-3 - (-7)]}{-2}$

10. $\dfrac{(2 - 5)^2}{3}$

11. $\dfrac{[-4 - (-5)]^2}{2}$

12. $\dfrac{2 - (-12)}{12 - 7(-2)}$

13. $\dfrac{8 - (-12)}{12(-2) + 3(-2)}$

14. $\dfrac{(3 - 5)}{-2} \times \dfrac{(7 - 1)}{-4}$

15. $\dfrac{(7 - 10)}{-5} \times \dfrac{[7 - (-3)]}{(12 - 13)}$

16. $\dfrac{(7 - 10)}{(3 - 7)} \times \dfrac{(8 - (-4)}{(14 - 17)}$

17. $(-3 - 3)^2(-2 + 4)^2$

18. $(-3 - 3)^2(-2 + 4)^2 \div (-4)^2$

19. $(-3 - 3)^2 \div (-2 + 6)^2 \div (-2)^2$

20. $\dfrac{-(10 - 1)^2}{(-2)(7)} \times \dfrac{(2)(-7)}{(-9) \times (9)}$

21. $\dfrac{(7 - 1)^2}{(-2)(11)} \times \dfrac{(3)(-11)}{-(-3) \times (-6)}$

22. $\dfrac{(7 - 1)^2}{(-5)(10)} \times \dfrac{(3)(-10)}{(-3) \times (-6)}$

23. $\dfrac{(7 - 1)^2}{(-3)(10)} \times \dfrac{(3)(-1)}{(1 - 7)^2}$

24. $\dfrac{(2 - 5)^2}{[3 - (-1)]^2} \times \dfrac{(3)(-1)}{-(2 - 5)^2}$

25. Critical Thinking Find a such that $\dfrac{(4 - a)^2}{100} \times \dfrac{[8 - (-2)]^2}{(5 - 10)^2} = 1$.

MID-CHAPTER REVIEW

For each number, find:
a. the opposite and   **b. the absolute value.** (2.1 Skill C)

26. -5.2

27. 0

28. 12.6

29. 2.3345

Find each sum or difference. (2.2 Skill B, 2.3 Skill A)

30. $-5.1 - (-12)$

31. $18 + (-26)$

32. $-11 + (-19)$

33. $4 - 6.3$

Find each product or quotient. (2.4 Skills A and B)

34. $3.5 \times (-4)$

35. $-2.5 \times (-6)$

36. $28.4 \div (-4)$

37. $30.6 \div (-6)$

2.5 LESSON

Solving Equations Involving Real Numbers

Using properties of equality, addition rules, and subtraction rules to solve equations

You used the following properties of equality to solve simple equations in earlier lessons. These properties apply to both positive and negative numbers.

> **Addition and Subtraction Properties of Equality**
>
> Let a, b, and c be real numbers.
>
> **Addition**   If $a = b$, then $a + c = b + c$.
>
> **Subtraction**   If $a = b$, then $a - c = b - c$.

EXAMPLE 1   Solve $d + 5 = -4$. Check your solution.

▶ **Solution**

$$d + 5 = -4$$
$$d + 5 - 5 = -4 - 5 \qquad \longleftarrow \text{Apply the Subtraction Property of Equality.}$$
$$d + 0 = -4 - 5$$
$$d = -9 \qquad \longleftarrow \text{Use the rule for addition with same signs; } -4 + (-5) = -9$$

Check: Substitute -9 for d in the original equation.

$$-9 + 5 \stackrel{?}{=} -4$$
$$-4 \stackrel{?}{=} -4 \ ✔$$

TRY THIS   Solve $a + 2 = -4$.

EXAMPLE 2   Solve $z - 4.5 = -12$. Check your solution.

▶ **Solution**

$$z - 4.5 = -12$$
$$z - 4.5 + 4.5 = -12 + 4.5 \qquad \longleftarrow \text{Apply the Addition Property of Equality.}$$
$$z = -7.5$$

Check: Substitute -7.5 for z in the original equation.

$$-7.5 - 4.5 \stackrel{?}{=} -12$$
$$-12 \stackrel{?}{=} -12 \ ✔$$

TRY THIS   Solve $v - 3.5 = -2$.

EXERCISES

a. Identify the property of equality needed to solve each equation.
b. Then state the number that you would add or subtract. Do not solve.

Sample: $x + (-4) = 7$ Solution: Change to $x - 4 = 7$.

a. Addition Property of Equality   b. 4

1. $d + 5 = -3$ **2.** $r + (-3) = 6$ **3.** $x + 4 = -11$ **4.** $y - 5 = 12$

PRACTICE

Solve each equation. Check your solution.

5. $x - 23 = -7$ **6.** $x - 35 = -18$ **7.** $-12 = 18 - y$

8. $x - (-10) = 11$ **9.** $a - (-2) = 3$ **10.** $t + (-2) = -1$

11. $b + (-7) = -7$ **12.** $-15 + a = 0$ **13.** $-5 = p - 5$

14. $-7 = p + (-3)$ **15.** $-2.2 + x = -5$ **16.** $-11.2 = w + 5.6$

17. $7.5 = v - (-3.5)$ **18.** $-4.3 - (-y) = -6.7$ **19.** $4.8 - x = -0.5$

20. $g - \frac{2}{3} = 18$ **21.** $d + \frac{1}{2} = 5$ **22.** $6\frac{1}{2} - x = 4$

23. $4\frac{2}{3} + x = 5\frac{2}{3}$ **24.** $-x - 4.5 = 10.7$ **25.** $-y + 2.7 = 8.8$

MIXED REVIEW

Write each fraction as a decimal. (2.1 Skill A)

26. $\frac{7}{8}$ **27.** $\frac{24}{25}$ **28.** $\frac{2}{3}$ **29.** $\frac{5}{12}$

Write each decimal as a fraction in simplest form. (2.1 Skill A)

30. 1.35   **31.** 0.04   **32.** 0.0404   **33.** 3.7575

Classify each number in as many ways as possible. (2.1 Skill B)

34. -0.4545 **35.** $\frac{30}{6}$ **36.** $2.\overline{435}$ **37.** $1.2334445555\ldots$

Evaluate each expression for $d = -6$. (2.1 Skill C)

38. d **39.** $-d$ **40.** $-(-d)$ **41.** $|d|$

42. $|-d|$ **43.** $-|-d|$ **44.** $-|d|$ **45.** $|-(-d)|$

You used the following properties of equality to solve simple equations in earlier lessons. These properties apply to both positive and negative numbers.

Multiplication and Division Properties of Equality

Let a, b, and c be real numbers.

Multiplication If $a = b$, then $ac = bc$.

Division If c is nonzero and $a = b$, then $\dfrac{a}{c} = \dfrac{b}{c}$.

When solving certain equations, changing a decimal to a fraction first may make the computations easier.

EXAMPLE 1

Solve $-6.5x = 26$. Check your solution.

▶ **Solution**

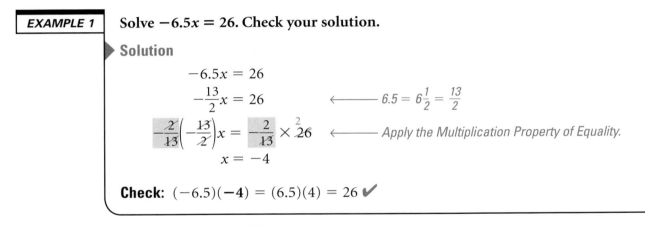

$$-6.5x = 26$$
$$-\frac{13}{2}x = 26 \qquad \longleftarrow \quad 6.5 = 6\frac{1}{2} = \frac{13}{2}$$
$$-\frac{2}{13}\left(-\frac{13}{2}\right)x = -\frac{2}{13} \times \overset{2}{26} \qquad \longleftarrow \text{ Apply the Multiplication Property of Equality.}$$
$$x = -4$$

Check: $(-6.5)(-4) = (6.5)(4) = 26$ ✔

TRY THIS   Solve $-5.5t = 22$. Check your solution.

Before applying a property of equality to solve an equation, check to see if you can first simplify any expressions. In the following example, replacing $3 + (-5)$ with its sum, -2, gives a simpler equation to solve.

EXAMPLE 2

Solve $\dfrac{z}{3 + (-5)} = 12$. Check your solution.

▶ **Solution**

$$\frac{z}{3 + (-5)} = 12$$
$$\frac{z}{-2} = 12$$
$$(-2)\left(\frac{z}{-2}\right) = (-2)(12) \qquad \longleftarrow \text{ Apply the Multiplication Property of Equality.}$$
$$z = -24$$

Check: $\dfrac{-24}{3 + (-5)} = \dfrac{-24}{-2} = 12$ ✔

TRY THIS   Solve $3a = -11 - 7$. Check your solution.

EXERCISES

KEY SKILLS

Simplify and rewrite each equation. Do not solve.

1. $(2 + 4)x = 12$

2. $25 = \dfrac{4 - 3}{1 + 1}d$

3. $(1.5 + 1.5)a = 12 + 6$

4. $-5 = \dfrac{2 - 3}{5}d$

5. $\dfrac{2 + 11}{13 + 13}c = 1 - 2$

6. $\dfrac{2 + 11}{13 + 13}c = \dfrac{2.6 + 3.4}{2}$

PRACTICE

Solve each equation. Check your solution.

7. $-6x = -42$

8. $-5a = -30$

9. $-\dfrac{1}{5}x = 7$

10. $-\dfrac{1}{7}g = -3$

11. $\dfrac{2}{3}y = -14$

12. $\dfrac{3}{4}x = 15$

13. $-2.5x = 5$

14. $\dfrac{d}{5 + 3} = -6$

15. $20 = -5z$

16. $-6 = \dfrac{d}{-6 + 3}$

17. $(-4.1 - 3.9)z = -64$

18. $\dfrac{d}{-6.2 - 3.8} = -2.6$

19. $(10 - 2.1 - 2.9)t = 100$

20. $-1 = \dfrac{(4 - 5)d}{-4.2 + 3.2}$

21. $2.9 = (100 - 28.1 - 61.9)t$

22. $\dfrac{d}{100 - 44.3 - 45.7} = 34.8$

23. $21 = \left(3 - \dfrac{1}{3} - \dfrac{1}{3}\right)r$

24. $\dfrac{d}{10 - \dfrac{22}{5} - \dfrac{33}{5}} = -2$

Is the solution positive or negative? Use mental math where possible. Do not solve.

25. $[4 + (-6)]x = -10$

26. $\dfrac{z}{10 - 5 + 2} = -2$

27. $(-2 - 2 - 2)r = 10$

28. $12 = [3 - 5 + (-7)]s$

29. $-2.1 = (-2 - 3 - 4)a$

30. $\dfrac{z}{1.2 + 1.7 + 2.2} = -2$

31. Critical Thinking Use the equation $-\dfrac{a}{b} = \dfrac{-a}{b}$ to show that $-\dfrac{a}{b} = \dfrac{a}{-b}$.

(*Hint:* Multiply each side of the equation by $\dfrac{-1}{-1}$, which equals 1.)

MIXED REVIEW

Give a counterexample to show that each statement is false. (1.3 Skill B)

32. The product of 1 and a number is 1 more than the number.

33. The quotient of a number divided by 1 equals the quotient of 1 divided by that number.

2.6
LESSON

The Distributive Property and Combining Like Terms

Using the Distributive Property to combine like terms

A **monomial** is a real number or the product of a real number and a variable raised to a whole-number power. For example, 6, $-4c$, and $3x^2$ are monomials. When there is both a number and a variable in the product, the number is called the **coefficient**. In the monomials $-4c$ and $3x^2$, -4 and 3 are the coefficients.

Monomials are **like terms** if the variable parts are the same.
    like terms: $3x$ and $-5x$ unlike terms: $3x$ and -5
    like terms: 3 and -5 unlike terms: $3x^2$ and $-5x$

Recall that a monomial with no visible coefficient, such as x, actually has a coefficient of 1. Similarly, $-x$ has a coefficient of -1.

You can use the Distributive Property to combine like terms.

EXAMPLE 1 **Simplify.**   **a. $3x + 5x$**   **b. $-4a - 6a$**   **c. $4n - n$**

▶ **Solution**

  a. $3x + 5x = (3 + 5)x$ **b.** $-4a - 6a = (-4 - 6)a$ **c.** $4n - n = (4 - 1)n$
           $= 8x$ $= -10a$ $= 3n$

TRY THIS   Simplify.   **a. $-2x + 9x$**   **b. $-4.5r - 2.5r$**   **c. $11m - 10m$**

EXAMPLE 2   **Simplify $4y - 2 - 3y$.**

▶ **Solution**

         $4y - 2 - 3y = 4y + (-3y) + (-2)$ ←——— *Apply the Commutative Property of Addition.*
                  $= [4 + (-3)]y + (-2)$
                  $= y - 2$

TRY THIS   Simplify $4x - 3 - 5x$.

When you have unlike terms in a sum or difference within parentheses, you may have to use the opposite of a sum or difference when simplifying.

EXAMPLE 3   **Simplify $7c - (5c + 5) - 3$.**

▶ **Solution**

     $7c - (5c + 5) - 3 = 7c - 5c - 5 - 3$ ←——— *Apply $-(a + b) = -a - b$.*
                     $= (7 - 5)c - 5 - 3$
                     $= 2c - 8$

TRY THIS   Simplify $6z - (z - 5) - 7$.

EXERCISES

Is each set of monomials like terms?

1. $3x$ and $-4x$

2. $3x$ and -4

3. $2x^2$ and $3x^2$

4. x and $-2x$

5. $6, 2x,$ and $3x^2$

6. 4 and 5

PRACTICE

Simplify each expression.

7. $-2x + 3x$

8. $4a + (-2a)$

9. $-8z + (-2z)$

10. $5x - 3x$

11. $3x - 5x$

12. $10a - 7a$

13. $7a - 10a$

14. $5.5d - 6.5d$

15. $7\frac{1}{2}t - 6\frac{1}{2}t$

16. $-3x - 1 + 5x$

17. $10a - 5 - 5a$

18. $12t + 3 - 5t$

19. $2y - 11 - 1 + 5y$

20. $-a - 5 - 5a - 3$

21. $12 + 5 - 8d + 7$

22. $(3y - 2y) - (4y + 5y)$

23. $(7w - 3w) - (4w + 6w)$

24. $-(3d + 7d) - (d + d)$

25. $10 - (3x + 5x) - (5x - 5x)$

26. $15 - (-3x + 7x) - (5x - 7x)$

27. $20 - (-3x - 8x) - (5x + 8x)$

**Let a, b, c, and d represent real numbers. Let x be the variable.
Rewrite each expression so that it contains only one x-term.**
Sample: $a(x + 3) + bx$ Solution: $(a + b)x + 3a$

28. $a(x + 3) - bx$

29. $a(2x + 1) - bx$

30. $a(3x - 1) + bx + 2$

31. $a(x - 1) + c(x + 3)$

32. $a(x - 2) + c(x + b)$

33. $a(x - 2) + c(x + b) + d(x + 1)$

MIXED REVIEW

Evaluate each expression. (2.3 Skill B)

34. $10 - (12 + 3)$

35. $-(3 + 9) - (3 + 10)$

36. $-(-2 - 5) + 3 - (-2)$

37. $12 - (-3 - 3) + (-9)$

38. $-10 - 12 - (25 - 13)$

39. $-(3 - 3) - (2 - 2) - (5 - 5)$

Simplify. (2.3 Skill B)

40. $6x - (x + 3x)$

41. $2x + 3x - (2x + 2x)$

42. $x + 2x + 3x - (6x - 3x)$

You can use the Distributive Property to simplify a product. The Distributive Property applies to both negative and positive numbers.

EXAMPLE 1

Simplify each expression.

a. $3(x + 2)$ b. $-4(z - 7)$

▶ **Solution**

a. $3(x + 2) = 3(x) + 3(2)$
$$= 3x + 6$$

b. $-4(z - 7) = (-4)(z) + (-4)(-7)$
$$= -4z + 28$$

TRY THIS   Simplify each expression.   a. $3(-2x - 2)$ b. $-4(z + 7)$

EXAMPLE 2

Simplify $3(a - 2) - 5(a + 1) + 12$.

▶ **Solution**

$$3(a - 2) - 5(a + 1) + 12 = 3(a) + 3(-2) + (-5)(a) + (-5)(1) + 12$$
$$= 3a - 6 - 5a - 5 + 12$$
$$= 3a - 5a - 6 - 5 + 12$$
$$= (3 - 5)a - 6 - 5 + 12$$
$$= -2a + 1$$

TRY THIS   Simplify $-4(t - 7) - 2(3t + 9) - 3$.

EXAMPLE 3

Simplify each expression.

a. $-\dfrac{4}{3}(-6t - 12)$ b. $\dfrac{10x + 25}{-10}$

▶ **Solution**

a. $-\dfrac{4}{3}(-6t - 12) = \left(-\dfrac{4}{3}\right)(-6t) + \left(-\dfrac{4}{3}\right)(-12)$ ⟵ *Apply the Distributive Property.*

$$= \left[\left(-\dfrac{4}{3}\right)(-6)\right]t + \left(-\dfrac{4}{3}\right)(-12)$$

$$= 8t + 16$$

b. $\dfrac{10x + 25}{-10} = -\dfrac{1}{10}(10x + 25)$

$$= \left(-\dfrac{1}{10}\right)(10x) + \left(-\dfrac{1}{10}\right)(25)$$ ⟵ *Apply the Distributive Property.*

$$= \left[\left(-\dfrac{1}{10}\right)(10)\right]x + \left(-\dfrac{1}{10}\right)(25)$$

$$= -x - 2.5$$

TRY THIS   Simplify each expression.   a. $-\dfrac{1}{3}(6z + 9)$ b. $\dfrac{10t - 5}{-5}$

EXERCISES

KEY SKILLS

In the following exercises, the first step(s) of simplifying the given expression are shown. Complete each simplification.

1. $-2(3g - 3) = (-2)(3g) + (-2)(-3)$

2. $4(-3b - 5) = 4(-3b) + 4(-5)$

3. $\frac{3}{5}(10s - 20) = \frac{3}{5}(10s) + \frac{3}{5}(-20)$

4. $\frac{-4y + 16}{2} = \frac{1}{2}(-4y + 16) = \frac{1}{2}(-4y) + \left(\frac{1}{2}\right)(16)$

PRACTICE

Simplify each expression.

5. $-2(x + 5)$

6. $-3(a + 4)$

7. $7(t - 2)$

8. $-3(d - 10)$

9. $6(-2q - 3)$

10. $-2(-2d - 2)$

11. $2(a - 3)$

12. $-3(4 - b)$

13. $2.5(4 - 2b)$

14. $2(h - 3) - 4(2h + 1) + 3$

15. $2(h + 3) - 2(3h - 1) + 7$

16. $-2(4h + 5) - (2h + 4) - 5$

17. $-(y - 3) - (2y + 1) + 1$

18. $2(n + 1) + 5(3n - 3) + 1$

19. $-9(-2m + 5) + (4m - 4) - 10$

20. $\frac{1}{3}(12r - 12)$

21. $-\frac{3}{4}(12s - 60)$

22. $-\frac{5}{6}(-9t + 18)$

23. $\frac{1}{2}(10g - 4)$

24. $\frac{3}{4}(12z - 4)$

25. $-\frac{1}{2}(2a + 4)$

26. $\frac{6x + 12}{3}$

27. $\frac{15z - 40}{-5}$

28. $\frac{-12x - 15}{-12}$

29. $3(2z - 4) + \frac{z}{2}$

30. $2(z + 5) - \frac{z}{2}$

31. $\frac{3}{2}a + 2(a - 5)$

32. $\frac{2}{5}b - 2(b + 5)$

33. $3(c + 5) - 2(c - 3)$

34. $-3(p + 5) + 2(p - 5)$

35. $-4(u + 1) + 2(u - 2) - \frac{u}{3}$

36. $\frac{2t}{3} - 2(t - 1) + 5(t - 2)$

37. $\frac{2k}{5} + \frac{3}{5}(10k - 15) + 5(k - 2)$

MIXED REVIEW

Evaluate each expression for the given value of the variable. (1.2 Skill C)

38. $6x + 5; x = \frac{2}{3}$

39. $14a + 7; a = \frac{3}{7}$

40. $45z + 20; z = \frac{3}{5}$

Solve each equation. (2.5 Skills A and B)

41. $x - 6.5 = -11$

42. $y + 0.5 = -1$

43. $a - 1\frac{1}{2} = 0$

44. $b + 1\frac{1}{2} = -3\frac{2}{3}$

45. $-3x = 120$

46. $\frac{-4b}{3} = 24$

47. $-2.5x = -10$

48. $\frac{4m}{-5} = -4$

2.7 Solving Two-Step Equations

LESSON

SKILL A *Solving an equation in two steps*

Often you will need to use two properties of equality to solve a single equation.

EXAMPLE 1 **Solve $-3x + 5 = 1$. Check your solution.**

▶ **Solution**

$$-3x + 5 = 1$$
$$-3x + 5 - 5 = 1 - 5 \quad \longleftarrow \text{Apply the Subtraction Property of Equality.}$$
$$-3x = -4$$
$$\frac{-3x}{-3} = \frac{-4}{-3} \quad \longleftarrow \text{Apply the Division Property of Equality.}$$
$$x = \frac{4}{3}, \text{ or } 1\frac{1}{3}$$

Check: Use the fraction form rather than the mixed number.

$$-3\left(\frac{4}{3}\right) + 5 \stackrel{?}{=} 1$$
$$-4 + 5 \stackrel{?}{=} 1 \checkmark$$

TRY THIS Solve $4x - 5 = -10$. Check your solution.

In Example 1, the equations $-3x + 5 = 1$, $-3x = -4$, and $x = \frac{4}{3}$ are *equivalent equations*. They are called **equivalent** because they all have the same solution.

EXAMPLE 2 **Solve $5x = 7x + 6$. Check your solution.**

▶ **Solution**

$$5x = 7x + 6$$
$$5x - 7x = 7x + 6 - 7x \quad \longleftarrow \text{Apply the Subtraction Property of Equality.}$$
$$(5 - 7)x = 6 \quad \longleftarrow \text{Apply the Distributive Property.}$$
$$-2x = 6$$
$$\frac{-2x}{-2} = \frac{6}{-2} \quad \longleftarrow \text{Apply the Division Property of Equality.}$$
$$x = -3$$

Check: Substitute -3 for x in each place where x occurs.

$$5(-3) \stackrel{?}{=} 7(-3) + 6$$
$$-15 \stackrel{?}{=} -15 \checkmark$$

TRY THIS Solve $-6x = -8x + 1$. Check your solution.

EXERCISES

The equations in each exercise are equivalent. Which property of equality can be used to show the equivalence?

1. $9x = 3x + 1$ and $6x = 1$

2. $3x = 18$ and $x = 6$

3. $6x + 10 = 7x - 3$ and $6x + 13 = 7x$

4. $6x + 10 = 7x - 3$ and $10 = x - 3$

PRACTICE

Solve each equation. Check your solution.

5. $2x + 5 = 9$

6. $3m - 4 = 15$

7. $2 = 5x - 13$

8. $2 = 4n - 14$

9. $-2v + 5 = 9$

10. $-3x + 5 = 2$

11. $2 = -5f - 18$

12. $-8 = -2z + 4$

13. $2p + 5 = 3p - 1$

14. $7a = -7a - 4$

15. $-3b + 5 = 3b$

16. $2x = -7x - 5$

17. $-3t = -2t + 1$

18. $-2w = -6w + 1$

19. $2 = -3d - 6$

20. $10x = -9x - 38$

21. $\frac{2x}{3} - 3 = \frac{x}{3}$

22. $\frac{-2c}{5} + 4 = \frac{c}{5}$

23. $10.2x = -9.8x - 40$

24. $-2.7h = 7.3h - 13$

25. $-5.4j = -4.6j - 15$

26. $3 - \frac{1}{2}w = 12$

27. $7 - \frac{3}{4}q = 14$

28. $-11 = 7 - \frac{3}{5}y$

29. $-32 = -5 - \frac{7}{2}n$

30. $1 - \frac{3}{2}r = 1$

31. $-6 - \frac{3}{2}d = -6$

32. Critical Thinking   Let a, b, and c represent real numbers, where $a \neq 0$.
Solve $ax + b = c$ for x.

33. Critical Thinking   Let a, b, and c represent real numbers, where $a \neq 0$.
Solve $a(x + b) = c$ for x.

MIXED REVIEW

Simplify each expression. (2.6 Skill A)

34. $-2m + 6m$

35. $-n + 7n$

36. $-3p - 7p$

37. $-5g + 6g$

38. $-2.9h - 7.1h$

39. $-5.4k - 14.6k$

Simplify each expression. (2.6 Skill B)

40. $2(x - 3) - (x - 5)$

41. $-2(d + 5) - 3(d - 5)$

42. $-2(3n - 5) + 3(2n + 5)$

43. $-2(k + 1) - (k - 6)$

44. $-(p + 1.5) + 3(p - 1)$

45. $7(2u + 3) - 3(2u + 3)$

Recall that a formula expresses a relationship between two or more variables in a mathematical or physical application.

When you solve a formula for one of the variables, the result will not be a number. Instead, you will have an algebraic expression that contains the other variable(s). This is called *solving for one variable in terms of another.*

EXAMPLE 1

The formula $C = \frac{5}{9}(F - 32)$ enables you to find a temperature, C, on the Celsius scale, given a temperature, F, on the Fahrenheit scale. Solve the formula for F in terms of C.

▶ **Solution**

$$C = \frac{5}{9}(F - 32)$$

$$\frac{9}{5} \cdot C = \frac{9}{5} \cdot \frac{5}{9}(F - 32) \quad \longleftarrow \text{ Apply the Multiplication Property of Equality.}$$

$$\frac{9}{5}C = F - 32$$

$$\frac{9}{5}C + 32 = F \quad \longleftarrow \text{ Apply the Addition Property of Equality.}$$

The formula is $F = \frac{9}{5}C + 32$.

TRY THIS Solve $F = \frac{9}{5}C + 32$ for C in terms of F.

If you deposit P dollars into an account paying simple interest, then the amount, A, that you will have after one year is given by $A = P + Pr$, where r is the annual interest rate as a decimal.

EXAMPLE 2

Solve $A = P + Pr$ for P.

▶ **Solution**

$$A = P + Pr$$

$$A = P(1 + r) \quad \longleftarrow \text{ Apply the Distributive Property.}$$

$$\frac{A}{1 + r} = \frac{P(1 + r)}{1 + r} \quad \longleftarrow \text{ Divide each side by } 1 + r.$$

$$\frac{A}{1 + r} = P$$

Thus, $P = \frac{A}{1 + r}$.

TRY THIS Solve $A = P + Pr$ for r.

EXERCISES

KEY SKILLS

Identify the properties of equality that you would use to solve each equation for the variable in parentheses.

1. $(x)y = 4$

2. $\frac{2(a)}{5} = d$

3. $A = \frac{1}{2}b(h)$

4. $P = 2l + 2(w)$

PRACTICE

Solve each literal equation for the specified variable.

5. $x + a = b$; a

6. $x - r = c$; x

7. $rt = d$; $r\ (t \neq 0)$

8. $\frac{x}{r} = v$; x

9. $2x + a = 4$; x

10. $2a - 3x = 4$; a

11. $y = mx + b$; $x\ (m \neq 0)$

12. $y = mx + b$; $m\ (x \neq 0)$

13. $2x + 3y = 12$; y

14. $2x + 3y = 12$; x

15. $A = P + Prt$; $t\ (Pr \neq 0)$

16. $w = P + P(1 + c)$; $P\ (c \neq -2)$

17. $2f = 3f + ag$; $g\ (a \neq 0)$

18. $4z = 2z + bd$; z

19. $2f = 3f + ag$; f

20. $4z = 2z + bd$; $d\ (b \neq 0)$

21. Twice the sum of x and y equals 12.
    a. Write an equation to represent this statement.   **b.** Solve for x.   **c.** Solve for y.

22. **Critical Thinking**   Let a, b, and c be real numbers whose sum is nonzero. Solve $ax + bx + cx = 1$ for x.

MIXED REVIEW APPLICATIONS

Find the net change in each bank balance. (2.3 Skill C)

23. deposits: $210.30, $100.56, $17.95, and $34.22
    withdrawals: $43.50, $250.56, and $145.00

24. deposits: $1000.50, $100.50, $187.95, and $314.20
    withdrawals: $247.50, $274.56, and $305.00

25. The Lincoln High football team completes 2 series of downs with the gains and losses in yards shown below. Find the net gain (or loss) after these 2 series.

Series 1	Down	1	2	3
	Gain/Loss	8	−4	15

Series 2	Down	1	2	3
	Gain/Loss	−3	−5	4

You can use a problem-solving plan to answer real-world questions.

A Problem-Solving Plan

1. **Understand the problem.** Read the problem carefully, perhaps more than once. Identify what information is given and what you are asked to find. Use a variable to represent any unknown quantity.
2. **Choose a strategy.** Make tables, diagrams, graphs, or other visual aids to help relate the known and the unknown quantities. Then express the relationships in an equation.
3. **Solve the problem.** Solve the equation using the properties of numbers and equality as you have learned. Express the answer using appropriate units.
4. **Check the results.** Check your calculations by substituting the solution in place of the variable in the equation. If the equation checks, then relate the solution to the statement of the original problem. If the solution is not reasonable, then repeat the four steps of the plan, looking for an error or faulty reasoning.

EXAMPLE

Jackie and Kevin want to design and build a rectangular garden. They have 120 feet of fencing for the border, and they want their garden's length to be twice its width. What will be the dimensions of their garden?

▶ **Solution**

1. Let w represent the width. Because the length is twice the width, let $2w$ represent the length.

Make a diagram.

2. Draw a diagram to represent the rectangular garden.

width, w | $2w$ | w
length, $2w$

length, $2w$
border = 2 (width) + 2 (length)
= $2w + 2(2w)$

Write an equation.

3. Write and solve an equation.

$$2w + 2(2w) = 120$$
$$2w + 4w = 120$$
$$6w = 120 \quad \longleftarrow \text{ Combine like terms.}$$
$$w = 20 \quad \longleftarrow \text{ Apply the Division Property of Equality.}$$

4. **Check:** $2(20) + 2[2(20)] = 40 + 80 = 120$ ✔

The solution is 20. So, the width of the garden will be 20 feet and the length will be twice the width, or 40 feet.

TRY THIS Suppose that Jackie and Kevin want to use 90 feet of fencing and want the length to be 1.5 times the width. What will be the dimensions of their garden?

EXERCISES

For each situation, choose a variable for the unknown quantity and write an equation. Do not solve.

1. Rochelle's age is twice Lin's age. If the sum of their ages is 27, how old is Lin?

2. A whole number plus one half itself less 2 equals 94. What is it?

PRACTICE

Solve each problem. Check your solution.

3. Ten ounces of water are removed from a container and two-thirds of the original amount remains. How much water did the container have originally?

4. Mickey has 160 yards of fencing to enclose a rectangular garden whose length is to be 4 times its width. What will be the dimensions of the garden?

5. A taxicab ride costs $1.10 plus $0.95 per mile. If the total cost of the trip is $12.50, how many miles did the passenger travel?

6. Karen has $120. If she saves $24 per week, how many weeks will it take her to have a total of $300?

7. In a *right triangle*, the sum of the two non-right angles is 90°. If the measure of one non-right angle is twice that of the other, what are the measures of the angles?

8. Maria's salary is three times Eileen's salary. They decide to share the cost of an $80 gift. Maria will contribute three times as much as Eileen. How much does each contribute?

9. **Critical Thinking**   Jamie has a total of 40 coins that are all nickels and dimes. She has 12 more nickels than dimes. How many of each does she have?

10. **Critical Thinking**   Show that two integers, one of which is three times the other, cannot add up to 13.

MIXED REVIEW APPLICATIONS

Solve the problem and answer the question. (2.3 Skill C)

11. The temperature was 15°F at 6:00 P.M. It dropped 4°F by midnight and then dropped another 6°F by 3:00 A.M. It then rose 12°F by 9:00 A.M. What was the temperature at 9:00 A.M.?

 **2.8**

LESSON

Solving Multistep Equations

SKILL A   *Solving an equation in multiple steps*

In many situations, the solution process has more than two steps.

EXAMPLE 1   **Solve $5x - 17 = 3x - 10$. Check your solution.**

▶ **Solution**

$$5x - 17 = 3x - 10$$
$$5x - 17 + 17 = 3x - 10 + 17 \quad \longleftarrow \text{Apply the Addition Property of Equality.}$$
$$5x = 3x + 7$$
$$5x - 3x = 3x + 7 - 3x \quad \longleftarrow \text{Apply the Subtraction Property of Equality.}$$
$$2x = 7$$
$$x = \frac{7}{2}, \text{ or } 3.5 \quad \longleftarrow \text{Apply the Division Property of Equality.}$$

Check: $5(\mathbf{3.5}) - 17 \overset{?}{=} 3(\mathbf{3.5}) - 10 \quad \rightarrow \quad 0.5 = 0.5$ ✔

TRY THIS   Solve $-2c + 8 = 5c - 10$. Check your solution.

Recall that the **least common multiple** (LCM) of two natural numbers is the smallest number divisible by the two numbers. You can eliminate fractions in an equation by multiplying both sides of the equation by the LCM of all the denominators (using the Multiplication Property of Equality). However, do *not* use this method to simplify an expression that is not part of an equation.

EXAMPLE 2   **Solve $\frac{1}{4}d + 3 = \frac{1}{3}d + 1$. Check your solution.**

▶ **Solution**

$$\frac{1}{4}d + 3 = \frac{1}{3}d + 1$$
$$12\left(\frac{1}{4}d + 3\right) = 12\left(\frac{1}{3}d + 1\right) \quad \longleftarrow \begin{array}{l}\text{Apply the Multiplication Property of Equality.}\\ \text{The LCM of 3 and 4 is 12.}\end{array}$$
$$3d + 36 = 4d + 12$$
$$3d + 36 - 3d = 4d + 12 - 3d \quad \longleftarrow \text{Apply the Subtraction Property of Equality.}$$
$$36 = 12 + d$$
$$36 - 12 = 12 + d - 12 \quad \longleftarrow \text{Apply the Subtraction Property of Equality.}$$
$$24 = d$$
$$d = 24 \quad \longleftarrow \text{Apply the Symmetric Property of Equality.}$$

Check: $\frac{1}{4}(\mathbf{24}) + 3 \overset{?}{=} \frac{1}{3}(\mathbf{24}) + 1 \quad \rightarrow \quad 9 = 9$ ✔

TRY THIS   Solve $\frac{1}{2}r - 1 = \frac{1}{3}r + 2$. Check your solution.

EXERCISES

Write the equation that results from the specified step.

1. $-3x + 5 = 2x + 7$; add -5 to each side of the equation.

2. $5x - 6 = 7x + 7$; add $-5x$ to each side of the equation.

3. $\frac{5}{8}t - 5 = \frac{2}{3}t$; multiply each side of the equation by 24.

4. $0.3x - 1 = 0.5x$; multiply each side of the equation by 10.

PRACTICE

Solve each equation. Check your solution.

5. $x + 3 = 2x + 5$

6. $3c - 5 = 2c + 5$

7. $10t - 6 = -2t - 6$

8. $7x - 9 = 3x + 19$

9. $6 + 10t = 8t + 12$

10. $3x + 7 = 16 + 6x$

11. $18 + 3y = 5y - 4$

12. $11a + 8 = -2 + 9a$

13. $9x - 5 = 6x + 13$

14. $6x - 5 = 2x - 21$

15. $8 - x = 5x - 4$

16. $-17 - 2x = 6 - x$

17. $5y - 0.3 = 4y + 0.6$

18. $10z + 1.3 = 5z - 1.6$

19. $z - 2.7 = -1.5z + 1$

20. $-0.6d + 1.1 = d + 1$

21. $\frac{3}{4}a + 1 = \frac{5}{4}a + 2$

22. $3 + \frac{1}{3}d = 5 + \frac{7}{3}d$

23. $\frac{1}{3}z - 1 = \frac{5}{3}z$

24. $\frac{3}{7}x + 3 = \frac{2}{7}x$

25. $\frac{1}{3}x + \frac{2}{3} = \frac{5}{3}x$

26. $\frac{3}{7}x + \frac{4}{7} = \frac{2}{3}x$

27. $\frac{10}{7}x = \frac{3}{4}x + \frac{1}{2}$

28. $\frac{5}{16}x = \frac{1}{5}x + \frac{3}{40}$

29. **Critical Thinking**   Solve for x in terms of a, b, and c: $\frac{1}{a}x + \frac{1}{b} = \frac{1}{c}x$

MIXED REVIEW

Use the order of operations to simplify each expression. (1.2 Skill B)

30. $\dfrac{3(5 - 2)}{9}$

31. $\frac{1}{5}(2 - 1) + \frac{4}{5}$

32. $\frac{4}{7}(12 - 5) - 4$

Evaluate each expression. (1.2 Skill C)

33. $-3(x + 5)$ given $x = -2$

34. $4.5(t - 2)$ given $t = 0$

35. $4.5\left(\dfrac{f + 1}{2}\right)$ given $f = 3$

Find the values in the replacement set $\{-3, -2, -1, 0, 1, 2, 3\}$ that are solutions to each equation. If the equation has no solution, write *none*. (1.2 Skill C)

36. $-2x + 3 = -1$

37. $x - 1 = -8$

38. $2(x + 3) = 12$

Sometimes the first step in solving an equation is simplifying the
expression(s) on one or both sides.

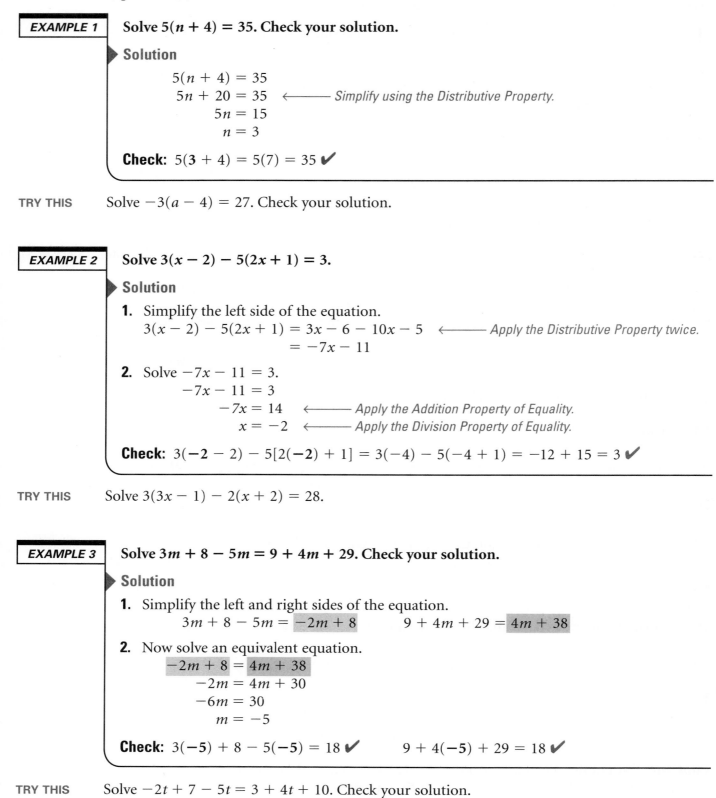

EXAMPLE 1   **Solve $5(n + 4) = 35$. Check your solution.**

▶ **Solution**

$$5(n + 4) = 35$$
$$5n + 20 = 35 \quad \longleftarrow \text{\textit{Simplify using the Distributive Property.}}$$
$$5n = 15$$
$$n = 3$$

Check: $5(\mathbf{3} + 4) = 5(7) = 35$ ✔

TRY THIS   Solve $-3(a - 4) = 27$. Check your solution.

EXAMPLE 2   **Solve $3(x - 2) - 5(2x + 1) = 3$.**

▶ **Solution**

1. Simplify the left side of the equation.
$$3(x - 2) - 5(2x + 1) = 3x - 6 - 10x - 5 \quad \longleftarrow \text{\textit{Apply the Distributive Property twice.}}$$
$$= -7x - 11$$

2. Solve $-7x - 11 = 3$.
$$-7x - 11 = 3$$
$$-7x = 14 \quad \longleftarrow \text{\textit{Apply the Addition Property of Equality.}}$$
$$x = -2 \quad \longleftarrow \text{\textit{Apply the Division Property of Equality.}}$$

Check: $3(\mathbf{-2} - 2) - 5[2(\mathbf{-2}) + 1] = 3(-4) - 5(-4 + 1) = -12 + 15 = 3$ ✔

TRY THIS   Solve $3(3x - 1) - 2(x + 2) = 28$.

EXAMPLE 3   **Solve $3m + 8 - 5m = 9 + 4m + 29$. Check your solution.**

▶ **Solution**

1. Simplify the left and right sides of the equation.
$$3m + 8 - 5m = \boxed{-2m + 8} \qquad 9 + 4m + 29 = \boxed{4m + 38}$$

2. Now solve an equivalent equation.
$$\boxed{-2m + 8} = \boxed{4m + 38}$$
$$-2m = 4m + 30$$
$$-6m = 30$$
$$m = -5$$

Check: $3(\mathbf{-5}) + 8 - 5(\mathbf{-5}) = 18$ ✔ $\qquad 9 + 4(\mathbf{-5}) + 29 = 18$ ✔

TRY THIS   Solve $-2t + 7 - 5t = 3 + 4t + 10$. Check your solution.

EXERCISES

Simplify each expression.

1. $2(x + 1) - 3$

2. $3 - 4(t + 1)$

3. $3(z + 1) + 4(z - 3)$

4. $-(a + 1) - 2(a + 1)$

5. $5t - 5 + t$

6. $-3y + 7 - 2y$

PRACTICE

Solve each equation. Check your solution.

7. $-3(h + 1) = 9$

8. $6(b - 5) = 10$

9. $2 = 4(z + 1)$

10. $-12 = 5(p - 2)$

11. $0 = 2(3 - q)$

12. $3 = 2(5 - r)$

13. $3(w + 1) - 2(w - 3) = 7$

14. $3(t - 1) - (t + 2) = 7$

15. $-2(3a + 1) - (a - 5) = -7$

16. $2(x - 1) - 3(2x + 2) = 10$

17. $14 = 3(d - 5) + 2(d - 3)$

18. $10 = -2(s + 2) + 3(s - 10)$

19. $4 + 2(w - 3) = 3w - 2(w + 5)$

20. $-3y + 2(y + 3) = -4 + 2(y - 3)$

21. $\frac{1}{2}y + \frac{1}{2}(y + 2) = 5 + 3(y - 3)$

22. $\frac{4}{5}a + \frac{1}{5}(a - 10) = 5 + 5(a - 5)$

23. $5n - 1 + n = 3 - 3n + 1$

24. $4k - 1 + k = 1 + 3k + 10$

25. $b + 1 + b = 10 - 5b - 12$

26. $3g - 5 - 3g = 7g + 5 - 6g$

27. $3r - 2 + 4r - 2 = 3r - 2r + 7r + 1$

28. $-2a + 3 + 3a - 2 = 5a + 2a - 7a + 9$

29. Critical Thinking Show that every real number is a solution to
$2(s + 1) + 4(s + 1) + 6(s + 1) = 12(s + 1)$.

MIXED REVIEW

Find each amount. (previous courses)

30. 6% of 1200

31. (8% of 10,000) − 400

32. 10% of (10,000 − 500)

Simplify each expression. (2.6 Skill B)

33. $0.05x + 0.05(10,000 - x)$

34. $0.06x + 0.04(8000 - x)$

35. $0.08x + 0.08(8000 - x)$

36. $0.05(12,000 - x) + 0.06x$

When you write an equation to solve a real-world problem, the equation you write may require a multistep solution process.

Recall from previous courses that simple interest on an investment is the product of the amount of money invested, P, the annual interest rate, r, and the time in years, t. The formula is $I = Prt$.

EXAMPLE

Write an equation.

Mr. Shaw has $10,000 to invest. He wants to divide the money between two investments that earn interest. One investment pays 6% simple interest annually, and the other pays 8% simple interest annually. How much money should he invest in each account so that he will earn $720 of interest in one year?

▶ **Solution**

1. Let x represent the amount Mr. Shaw will invest at 6%. He will invest a total of $10,000, so $(10,000 - x)$ represents the amount that he will invest at 8%.

2. Write an equation. The *total* interest earned equals the sum of the interest earned from each investment.

interest earned at 6%		interest earned at 8%		total interest
↓		↓		↓
6% of x	+	8% of $(10,000 - x)$	=	720
$0.06x$	+	$0.08(10,000 - x)$	=	720

3. Simplify the left side of the equation.
$$0.06x + 0.08(10,000 - x) = 0.06x + 800 - 0.08x \quad \longleftarrow \text{Apply the Distributive Property.}$$
$$= 0.06x - 0.08x + 800$$
$$= -0.02x + 800$$

4. Solve the new equation.
$$-0.02x + 800 = 720$$
$$-0.02x + 800 - 800 = 720 - 800 \quad \longleftarrow \text{Apply the Subtraction Property of Equality.}$$
$$-0.02x = -80$$
$$\frac{-0.02}{-0.02}x = \frac{-80}{-0.02} \quad \longleftarrow \text{Apply the Division Property of Equality.}$$
$$x = 4000$$

Mr. Shaw should invest $4000 at 6% and $6000 at 8%.

TRY THIS

Ms. Moore has $10,000 to invest. She wants to divide the money between two investments that earn interest. One investment pays 6% simple interest annually, and the other pays 8% simple interest annually. How much money should she invest in each account so that she will earn $740 of interest in one year?

EXERCISES

KEY SKILLS

Write an equation to represent each situation. Do not solve.

1. You have $1500 to divide between two accounts, one paying 5% simple interest and the other paying 6% simple interest. You want to earn $85 of interest in one year. How much should you invest in each account?

2. Twenty-four children bought cookies. Some children bought plain cookies for $1 each. The others bought frosted cookies for $2 each. Together, they spent $30. How many of each kind of cookie did they buy?

PRACTICE

Solve each problem.

3. Suppose that you distribute $2000 between two interest-paying accounts. One account pays 5% simple interest, and the other pays 6% simple interest. How much should you invest in each account to earn $113 of interest in one year?

4. How many nickels and how many dimes does Shannon have if she has 40 coins whose total value is $4.00?

5. Mr. Suzuki drove 365 miles in 7 hours. During one part of the trip, he drove 50 miles per hour and during the other part, he drove 55 miles per hour. How far did he drive at each speed?

6. A chemist must divide 500 milliliters of a solution into two containers. One container will contain 50 milliliters more than twice the amount in the other container. How much will each container have?

7. A drama club sold 200 tickets and collected $640. If a child's ticket cost $2 and an adult ticket cost $5, how many of each ticket did the club sell?

8. **Critical Thinking** Suppose Mr. Shaw wants to divide $10,000 between two accounts that earn interest at 6% and 8%. Show that it is impossible for him to earn $850 of interest in one year.

MIXED REVIEW

Solve each equation. Identify the property used in each step. (2.5 Skills A and B)

9. $2(x - 5) = 11$

10. $2x - 5 = 12 + x$

11. $2x + x = -2$

2.9 LESSON

Deductive and Indirect Reasoning in Algebra

When you read or write a mathematical argument, you should check it carefully for mistakes in reasoning.

EXAMPLE 1

Find the error in the simplification below. Then give the correct value of $\frac{14 - 3}{2} - 4$.

$$\frac{14 - 3}{2} - 4 = \frac{14}{2} - 3 - 4 = 7 - 3 - 4 = 4 - 4 = 0 \qquad \text{Therefore, } \frac{14 - 3}{2} - 4 = 0. \ ✗$$

▶ **Solution**

In the first step, the Distributive Property was not applied correctly. The value of $\frac{14 - 3}{2}$ should be $\frac{14}{2} - \frac{3}{2}$, or $5\frac{1}{2}$.

The correct value of $\frac{14 - 3}{2} - 4$ is $1\frac{1}{2}$.

TRY THIS

Find the error in the simplification below. Then give the correct value of $\frac{4 - 10}{2} - 3$.

$$\frac{4 - 10}{2} - 3 = \frac{10 - 4}{2} - 3 = \frac{6}{2} - 3 = 3 - 3 = 0 \qquad \text{Therefore, } \frac{4 - 10}{2} - 3 = 0. \ ✗$$

EXAMPLE 2

Find the error in the solution to $3(x - 1) = 12$ shown at right. Then give the correct solution.

$$3(x - 1) = 12$$
$$3x - 3 = 12$$
$$3x = 9$$
$$x = 3 \ ✗$$

▶ **Solution**

From the second step to the third step ($3x - 3 = 12$ to $3x = 9$), the Addition Property of Equality was applied incorrectly; the number 3 was added to the left side of the equation but subtracted from the right side of the equation.

$$3x - 3 = 12$$
$$3x - 3 + 3 = 12 + 3$$
$$3x = 15$$
$$x = 5$$

The correct solution is 5.

TRY THIS

Find the error in the solution to $-2(x - 3) = 18$ shown at right. Then give the correct solution.

$$-2(x - 3) = 18$$
$$-2x - 6 = 18$$
$$-2x = 24$$
$$x = -12 \ ✗$$

EXERCISES

KEY SKILLS

Substitute the value of the variable from the last line into each of the equations above it. Which equations does it satisfy?

1. $3(x + 4) = -11$
   $3x + 4 = -11$
   $3x = -15$
   $x = -5$

2. $0 = 4(a - 2) + 12$
   $12 = 4(a - 2)$
   $12 = 4a - 8$
   $20 = 4a$
   $5 = a$

PRACTICE

Find the error(s) in each simplification. Then give the correct value.

3. $-(3 + 4) + 5 = -3 + 4 + 5$
   $= 1 + 5$
   $= 6$

4. $-2(3 + 4) + 5 = -6 + 4 + 5$
   $= -2 + 5$
   $= 3$

5. $\dfrac{3}{7} + \dfrac{2}{7} = \dfrac{3 + 2}{7 + 7} = \dfrac{5}{14}$

6. $\dfrac{48 - 5}{7} = \dfrac{48}{7} - 5 = 6\dfrac{6}{7} - 5 = 1\dfrac{6}{7}$

Find the error(s) in each solution. Then write the correct solution.

7. $3x + 5 - 4x = 19$
   $3x + 4x - 5 = 19$
   $7x + 5 = 19$
   $7x = 14$
   $x = 2$

8. $2(d - 3) + 3(d + 5) = 10$
   $2d - 6 + 3d + 5 = 10$
   $2d - 3d + 6 + 5 = 10$
   $-d + 11 = 10$
   $-d = -1$
   $d = -1$

Give a reason for each step in the proofs below.

9. $(a + b) + (-a) = b$
   $(a + b) + (-a) = a + [b + (-a)]$
   $= a + [(-a) + b]$
   $= [a + (-a)] + b$
   $= 0 + b$
   $= b$

10. $a(b + c + d) = ab + ac + ad$
    $a(b + c + d) = a[(b + c) + d]$
    $= a(b + c) + ad$
    $= ab + ac + ad$

MIXED REVIEW

Solve each equation for the specified variable. (2.7 Skill B)

11. $y = -2x + 5; x$

12. $y = -1.5x - 3; x$

13. $-3x + 5y = 12; x$

14. $7x - 5y = 15; y$

15. $y = -3(x - 2) + 3; x$

16. $y = 4(x + 5) + 3; x$

17. $-3x = -7y + 24; y$

18. $5x = -7y - 14; y$

19. $7y + 8x = 20; y$

In everyday conversation, we often make *if-then* statements. Some examples are "If it is raining, then I carry an umbrella." and "If it is winter, then birds fly south."

In mathematics, an if-then statement is called a **conditional statement**. The part following the word *if* is the **hypothesis**, and the part following the word *then* is the **conclusion**. The Symmetric Property of Equality is an example of a conditional statement.

$$\overbrace{\text{If } a \text{ and } b \text{ are real numbers and } a = b,}^{\text{conditional statement}} \underbrace{\text{then } b = a.}$$

$$\underbrace{\text{If } a \text{ and } b \text{ are real numbers and } a = b}_{\text{hypothesis}}, \text{then } \underbrace{b = a}_{\text{conclusion}}.$$

The hypothesis of a statement is often represented by the letter p. The conclusion is often represented by q. A conditional statement is often abbreviated as $p \Rightarrow q$. You read $p \Rightarrow q$ as "If p then q" or "p implies q."

The process of beginning with a hypothesis and following a sequence of logical steps to reach a conclusion is called **deductive reasoning**. An important rule in deductive reasoning is the Transitive Property of Deductive Reasoning.

Transitive Property of Deductive Reasoning
Let p, q, and r represent statements. If $p \Rightarrow q$ and $q \Rightarrow r$, then $p \Rightarrow r$.

Solving an equation is actually a form of deductive reasoning.

EXAMPLE

Solve $2x + 5 = 11$. Write the steps in the solution as a logical argument. Write the solution as a conditional statement.

▶ **Solution**

Conditional Statement	Justification
If $\underbrace{2x + 5 = 11}_{p}$, $\Rightarrow$ then $\underbrace{2x = 6}_{q}$.	Subtraction Property of Equality Subtract 5 from each side.
If $\underbrace{2x = 6}_{q}$, $\Rightarrow$ then $\underbrace{x = 3}_{r}$.	Division Property of Equality Divide each side by 2.
If $\underbrace{2x + 5 = 11}_{p}$, $\Rightarrow$ then $\underbrace{x = 3}_{r}$.	Transitive Property of Deductive Reasoning If $p \Rightarrow q$ and $q \Rightarrow r$, then $p \Rightarrow r$.

TRY THIS

Solve $\dfrac{x}{3} - 5 = 11$. Write the steps in the solution as a logical argument.

Write the solution as a single conditional statement.

EXERCISES

Identify p, q, and r in each pair of conditional statements.

1. If $-2d + 7 = 11$, then $-2d = 4$.
   If $-2d = 4$, then $d = -2$.

2. If $4y - 6 = 11$, then $4y = 17$.
   If $4y = 17$, then $y = \dfrac{17}{4}$

PRACTICE

Justify each step in each logical argument.

3. If $2(3x - 1) = 7$, then $6x - 2 = 7$.
   If $6x - 2 = 7$, then $6x = 9$.
   If $6x = 9$, then $x = \dfrac{3}{2}$.
   If $2(3x - 1) = 7$, then $x = \dfrac{3}{2}$.

4. If $2v - 5 = 3v + 1$, then $-v - 5 = 1$.
   If $-v - 5 = 1$, then $v + 5 = -1$.
   If $v + 5 = -1$, then $v = -6$.
   If $2v - 5 = 3v + 1$, then $v = -6$.

5. If $3a - 4 = 7 + 5a$, then $-4 = 7 + 2a$.
   If $-4 = 7 + 2a$, then $-11 = 2a$.
   If $-11 = 2a$, then $-\dfrac{11}{2} = a$.
   If $-\dfrac{11}{2} = a$, then $a = -\dfrac{11}{2}$.
   If $3a - 4 = 7 + 5a$, then $a = -\dfrac{11}{2}$.

6. If $2w - 3w = 4w + 10$, then $-w = 4w + 10$.
   If $-w = 4w + 10$, then $-5w = 10$.
   If $-5w = 10$, then $w = -2$.
   If $2w - 3w = 4w + 10$, then $w = -2$.

Write the logical steps in the solution to each equation. Write the solution as a conditional statement.

7. $3(x - 5) = 5x$

8. $7h - (3h - 5) = 9$

9. $6(x - 3) + 4(x + 2) = 6$

10. **Critical Thinking** Find the missing step in the logical argument below.
    If $3(p + 1) - (p - 1) = 6$, then $3p + 3 - p + 1 = 6$.
    If $2p + 4 = 6$, then $2p = 2$. If $2p = 2$, then $p = 1$.

MIXED REVIEW

Write each set of numbers in order from least to greatest.
(previous courses)

11. $-2, 9.2, 0, -6.1, 0.5$

12. $0, 0.1, 0.2, -0.1, -0.2, 5$

13. $3.7, 3.75, 3.754, 3.6, 3.65$

**Replace each ___?___ with < (is less than), > (is greater than), or =
to make the statement true.** (previous courses)

14. $4\dfrac{1}{4}$ ___?___ 4.25

15. 6.33 ___?___ 6.43

16. $\dfrac{9}{5}$ ___?___ 1.79

17. $-3\dfrac{1}{3}$ ___?___ $3\dfrac{1}{3}$

There are various ways to determine whether a statement is sometimes, always, or never true.

EXAMPLE 1	**Given that a is a real number, is the equation $a^2 = 2a$ sometimes, always, or never true? Justify your response.**

▶ **Solution**

Guess and check.

Test values of a. $a = 0$: $0^2 = 2(0)$ ✔ True $a = 1$: $1^2 \neq 2(1)$ ✘ False

Because $a^2 = 2a$ is true for $a = 0$ but false for $a = 1$, it is sometimes true.

TRY THIS   Given that a is a real number, is the equation $(a - 3)^2 = 2(a - 3)$ sometimes, always, or never true? Justify your response.

EXAMPLE 2	**Given that x is a real number, is the equation $2(2x + 3) = 5 + 4x + 1$ sometimes, always, or never true? Justify your response.**

▶ **Solution**

$$2(2x + 3) = 4x + 6 \qquad \longleftarrow \text{ Distributive Property}$$
$$4x + 6 = 4x + 5 + 1 \qquad \longleftarrow \text{ } 6 = 5 + 1$$
$$2(2x + 3) = 5 + 4x + 1 \qquad \longleftarrow \text{ Transitive Property of Equality}$$

The statement is always true.

TRY THIS   Given that b is a real number, is $2(b - 1) - 7(b + 1) = -3b - 9 - 2b$ sometimes, always, or never true? Justify your response.

If p represents a statement, then $\sim p$ ("not p") represents the **negation** of p. If p is true, then $\sim p$ is never true. So, another way to show that a statement, p, is true is to show that $\sim p$ is never true. When you do this, you are using a process called **indirect reasoning**.

EXAMPLE 3	**Use indirect reasoning to show that $2(a + 1) \neq 2a + 1$.**

▶ **Solution**

1. Identify p and $\sim p$. p: $2(a + 1) \neq 2a + 1$ $\sim p$: $2(a + 1) = 2a + 1$

2. Begin with $\sim p$ and use properties to arrive at a false statement.

$$2(a + 1) = 2a + 1$$
$$2a + 2 = 2a + 1 \qquad \longleftarrow \text{ Distributive Property}$$
$$2 = 1 \text{ ✘} \qquad \longleftarrow \text{ Subtraction Property of Inequality}$$

When you arrive at a false statement, the statement you began with must also be false. Since $2 = 1$ is false, $\sim p$ is never true.

3. Conclude that p is true: $2(a + 1) \neq 2a + 1$

TRY THIS   Use indirect reasoning to show that $3(t - 1) \neq 3t$.

EXERCISES

1. Give one example and one counterexample to show that $n^2 = 3n$ is sometimes true.

2. Use properties to show that $n + (n + 1) = 2n + 1$ is always true.

3. Use indirect reasoning to show that $3(x + 5) + 2 \neq 3x + 20$.

PRACTICE

Is the given statement sometimes true or always true? Justify your response. (Assume that all variables represent real numbers.)

4. $3(n + 1) + (n + 1) = 4(n + 1)$

5. $m^2 = 4m$

6. $4(k - 3) = 4k - 12$

7. $4(k - 3) = 4(3 - k)$

8. $3x - 5 + 7x = -5 + 10x$

9. $x(y + 5) - 5 = xy$

10. $x(y + 5) - 5x = xy$

11. $zxy + zx + zy = z[x(y + 1) + y]$

12. Use indirect reasoning to show that $4(3 + y) \neq 4y - 4$.

13. Use indirect reasoning to show that $3x + 6 \neq 3(x + 4) + 2$.

14. Use indirect reasoning to show that $4y + 2 + 3(5 - y) \neq y + 1$.

15. **Critical Thinking** Refer to the geometry definitions below. Then decide whether the statement following them is sometimes, always, or never true.

   scalene triangle: no two sides have the same length.
   isosceles triangle: at least two sides have the same length.
   equilateral triangle: all three sides have the same length.

   Statement: No scalene triangle is isosceles or equilateral and all equilateral triangles are isosceles.

MIXED REVIEW

Solve each equation. Check your solution. (2.8 Skill B)

16. $-2(x - 1) + 4x = 5x + 3$

17. $x - 1 - (4x + 3) = 5x$

18. $\frac{1}{2}(4x + 3) = \frac{3}{2}(5x - 1)$

19. $\frac{1}{2}(4x + 4) + \frac{1}{2}(4x - 8) = 0$

20. $c - 2c - 3c + 2 = 8$

21. $(r - 1) - (r - 2) + (r - 3) = 1$

22. $4(2n - 1) - 3(2n - 1) = 3(n - 1)$

23. $2m + 1 + 2(2m + 1) = -2(2m + 1)$

Write each fraction as a decimal and each decimal as a fraction.

1. $\frac{13}{25}$ **2.** $\frac{5}{12}$ **3.** 0.58   **4.** 3.26

Classify each number in as many ways as possible.

5. $-\frac{7}{15}$ **6.** 13   **7.** -180 **8.** 5.3030030003...

Evaluate each expression given $b = 12$.

9. $-|b + 20|$ **10.** $|b + 3| + |b - 3|$

11. $-(-|b|)$ **12.** $-(-|b - 3|)$

Use a number line to find each sum.

13. $-4 + 7$ **14.** $-1 + (-5)$

Find each sum.

15. $-5 + 12$ **16.** $-2\frac{1}{4} + 3\frac{3}{4}$

17. $-6.4 + (-7.3)$ **18.** $1.2 + (-0.5) + (-3)$

Find each difference.

19. $-7 - 13$ **20.** $12.4 - 18.1$

21. $-13 - (-19)$ **22.** $-2\frac{1}{2} - 3\frac{1}{2}$

Simplify each expression.

23. $-(5 - 3) - (4 + 8)$ **24.** $10 - (3 - 8) + (2 - 10)$

25. A bank customer deposited checks in the amounts of $45.20, $56.75, and $110.39. The customer also withdrew $36.75, $120.25, and $110.45. What was the net change in the account balance?

Find each product. Write fractions in simplest form.

26. $(-12)(12)$ **27.** $\left(-2\frac{1}{2}\right)\left(-4\frac{1}{2}\right)$ **28.** $\frac{-8}{15} \cdot \frac{3}{32}$

Find each quotient. Write fractions in simplest form.

29. $(-19) \div (19)$ **30.** $\left(-3\frac{1}{2}\right) \div \left(-2\frac{2}{3}\right)$ **31.** $\frac{-8}{15} \div \frac{4}{-35}$

Simplify each expression.

32. $\dfrac{(3-5)}{-2} \times \dfrac{-(9-3)}{-2}$

33. $\dfrac{9-(-15)}{12(-2)-3(2)}$

Solve each equation. Check your solution.

34. $y + 5.6 = 2.5$

35. $d - 3\frac{1}{3} = -1\frac{1}{3}$

36. $-11.5 = d - \frac{1}{2}$

37. $\dfrac{x}{-5} = 2.5$

38. $4g = -4.8$

39. $\dfrac{z}{-2-5} = -3$

Simplify each expression.

40. $3a - 6 - 8a$

41. $(4x - 5x) - (3x - x)$

42. $-3(3a - 5) - 8a + 12$

43. $-2(6c - 8c) - 3(3c - 5c)$

Solve each equation. Check your solution.

44. $-2 - 5r = -7r$

45. $-3a + 7 = -4a + 12$

46. Solve $3x - 5y = 12$ for y.

47. Tara has \$200. If she can save \$22 per week, how many weeks will it take her to save \$420?

Solve each equation. Check your solution.

48. $z + 8.1 - 5z = -7z$

49. $-3a - 4 = 5a + 1$

50. $12 = -3(x + 4) + 5(x - 1)$

51. $4n - 6 - 7n = 2n - 5 + 7n$

52. A chemist must separate 770 milliliters of solution into two containers. One container will have 10 milliliters more than 3 times the amount of solution in the other container. How much will each container have?

53. Find the error(s) in the solution at right. Then write a correct solution.

$$-5(2 - 3) + 4 = -10 - 15 + 4$$
$$= -25 + 4$$
$$= 21$$

54. Write the steps in the solution of $4x = 2(x - 5)$ as a logical argument. Then write the solution as a single conditional statement.

55. Is the statement at right sometimes true or always true? Justify your response.

$$5(h + 3) - 2 = 5h + 13$$

Linear Equations in Two Variables

▶ **What You Already Know**

In Chapters 1 and 2, you learned how to represent real numbers on a number line and how to solve linear equations in one variable. The connection between real numbers, solutions to linear equations in one variable, and points on a number line will help you take the next step in your study of equations.

▶ **What You Will Learn**

In Chapter 3, you will first learn about the coordinate plane, linear equations in two variables, and graphs of ordered pairs. You will see that many real-world relationships involving ratio and rate can be represented by linear equations. You will also see that many linear relationships are examples of a more general concept, that of a function.

Your study of linear equations in two variables then takes a geometric point of view. You will learn how to sketch the graph of a linear equation in two variables and how to find an equation for a specified line.

Finally, you will use algebra to find a geometric relationship between two lines in the coordinate plane and you will model geometric patterns with linear equations.

VOCABULARY

collinear	mathematical model	right triangle
constant of variation	ordered pair	sequence
continuous graph	origin	slope
coordinate axes	parallel lines	slope-intercept form
coordinate plane	parallelogram	solution to an equation
deductive reasoning	parametric equations	standard form
dependent variable	perpendicular	term of a sequence
direct variation	point-slope form	trapezoid
discrete graph	proportion	vertical lines
domain	quadrilateral	vertical-line test
function	quadrants	x-axis
horizontal line	range	x-coordinate
independent variable	rate	x-intercept
inductive reasoning	rate of change	y-axis
linear equation	ratio	y-coordinate
linear function	relation	y-intercept
linear pattern		

The diagram below shows how mathematical skills and mathematical reasoning are interrelated with the skills and concepts in Chapter 3. Notice that this chapter has two major topics: 1) functions and relations, and 2) linear equations.

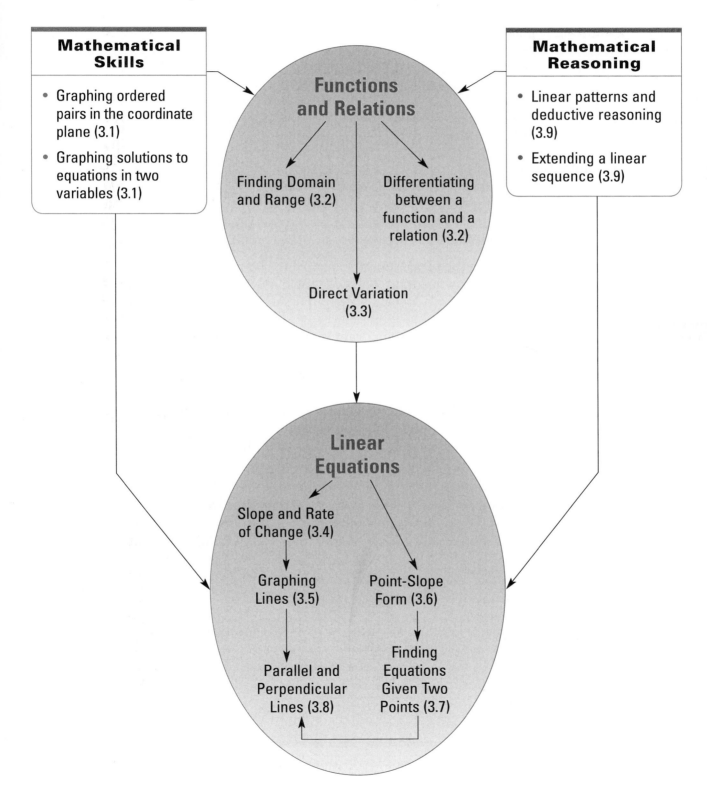

3.1

LESSON

The Coordinate Plane

Graphing ordered pairs in the coordinate plane

The **coordinate plane** is formed by placing two number lines called **coordinate axes** so that one is horizontal (**x-axis**) and one is vertical (**y-axis**). These axes intersect at a point called the **origin,** which is often labeled *O.*

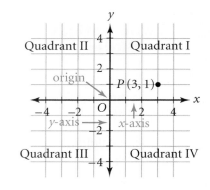

An **ordered pair** (*x, y*) is a pair of real numbers that correspond to a point in the coordinate plane. The first number in an ordered pair is the **x-coordinate** and the second number is the **y-coordinate**. The *x*-coordinate gives the distance right (positive) or left (negative) from the *y*-axis. The *y*-coordinate gives the distance up (positive) or down (negative) from the *x*-axis. In the figure shown above, point *P* has *x*-coordinate 3 and *y*-coordinate 1. The coordinates of the origin are (0, 0).

The coordinate axes divide the plane into four **quadrants**. Points on the coordinate axes are not in any quadrant. Point *P*, shown above, is in Quadrant I.

EXAMPLE

Graph each ordered pair. State the quadrant in which the point lies.

$$A(3, 4) \qquad B(-4, -4) \qquad C(0, -3)$$

▶ **Solution**

Begin at the origin, (0, 0). Count right or left (positive or negative) on the *x*-axis to locate the desired *x*-value. Then count up or down (positive or negative) to locate the desired *y*-value. Label this point with its *x*- and *y*-value.

A(3, 4): Count 3 units right from *O* and then count 4 units up. Point *A* is in Quadrant I.

B(−4, −4): Count 4 units left from *O* and then count 4 units down. Point *B* is in Quadrant III.

C(0, −3): Because the *x*-coordinate is 0, do not move to the right or left. Count 3 units down from *O*. Point *C* is on the *y*-axis, so it is not in any quadrant.

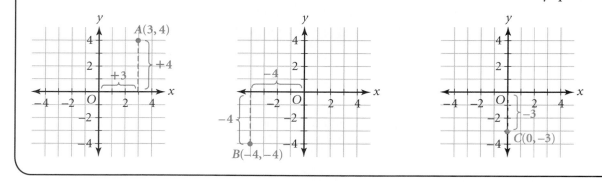

TRY THIS   Graph each ordered pair below. State the quadrant in which each point lies.
$$D(-2, -5), E(-1, -3), F(0, -1), G(1, 1), H(2, 3), J(3, 5)$$

EXERCISES

Examine the coordinates for each ordered pair. Is the corresponding point in a quadrant or on an axis?

1. $P(3, -2)$ **2.** $Q(0, 0)$ **3.** $R(3, 3)$ **4.** $B(0, -10)$

Write the ordered pair that corresponds to each point.

5. point F **6.** point C **7.** point X

8. point W **9.** point Z **10.** point A

11. point H **12.** point J **13.** point L

14. Write the coordinates of a point on the x-axis that lies to the right of point H and to the left of point F.

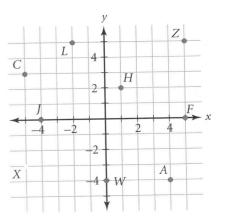

Graph and label each point on the same coordinate plane.

15. $A(-2, 0)$ **16.** $B(-2, -2)$ **17.** $C(0, -2)$ **18.** $D(4, -1)$

19. $W(-3, 5)$ **20.** $X(3, 5)$ **21.** $Y(-6, 2)$ **22.** $Z(-3, 0)$

Solve each problem.

23. If $N(a, 1)$ is in Quadrant II, what must be true about a?

24. If $C(a, b)$ is on the y-axis and above the x-axis, what must be true about a and b?

25. How are the locations of $P(a, b)$ and $Q(a, -b)$ related if $b \neq 0$?

26. How are the locations of $R(a, b)$ and $S(-a, b)$ related if $a \neq 0$?

27. Critical Thinking   Let r and s be nonzero real numbers. What must be true about r and s if $X(|r|, |s|)$ and $Y(r, s)$ are the same point? In which quadrant is X found?

Find the values in the given replacement set that are solutions to the given equation. (1.3 Skill A)

28. $-2x + 5 = 12; \{-3, -2, -1, 0, 1, 2, 3\}$ **29.** $-2(t + 3) = -12; \{-3, -2, -1, 0, 1, 2, 3\}$

30. $2(a + 3) + 2a = 10; \{-3, -2, -1, 0, 1, 2, 3\}$ **31.** $-2(r + 3) = 0; \{-3, -2, -1, 0, 1, 2, 3\}$

In Lesson 2.5, you solved equations in one variable. For example, given $x + 5 = -4$, you can find $x = -9$ as the solution. You can also graph this solution on a number line.

You will now learn how to solve equations in two variables, such as $y = 2x - 3$. The **solution to an equation in two variables** is all ordered pairs of real numbers (x, y) that satisfy the equation.

You can graph the solution to an equation in two variables on the coordinate plane, as shown below.

| EXAMPLE 1 | **Let $y = 2x - 3$ and let $x = -3, -2, -1, 0, 1, 2, 3$. Graph the solutions.** |

▶ **Solution**

Substitute each value of x into $2x - 3$, evaluate the expression, and then make a table of ordered pairs.

Make a table.

x	-3	-2	-1	0	1	2	3
y	-9	-7	-5	-3	-1	1	3

Graph each ordered pair in the table. The points appear to lie on a line.

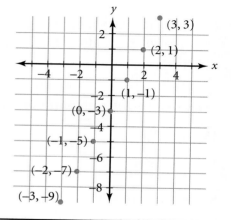

TRY THIS   Let $y = -2x + 1$ and let $x = -2, -1, 0, 1, 2$. Graph the solutions.

| EXAMPLE 2 | **Let $y = (x + 1)^2$ and let $x = -4, -3, -2, -1, 0, 1, 2$. Graph the solutions.** |

▶ **Solution**

Substitute each value of x into $(x + 1)^2$, evaluate the expression, and then make a table of ordered pairs.

Make a table.

x	-4	-3	-2	-1	0	1	2
y	9	4	1	0	1	4	9

Graph each ordered pair in the table. The points do not lie along a line.

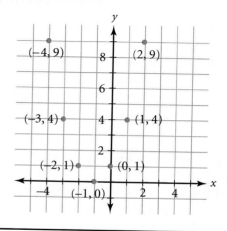

TRY THIS   Let $y = (x - 2)^2$ and let $x = -1, 0, 1, 2, 3, 4, 5$. Graph the solutions.

EXERCISES

KEY SKILLS

Make a table of ordered pairs for each equation.

1. $y = -3x + 5$; $x = -3, -2, -1, 0, 1, 2, 3$

2. $y = 3x - 1$; $x = -3, -2, -1, 0, 1, 2, 3$

3. $y = 2x^2 + 5$; $x = 0, 1, 2, 3, 4, 5, 6$

4. $y = 2x^2 - 3$; $x = -6, -4, -2, 0, 2, 4, 6$

PRACTICE

Graph the ordered pairs (x, y) in each table.

5.

x	-3	-2	-1	0	1	2	3
y	4	0	4	0	4	0	4

6.

x	-3	-2	-1	0	1	2	3
y	-5	-3	-1	3	5	7	9

7.

x	-3	-2	-1	0	1	2	3
y	8	6	4	2	0	-2	-4

8.

x	-3	-2	-1	0	1	2	3
y	4	4	4	4	4	4	4

Make a table of ordered pairs for each equation. Then graph the ordered pairs.

9. $y = x + 3$; $x = -3, -2, -1, 0, 1, 2, 3$

10. $y = x - 2$; $x = -3, -2, -1, 0, 1, 2, 3$

11. $y = 2x$; $x = 0, 1, 2, 3, 4, 5$

12. $y = -2x$; $x = -3, -2, -1, 0, 1, 2, 3$

13. $y = -x - 2$; $x = -3, -2, -1, 0, 1, 2, 3$

14. $y = -x - 1$; $x = -3, -2, -1, 0, 1, 2, 3$

15. $y = 2x^2 - 1$; $x = -2, -1, 0, 1, 2$

16. $y = x^2 - 4x$; $x = 0, 1, 2, 3, 4, 5$

17. $y = -2$; $x = 0, 1, 2, 3, 4, 5, 6$

18. $y = 3$; $x = -4, -2, 0, 2, 4, 6$

Let $x = -3, -2, -1, 0, 1, 2, 3$. For each value of x, find y and make a table of ordered pairs. Graph the ordered pairs.

19. $x + y = 3$

20. $x - y = 4$

21. $2x + y = 2$

22. $x - 2y = 6$

MIXED REVIEW

Write each phrase as a mathematical expression. Do not calculate.
(1.1 Skill A)

23. 2 subtracted from the product of 5 and 9

24. the product of 3 and the difference of 12 minus 5

25. the difference of 5 minus the product of 2 and 5

3.2
LESSON

Functions and Relations

SKILL A *Differentiating between functions and relations*

The equation $y = 2x - 3$ has an important characteristic. For each value of x, you will find *exactly one* value of y.

Definition of Relation and Function

A **relation** is a pairing between two sets. A **function** is a relation in which each member of the first set, the **domain**, is assigned exactly one member of the second set, the **range**.

A relation can be represented by a set of ordered pairs (x, y). The first number, x, is a member of the domain and the second number, y, is a member of the range.

EXAMPLE 1 Determine whether each relation is a function. Explain.

  a. $\{(-1, 7), (0, 3), (1, 5), (0, -3)\}$ **b.** $\{(-1, 7), (0, 3), (1, 5), (2, -3)\}$

▶ **Solution**

  a. Because 0 is paired with both 3 and -3, the relation is not a function.
  b. For each value of x, $\{-1, 0, 1, 2\}$, there is exactly one value of y, $\{7, 3, 5, -3\}$. This set of ordered pairs represents a function.

TRY THIS Determine whether each relation is a function. Explain.

  a. $\{(0, 2), (2, 4), (4, 8), (8, 10)\}$ **b.** $\{(0, 2), (2, 4), (4, 8), (2, 10)\}$

Use the following test to determine whether a graph represents a function.

Vertical-Line Test for a Function

If no vertical line in the coordinate plane intersects a graph in more than one point, then the graph represents a function.

EXAMPLE 2 **Does the graph at right represent a function?**

▶ **Solution**

Because it is impossible to draw a vertical line that will intersect the graph in more than one point, the graph represents a function.

TRY THIS Does the graph at right represent a function?

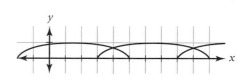

EXERCISES

Write the domain and range of each relation.

1. $\{(3, 0), (-4, 1), (9, 2)\}$

2. $\{(2, 1), (3, 2), (7, 3)\}$

Does each situation describe a function or a relation that is not a function?

3. Each car is assigned one license plate number.

4. Each telephone owner is assigned two telephone numbers.

PRACTICE

Does each set of ordered pairs represent a function? Explain.

5. $\{(1, 10), (2, 8), (3, 6), (4, 4), (5, 2)\}$

6. $\{(1, 10), (2, 8), (3, 6), (4, 8), (5, 2)\}$

7. $\{(1, 1), (2, 1), (3, 1), (4, 1), (5, 1)\}$

8. $\{(1, 1), (1, 2), (1, 3), (1, 4), (1, 5)\}$

9. $\{(1, 1), (4, 2), (9, 2), (16, 4), (16, -4)\}$

10. $\{(1, 1), (4, 2), (9, 3), (16, 4), (25, 5)\}$

Does each graph represent a function? Explain.

11.

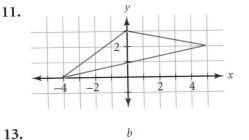

12.

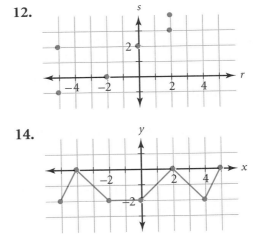

13.

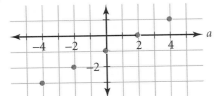

14.

15. Critical Thinking For what value(s) of y will the list of ordered pairs $\{(2, y), (2, 4), (3, 5), (2, y + 1), (10, -3)\}$ be a function?

MIXED REVIEW

Graph all of the following points on the same coordinate plane. (3.1 Skill A)

16. $A(-2, 3)$

17. $B(6, 0)$

18. $C(0, -3)$

19. $D(5, 2)$

Let $x = -3, -2, -1, 0, 1, 2, 3$. Graph the solutions to each equation that correspond to these values of x. (3.1 Skill B)

20. $y = -x - 1.5$

21. $y = 2x + 1.5$

A relation or a function can be represented by a graph. If the graph contains sufficient information, you can use it to find the domain and range.

EXAMPLE 1

Find the domain and range of each relation.

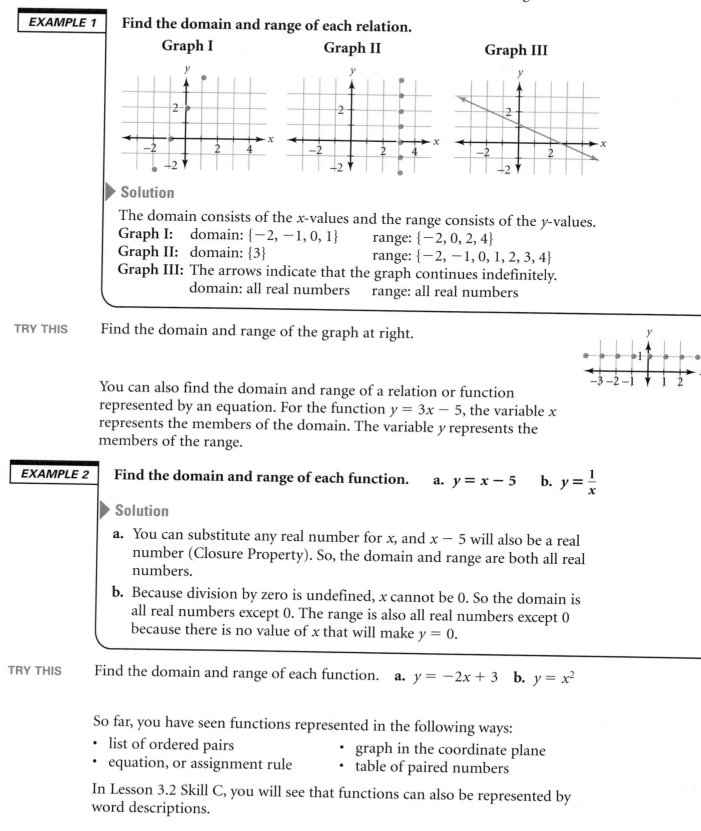

Graph I   Graph II   Graph III

▶ **Solution**

The domain consists of the *x*-values and the range consists of the *y*-values.
Graph I: domain: {−2, −1, 0, 1}   range: {−2, 0, 2, 4}
Graph II: domain: {3}   range: {−2, −1, 0, 1, 2, 3, 4}
Graph III: The arrows indicate that the graph continues indefinitely.
              domain: all real numbers   range: all real numbers

TRY THIS   Find the domain and range of the graph at right.

You can also find the domain and range of a relation or function represented by an equation. For the function $y = 3x - 5$, the variable x represents the members of the domain. The variable y represents the members of the range.

EXAMPLE 2

Find the domain and range of each function.   **a.** $y = x - 5$ **b.** $y = \frac{1}{x}$

▶ **Solution**

a. You can substitute any real number for *x*, and $x - 5$ will also be a real number (Closure Property). So, the domain and range are both all real numbers.

b. Because division by zero is undefined, *x* cannot be 0. So the domain is all real numbers except 0. The range is also all real numbers except 0 because there is no value of *x* that will make $y = 0$.

TRY THIS   Find the domain and range of each function.   **a.** $y = -2x + 3$ **b.** $y = x^2$

So far, you have seen functions represented in the following ways:
• list of ordered pairs   • graph in the coordinate plane
• equation, or assignment rule   • table of paired numbers

In Lesson 3.2 Skill C, you will see that functions can also be represented by word descriptions.

EXERCISES

KEY SKILLS

Refer to the diagram at right. Are the following statements true or false?

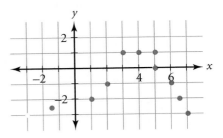

1. All values of the range are less than 2.

2. All values of the domain are integers.

3. The domain contains 0, but the range does not.

PRACTICE

Find the domain and range of each relation.

4.

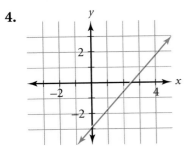

5.

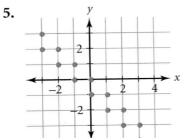

6.

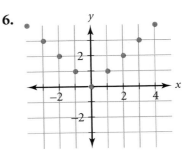

7.

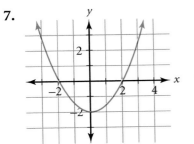

8.

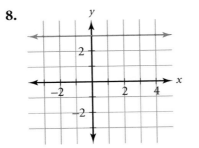

9.

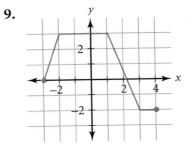

Find the domain and range of each function.

10. $y = \frac{5}{2}x$

11. $y = -\frac{2}{3}x$

12. $y = 3x + 4$

13. $y = -2x - 7$

14. $y = x^2 - 1$

15. $y = -2x^2$

16. **Critical Thinking** Let x be any real number, let a be a fixed real number, and let $y = a|x|$. What is the range of the function?

MIXED REVIEW

Simplify each expression. (2.4 Skill C)

17. $\dfrac{8 + (-2)}{3 - 7} \times \dfrac{(-4)(-2)}{6(-5)}$

18. $\dfrac{-2 + (-2)}{5 - 7} \times \dfrac{(-2)(-2)}{3(-8)}$

19. $\dfrac{-2 + (-3)}{5(-5)} \times \dfrac{(-2)(-2)}{2(-2)}$

20. $\dfrac{(-2)(-3)}{5(-5)} \times \dfrac{(-1)(-1)}{5(-2)} \times \dfrac{-5}{3}$

21. $\dfrac{(-5)(10)}{5(-10)} \times \dfrac{(-1)(-1)}{(-2)(-2)} \times \dfrac{-4}{7}$

22. $\dfrac{(-5)(10)}{-1} \times \dfrac{(-1)(-1)}{(-6)(6)} \times \dfrac{-5}{5(10)}$

In a function such as $y = 2x$, each x-value is paired with exactly one y-value. Thus, the value of y *depends* on the value of x. In a function, the variable of the domain is called the **independent variable** and the variable of the range is called the **dependent variable**. On a graph, the independent variable is represented on the horizontal axis and the dependent variable is represented on the vertical axis.

EXAMPLE 1

The value v (in cents) of n nickels is given by $v = 5n$.
a. Identify the independent and dependent variables.
b. Represent this function in a table for $n = 0, 1, 2, 3, 4, 5, 6$.
c. Represent this function in a graph.

▶ **Solution**

a. The value, v, *depends* on the number of nickels, n. Therefore the independent variable is n and the dependent variable is v.

b.

n	0	1	2	3	4	5	6
v	0	5	10	15	20	25	30

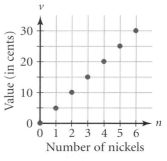

c. Label the horizontal axis with the independent variable, n, and the vertical axis with the dependent variable, v. Graph the ordered pairs (n, v).

(0, 0), (1, 5), (2, 10), (3, 15), (4, 20), (5, 25), (6, 30)

TRY THIS

The number of wheels, w, on n tricycles is given by $w = 3n$. Identify the independent and dependent variables. Represent the function in a table and in a graph for $n = 0, 1, 2, 3, 4, 5$.

EXAMPLE 2

Water is pumped through a pipe at the rate of 50 gallons per minute. The number of gallons pumped is a function of the number of minutes the pump operates. Find the domain and range of this function.

▶ **Solution**

Make a graph.

The domain, number of minutes, must be greater than or equal to zero. The range, number of gallons of water pumped, must also be greater than or equal to zero. Because time and water flow are continuous, the domain and range are all real numbers greater than or equal to zero.

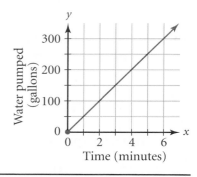

TRY THIS

Find the domain and range of the function described in Example 1.

EXERCISES

A 1200-gallon tank is empty. A valve is opened and water flows into the tank at the constant rate of 25 gallons per minute.

1. How would you find the amount of water in the tank after 5 minutes, 6 minutes, and 7 minutes?

2. How would you find the time it takes for the tank to contain 500 gallons, 600 gallons, and 700 gallons of water?

Use both a table and a graph to represent each function for the given values of the independent variable.

3. The value v of 0, 1, 2, 3, 4, 5, and 6 dimes is a function of the number of dimes, n.

4. The distance d traveled by a jogger who jogs 0.1 mile per minute over 0, 1, 2, 3, 4, and 5 minutes is a function of time t.

5. One hat costs $24. The total cost c of 1, 2, 3, 4, 5, and 6 hats is a function of the number n of hats purchased.

Find the domain and range of each function.

6. independent variable: number of dimes, d
   dependent variable: total value of the coins, v

7. One package of cookies contains 6 cookies.
   independent variable: number of packages of cookies, n
   dependent variable: number of cookies, c

8. A conveyor belt adds 300 pounds of sand to a pile each minute.
   independent variable: elapsed time, t
   dependent variable: total weight of the sand, w

Solve each problem. (2.8 Skill C)

9. How much of $2000 should be put into an account paying 6% and how much should go into an account paying 12% simple interest to earn $200 interest in one year?

10. A chemist divides 400 milliliters of a solution into two beakers. One beaker contains 70 milliliters more than the other beaker. How much will each beaker contain?

3.3 Ratio, Rate, and Direct Variation

LESSON

A **ratio** is a quotient that compares two quantities. A **rate** is a quotient that compares two different types of quantities.

ratio: 80 people out of 100 people

$$\frac{80}{100}, \text{ or } \frac{2}{5}$$

rate: 55 miles per hour

$$\frac{55 \text{ miles}}{1 \text{ hour}}$$

Both ratios and rates are the basis for many simple but important functions. One type of function based on ratio or rate is *direct variation.*

Direct Variation

If there is a fixed nonzero number k such that $y = kx$, then you can say that y *varies directly as x.* The relationship between x and y is called a **direct-variation** relationship and k is called the **constant of variation.**

If you write $y = kx$ as $\frac{y}{x} = k$, you can see that k is a constant ratio or a constant rate. For example, if you drive for t hours at 55 miles per hour, then the distance d (in miles) that you travel is given by $d = 55t$. In other words, $\frac{d}{t} = 55$ miles per hour, which is a constant rate.

EXAMPLE

In a certain state, the sales tax rate is $0.05 per $1 of purchase.
  a. Use an equation to express the total cost of a purchase as a function of the purchase price.
  b. Identify the constant of variation.
  c. Find the domain and range.
  d. Find the total cost of a purchase of $14.39.

▸ **Solution**

  a. Let p represent the purchase price and let t represent the total cost.
      total cost = purchase price + 0.05 × purchase price
      $t = p + 0.05 \times p$
      $t = 1p + 0.05p = (1 + 0.05)p = 1.05p$ ⟵——— Apply the Distributive Property.
  b. The constant of variation is 1.05.
  c. The domain is any nonnegative money amount in dollars and cents.
      The range is any nonnegative money amount in dollars and cents.
  d. $t = 1.05p$
      $= 1.05(14.39)$ ⟵——— *Substitution*
      ≈ 15.11 ⟵——— *Round to the nearest cent.*
   The total cost is $15.11.

TRY THIS   Rework the Example with a sales tax of 6 cents per dollar.

EXERCISES

KEY SKILLS

Write the unit that would result from each product.

1. $\dfrac{5 \text{ dollars}}{1 \text{ package}} \times 6 \text{ packages}$

2. $\dfrac{55 \text{ miles}}{1 \text{ hour}} \times 8 \text{ hours}$

PRACTICE

For Exercises 3–5,
   a. **use an equation to express each direct variation as a function.**
   b. **identify the constant of variation.**
   c. **find the domain and range.**
   d. **find the specified quantity.**

3. The total price of a dinner, p, is the cost of the meal, m, plus a 15% tip. Find the total price of a dinner if the meal costs $19.50.

4. In a certain store, the retail price, p, of an item is the wholesale cost, c, of the item plus a 50% markup. Find the retail price of a coat whose wholesale cost is $98.00.

5. A pump fills an empty tank at the rate of 30 gallons per minute. Find the amount of liquid in the tank after 5.5 minutes.

Use a direct-variation equation to solve each problem.

6. A certain automobile can travel 22 miles on each gallon of gas.
   a. How far will it travel on 36 gallons of gas?
   b. How many gallons of gasoline are needed for an 800-mile trip?

7. On a certain map, a map distance of 1 inch corresponds to an actual distance of 50 miles.
   a. What actual distance corresponds to a map distance of 2.5 inches?
   b. What map distance would correspond to an actual distance of 300 miles?

8. A store has a sale and marks each item down to 80% of its original price.
   a. Find the sale price of an appliance that cost $84 before the discount was taken.
   b. If the sale price of an appliance is $64, what was its original price?

9. An employee's wage is $12.45 per hour.
   a. How much money will the employee earn for 40 hours of work?
   b. How many hours must the employee work to earn $240?

MIXED REVIEW APPLICATIONS

Solve each problem. (previous courses)

10. Find 60% of $18.

11. What percent of $140 is $210?

12. Find 20% of $20.

13. Find 150% of $200.

A **proportion** is a statement that two ratios are equal. If y varies directly as x, then you can say that *y is proportional to x.*

Alternate Form of Direct Variation

If (x_1, y_1) and (x_2, y_2) satisfy the direct-variation relationship $y = kx$, then
$$\frac{y_1}{x_1} = \frac{y_2}{x_2}.$$

Proof: Suppose that (x_1, y_1) and (x_2, y_2) satisfy $y = kx$.

$$y_1 = kx_1 \qquad\qquad y_2 = kx_2 \qquad \longleftarrow \text{\textit{Definition of solution}}$$

$$\frac{y_1}{x_1} = k \qquad\qquad \frac{y_2}{x_2} = k \qquad \longleftarrow \text{\textit{Multiplication Property of Equality}}$$

$$\frac{y_1}{x_1} = k = \frac{y_2}{x_2} \qquad\qquad \longleftarrow \text{\textit{Transitive Property of Equality}}$$

EXAMPLE 1

The cost of lunch, c, varies directly as the number of people, n. If lunch for 3 people costs \$13, find the cost of lunch for 6 people.

▸ **Solution**

Write and solve a proportion.

$$\frac{13}{3} = \frac{c}{6} \qquad \longleftarrow \text{\textit{Set the ratios equal to each other.}}$$

$$6\left(\frac{13}{3}\right) = 6\left(\frac{c}{6}\right) \qquad \longleftarrow \text{\textit{Apply the Multiplication Property of Equality.}}$$

$$26 = c$$

The cost of lunch for 6 people is \$26.

TRY THIS   Refer to Example 1. Find the cost of lunch for 9 people.

EXAMPLE 2

An empty water tank is being filled by a pump. After 2 minutes, the tank contains 125 gallons. If the volume of water in the tank varies directly as the time the water is pumped, how much water is in the tank after 5 minutes?

▸ **Solution**

Let v represent the volume of water in the tank after 5 minutes.
Write and solve a proportion.

$$\frac{volume}{time} \rightarrow \frac{125}{2} = \frac{v}{5}$$

$$5\left(\frac{125}{2}\right) = 5\left(\frac{v}{5}\right) \qquad \longleftarrow \text{\textit{Apply the Multiplication Property of Equality.}}$$

$$312.5 = v$$

After 5 minutes, the tank contains 312.5 gallons of water.

TRY THIS   Refer to Example 2. How much water is in the tank after 9 minutes?

EXERCISES

Solve each proportion.

1. $\dfrac{12}{36} = \dfrac{n}{6}$

2. $\dfrac{a}{42} = \dfrac{3}{14}$

3. $\dfrac{10}{25} = \dfrac{n}{12}$

4. $\dfrac{m}{21} = \dfrac{3}{10}$

PRACTICE

Given that y varies directly as x, use a proportion to find y.

5. If $y = 33$ when $x = 3$, find y when $x = 22$.

6. If $y = 4.6$ when $x = 2.3$, find y when $x = 2$.

7. If $y = 8.4$ when $x = 3.5$, find y when $x = 12$.

8. If $y = 35.1$ when $x = 13.5$, find y when $x = 5.5$.

9. If $y = 109.56$ when $x = 8.3$, find y when $x = 2.2$.

Solve each direct-variation problem using:
   a. a direct-variation equation.   b. a proportion.

10. Simple interest in a period of time varies directly as the amount of the deposit. If a $2000 deposit earns $120, how much interest will $2400 earn in the same amount of time?

11. The number of gallons of paint needed is proportional to the area of the surface to be painted. If one gallon covers 250 square feet, how many gallons will be needed to cover 540 square feet? (Give one decimal answer and one whole number answer.)

12. The number of pages read is proportional to the time spent reading. If a student can read 3 pages in 10 minutes, how many pages will the student read in 30 minutes?

13. The number of students served in a cafeteria varies directly as time. If 10 students can be served in 6 minutes, how many students will be served in 48 minutes?

MIXED REVIEW APPLICATIONS

Solve each problem using an equation. (1.9 Skill B)

14. Six workers together receive $111 for one hour of work. What is each worker's wage given that they all earn the same amount?

15. A stamp collector separated stamps into 5 equal groups of 31 stamps each. How many stamps did the collector have to sort?

Slope and Rate of Change

SKILL A *Finding the slope of a line*

Every nonvertical line in the coordinate plane has a steepness, or **slope**. The ratio $\frac{\text{rise}}{\text{run}}$, illustrated at right, is a measure of that steepness. If two points lie on the same line, then you can use the coordinates of those points to calculate the slope.

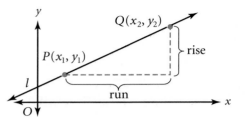

The Slope of a Line

If $P(x_1, y_1)$ and $Q(x_2, y_2)$ lie along a nonvertical line in the coordinate plane, the line has slope m, given by $m = \frac{\text{rise}}{\text{run}} = \frac{y_2 - y_1}{x_2 - x_1}$.

EXAMPLE

Find the slope of the line that contains each pair of points.
a. $P(-3, -1)$ and $Q(5, 6)$ **b. $R(-2, 8)$ and $S(7, 3)$**

▶ **Solution**

Choose (x_1, y_1) and (x_2, y_2) and apply the formula for slope.

a. $m = \frac{y_2 - y_1}{x_2 - x_1} \rightarrow \frac{6 - (-1)}{5 - (-3)} = \frac{7}{8}$ **b.** $m = \frac{y_2 - y_1}{x_2 - x_1} \rightarrow \frac{3 - 8}{7 - (-2)} = \frac{-5}{9} = -\frac{5}{9}$

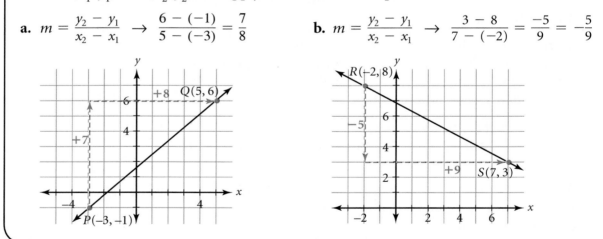

TRY THIS

Find the slope of the line that contains each pair of points.
a. $A(-2, 3)$ and $B(5, 4)$ **b.** $A(-2, 3)$ and $C(5, -4)$

Notice that the line graphed in part **a** *rises* from left to right and the line graphed in part **b** *falls* from left to right.

Slope and Orientation of a Line

In the coordinate plane, a line with positive slope *rises* from left to right. A line with negative slope *falls* from left to right. A line with 0 slope is horizontal. The slope of a vertical line is undefined.

EXERCISES

Graph the line that contains each pair of points. Does the line rise from left to right, fall from left to right, or is it horizontal?

1. $A(2, 6)$ and $B(6, 0)$

2. $M(-3, 4)$ and $N(5, 4)$

3. $X(-1, -4)$ and $Y(4, 6)$

PRACTICE

Find the slope of the line that contains each pair of points.

4. $R(0, 0)$ and $S(6, 2)$

5. $G(0, 0)$ and $H(5, -1)$

6. $W(0, 0)$ and $Z(4, 0)$

7. $A(-2, 0)$ and $B(2, 0)$

8. $F(0, 4)$ and $G(-4, 0)$

9. $B(-1, -4)$ and $C(3, 9)$

10. $P(6, 2)$ and $Q(-6, 1)$

11. $M(7, -2)$ and $N(-5, 4)$

12. $X(4, -4)$ and $Y(-4, -4)$

13. $A(2.3, 0)$ and $B(6.6, 0)$

14. $D(3.5, 2)$ and $E(5.5, 7)$

15. $L(-1, -3)$ and $K(-4, -3)$

Use the figure at right to find the slope of each line.

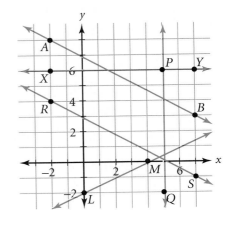

16. the line that contains A and B

17. the line that contains X and Y

18. the line that contains R and S

19. the line that contains P and Q

20. the line that contains L and M

21. Points $P(5, 2)$ and $Q(-3, y)$ lie on a line with slope $\frac{3}{2}$. What is the y-coordinate of Q?

22. **Critical Thinking** You are given $A(a, b)$ and $R(r, s)$. What must be true of a, b, r, and s if the line that contains A and R rises from left to right?

MIXED REVIEW

Solve each equation. (2.9 Skill B)
  a. Write the steps in the solution as a logical argument.
  b. Write the solution as a conditional statement.

23. $-2(x + 5) = 16$

24. $3a - 2.6 = 4$

25. $-4 = 3x + 5x$

26. $-3r - r + 5 = 6$

27. $-2(x - 1) + 3 = 7$

28. $4b - 1 = 2b + 8$

EXAMPLE 1 A line with slope $\frac{3}{5}$ contains $P(1, -2)$.

a. **Sketch the line.**　　b. **Find the coordinates of a second point on the line.**

▶ **Solution**

a. Graph $P(1, -2)$.

Because the slope is $\frac{3}{5}$, you can find a second point on the line by counting up 3 units and then right 5 units. Draw a line through $P(1, -2)$ and the new point. The graph is shown at right.

b. **Method 1:** Use the graph from part **a.**
A second point on the line is $Q(1 + 5, -2 + 3)$, or $Q(6, 1)$.

Method 2: Use the definition of slope.

Let $Q(x_2, y_2)$ be a second point on the line. Then $\dfrac{y_2 - (-2)}{x_2 - 1} = \dfrac{3}{5}$.

Choose a number other than 1 for x_2, because division by 0 is undefined. For example, choose 0 for x_2. Then solve for y_2.

$$\frac{y_2 - (-2)}{0 - 1} = \frac{3}{5} \quad \rightarrow \quad \frac{y_2 + 2}{-1} = \frac{3}{5} \quad \rightarrow \quad y_2 = -\frac{13}{5}$$

Point $Q\left(0, -\dfrac{13}{5}\right)$ is another point on the line.

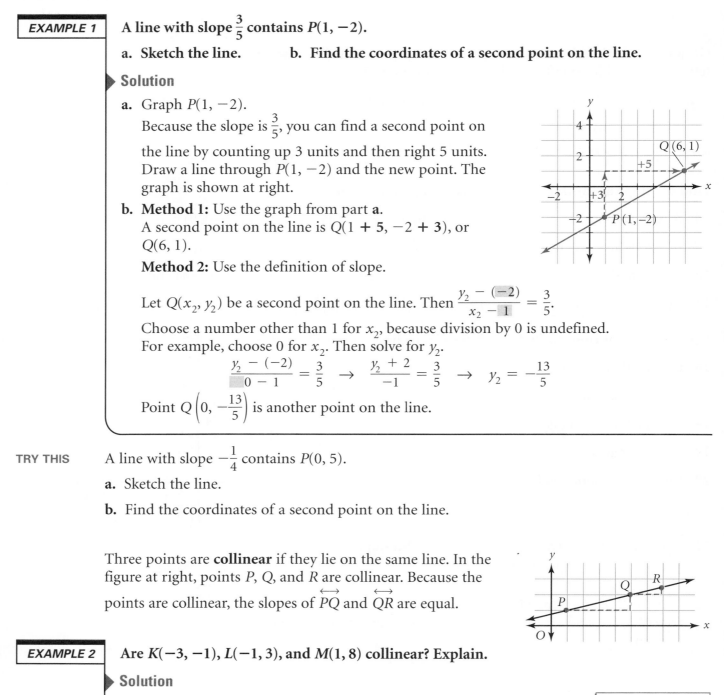

TRY THIS　A line with slope $-\dfrac{1}{4}$ contains $P(0, 5)$.

a. Sketch the line.

b. Find the coordinates of a second point on the line.

Three points are **collinear** if they lie on the same line. In the figure at right, points P, Q, and R are collinear. Because the points are collinear, the slopes of $\overleftrightarrow{PQ}$ and $\overleftrightarrow{QR}$ are equal.

EXAMPLE 2 Are $K(-3, -1)$, $L(-1, 3)$, and $M(1, 8)$ collinear? Explain.

▶ **Solution**

1. Calculate the slopes.

$$\text{slope of } \overleftrightarrow{LM}: \frac{8 - 3}{1 - (-1)} = \frac{5}{2} \qquad \text{slope of } \overleftrightarrow{KL}: \frac{3 - (-1)}{-1 - (-3)} = 2$$

> $\overleftrightarrow{KL}$ refers to the line that contains points K and L.

2. Compare the slopes.

The slopes, 2 and $\dfrac{5}{2}$, are not equal. Thus, the line that contains K and L is not the same line that contains L and M. The three points are not collinear.

TRY THIS　Are $A(-3, -1)$, $B(-1, 3)$, and $C(3, 7)$ collinear? Explain.

EXERCISES

Graph the three given points. Are they collinear?

1. $P(0, 0)$, $B(3, 3)$, and $D(-2, -2)$

2. $R(2, 1)$, $E(6, 3)$, and $F(3, -2)$

For Exercises 3–6,
   a. sketch each line.
   b. find a second point on the line.

3. containing $H(0, 0)$; slope: 2

4. containing $S(3, 3)$; slope: -3

5. containing $J(2, 5)$; slope: $-\dfrac{1}{2}$

6. containing $T(-3, 1)$; slope: $-\dfrac{2}{5}$

Are the given points collinear? Explain.

7. $A(0, 0)$, $B(2, 2)$, $C(4, 4)$

8. $D(3, 1)$, $E(6, 5)$, $F(9, 8)$

9. $P(6, 5)$, $Q(2, 2)$, $R(0, -2)$

10. $X(3, 7)$, $Y(4, 9)$, $Z(5, 11)$

Two distinct lines are *parallel* if they have the same slope. Determine if the given lines are parallel.

11. Line s contains $(0, 5)$ and $(2, 7)$;
    line t contains $(-1, 6)$ and $(2, 9)$.

12. Line q contains $(2, -3)$ and $(4, -7)$;
    line r contains $(-1, 3)$ and $(0, 2)$.

13. **Critical Thinking**   A line has slope $\dfrac{m}{n}$ and contains $P(a, b)$.

    Show that $Q(a + n, b + m)$ is also on the line.

14. Graph the solutions to $y = -\dfrac{3}{2}x + 2$ that correspond to

    $x = -4, -2, 0, 2, 4, 6.$ (3.1 Skill B)

15. Does $\{(1, 5), (3, 2), (6, 2), (0, -2)\}$ represent a function? Explain.
    (3.2 Skill A)

16. Find the domain and range of $y = \dfrac{2}{5}x + 3.$ (3.2 Skill B)

17. The perimeter of a square is 4 times the length of one side. Find the
    domain and range of the perimeter function. (3.2 Skill C)

18. Find the slope of the line that contains $A(-2, 1)$ and $B(5, 7)$.
    (3.4 Skill A)

19. If y varies directly as x, and $y = 28$ when $x = 4$, find y when $x = 11$.
    (3.3 Skill B)

The slope of a line is called *constant* because it is the same regardless of the points chosen to calculate it. This is illustrated at right by using the definition of slope as $\frac{\text{rise}}{\text{run}}$.

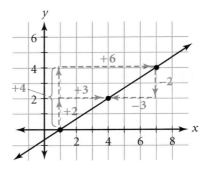

$$\frac{\text{rise}}{\text{run}} = \frac{2}{3} = \frac{4}{6} = \frac{-2}{-3}$$

Notice that rise represents the *change* in *y* and run represents the corresponding *change* in *x*. Because slope is the ratio $\frac{\text{rise}}{\text{run}}$, slope is sometimes called *rate of change*.

$$\text{slope} = m = \frac{\text{rise}}{\text{run}} = \frac{\text{change in } y}{\text{change in } x} = \text{rate of change}$$

Rate of change describes the relationship between two different quantities that are changing. An example of a rate of change is speed, or $\frac{\text{change in distance}}{\text{change in time}}$. Rates of change are usually expressed as *unit rates*. A unit rate has a denominator of one unit, such as 55 miles per hour, or $\frac{55 \text{ miles}}{1 \text{ hour}}$.

So, if you graph the time and distance data for a person driving 55 miles per hour, the slope of the line is 55.

EXAMPLE

After stopping to buy gas, a bus driver drives at a constant rate as indicated in the table at right. Find the speed or rate of change for the bus and graph the situation.

Time *t*	0 hr	2 hr
Distance *d*	5 mi	120 mi

▸ **Solution**

Plot and connect the points in the table. The speed or rate of change is the slope of the line.

$$\text{slope} = \frac{\text{change in distance traveled}}{\text{corresponding time change}}$$

$$= \frac{d_2 - d_1}{t_2 - t_1} = \frac{120 - 5}{2 - 0} = \frac{115}{2} = 57.5$$

The bus traveled at a constant speed of 57.5 miles per hour.

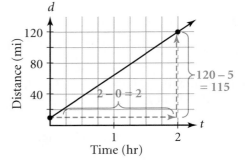

TRY THIS

A small aircraft begins a descent to land as indicated in the table at right. If the plane descends at a constant rate, find its rate of descent and graph the situation.

Time *t*	1.5 min	3 min
Altitude *a*	5450 ft	4700 ft

EXERCISES

Is the rate of change positive or negative?

1. $\dfrac{32 - 12}{10 - 5}$

2. $\dfrac{28 - 60}{100 - 80}$

3. $\dfrac{117.4 - 126.5}{7.0 - 6.9}$

4. $\dfrac{17.5 - 16.5}{3 - 2}$

Calculate each rate of change.

5.

elapsed time t	3 hr	4.4 hr
distance d	140 mi	280 mi

6.

elapsed time t	3 min	4.6 min
altitude a	3450 ft	2350 ft

7.

elapsed time t	5 min	6.5 min
temperature T	35°F	32°F

8.

elapsed time t	1 min	5.4 min
temperature T	80°F	212°F

In Exercises 9–12, calculate each rate of change.

9. A hose flows 250 gallons of water in 2 minutes and 875 gallons in 7 minutes.

10. 32 people exit the stadium in 2.5 minutes and 384 people leave the stadium in 30 minutes.

11. After 3.5 minutes of descent, a small plane's altitude is 3425 feet. After 6.0 minutes, the plane's altitude is 2375 feet.

12. A racer drives 2 miles in 0.02 hours and 305 miles in 3.02 hours.

13. Does the data below indicate a constant speed? Explain.
2 miles/4 minutes   8 miles/16 minutes   10 miles/24 minutes

14. **Critical Thinking**   Let m and b represent real numbers. Find the rate of change given $(x_1, mx_1 + b)$ and $(x_2, mx_2 + b)$.

Solve each direct-variation problem. (3.3 Skill B)

15. Pizza costs vary directly as quantity. Two pizzas costs $12.96 and 7 pizzas cost $45.36. How much will 10 pizzas cost?

16. The amount of an employee's paycheck, p, varies directly as the number of hours worked, h. If $p = 400$ when $h = 20$, find p when $h = 15$.

17. Interest earned varies directly as money invested. If $450 earns $15.75 and $2400 earns $84, how much interest will $6000 earn in the same amount of time?

Graphing Linear Equations in Two Variables

SKILL A *Using intercepts to graph a linear equation*

A linear equation is an equation whose graph is a line. The following equations are examples of *linear equations.*

$$y = 3.5x - 5 \qquad 2x - 3y = 12 \qquad y = 5 \qquad x = -5.6 \qquad y + 2 = -3(x + 7)$$

> ### Linear Equations in Two Variables
> A **linear equation in x and y** is any equation of the form $Ax + By = C$, where A, B, and C are real numbers, and A and B are not both 0. The graph of $Ax + By = C$ is a straight line in the coordinate plane.

The **solution to a linear equation** in two variables is all ordered pairs of real numbers that satisfy the equation. One way to graph a linear equation is to use *intercepts* to find two points and draw a line through them.

The **y-intercept** is the y-coordinate of the point where a line crosses the y-axis. The **x-intercept** is the x-coordinate of the point where a line crosses the x-axis.

EXAMPLE

Let $2x + y = 4$. Use the x- and y-intercepts to graph the equation.

▶ **Solution**

To find the y-intercept, let $x = 0$ and solve for y.
$$x = 0 \quad \rightarrow \quad 2(0) + y = 4 \quad \rightarrow \quad y = 4$$
The y-intercept is 4, so $(0, 4)$ is on the graph.

To find the x-intercept, let $y = 0$ and solve for x.
$$y = 0 \quad \rightarrow \quad 2x + 0 = 4 \quad \rightarrow \quad x = 2$$
The x-intercept is 2, so $(2, 0)$ is on the graph.

Graph $(0, 4)$ and $(2, 0)$. Draw the line through $(0, 4)$ and $(2, 0)$.

TRY THIS   Let $-3x + y = 6$. Use the x- and y-intercepts to graph the equation.

Lines that rise or fall from left to right intersect both axes, as shown in the example above. Vertical lines only intersect the x-axis, and horizontal lines only intersect the y-axis.

> ### Vertical and Horizontal Lines
> The graph of $x = r$ is a vertical line that contains $(r, 0)$.
> The graph of $y = s$ is a horizontal line that contains $(0, s)$.

EXERCISES

KEY SKILLS

Find the *x*- and *y*-intercepts of the graph of each equation. If the graph has no *x*- or no *y*-intercept, write *none*.

1. $3x - 5y = 12$ **2.** $x = -2.5$ **3.** $y = 12\frac{3}{4}$ **4.** $3x + 5y = 25$

Tell whether the line is vertical, horizontal, or neither.

5. $6x + 2y = 8$ **6.** $y = -2$ **7.** $x = 5$ **8.** $4x - 3y = 11$

PRACTICE

Find the intercepts for the graph of each equation. Then graph the line.

9. $x + y = 3$ **10.** $x - y = 3$ **11.** $2x + y = 8$ **12.** $-2x + y = 8$

13. $3x + 4y = 12$ **14.** $2x + 3y = 12$ **15.** $y = -1$ **16.** $x = -1$

17. $y = \frac{5}{2}$ **18.** $x = \frac{5}{2}$ **19.** $x + y = 4$ **20.** $x - y = 4$

21. $2.5x + 2.5y = 5$ **22.** $1.5x + y = 9$ **23.** $3x + 5y = 15$ **24.** $3x - 6y = 15$

25. $y = 3x + 4$ **26.** $y = -2x - 3$ **27.** $y - 2 = 3(x + 1)$ **28.** $y + 1 = -2(x - 1)$

29. A line has *x*-intercept 3 and *y*-intercept 7. Is the line horizontal, vertical, or neither?

30. A line has *x*-intercept 2 and *y*-intercept 4. Write an equation for the line.

31. A line has *x*-intercept 3. Write an equation for the line given that
   a. it rises from left to right.   **b.** it is vertical.

32. Critical Thinking   Let *a* and *b* be nonzero real numbers. Find the *x*- and *y*-intercepts of the graph of $\frac{x}{a} + \frac{y}{b} = 1$.

MIXED REVIEW

Write each decimal as a fraction. (previous courses)

33. 3.5 **34.** -1.5 **35.** 0.4 **36.** -0.6

Find the slope of the line that contains the given points. (3.4 Skill A)

37. $A(1, 2); B(3, 5)$ **38.** $K(1, -2); L(3, -6)$ **39.** $R(1, 2); S(5, 2)$ **40.** $X(5, -1); Y(-3, 0)$

One useful form of a linear equation is $y = mx + b$, the *slope-intercept form.*

Slope-Intercept Form

If $y = mx + b$, the slope of the line is m and the y-intercept is b.

When the equation of a line is solved for y, you can quickly identify the slope and the y-intercept of the line.

EXAMPLE 1 Let $y = -\frac{3}{5}x + 5$. Use the slope and the y-intercept to graph the equation.

Solution

1. Because the y-intercept is 5, plot $(0, 5)$.
2. Because the slope is $-\frac{3}{5}$, or $\frac{-3}{5}$, start at $(0, 5)$, count 3 units down, and then count 5 units right to arrive at $(5, 2)$.
3. Draw the line through $(0, 5)$ and $(5, 2)$.

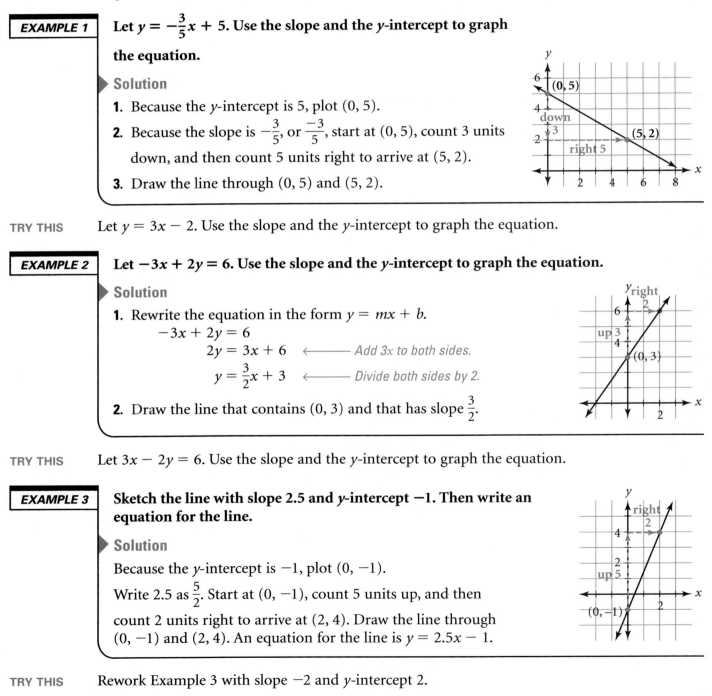

TRY THIS Let $y = 3x - 2$. Use the slope and the y-intercept to graph the equation.

EXAMPLE 2 Let $-3x + 2y = 6$. Use the slope and the y-intercept to graph the equation.

Solution

1. Rewrite the equation in the form $y = mx + b$.

$$-3x + 2y = 6$$
$$2y = 3x + 6 \quad \longleftarrow \textit{Add 3x to both sides.}$$
$$y = \frac{3}{2}x + 3 \quad \longleftarrow \textit{Divide both sides by 2.}$$

2. Draw the line that contains $(0, 3)$ and that has slope $\frac{3}{2}$.

TRY THIS Let $3x - 2y = 6$. Use the slope and the y-intercept to graph the equation.

EXAMPLE 3 Sketch the line with slope 2.5 and y-intercept -1. Then write an equation for the line.

Solution

Because the y-intercept is -1, plot $(0, -1)$.

Write 2.5 as $\frac{5}{2}$. Start at $(0, -1)$, count 5 units up, and then count 2 units right to arrive at $(2, 4)$. Draw the line through $(0, -1)$ and $(2, 4)$. An equation for the line is $y = 2.5x - 1$.

TRY THIS Rework Example 3 with slope -2 and y-intercept 2.

EXERCISES

KEY SKILLS

Start from $(0, -2)$ and find the coordinates of the point that results.

1. Count 2 units up and 3 units right.

2. Count 2 units up and 3 units left.

3. Count 2 units down and 3 units right.

4. Count 2 units down and 3 units left.

Describe each slope in terms of rise and run.

5. $\frac{3}{2}$

6. $-\frac{3}{2}$

7. $\frac{1}{3}$

8. $-\frac{1}{3}$

PRACTICE

Use the slope and the y-intercept to graph each equation.

9. $y = -x + 2$

10. $y = x + 3$

11. $y = 3x + 1$

12. $y = -2x + 4$

13. $y = -\frac{1}{3}x$

14. $y = \frac{1}{3}x$

15. $y = \frac{5}{3}x + 2$

16. $y = -\frac{5}{3}x + 3$

17. $y = -1.5x$

18. $y = 2.5x$

19. $y = 0.5x + 1$

20. $y = -0.6x - 1$

Write each equation in slope-intercept form. Then use the slope and the
y-intercept to graph the equation.

21. $-\frac{1}{2}x + y = 4$

22. $\frac{2}{3}x + y = 2$

23. $2x - 3y = 6$

24. $-3x + 5y = 15$

Graph the line with the given slope and y-intercept. Then write an
equation for the line in slope-intercept form. (*Hint:* Write the slopes as
fractions where necessary. Example: $6 = \frac{6}{1}$)

25. slope 3; y-intercept 0

26. slope -2; y-intercept 0

27. slope 1.5; y-intercept -2

28. slope -2.5; y-intercept 4

29. slope $\frac{7}{5}$; y-intercept 1

30. slope $-\frac{7}{4}$; y-intercept -2

MIXED REVIEW

Write each mixed number as a fraction and as a decimal. (previous courses)

31. $2\frac{1}{3}$

32. $-3\frac{3}{5}$

33. $10\frac{1}{2}$

34. $-9\frac{1}{3}$

Solve each equation. (2.8 Skill B)

35. $2(x + 5) - 3 = 11$

36. $2x + 5 = 11x + 32$

37. $2(x + 3) + 5 = -2(x + 1) - 6$

EXAMPLE 1

A storage tank contains 300 gallons of water. A valve is opened and water is drained out at the rate of 60 gallons per minute. Represent the amount of water, V, in the tank after t minutes of drainage as a linear equation in V and as a graph.

▶ Solution

1. When the valve is opened, the volume is 300 gallons.
   discharge rate: $\dfrac{-60 \text{ gallons}}{1 \text{ minute}}$

2. Write a function for volume V in terms of elapsed time t.
   $$V = -60t + 300$$

3. When $t = 0$, $V = 300$. When $t = 5$, $V = 0$. Connect the points $(0, 300)$ and $(5, 0)$ with a line. Notice that the graph ends where it meets the axes.

TRY THIS

Rework Example 1 given a 600-gallon tank that is being drained at the rate of 120 gallons per minute.

The graph in Example 1 is a *continuous graph,* a graph that is unbroken.

EXAMPLE 2

Diego has $1.00 in nickels and dimes. Represent this with an equation and with a graph.

▶ Solution

1. Let n represent the number of nickels and let d represent the number of dimes. Make a table with amounts in cents.

Make a table.

Coins	n	d	$n + d$
Value	$5n$	$10d$	100

2. Write an equation for total value.
   $5n + 10d = 100$, or $n = 20 - 2d$
   d: whole number not more than 10
   n: positive even number 20 or less

3. Graph the intercepts, $(0, 20)$ and $(10, 0)$. Points (d, n) are solutions if d and n meet the conditions stated in Step **2**.

TRY THIS

Rework Example 2 given that Diego has $2.00.

The graph in Example 2 is a *discrete graph,* a graph of separate points. Because the variables in Example 2 represent coins and therefore must be whole numbers, we cannot connect the points with a line.

EXERCISES

KEY SKILLS

Write an algebraic expression for each situation.

1. The total value v of d dimes and n nickels

2. Due to a leak, a tank containing 1000 gallons loses 2 gallons per minute.

3. An airplane flying at an altitude of 1200 feet ascends 130 feet per minute.

PRACTICE

Represent each situation with an equation and with a graph. Is the graph discrete or continuous? Explain your response.

4. A box contains 15 marbles. Each minute, 3 marbles are removed from the box.

5. An airplane is flying at 1200 feet. Each minute, the plane descends 300 feet.

6. Jamie has $0.25 in pennies and nickels.

7. A tank contains 200 gallons of water when a discharge pipe is opened. Water is drained from the tank at the rate of 40 gallons per minute.

8. A clerk earns $12.50 per hour and works a maximum of 8 hours per day. Time worked is a whole number of hours.

9. The sale price is 75% of the original price.

10. **Critical Thinking**   An empty tub is filled with water at the rate of 3 gallons per minute for 5 minutes. Then the water is turned off and the water is allowed to sit for 5 minutes. At that time, the drain is opened and water is allowed to drain at the rate of 2 gallons per minute.

MIXED REVIEW APPLICATIONS

Represent volume as a function of time in a direct-variation relationship. Find the volume after 5.5 minutes. (3.3 Skill B)

11. A bin is initially empty. A conveyor belt pours 10 cubic feet of sand into the bin each minute.

12. A swimming pool is initially empty. A pump pours 125 gallons of water into the pool each minute.

The Point-Slope Form of an Equation

SKILL A *Using slope and a given point to find the equation of a line*

If you know the coordinates of one point on a line and the slope of the line, you can write an equation for the line.

> **Point-Slope Form**
>
> If $P(x_1, y_1)$ is a point on a nonvertical line with slope m, then the line can be represented by $y - y_1 = m(x - x_1)$.
>
> This is called the *point-slope form* of a linear equation.

Proof: Suppose $P_1(x_1, y_1)$ and $P(x, y)$ lie on a line with slope m.

Then $\dfrac{y - y_1}{x - x_1} = m$. Multiplying by $x - x_1$ results in $y - y_1 = m(x - x_1)$.

EXAMPLE 1 Point $Q(-3, -5)$ is on a line with slope $-\dfrac{5}{3}$. Write an equation for the line in both point-slope form and slope-intercept form.

▶ Solution

$$y - \boxed{y_1} = \boldsymbol{m}(x - \boxed{x_1}) \qquad \longleftarrow \text{ Use the point-slope form.}$$

$$y - (-5) = -\frac{5}{3}[x - (\boxed{-3})] \qquad \longleftarrow \text{ Substitute } -3 \text{ for } x_1, -5 \text{ for } y_1, \text{ and } -\frac{5}{3} \text{ for } m.$$

$$y = -\frac{5}{3}x - 10 \qquad \longleftarrow \text{ Write in slope-intercept form.}$$

TRY THIS Point $Q(2, -3)$ is on a line with slope $\dfrac{3}{5}$. Write an equation for the line in both point-slope form and slope-intercept form.

Recall from Lesson 3.5 Skill A that the graph of an equation of the form $y = s$ is a horizontal line. Also recall that all horizontal lines have slope 0.

EXAMPLE 2 Point $Q(-3, -5)$ is on a line with slope 0. Write the equation for the line in slope-intercept form.

▶ Solution

$$y - \boxed{y_1} = \boldsymbol{m}(x - \boxed{x_1}) \qquad \longleftarrow \text{ Use the point-slope form.}$$
$$y - (-5) = 0[x - (\boxed{-3})] \qquad \longleftarrow \text{ Substitute } -3 \text{ for } x_1, -5 \text{ for } y_1, \text{ and } 0 \text{ for } m.$$
$$y + 5 = 0 \qquad \longleftarrow \text{ Simplify.}$$
$$y = -5 \qquad \longleftarrow \text{ Write in slope-intercept form.}$$

TRY THIS Point $Q(7, 4)$ is on a line with slope 0. Write the equation for the line in slope-intercept form.

EXERCISES

1. Graph several different lines that all contain $Q(3, -3)$. Use the graph to show that one point alone is not enough to determine a unique line.

2. Graph several different lines that all have slope $-\frac{5}{3}$. Use the graph to show that slope alone is not enough to determine a unique line.

PRACTICE

Using the given information, write an equation in point-slope form and in slope-intercept form.

3. $A(-4, 2)$ and slope 1

4. $B(5, -2)$ and slope -1

5. $V(5, 3)$ and slope 2

6. $K(-4, 0)$ and slope -3

7. $X(0, -3)$ and slope -0.5

8. $R(5, 1)$ and slope 0.5

9. $F(-4, -2)$ and slope $\frac{3}{7}$

10. $B(-2, -2)$ and slope $-\frac{1}{7}$

11. $A(5, -9)$ and slope 2.5

12. $C(7, 2)$ and slope 0

13. $P(6, -1)$ and slope 0

14. $N(0, 0)$ and slope 0

15. $S(0, 2)$ and slope 0

16. $Q(6, -1)$ and slope 1.6

17. $U(4, 10)$ and slope $\frac{11}{12}$

18. A bug begins traveling at $P(-5, 1)$ along a line with slope $-\frac{5}{7}$. The bug travels in the positive x-direction. Where will the bug cross the y-axis?

19. **Critical Thinking** Use the point-slope form to show that the line containing $P(a, b)$ with slope 0 has the form $y = b$.

20. **Critical Thinking** Use the point-slope form to show that the line containing $P(0, 0)$ with slope m has the form $y = mx$.

MIXED REVIEW

Find the slope of the line that contains the given points. (3.4 Skill B)

21. $T(0, 2.4)$ and $C(2.4, 0)$

22. $M(5, -1)$ and $Y(6, -5)$

23. $N(4.2, 8)$ and $X(4.4, 10)$

Use the x- and y-intercepts to graph each equation. (3.5 Skill A)

24. $-4.5x + 4.5y = 9$

25. $5x - 8y = 40$

26. $x - 8y = 2$

27. Is the graph of $x = r$ a horizontal or a vertical line? What point does the line contain? What can you say about the slope of the line?

The point-slope form of an equation can be used to write equations that represent real-world problems.

EXAMPLE 1

After driving for 2.5 hours, Grace and Jamal had traveled 180 miles from their starting point. If they drive 55 miles per hour from that point on, how far from their starting point will they be after 5 total hours of driving?

▶ **Solution**

1. Interpret the data. Let d represent the distance traveled and let t represent the elapsed time.

   (elapsed time, distance traveled)   → (t_1, d_1) → $(2.5, 180)$

   constant speed of 55 mph   →   slope is 55   → $m = 55$

2. Write and simplify an equation. Because you know the coordinates of a point on the line and the slope, use the point-slope form.

$$d - d_1 = m(t - t_1)$$
$$d - 180 = 55(t - 2.5)$$
$$d = 55t + 42.5$$

3. If $t = 5$, then $d = 55(5) + 42.5 = 317.5$.

   After 5 hours, they will be 317.5 miles from their starting point.

TRY THIS   Rework Example 1 given that they were 120 miles from their starting point after 1.5 hours and they drove 52 miles per hour.

EXAMPLE 2

A machine operator counted 26 full bottles in stock at the start of the business day. A filling machine is turned on and can fill 12 bottles every minute. Write an equation for the function. How many bottles will be filled after 8 hours?

▶ **Solution**

1. Interpret the data. Let n represent the number of bottles on hand and let t represent the elapsed time in minutes.

   (elapsed time, number of bottles)   → (t_1, n_1) → $(0, 26)$

   12 bottles filled every minute   →   slope is 12   → $m = 12$

2. Write and simplify an equation.

$$n - n_1 = m(t - t_1)$$
$$n - 26 = 12(t - 0)$$
$$n = 12t + 26$$

3. Because 8 hours = 480 minutes, let $t = 480$.

$$n = 12t + 26 \quad → \quad n = 12(480) + 26 = 5786$$

There will be 5786 bottles filled after 8 hours (480 minutes).

TRY THIS   Suppose the filling rate is 15 bottles per minute and the day starts with 120 full bottles in stock. How many full bottles will be in stock after 4 hours?

EXERCISES

KEY SKILLS

Identify the point and slope found in each verbal description.

1. A small aircraft ascends at 350 feet per minute. After 4 minutes, the plane's altitude is 1900 feet.

2. Twenty students enter an auditorium each minute. After 5 minutes, there are 135 students in the room.

PRACTICE

Write an equation in point-slope form and solve each problem.

3. A bakery has 144 sugar cookies on hand. The baker makes 24 cookies every hour. How many cookies will be on hand after 6 hours?

4. After 2 years, a computer system is valued at $6200. Its value will diminish at the rate of $500 per year. What will its value be after 5 years?

5. A T-shirt company charges $50 for the first 5 T-shirts ordered, plus $6 for every additional shirt order after that. How much would ordering 20 shirts cost?

Find the domain and range of each function. Write an equation for the function. Then answer the question.

6. A wood-cutting machine has already cut 120 boards. If the machine can cut 12 boards per minute, how many cut boards will be on hand after 18 minutes?

7. After driving for 3 hours, Dave and Jenna have traveled 520 miles. If they drive at a constant speed of 50 miles per hour from that point on, how far will they have traveled after driving a total of 9 hours?

MIXED REVIEW APPLICATIONS

Find the domain and range of each function. (3.2 Skill C)

8. the cost of n dozen bagels if one dozen bagels costs $4.50

9. the quantity of r inches of rainfall in a storm when it rains 0.6 inch per minute

10. the cost of n nights in a hotel that costs $195 per night

11. the cost of p postage stamps if each stamp is valued at $0.33

3.7 LESSON
Finding the Equation of a Line Given Two Points

SKILL A *Using two points to find the equation of a line*

If the coordinates of two points are given, you can write an equation for the line that passes through them.

EXAMPLE 1   **Find an equation of the line that contains $P(-4, -2)$ and $Q(5, 3)$. Write the equation in slope-intercept form.**

▶ **Solution**

1. Find the slope of $\overleftrightarrow{PQ}$. $m = \dfrac{3 - (-2)}{5 - (-4)} = \dfrac{5}{9}$

2. Use the point-slope form with either point P or point Q.

$$y - y_1 = m(x - x_1) \quad \rightarrow \quad y - 3 = \frac{5}{9}(x - 5)$$

$$y = \frac{5}{9}x + \frac{2}{9} \qquad \longleftarrow \text{Slope-intercept form}$$

TRY THIS   Find the equation of the line that contains $A(4, -2)$ and $B(-5, 1)$. Write the equation in slope-intercept form.

When two points are given, the line that passes through them might be vertical or horizontal. Recall from Lesson 3.5 that if a line is vertical, its equation has the form $x = a$. If it is horizontal, its equation has the form $y = b$.

EXAMPLE 2   **Find an equation for the line that contains each pair of points.**
a. $A(4, 3)$ and $B(4, -5)$ **b.** $R(2.5, 3.5)$ and $S(-1, 3.5)$

▶ **Solution**

a. Because the x-coordinates of A and B are equal, we know that the line is vertical. An equation for $\overleftrightarrow{AB}$ is $x = 4$. The slope is undefined.

b. Because the y-coordinates of R and S are equal, we know that the line is horizontal. An equation for $\overleftrightarrow{RS}$ is $y = 3.5$. The slope is 0.

TRY THIS   Find an equation for the line that contains each pair of points.
a. $K(4, 3)$ and $L(-4, 3)$ **b.** $D(-1.4, 10)$ and $E(-1.4, -10)$

Summary of the Forms for the Equation of a Line

Name	Form	Example
Standard	$Ax + By = C$	$2x + 3y = 7$
Slope-intercept	$y = mx + b$	$y = 5x + 1$
Point-slope	$y - y_1 = m(x - x_1)$	$y - 4 = -2(x - 3)$

EXERCISES

A line contains points $(2, 5)$ and (a, b), where a and b are real numbers.
Tell whether the line is horizontal, vertical, or neither.

1. $a = 2$

2. $b = 5$

3. $a \neq 2$ and $b \neq 5$

PRACTICE

Find an equation for the line that contains each pair of points. If the line
is not vertical, write the equation in slope-intercept form.

4. $A(0, 0)$ and $B(4, 5)$

5. $K(0, 0)$ and $L(-4, 5)$

6. $X(4.5, 0)$ and $Y(0, 4.5)$

7. $P(-5.5, 0)$ and $Q(0, -5.5)$

8. $G(4.5, 0)$ and $H(0, 4)$

9. $A(-2, 0)$ and $C(0, 7)$

10. $D(-2, 3)$ and $K(-2, 5)$

11. $W(-1, 2.4)$ and $S(5, 2.4)$

12. $X(-2, 3)$ and $Y(3, 6)$

13. $X(-1, 7)$ and $Y(3, -1)$

14. $Q(1, 3)$ and $B(10, 24)$

15. $T(-1, -9)$ and $R(5, 18)$

Tell whether the given point lies on, above, or below the line defined by
A and C.

16. $P(0, 5)$; $A(-1, 3)$ and $C(2, 7)$

17. $Q(-2, 5)$; $A(-3, 3)$ and $C(-1, 3)$

18. $Z(-5, -3)$; $A(0, 3)$ and $C(2, 4)$

19. $D(7, 2)$; $A(-4, 7)$ and $C(5, -4)$

20. Critical Thinking Show that $y = \frac{y_2 - y_1}{x_2 - x_1}(x - x_1) + y_1$ represents the
line that contains (x_1, y_1) and (x_2, y_2), where $x_1 \neq x_2$.

MIXED REVIEW

Use the x- and y-intercepts to graph each equation. (3.5 Skill A)

21. $-x + 2y = 4$

22. $-x = 3$

23. $4x + 6y = 24$

Use the slope and the y-intercept to graph each equation. (3.5 Skill B)

24. $y = 4x - 2$

25. $y = 2x - 6$

26. $2x + 3y = 9$

Find the equation in slope-intercept form for each line. (3.6 Skill A)

27. slope $-\frac{3}{2}$; contains $Q(-3, 7)$

28. slope $\frac{9}{20}$; contains $R(2, 12)$

A **mathematical model** can describe the relationship between *data* taken from real-world situations. Some real-world situations can be modeled by linear equations, as shown below.

EXAMPLE 1

A T-shirt company charges a certain amount per T-shirt plus a handling fee for each order. Raul ordered 3 shirts and paid $32, and Maria ordered 5 shirts and paid $50.

   a. How much does the company charge per shirt?
   b. How much is the handling fee?
   c. How much would 7 T-shirts cost?

▶ **Solution**

   a. Use the given information to write data points.

   (number of shirts x, total cost y)   →   (3, 32) and (5, 50)

   Use the data points to find the slope and write an equation.

   $$\text{slope} = \frac{50 - 32}{5 - 3} = \frac{18}{2} = 9 \quad y - y_1 = m(x - x_1) \quad \to \quad y - 50 = 9(x - 5)$$
   $$y = 9x + 5$$

   The slope, 9, gives the cost per shirt, $9.

   b. The y-intercept, 5, gives the handling fee for each order, $5.

   c. If $x = 7$, then $y = 9(7) + 5$, or 68. It costs $68 for an order of 7 T-shirts.

TRY THIS   Refer to Example 1. Find the cost of 8 shirts if 3 shirts cost $25 and 5 shirts cost $39.

EXAMPLE 2

It costs $570 to cater an outdoor party for a group of 24 people and $1035 to cater the same party for a group of 55 people.

a. Find the slope and explain what it represents.
b. Find the cost to cater a party for 60 people.

▶ **Solution**

   a. Use the given information to write two data points.

   (number of people x, party cost y)   →   (24, 570) and (55, 1035)

   Use the data points to write an equation for the line.

   $$y - 570 = \frac{1035 - 570}{55 - 24}(x - 24) \text{ , or } y = 15x + 210$$

   The slope, 15, gives the cost per person, $15.

   b. If $x = 60$, then $y = 15(60) + 210$, or 1110. It costs $1110 for a party of 60 people.

TRY THIS   After driving for 5 hours, Ben is 380 miles from home. After driving for 8 hours, Ben is 560 miles from home. Assume Ben's driving speed during this three-hour period is constant.

   a. Find the slope and explain what it represents.

   b. How far from home will Ben be after 10 hours?

EXERCISES

Use the given information to create data points.

1. Four pounds of apples cost $3.60 and seven pounds of apples cost $6.30.

2. The cost of renting a garden tiller for 2 days is $55.00. The cost of renting a garden tiller for 3 days is $75.00.

PRACTICE

Solve each problem by finding a linear equation that models the data. Interpret the slope and the *y*-intercept of the line.

3. A home-health aide charges a fixed fee for each visit plus an hourly charge. A 2-hour visit costs $70 and a 4-hour visit costs $120. How much would a 5-hour visit cost?

4. The monthly charge for a cellular phone consists of a fixed fee plus a fee for every minute of calling time used. Vicki's monthly bill for 300 minutes of calling time was $35. Marty's monthly bill for 400 minutes of calling time was $40. How much would a monthly bill be for 500 minutes of calling time?

5. After a candle has been burning for 1 hour, its height is 10 cm. After three hours, the candle's height is 6 cm. What will the height of the candle be after it has been burning for 5 hours?

Solve each problem. Interpret the slope and the *y*-intercept of the line.

6. After driving for 2 hours, Lee is 220 from his starting point. After driving a total of 6 hours, Lee is 420 miles from his starting point. Assuming his speed is constant, how far is Lee from his starting point after a total of 8 hours?

7. It costs $77 to rent a car for 3 days and $115 to rent a car for 5 days. How much does it cost to rent a car for one week? Explain why the cost for 7.5 days might be the same as the cost for 8 days.

MIXED REVIEW APPLICATIONS

Solve each problem. (2.8 Skill C)

8. Jarrod needs 4 pounds of peanuts and cashews to put into a mix. Cashews cost $2.19 per pound and peanuts cost $1.98 per pound. How much of each will he buy if he spends $8.13?

9. Chen drives 375 miles in 7 hours. For the first part of the trip, he drives at 55 miles per hour. Then he drives at 50 miles per hour. How far did he drive at each speed?

Relating Two Lines in the Plane

SKILL A *Working with parallel lines in the coordinate plane*

Two lines in the plane are *parallel* if they never intersect. In the diagram at right, lines m and n are parallel and have the same slope. You can use slope to determine if two lines are parallel. Note: If two equations can be simplified to the same equation, then they will define the same line, not two distinct parallel lines.

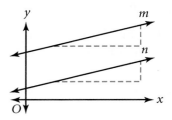

> **Parallel Lines**
>
> All vertical lines are parallel.
>
> If two distinct nonvertical lines are parallel, then they have equal slopes.
>
> If two distinct lines have equal slopes, then they are parallel.

EXAMPLE 1 Are the lines with equations $2x - 3y = 10$ and $-3x + 2y = 2$ parallel? Justify your response.

▶ **Solution**

1. Write both equations in slope-intercept form.

$$2x - 3y = 10 \quad \rightarrow \quad y = \frac{2}{3}x - \frac{10}{3} \qquad -3x + 2y = 2 \quad \rightarrow \quad y = \frac{3}{2}x + 1$$

2. Compare the slopes. $\qquad y = \boxed{\frac{2}{3}}\, x - \frac{10}{3} \qquad y = \boxed{\frac{3}{2}}\, x + 1 \qquad \frac{2}{3} \neq \frac{3}{2}$

Because the slopes, $\frac{2}{3}$ and $\frac{3}{2}$, are not equal, the lines are not parallel.

TRY THIS Are the lines with equations $-5x + 3y = 15$ and $-3x + 5y = 12$ parallel?

EXAMPLE 2 Find the equation in slope-intercept form for the line parallel to $y = \frac{3}{2}x + 7$ that contains $P(4, 5)$.

▶ **Solution**

Any line parallel to $y = \frac{3}{2}x + 7$ must also have slope $\frac{3}{2}$. Use the slope and point $P(4, 5)$ to write an equation for the line.

$$y - y_1 = m(x - x_1) \quad \rightarrow \quad y - 5 = \frac{3}{2}(x - 4)$$

Simplify and write the equation in slope-intercept form: $y = \frac{3}{2}x - 1$.

TRY THIS Find the equation in slope-intercept form for the line parallel to $y = \frac{5}{4}x - 3$ that contains $P(-3, 1)$.

EXERCISES

KEY SKILLS

Compare the slopes and tell whether the lines are parallel.

1. $y = x + 2$ and $y = -x - 3$

2. $y = -\frac{2}{7}x - 1$ and $y = \frac{7}{2}x - 2$

3. $y = 2.5x + 2$ and $y = \frac{5}{2}x - 3$

4. $y = -\frac{1}{10}x + 5$ and $y = -0.1x + 5$

PRACTICE

Are the given lines parallel? Justify your response.

5. $x + 2y = 3$ and $x - 2y = 5$

6. $6x - 5y = 14$ and $6x - 5y = 13$

7. $y = 4$ and $x = 6$

8. $x = -2.6$ and $y = 7$

Find the equation in slope-intercept form for the line that contains the given point and that is parallel to the given line.

9. $P(0, 0)$; $7x + 2y = 14$

10. $Q(0, 0)$; $2x - 7y = 5$

11. $M(-5, 2)$; $-3x + 8y = 24$

12. $N(7, 3)$; $8x - 3y = 15$

13. $T(7, 12)$; $-7x - 8y = 15$

14. $Z(-3, 1)$; $11x - 3y = 1$

15. $V(0, 6)$; $2.5x - 1.5y = 1$

16. $J(3, 0)$; $10x + 15y = 50$

17. Find the equation in slope-intercept form for the line that passes through $(1, 6)$ and that is parallel to the line containing $(4, 3)$ and $(5, 1)$.

18. Find the equation in slope-intercept form for the line that passes through $(5, 2)$ and that is parallel to the line containing $(4, 2)$ and $(7, 1)$.

19. Line $\overleftrightarrow{PQ}$ contains $P(4, 3)$ and $Q(7, 5)$. Find the x-intercept of the line that is parallel to $\overleftrightarrow{PQ}$ and that contains $Z(5, 10)$.

20. **Critical Thinking**   Are the lines with equations $35x + 14y = 21$ and $5x + 2y = 3$ parallel? Explain your response.

MIXED REVIEW

List in order from least to greatest. (previous courses)

21. 2.63, 2.62, 2.36, 2.26, 0, and 2.24

22. 10.5, 10, 11.1, 0.3, 0.03, and 1.05

Determine if each relation represents a function. (3.2 Skill A)

23. $\{(1, 2), (1, 3), (1, 4), (1, 5), (1, 6)\}$

24. $\{(3, 4), (4, 4), (5, 4), (6, 4), (7, 4)\}$

25. $\{(3, 5), (5, 8), (7, 11), (9, 14), (11, 17)\}$

26. $\{(1, 1), (2, 4), (3, 9), (2, 8), (3, 27)\}$

If two lines in the coordinate plane are not parallel, then they intersect at a point. If the lines intersect in such a way that they form a right angle, then the lines are *perpendicular*.

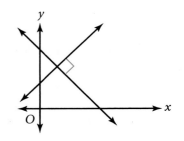

Perpendicular Lines

Every vertical line is perpendicular to every horizontal line.

Two lines are perpendicular if the product of their slopes is -1.

EXAMPLE 1   **Are the lines with equations $3x - 5y = 5$ and $5x + 3y = -21$ perpendicular? Justify your response.**

▶ **Solution**

1. Write both equations in slope-intercept form.

$$3x - 5y = 5 \quad \rightarrow \quad y = \frac{3}{5}x - 1 \qquad 5x + 3y = -21 \quad \rightarrow \quad y = -\frac{5}{3}x - 7$$

2. Compare the slopes. $y = \boxed{\frac{3}{5}} x - 1 \qquad y = \boxed{-\frac{5}{3}} x - 7 \qquad \frac{3}{5} \times \left(-\frac{5}{3}\right) = -1$

Because the product of the slopes is -1, the lines must be perpendicular.

TRY THIS   Are the lines with equations $x - y = 5$ and $x + y = 3$ perpendicular? Justify your response.

EXAMPLE 2   **Find the equation in slope-intercept form for the line perpendicular to $y = \frac{5}{2}x - 2$ that contains $Q(-10, 1)$.**

▶ **Solution**

All lines perpendicular to $y = \frac{5}{2}x - 2$ must have slope $-\frac{2}{5}$, the negative reciprocal of $\frac{5}{2}$. Use the slope and point $Q(-10, 1)$ to write an equation.

$$y - y_1 = m(x - x_1) \quad \rightarrow \quad y - 1 = -\frac{2}{5}[x - (-10)]$$

Simplify and write the equation in slope-intercept form: $y = -\frac{2}{5}x - 3$.

TRY THIS   Find the equation in slope-intercept form for the line that contains $Z(3, -5)$ and that is perpendicular to $y = -\frac{7}{2}x + 6$.

EXERCISES

Compare the slopes and tell whether the lines are perpendicular.

1. $y = \frac{1}{2}x - 5$ and $y = \frac{1}{2}x + 5$

2. $y = -\frac{4}{3}x - 2$ and $y = \frac{3}{4}x + 4$

3. $y = -\frac{1}{9}x + 8$ and $y = 9x - 2$

4. $y = 3x + 5$ and $y = -3x - 1$

Are the given lines perpendicular? Justify your response.

5. $x = 4$ and $y = -3$

6. $x = -4$ and $y = 3.2$

7. $y = -\frac{1}{2}x + 3$ and $y = 2x + 3$

8. $y = 1.3x + 3$ and $y = -3.1x - 2$

9. $-7x + 2y = 7$ and $-4x - 7y = 10$

10. $3x + 4y = 13$ and $-4x + 3y = 10$

Find the equation in slope-intercept form for the line that is perpendicular to the given line and that contains the given point.

11. $y = -2$; $P(0, 0)$

12. $x = 5$; $Q(0, 0)$

13. $3x - y = 12$; $R(1, 0)$

14. $-2x - 5y = 10$; $Z(0, 1)$

15. $-6x - 5y = 3$; $C(1, 4)$

16. $9x + 2y = 9$; $D(4, 5)$

17. For what values of a and b will $ax + y = 10$ and $x + by = 10$ be perpendicular?

18. Write the equation of the line that contains $A(4, 2)$ and $B(8, 4)$. Then find the equation of the line that is perpendicular to $\overleftrightarrow{AB}$ and that contains $C(6, 3)$.

19. Write the equation of the line perpendicular to $3x - 11y = 12$ that contains $P(5, 2)$. Where does the line cross the x-axis?

20. **Critical Thinking** Without using pencil and paper, predict where the lines $y = \frac{7}{9}x + 4$ and $-4 = -\frac{9}{7}x - y$ will intersect.

Graph each set of numbers on a number line. (2.1 Skill C)

21. $-4, 3, 0, -2.5, 4$

22. $-4, 4, -2, 2, 0, 3$

23. $0, 1.5, 2.5, -3.5. -4.5, -5.5$

Let $x = -2, -1, 0, 1, 2$. Graph the solutions to each equation. (3.1 Skill B)

24. $y = -2.5x + 4$

25. $y = 2x - 2.5$

26. $y = -2.5x + 1$

A **right triangle** is a triangle with one right angle. The sides of the triangle that form the right angle are perpendicular to each other.

EXAMPLE 1

Do $A(-3, 2)$, $B(-1, -2)$, and $C(3, 0)$ determine a right triangle? Justify your response.

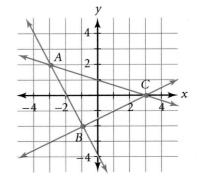

▶ **Solution**

Graph points A, B, and C and connect the points with lines.

From the diagram, it looks like the angle between $\overleftrightarrow{AB}$ and $\overleftrightarrow{BC}$ may be a right angle, but to be certain, you must calculate the slopes.

$$\text{slope } \overleftrightarrow{AB} = \frac{-2 - 2}{-1 - (-3)} = \frac{-4}{2} = -2$$

$$-2 \times \frac{1}{2} = -1$$

$$\text{slope } \overleftrightarrow{BC} = \frac{0 - (-2)}{3 - (-1)} = \frac{2}{4} = \frac{1}{2}$$

Because the product of their slopes is -1, $\overleftrightarrow{AB}$ and $\overleftrightarrow{BC}$ are perpendicular and thus form a right angle. Therefore, the triangle is a right triangle.

TRY THIS

Do $A(-5, 2)$, $B(0, 0)$, and $C(3, 5)$ determine a right triangle? Justify your response.

Any four-sided figure is a **quadrilateral**. Quadrilaterals that have certain properties are called *special quadrilaterals*. One type of special quadrilateral is a *trapezoid*. A **trapezoid** is a four-sided figure with exactly one pair of parallel sides.

EXAMPLE 2

Do $A(2, 2)$, $B(4, 2)$, $C(5, 0)$, and $D(1, 0)$ determine a trapezoid? Justify your response.

▶ **Solution**

Graph points A, B, C, and D and connect the points with lines.

From the diagram, it looks like $\overleftrightarrow{AB}$ and $\overleftrightarrow{CD}$ may be parallel. To be certain, calculate their slopes.

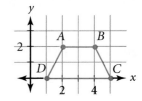

$$\overleftrightarrow{AB} = \frac{2 - 2}{4 - 2} = \frac{0}{2} = 0 \qquad \overleftrightarrow{BC} = \frac{0 - 2}{5 - 4} = \frac{-2}{1} = -2$$

$$\overleftrightarrow{CD} = \frac{0 - 0}{5 - 1} = 0 \qquad \overleftrightarrow{AD} = \frac{0 - 2}{1 - 2} = \frac{-2}{-1} = 2$$

Because the slopes of $\overleftrightarrow{AB}$ and $\overleftrightarrow{CD}$ are equal and the slopes of $\overleftrightarrow{AD}$ and $\overleftrightarrow{BC}$ are not equal, this four-sided figure has exactly one pair of parallel sides and thus is a trapezoid.

TRY THIS

Do $A(-3, -1)$, $B(-1, 2)$, $C(3, 2)$, and $D(4, -1)$ determine a trapezoid? Justify your response.

EXERCISES

KEY SKILLS

Refer to the diagram at right.

1. Find the slopes of $\overleftrightarrow{CF}$, $\overleftrightarrow{AC}$, and $\overleftrightarrow{LX}$.

How are each pair of lines below related?

2. $\overleftrightarrow{CF}$ and $\overleftrightarrow{AC}$

3. $\overleftrightarrow{CF}$ and $\overleftrightarrow{LX}$

4. $\overleftrightarrow{AC}$ and $\overleftrightarrow{LX}$

PRACTICE

Do the given points determine a right triangle? Justify your response.

5. $G(-4, -8)$, $H(6, -1)$, and $J(-1, 9)$

6. $X(10, 2)$, $Y(7, 0)$, and $Z(-4, 1)$

7. $C(-3, 2)$, $L(7, 0)$, and $E(5, 1)$

8. $V(3, 1)$, $T(9, 6)$, and $W(4, 12)$

Do the given points determine a trapezoid? Justify your response.

9. $A(6, 4)$, $B(9, 4)$, $C(12, 0)$, and $D(4, 0)$

10. $A(-4, 2)$, $B(2, 2)$, $C(6, -3)$, and $D(-6, -3)$

11. $A(-4, 2)$, $B(2, 4)$, $C(5, 2)$, and $D(-7, -2)$

12. $A(5, 5)$, $B(8, 1)$, $C(10, -4)$, and $D(1, 8)$

Another *special quadrilateral* is a parallelogram. A parallelogram is a four-sided figure in which both pairs of opposite sides are parallel. Use the slopes of the sides to show that each figure is a parallelogram.

13.

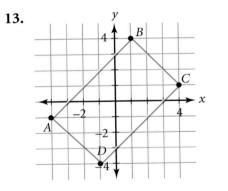

14.

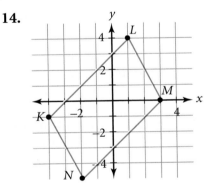

15. **Critical Thinking** What geometric figure in the coordinate plane is determined by $P(-6, 4)$, $Q(-3, 7)$, $R(6, -2)$, and $S(3, -5)$? Justify your answer.

MIXED REVIEW APPLICATIONS

Represent each situation with an equation and a graph. (3.5 Skill C)

16. Kim has $0.30 in nickels and dimes.

17. The temperature is 25°C and is increasing 5°C per minute.

Linear Patterns and Inductive Reasoning

SKILL A *Extending a linear sequence given a table*

A **sequence** is an ordered set of real numbers. Each number in the sequence is called a **term of the sequence.** An example of a sequence is 2, 8, 14, 20, 26 ... This can be represented in table form, as shown below.

term number x	1	2	3	4	5	... n
term y	2	8	14	20	26	... ?

EXAMPLE 1 | **Find a pattern in the table above and use it to predict the 7th term.**

▶ **Solution**

For each increase of 1 in the term number, the terms increase by 6.

6th term: 26 + 6, or 32
7th term: 32 + 6, or 38

x	1	2	3	4	5
y	2	8	14	20	26

$+1 \quad +1 \quad +1 \quad +1$

$+6 \quad +6 \quad +6 \quad +6$

TRY THIS | Rework Example 1 using the table at right.

x	1	2	3	4	5
y	3	-1	-5	-9	-13

Linear Patterns and Linear Functions

Given a sequence of ordered pairs (x, y), if the differences between successive x-terms are the same and the differences between successive y-terms are the same, then the data can be modeled by a linear function.

EXAMPLE 2 | **Express the pattern in the table as a function with x as the independent variable. Predict the 80th term.**

x	1	2	3	4	5
y	3	10	17	24	31

▶ **Solution**

The values of x increase by 1 and the values of y increase by 7. The pattern can be modeled by a linear function. Use data pairs (1, 3) and (2, 10) to write an equation.

x	1	2	3	4	5
y	3	10	17	24	31

$+1 \quad +1 \quad +1 \quad +1$

$+7 \quad +7 \quad +7 \quad +7$

$$y - y_1 = m(x - x_1) \rightarrow y - 3 = \frac{7}{1}(x - 1), \text{ or } y = 7x - 4$$

To predict the 80th term, let $x = 80$. $y = 7(80) - 4 = 556$

A prediction for the 80th term is 556.

TRY THIS | Rework Example 2 using the table at right.

x	1	2	3	4	5
y	6	11	16	21	26

EXERCISES

KEY SKILLS

Tell whether the data in the table shows a linear pattern or a nonlinear pattern. Justify your response.

1.

x	1	2	3	4	5
y	−4	4	12	20	28

2.

x	1	2	3	4	5
y	−4	9	22	33	46

3. Verify that $y + 11 = 6(x − 1)$ models the data at right by verifying that each ordered pair is a solution to the equation.

x	1	2	3	4	5
y	−11	−5	1	7	13

PRACTICE

Find the specified terms in each linear sequence.

4. 7th and 8th terms

x	1	2	3	4	5
y	3	11	19	27	35

5. 7th, 8th, and 9th terms

x	1	2	3	4	5
y	19	9	−1	−11	−21

6. 7th and 9th terms

x	1	2	3	4	5
y	2.5	0.5	−1.5	−3.5	−5.5

7. 8th and 10th terms

x	1	2	3	4	5
y	0.3	2.8	5.3	7.8	10.3

Express each pattern as a function with x as the independent variable. Then predict the specified term of the sequence.

8. the table in Exercise 4; 90th term

9. the table in Exercise 5; 100th term

10. the table in Exercise 6; 29th term

11. the table in Exercise 7; 44th term

12. Critical Thinking The table at right contains a linear pattern. Find the missing terms.

x	1	2	3	4	5	6	7	8	9	10
y	4	?	?	?	16	?	?	25	?	?

MIXED REVIEW

Find the equation in slope-intercept form for the line that contains the given points. (3.7 Skill A)

13. $A(2.5, −4)$ and $B(5.5, 7)$

14. $X(0, −7)$ and $Y(−4, −7)$

15. $P(3.5, 0)$ and $Q(3.5, −8)$

16. $K(6.5, −4)$ and $L(−0.5, −4)$

17. $D(24, −2)$ and $E(27, 28)$

18. $R(98, −3)$ and $S(8, −13)$

You use *inductive reasoning* when you look for a pattern in particular situations and then develop a generalization that represents that pattern.

EXAMPLE 1

Write a function to predict the number of squares, *s*, in the *n*th diagram. Then predict the number of squares in the 100th diagram.

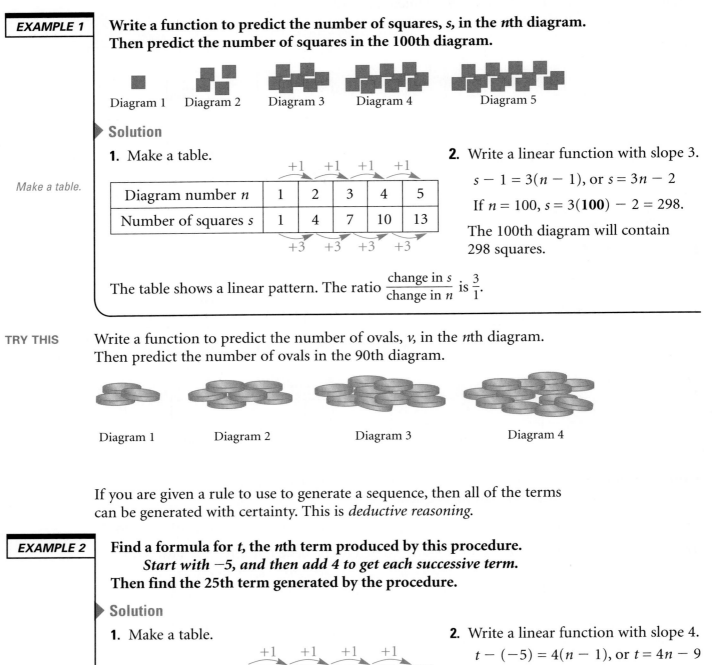

Diagram 1   Diagram 2   Diagram 3   Diagram 4   Diagram 5

▶ **Solution**

1. Make a table.

Make a table.

	+1	+1	+1	+1	
Diagram number *n*	1	2	3	4	5
Number of squares *s*	1	4	7	10	13
	+3	+3	+3	+3	

The table shows a linear pattern. The ratio $\dfrac{\text{change in } s}{\text{change in } n}$ is $\dfrac{3}{1}$.

2. Write a linear function with slope 3.

$s - 1 = 3(n - 1)$, or $s = 3n - 2$

If $n = 100$, $s = 3(\mathbf{100}) - 2 = 298$.

The 100th diagram will contain 298 squares.

TRY THIS

Write a function to predict the number of ovals, *v*, in the *n*th diagram. Then predict the number of ovals in the 90th diagram.

Diagram 1   Diagram 2   Diagram 3   Diagram 4

If you are given a rule to use to generate a sequence, then all of the terms can be generated with certainty. This is *deductive reasoning*.

EXAMPLE 2

Find a formula for *t*, the *n*th term produced by this procedure.
     Start with −5, and then add 4 to get each successive term.
Then find the 25th term generated by the procedure.

▶ **Solution**

1. Make a table.

Make a table.

	+1	+1	+1	+1	
term number *n*	1	2	3	4	5
term *t*	−5	−1	3	7	11
	+4	+4	+4	+4	

2. Write a linear function with slope 4.

$t - (-5) = 4(n - 1)$, or $t = 4n - 9$

If $n = 25$, then $t = 4(\mathbf{25}) - 9 = 91$.

The 25th term is 91.

TRY THIS

Find a formula for *t*, the *n*th term produced by this procedure.
     Start with 12, and then subtract 7 to get each successive term.
Then find the 35th term generated by the procedure.

EXERCISES

Write the first five numbers generated by each procedure.

1. Start with −7, and then add 6 to get each successive term.

2. Start with 2.4, and then subtract 0.4 to get each successive term.

PRACTICE

Write a function to predict the number of objects, *t*, in the *n*th diagram. Then predict the number of objects in the 100th diagram.

3.

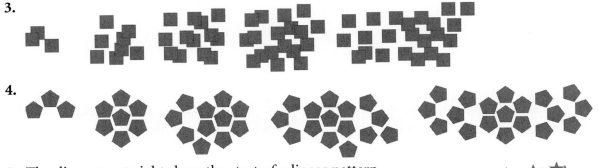

4.

5. The diagrams at right show the start of a linear pattern. After how many steps will all the stars disappear?

Step 1   Step 2

Find a formula for *t*, the *n*th term. Then find the 40th term.

6. Start with −13, and then add 8 to get each successive term.

7. Start with −12.4, and then subtract 3 to get each successive term.

8. Which type of reasoning, inductive or deductive, is needed to solve Exercises 3 and 4? Which type of reasoning is needed to solve Exercises 6 and 7? Explain your responses.

9. **Critical Thinking**   Suppose you apply a procedure that determines a linear pattern, and the 4th number you get is 7 and the 8th number is 2. What is the starting number and how much do you add or subtract to get each successive term?

MIXED REVIEW

Find the slope of the line whose equation is given. (3.5 Skill B)

10. $-8x - y = 4$ 11. $7x - 8y = 56$ 12. $9x - 7y = 63$ 13. $x - y = 4.3$

Find the equation in slope-intercept form for the line that contains each pair of points. (3.7 Skill A)

14. $A(-9, -7)$ and $B(7, 9)$ 15. $J(-1, 11)$ and $K(1, -11)$ 16. $P(13, 2)$ and $Q(7, 10)$

You can also use inductive reasoning when you look for a pattern in graphical data. In the graph at right, you can see a set of points that suggest a pattern. In the Example, you will see how to represent the *x*- and *y*-coordinates using linear functions.

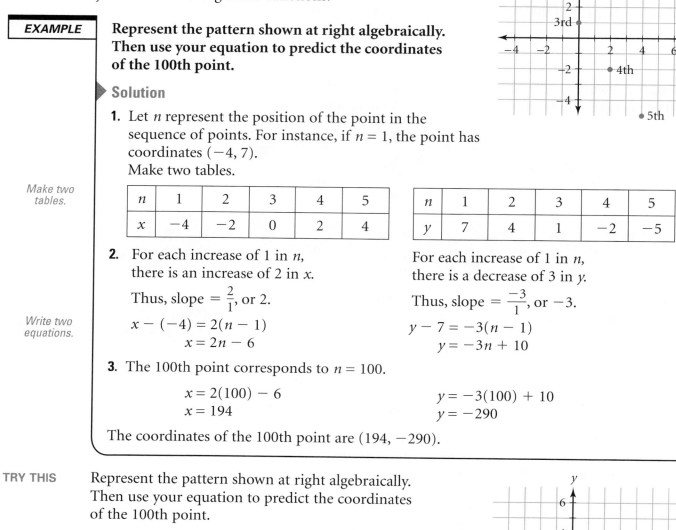

EXAMPLE

Represent the pattern shown at right algebraically. Then use your equation to predict the coordinates of the 100th point.

▸ **Solution**

Make two tables.

1. Let *n* represent the position of the point in the sequence of points. For instance, if $n = 1$, the point has coordinates $(-4, 7)$.
   Make two tables.

n	1	2	3	4	5
x	−4	−2	0	2	4

n	1	2	3	4	5
y	7	4	1	−2	−5

Write two equations.

2. For each increase of 1 in *n*, there is an increase of 2 in *x*.

   Thus, slope $= \frac{2}{1}$, or 2.

   $$x - (-4) = 2(n - 1)$$
   $$x = 2n - 6$$

   For each increase of 1 in *n*, there is a decrease of 3 in *y*.

   Thus, slope $= \frac{-3}{1}$, or −3.

   $$y - 7 = -3(n - 1)$$
   $$y = -3n + 10$$

3. The 100th point corresponds to $n = 100$.

   $$x = 2(100) - 6$$
   $$x = 194$$

   $$y = -3(100) + 10$$
   $$y = -290$$

   The coordinates of the 100th point are $(194, -290)$.

TRY THIS

Represent the pattern shown at right algebraically. Then use your equation to predict the coordinates of the 100th point.

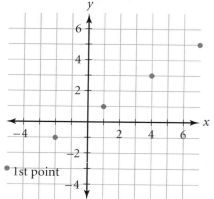

In the Example above, the equations for the *x*- and *y*-coordinates of the *n*th point are written in terms of the variable *n*.

$$x = 2n - 6 \qquad y = -3n - 4$$

Notice that, in the Example above, *n* is the independent variable. Both variables *x* and *y* are dependent variables.

Choose *n*. ⟶ This determines *x*.
         ⟶ This determines *y*.

Equations with this relationship are called *parametric equations*. The independent variable, in this case *n*, is called the *parameter*.

EXERCISES

Write a linear function to represent *x* in terms of *n*. Then write a second
linear function to represent *y* in terms of *n*.

1.

n	1	2	3	4	5
x	0	3	6	9	12

n	1	2	3	4	5
y	0	−4	−8	−12	−16

2.

n	1	2	3	4	5
x	−5	−1	3	7	11

n	1	2	3	4	5
y	11	3	−5	−13	−21

PRACTICE

Write a linear function to represent *x* in terms of *n*. Write a second
linear function to represent *y* in terms of *n*. Then find the coordinates
of the 60th point in the pattern.

3.
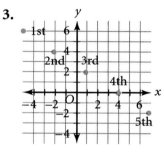

4.
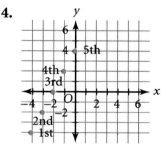

5.
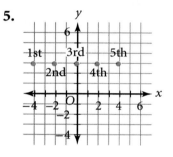

6. The following table represents a graphical pattern in a set of points.

Point position	first	second	third	fourth	fifth
Coordinates	(−15, 12)	(−11, 9)	(−7, 6)	(−3, 3)	(1, 0)

  a. Write a linear function to represent *x* in terms of *n*. Then write
     a second linear function to represent *y* in terms of *n*.
  b. Find the 70th point.   c. Is (481, −347) part of the pattern?

7. Find the coordinates of the unknown points in this linear pattern.

Point position	first	second	third	fourth	fifth
Coordinates	(3, 17)	?	?	?	(23, 37)

MIXED REVIEW APPLICATIONS

Solve the problem by finding a linear equation that models the data.
Interpret the slope and the *y*-intercept of the line. (3.7 Skill B)

8. It costs $810 to make 30 vases and $1242 to make 54 vases. What will
   it cost to make 120 vases? What will it cost to make 150 vases?

Write the ordered pair that corresponds to each point.

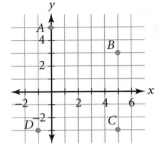

1. *A*   **2.** *B*   **3.** *C*   **4.** *D*

Graph each point on the same coordinate plane.

5. $W(3, -4)$ **6.** $X(-2, 0)$ **7.** $Y(4, 1)$ **8.** $Z(-4, -4)$

Let $x = -3, -2, -1, 0, 1, 2, 3$. For each value of x, find y and then make a table of ordered pairs. Then graph the ordered pairs.

9. $y = -2x + 3$ **10.** $y = x^2 - 2x + 1$

11. Does $\{(1, 6), (2, 1), (3, 1), (4, 6)\}$ represent a function? Explain.

12. Find the domain and range of the relation graphed at right.

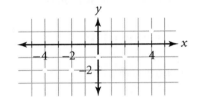

13. Represent the situation below in a table and in a graph for the given values of the independent variable.

   One notebook costs $1.50. The total cost, c, of 1, 2, 3, 4, 5, and 6 notebooks is given by $c = 1.5n$, where n the number of notebooks.

14. Gasoline costs $1.39 per gallon. Use direct-variation equations to answer the following questions.
   a. What is the cost of 20 gallons?
   b. How many gallons can be purchased for $25?

15. If y varies directly as x and if $y = 40$ when $x = 2.5$, find y when $x = 9$.

16. A small plane begins its descent. If the plane has an altitude of 4300 feet after 2 minutes and an altitude of 3100 feet after 5 minutes, what is the plane's constant rate of descent?

Find the slope of the line that contains each pair of points.

17. $P(-5, 0)$ and $B(6, 3)$ **18.** $X(5, 9)$ and $Y(7, -4)$

19. A line has slope $-\dfrac{3}{5}$ and contains $K(-4, 7)$. Graph the line and write the coordinates of a second point on it.

Use the intercepts to graph each equation.

20. $-4x + 3y = 24$ **21.** $y = 3.6$

Use the slope and the *y*-intercept of the line to graph each equation.

22. $y = -\dfrac{3}{2}x + 3$

23. $-2x + 5y = 10$

24. Represent the following situation in an equation and in a graph.

Michael has $0.40 in nickels and dimes.

Use the given information to write an equation in point-slope form and in slope-intercept form.

25. $K(-3, -5)$; slope $\dfrac{3}{7}$

26. $M(-5, -3)$; slope 0

27. Write a function to solve the problem below. Then solve.

After 1.5 years, a copier's value is $4900, and after 2 years its value is $4700. What will its value be after 4 years?

Find the equation in slope-intercept form for $\overleftrightarrow{PQ}$.

28. $P(3, -2)$ and $Q(6, 7)$

29. $P(-2, 5)$ and $Q(7, 15)$

30. Write an equation to solve the problem below. Interpret the slope and the *y*-intercept.

It costs $87 to rent a car for 3 days and $125 to rent the car for 5 days. What will the cost of renting the car for 7 days be?

31. Write the equation in slope-intercept form for the line that contains $Q(5, 2)$ and that is parallel to the graph of $3x - 5y = 15$.

32. Write the equation in slope-intercept form for the line that contains $S(3, 5)$ and that is perpendicular to the graph of $-3x - 2y = 4$.

33. Do $K(-3, 4)$, $L(0, 0)$, and $M(5, 2)$ determine a right triangle? Explain.

34. Express the pattern in the table at right as a function with *x* as the independent variable. Then predict the 50th term.

x	1	2	3	4	5
y	13	6	-1	-8	-15

35. Suppose you start with 10 and then subtract 2.5 to get each successive term. Find a formula for *t*, the *n*th term. Then predict the 55th term.

36. Refer to the pattern at right. Write two linear functions to represent *x* and *y* in terms of *n*, the point number. Then predict the coordinates of the 60th point.

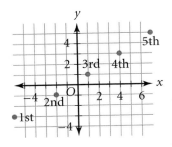

CHAPTER

4

Solving Inequalities in One Variable

▶ What You Already Know

In earlier mathematics courses, you learned how to compare and order numbers. The skills you learned will be helpful in your study of inequalities involving variables. Your study of properties of equality from earlier chapters will be especially helpful.

▶ What You Will Learn

In Chapter 4, you will extend your equation-solving skills to solve linear inequalities in one variable. Just as you learned properties of equality to help solve equations, you will learn properties of inequality that help you solve inequalities.

In Chapter 4, you will gradually move from solving simple inequalities in one variable to more complicated ones. You will see that there are infinitely many solutions to a linear inequality in one variable and that its solution on a number line is a ray.

You will then have the opportunity to solve a pair of inequalities joined by *and* or *or,* as well as inequalities involving absolute value. Lastly, you will encounter mathematical statements involving inequalities that are sometimes, always, or never true.

VOCABULARY	
absolute-value equation	maximum
absolute-value inequality	minimum
Addition Property of Inequality	Multiplication Property of Inequality
compound inequality	solution to an inequality
conjunction	Subtraction Property of Inequality
disjunction	tolerance
Division Property of Inequality	Transitive Property of Inequality
inequality	

The diagram below shows how mathematical skills and mathematical reasoning are interrelated with the skills and concepts in Chapter 4. Notice that this chapter focuses on solving inequalities in one variable.

CHAPTER 4

Mathematical Skills

- Representing inequalities (4.1)
- Graphing inequalities (4.1)
- Simplifying expressions (4.5)

Mathematical Reasoning

- Showing an inequality is sometimes, always, or never true

Solving Inequalities in One Variable

Using Addition or Subtraction (4.2)

Using Multiplication or Division (4.3)

Solving One-step Inequalities (4.4)

Solving Multistep Inequalities (4.5)

Solving Compound Inequalities (4.6)

Solving Absolute-value Inequalities (4.7)

Solving application problems (4.2, 4.3, 4.4, 4.5, 4.6, 4.7)

Choosing a solution method (4.4)

4.1

LESSON

Introduction to Inequalities

Representing inequalities with mathematical symbols

Recall that an equation is a mathematical statement that two expressions are equal. An **inequality**, such as $3 < 5$, is a statement that two expressions are *not* equal but are related to one another in one of these four ways:

> $<$ is less than
> $\leq$ is less than or equal to (is not more than, is at most)
> $>$ is greater than
> $\geq$ is greater than or equal to (is no less than, is at least)

Some examples of inequalities are:

$$-4 < -1 \qquad [3(5) - 10]^2 \leq 25 \qquad -2x + 5 \geq 10$$

Recall that on a number line the smaller of two numbers appears to the left of the larger number. You can represent the relationship between numbers with an inequality.

EXAMPLE 1
Graph -6 and 2 on a number line and write two inequalities that describe their relationship.

▶ **Solution**

$$\begin{array}{c}\xleftarrow{\hspace{0.3em}}\!\!+\!\!\!\underset{-7}{+}\!\!\!\bullet\!\!\!\underset{-6}{+}\!\!\!\underset{-5}{+}\!\!\!\underset{-4}{+}\!\!\!\underset{-3}{+}\!\!\!\underset{-2}{+}\!\!\!\underset{-1}{+}\!\!\!\underset{0}{+}\!\!\!\underset{1}{+}\!\!\!\underset{2}{\bullet}\!\!\!\underset{3}{+}\!\!\!\underset{4}{+}\!\!\!\xrightarrow{\hspace{0.3em}}\end{array}$$

Because -6 is to the left of 2, $-6 < 2$.

Because 2 is to the right of -6, $2 > -6$.

TRY THIS
Graph -7 and -2 on a number line and write two inequalities that describe their relationship.

EXAMPLE 2
Write an inequality that represents the following sentence.
The cost of a purchase, c, is more than \$5.

▶ **Solution**

Look for key words.

$$\begin{array}{ccc}\textit{The cost of a purchase, c, } & \textbf{\textit{is more than}} & \textit{\$5.} \\ c & > & 5\end{array}$$

The inequality is $c > 5$.

TRY THIS
Write an inequality that represents the following sentence.

The cost, c, is not more than \$12.95.

EXERCISES

> **KEY SKILLS**

Write the number(s) from the set {−3, −2, −1, 0, 1, 2, 3} that match each description.

1. the number(s) less than 2

2. the positive number(s)

3. the number(s) not more than −2

4. the number(s) not less than 0

> **PRACTICE**

Graph each pair of numbers on a number line and write two inequalities that describe their relationship.

5. 0 and 3

6. 0 and −3

7. −3 and 3

8. 4 and −4

9. 4 and 3

10. −4 and −3

11. −4 and 3

12. 4 and −3

13. −2 and 3.5

14. −1.5 and 3

15. $-\frac{1}{2}$ and 4

16. $-\frac{1}{2}$ and $2\frac{1}{2}$

Write an inequality that represents each sentence.

17. A motorist drives at a speed, s, that is more than 55 miles per hour.

18. The amount of sugar, a, is no more than 2 ounces.

19. A child's height, h, must be at least 42 inches.

20. A voter's age, a, must be at least 18.

21. An athlete's weight, w, is more than 125 pounds.

In Exercises 22–25, determine whether each statement is true or false.

22. $|3.2| > 0$

23. $-|4.5| \geq 0$

24. $-|6| \leq -6$

25. $|9| \leq -(-8)$

26. Let x be the average of two numbers a and b. On a number line, where is x located with respect to a and b?

> **MIXED REVIEW**

Solve each equation. Check your solution. (2.5 Skills A and B)

27. $x + 8 = 10$

28. $y - 5 = 4$

29. $z + 7 = -1$

30. $4 = y - 13$

31. $8x = 64$

32. $\frac{y}{4} = -5$

The **solution to an inequality** in one variable is the set of all numbers that make the inequality true. For example, given the replacement set {1, 2, 3, 4}, the solutions to $5x + 2 \geq 12$ are 2, 3, and 4 as shown below.

$x = 1 \rightarrow 5(1) + 2 \overset{?}{\geq} 12$
$\qquad\qquad\quad 7 \geq 12$ ✗

$x = 2 \rightarrow 5(2) + 2 \overset{?}{\geq} 12$
$\qquad\qquad\quad 12 \geq 12$ ✔

$x = 3 \rightarrow 5(3) + 2 \overset{?}{\geq} 12$
$\qquad\qquad\quad 17 \geq 12$ ✔

$x = 4 \rightarrow 5(4) + 2 \overset{?}{\geq} 12$
$\qquad\qquad\quad 22 \geq 12$ ✔

If no replacement set is specified, you may assume that the replacement set is the set of real numbers.

You can graph the solution to an inequality in one variable on a number line.

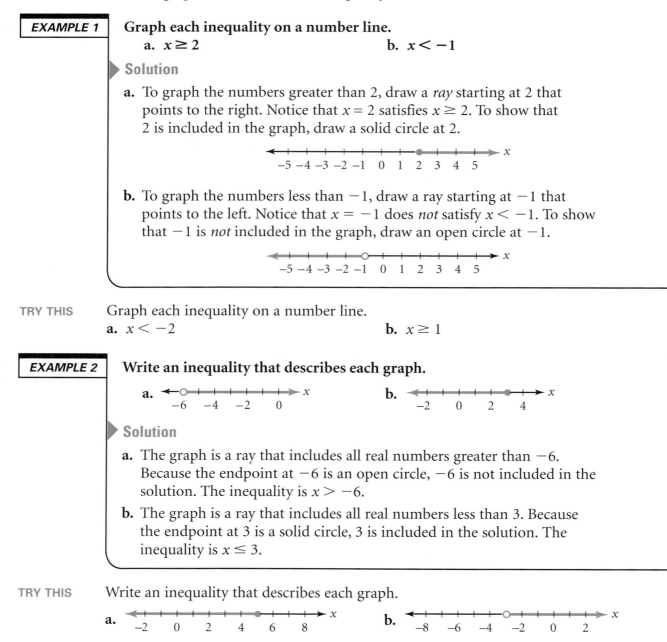

EXAMPLE 1

Graph each inequality on a number line.
a. $x \geq 2$
b. $x < -1$

▶ **Solution**

a. To graph the numbers greater than 2, draw a *ray* starting at 2 that points to the right. Notice that $x = 2$ satisfies $x \geq 2$. To show that 2 is included in the graph, draw a solid circle at 2.

b. To graph the numbers less than -1, draw a ray starting at -1 that points to the left. Notice that $x = -1$ does *not* satisfy $x < -1$. To show that -1 is *not* included in the graph, draw an open circle at -1.

TRY THIS

Graph each inequality on a number line.
a. $x < -2$
b. $x \geq 1$

EXAMPLE 2

Write an inequality that describes each graph.

a.
b.

▶ **Solution**

a. The graph is a ray that includes all real numbers greater than -6. Because the endpoint at -6 is an open circle, -6 is not included in the solution. The inequality is $x > -6$.

b. The graph is a ray that includes all real numbers less than 3. Because the endpoint at 3 is a solid circle, 3 is included in the solution. The inequality is $x \leq 3$.

TRY THIS

Write an inequality that describes each graph.

a.
b.

EXERCISES

Match each inequality on the left with its graph on the right.

1. $x < 1$

2. $x \geq 1$

3. $x \leq 1$

4. $x > 1$

a.

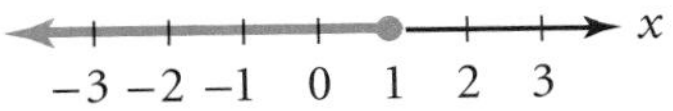

b.

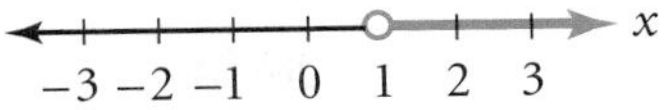

c.

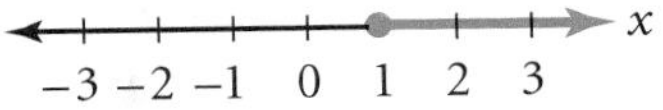

d.

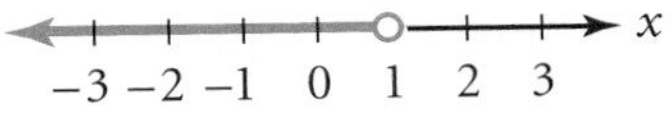

PRACTICE

Graph each inequality on a number line.

5. $x \geq 4$ 6. $t \leq 4$ 7. $s \geq -2$ 8. $s < 2$

9. $d < 5$ 10. $k > 2$ 11. $a > 3$ 12. $x \leq 3$

13. $k \geq -3$ 14. $b > 4$ 15. $w < -4$ 16. $y < 7$

Write an inequality that describes each graph.

17.

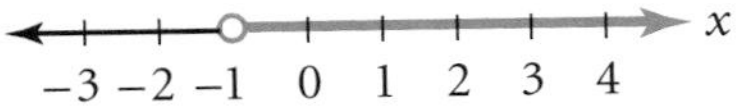

18.

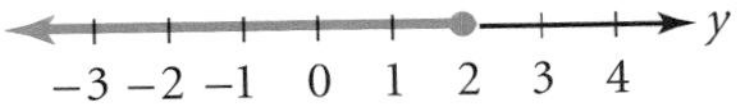

19.

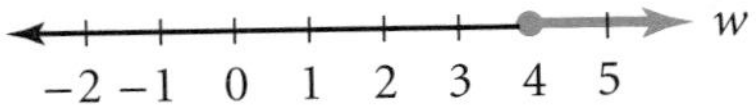

20.

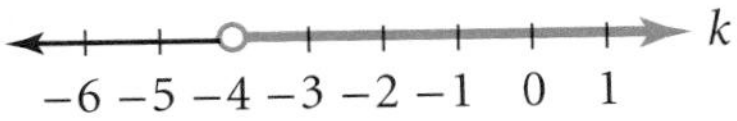

21.

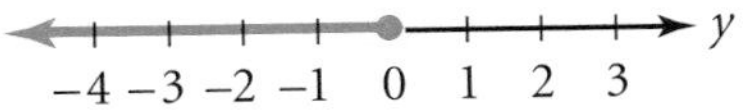

22.

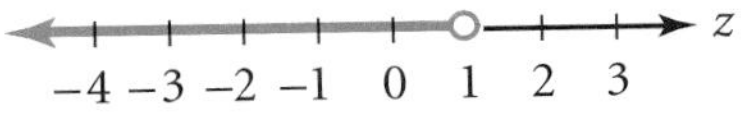

23.

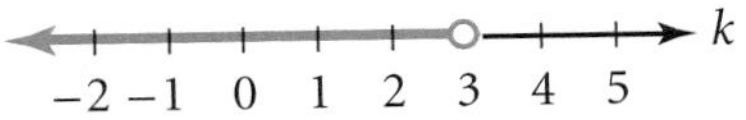

24.

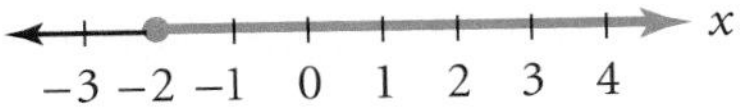

MIXED REVIEW

Solve each equation. Check your solution. (2.5 Skill A)

25. $x - 3.5 = -3$

26. $d - 4.5 = -7$

27. $v + 3\frac{1}{2} = -2\frac{1}{2}$

28. $-4.25 + q = -0.5$

29. $0 = -4\frac{1}{3} + p$

30. $-5.5 = -4.25 + q$

31. $-0.75 + x = -0.5$

32. $-4.5 = t - 1.25$

A **compound inequality** is a pair of inequalities joined by *and* or *or*. An example of each type of compound inequality follows.

$$x > 2 \ and \ x \leq 10 \qquad x < -5 \ or \ x > 3$$

A compound inequality joined by *and* is true only if *both* inequalities are true as shown in Example 1.

EXAMPLE 1   **Graph $x \geq -1$ *and* $x < 3$.**

▶ **Solution**

First graph each inequality separately.

Because the solutions to the compound inequality must satisfy both inequalities, identify the region where the two graphs overlap.

TRY THIS   Graph $x > -5$ *and* $x \leq 2$.

The Transitive Property of Inequality defined below allows us to write compound inequalities joined by *and* in a shortened form. We use this property to write $a < b$ *and* $b < c$ as $a < b < c$. Using this form, the compound inequality $x \geq -1$ *and* $x < 3$ in Example 1 can be shortened to $-1 \leq x < 3$.

Transitive Property of Inequality

If a, b, and c are real numbers, $a < b$, and $b < c$, then $a < c$.
This statement is also true if $<$ is replaced by $>$, $\leq$, or $\geq$.

A compound inequality joined by *or* is true if *at least one* of the inequalities is true.

EXAMPLE 2   **Graph $x < -1$ *or* $x \geq 2$.**

▶ **Solution**

First graph each inequality separately.

Because the solutions to the compound inequality must satisfy at least one of the inequalities, identify the combination of the two graphs.

TRY THIS   Graph $x \leq -4$ *or* $x > 0$.

EXERCISES

Match each inequality on the left with its graph on the right.

1. $x > -2$ and $x < 1$

a.

2. $x \leq -2$ or $x > 1$

b.

3. $x < -2$ or $x \geq 1$

c.

4. $-2 \leq x < 1$

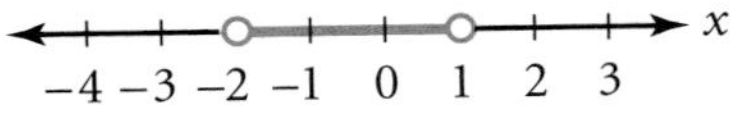

d.

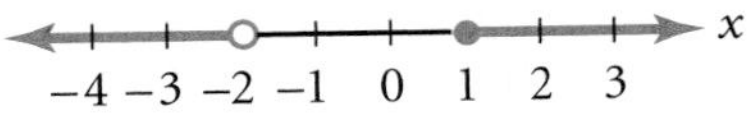

PRACTICE

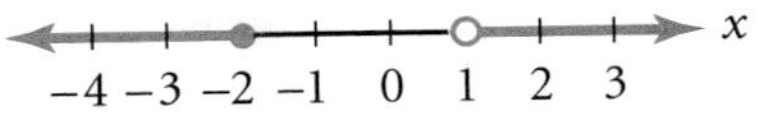

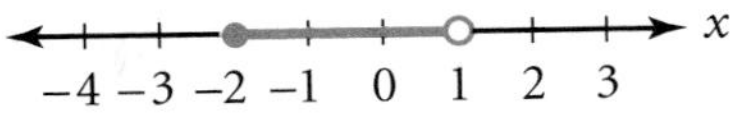

Graph each compound inequality.

5. $x \geq 1$ or $x < -1$ **6.** $g \geq 3$ or $g < -2$ **7.** $0 \leq a \leq 4$

8. $-4 \leq b \leq 4$ **9.** $c \geq 3$ or $c < 0$ **10.** $c \geq 0$ or $c < -2$

11. $q \geq 3$ or $q < 1$ **12.** $r \geq 4$ or $r \leq 2$ **13.** $w \leq 4$ and $w \geq 1$

14. $v \leq 0$ and $v \geq -3$ **15.** $z > 2$ and $z \leq 4$ **16.** $h < -2$ or $h \geq 2$

17. $u > 2$ or $u > 4$ **18.** $k \leq 0$ and $k \leq -3$ **19.** $d \leq 0$ and $d \geq -4$

Write a compound inequality that describes each graph.

20.

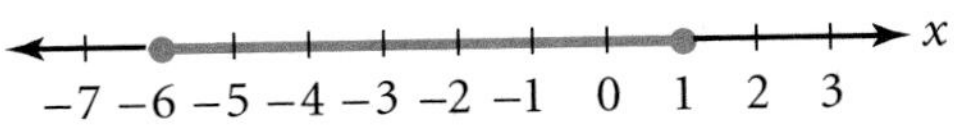

21.

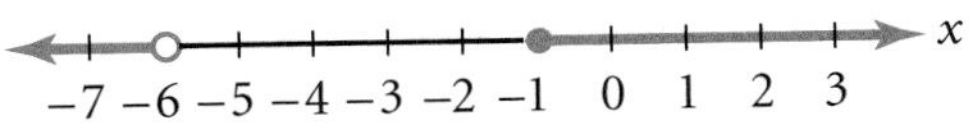

22. Critical Thinking What must be true of a and b if the compound inequality $x \geq a$ and $x < b$ has a solution?

23. Critical Thinking What must be true of r and s if the compound inequality $x > s$ and $x < r$ has no solution?

MIXED REVIEW

Solve each equation. Check your solution. (2.5 Skill B)

24. $-3z = -72$ **25.** $\frac{4}{3}b = -\frac{1}{2}$ **26.** $\frac{5d}{7} = -\frac{3}{7}$

27. $-\frac{5y}{7} = -4.5$ **28.** $4\frac{1}{2} = -\frac{2g}{7}$ **29.** $7.2 = -0.2w$

30. $2.5p = -10$ **31.** $0 = -4.875q$ **32.** $\frac{x}{-7} = 1$

Solving Inequalities Using Addition or Subtraction

LESSON

You solve an inequality by rewriting the inequality in a form whose solution set is easy to see. You can use the Addition Property of Inequality below to solve an inequality that involves subtraction, such as $x - 4.5 < -5$.

Addition Property of Inequality

If a, b, and c are real numbers and $a < b$, then $a + c < b + c$.
This statement is also true if $<$ is replaced by $>$, $\leq$, or $\geq$.

EXAMPLE 1 **Solve $x - 4.5 < -5$. Graph the solution on a number line.**

▶ **Solution**

$$x - 4.5 < -5$$
$$x - 4.5\ \boxed{+\ 4.5} < -5\ \boxed{+\ 4.5}\quad \longleftarrow \textit{Apply the Addition Property of Inequality.}$$
$$x < -0.5$$

The solution is all real numbers less than -0.5.

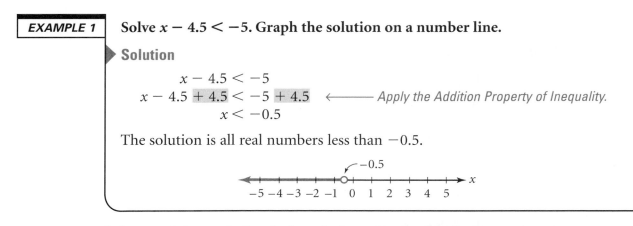

TRY THIS Solve $x - 2.5 > -4$. Graph the solution on a number line.

You can write any inequality in two ways. For example, $-3 \leq a$ is the same as $a \geq -3$. This is useful when you graph solutions.

EXAMPLE 2 **Solve $-8 \leq a - 5$. Graph the solution on a number line.**

▶ **Solution**

$$-8 \leq a - 5$$
$$-8\ \boxed{+\ 5} \leq a - 5\ \boxed{+\ 5}\quad \longleftarrow \textit{Apply the Addition Property of Inequality.}$$
$$-3 \leq a$$
$$a \geq -3\qquad\qquad \longleftarrow \textit{Rewrite the inequality so that } a \textit{ is on the left.}$$

The solution is all real numbers greater than or equal to -3.

TRY THIS Solve $3 < w + 7$. Graph the solution on a number line.

EXERCISES

Rewrite each inequality so that the variable is on the left.

1. $-4 \le x$
2. $3 < y$
3. $-2 \ge d$
4. $0 > w$

Solve each inequality. Graph the solution on a number line.

5. $x - 6 < 5$
6. $y - 2 \ge 10$
7. $a - 1 \le 9$

8. $x - 3 > -5$
9. $h - 3 > -1$
10. $5 \le k - 3$

11. $-2 > m - 4$
12. $-3.5 + x > -4$
13. $-1.5 + b < -3.5$

14. $-0.5 < p - 3.5$
15. $-1.25 \le q - 3.0$
16. $-\frac{1}{3} < z - \frac{7}{3}$

17. $-\frac{17}{4} \ge a - \frac{13}{4}$
18. $y - 3.5 > -6.25$
19. $s - 0.75 > -5.5$

Without solving, tell whether the solution contains only positive numbers, only negative numbers, or both positive and negative numbers.

20. $x - 4 > -10$
21. $s - 10 \le -12$
22. $d - 5 > 2$
23. $x - 3 \le -2$

List all integers that meet each given restriction.

24. $d - 3.5 > 4$; solutions less than 10
25. $b - 2 \le 6$; solutions greater than 0

26. $n - 3.5 < 4$; solutions greater than 4
27. $d - 1 \ge 6$; solutions less than 8

Find and graph all solutions.

28. $y - 3 > 5$ and $y - 3 < 6$
29. $x - 3 > 5$ or $x - 3 < 3$

30. $h - 4 \ge 1$ or $h - 3 < 0$
31. $n - 3 > -3$ and $n - 3 \le 5$

32. **Critical Thinking** Solve $x + a > b$ for x. What must be true about a and b if the solution contains only positive numbers?

Find each difference. (2.3 Skill A)

33. $-4 - 7$
34. $5 - 12$
35. $-4 - 3.6$

36. $-4.25 - 0.75$
37. $4.2 - 7.6$
38. $-2\frac{4}{5} - 2\frac{3}{5}$

Using subtraction to solve an inequality in one step

You can use the Subtraction Property of Inequality to solve an inequality that involves addition, such as $x + 6.5 > 5$.

Subtraction Property of Inequality

If a, b, and c are real numbers and $a < b$, then $a - c < b - c$. This statement is also true if $<$ is replaced by $>$, $\leq$, or $\geq$.

EXAMPLE 1

Solve $x + 6.5 > 5$. Graph the solution on a number line.

▶ **Solution**

$$x + 6.5 > 5$$
$$x + 6.5 - 6.5 > 5 - 6.5 \quad \longleftarrow \text{ Apply the Subtraction Property of Inequality.}$$
$$x > -1.5$$

The solution is all real numbers greater than -1.5. Because -1.5 is not included in the solution, draw an open circle at -1.5.

TRY THIS

Solve $x + 3 \geq 5$. Graph the solution on a number line.

As you saw in Skill A, it is usually easier to graph an inequality when the variable is on the left. In Example 2, you can rewrite $6 \leq t + 4.2$ as $t + 4.2 \geq 6$ and then solve the inequality.

EXAMPLE 2

Solve $6 \leq t + 4.2$. Graph the solution on a number line.

▶ **Solution**

$$6 \leq t + 4.2$$
$$t + 4.2 \geq 6 \quad \longleftarrow \text{ Rewrite the inequality so that } t \text{ is on the left.}$$
$$t + 4.2 - 4.2 \geq 6 - 4.2 \quad \longleftarrow \text{ Apply the Subtraction Property of Inequality.}$$
$$t \geq 1.8$$

The solution is all real numbers greater than or equal to 1.8. Because 1.8 is included in the solution, draw a closed circle at 1.8.

TRY THIS

Solve $5 > a + 3.7$. Graph the solution on a number line.

EXERCISES

Rewrite each inequality so that the variable is on the left. Do not solve.

1. $-14 < n + 3.2$

2. $-2 \geq 9 + n$

3. $4.7 > -2 + d$

4. $-10 \leq -10 + w$

PRACTICE

Solve each inequality. Graph the solution on a number line.

5. $n + 5 \geq 5$

6. $p + 4 \geq 8$

7. $4 < -2 + m$

8. $-7 > -5 + k$

9. $-5 < -5 + k$

10. $3 < -3 + f$

11. $c + 5.1 \geq 10$

12. $b + 3.1 \leq 1$

13. $0 < -2 + h$

14. $-7 < -3 + z$

15. $z - 2.6 \geq 4.4$

16. $z - (1.2 + 2.2) \leq -2\frac{2}{5}$

17. $x - \left(4\frac{1}{3} - 2\frac{1}{3}\right) \geq -2\frac{2}{3}$

18. $-6 > a - \left(2\frac{4}{5} + 2\frac{1}{5}\right)$

19. $x - 1\frac{1}{3} \geq -4\frac{2}{3}$

Without solving, tell whether the solution contains only positive numbers, only negative numbers, or both positive and negative numbers.

20. $v + 5 \leq -9$

21. $a + 11 \leq -12$

22. $q + 6 > 2$

23. $n + 4 > -2$

List all integers that meet each given restriction.

24. $b - 6 \leq -2$; solutions no less than -5

25. $n + 1 > 6$; solutions no more than 8

26. $c + 2.6 \geq -4$; solutions no more than 0

27. $m + 3 > -5$; solutions no more than 2

28. **Critical Thinking** Find and graph the integer solution(s) to the set of inequalities below.

$$x + 2 > -7 \text{ and } x - 3 \leq 7 \text{ and } x + 2 > 5 \text{ and } x - 4 < 1$$

MIXED REVIEW

Find each product or quotient. (2.4 Skill A)

29. $(-3)(-11)$

30. $(-4)(14)$

31. $(7)(-1)$

32. $4(4)$

33. $28 \div (-14)$

34. $-108 \div (12)$

35. $5 \div (-25)$

36. $(-15) \div (15)$

37. $\left(-\frac{3}{5}\right)\left(-\frac{5}{21}\right)$

38. $\left(-\frac{8}{3}\right)\left(\frac{15}{24}\right)$

39. $\left(-\frac{14}{5}\right) \div \left(\frac{28}{25}\right)$

40. $\left(\frac{35}{32}\right) \div \left(-\frac{7}{16}\right)$

In some real-world problems, you will need to modify your solution so that the answer is expressed in quantities that are actually possible.

EXAMPLE 1

Greg plans to enter a benefit walk. He will donate $16.40 plus one dollar for each whole mile that he walks. In all, he wants to donate at least $30. How many miles must he walk?

▶ **Solution**

Look for key words.

1. Let m represent the number of miles Greg must walk.

   (starting amount + dollars per mile × miles) **is at least** $30

Write an inequality.

2. Write and solve an inequality.

$$16.40 + 1m \geq 30$$
$$16.40 + m - 16.40 \geq 30 - 16.40$$
$$m \geq 13.6$$

*Because he receives one dollar for every **whole** mile walked, round 13.6 to 14.*

Greg must walk 14 miles or more.

TRY THIS

Rosa's long-distance phone calls cost $0.24 for each whole minute. If she wants her long-distance bill to be less than $15, how many minutes can she talk?

In a real-world problem, sometimes the solution is not a complete answer to the question. For example, if the question is about height, and your solution is $h < 50$, you should modify it to $0 \leq h < 50$, because height cannot be negative. A question sometimes gives additional information that must be included in the answer.

EXAMPLE 2

A store manager wants to have no more than 20 pairs of tennis shoes left at the end of the week. She estimates that 25 pairs of tennis shoes will be sold during the week. How many pairs should she have in stock at the beginning of the week?

▶ **Solution**

Look for key words.

1. Let s represent the number of pairs of tennis shoes in stock at the beginning of the week.

   (starting number – pairs of shoes sold) **is no more than** 20 pairs

Write an inequality.

2. Write and solve an inequality.

$$s - 25 \leq 20$$
$$s - 25 + 25 \leq 20 + 25$$
$$s \leq 45$$

Recall from the question that 25 pairs will be sold during the week.

The manager must have at least 25 pairs of tennis shoes and not more than 45 pairs of tennis shoes.

TRY THIS

How much grain should be in a silo so that when 40.2 tons are removed, there will be less than 42 tons remaining?

KEY SKILLS

EXERCISES

KEY SKILLS

1. Rework Example 1 given that Greg already has $15 to donate and wants to donate at least $35.

2. Rework Example 2 so that the manager wants to have no more than 5 pairs of tennis shoes at the end of the week.

PRACTICE

Use an inequality to solve each problem. Be sure to include any restrictions.

3. Russell must have at least 510 points in his science class to receive an A. If he has 375 points, at least how many more points does he need to receive an A?

4. Casey bought a sandwich and a drink for $3.75. If she has $6.00 to spend, what is the most that she can spend on dessert?

5. Jenna has entered a benefit walk. Her parents gave her $15, and she has saved $24.25. If she receives one dollar for each whole mile that she walks, how far must she walk to donate at least $50?

6. A manager believes that his ski shop will sell 35 ski caps, 24 ski jackets, and 30 pairs of goggles. How many of each should he have at the beginning of the week so that he will have no more than 20 of each item at the end of the week?

7. A baker needs between 13 and 18 pounds of flour, and she already has 5.5 pounds of flour. How many 1-pound bags does she need to buy?

8. The Barlow family has less than $540 to spend for food and entertainment during their vacation. If they spend $365 for food, what is the most that they can spend for entertainment?

MIXED REVIEW APPLICATIONS

Use an equation to solve each problem. (2.7 Skill C)

9. Find the dimensions of a rectangular garden whose perimeter is 180 feet and whose length is 5 times its width.

10. Find two consecutive numbers whose sum is 47.

Solving Inequalities Using Multiplication or Division

LESSON

Using multiplication to solve an inequality in one step

You may need to multiply to solve an inequality.

Multiplication Property of Inequality

If a and b are real numbers, c is positive, and $a < b$, then $ac < bc$.

If a and b are real numbers, c is negative, and $a < b$, then $ac > bc$.

Similar statements can be written for $a > b$, $a \leq b$, and $a \geq b$.

Notice that you must reverse the inequality symbol if you multiply both sides of an inequality by a negative number.

EXAMPLE 1 **Solve each inequality. Graph the solution on a number line.**

a. $\dfrac{x}{3} \leq 12$

b. $2 > \dfrac{d}{-2}$

▶ **Solution**

a. $\quad \dfrac{x}{3} \leq 12$

$3 \cdot \dfrac{x}{3} \leq 3 \cdot 12$

$x \leq 36$

number line: $-24\ -12\ \ 0\ \ 12\ \ 24\ \ 36\ \ 48$ — x

b. $\quad 2 > \dfrac{d}{-2}$

$(-2)2 < (-2)\left(\dfrac{d}{-2}\right)$ ← *Reverse the inequality symbol.*

$-4 < d$, or $d > -4$

number line: $-6\ -5\ -4\ -3\ -2\ -1\ \ 0\ \ 1$ — d

TRY THIS Solve each inequality. Graph the solution on a number line.

a. $\dfrac{x}{4} > -\dfrac{1}{2}$

b. $-\dfrac{1}{3} > \dfrac{t}{-6}$

A multiplier can be a rational number other than an integer.

EXAMPLE 2 **Solve $\dfrac{2}{5}y < -1$. Graph the solution on a number line.**

▶ **Solution**

$\dfrac{2}{5}y < -1$

$\dfrac{5}{2} \cdot \dfrac{2}{5}y < \dfrac{5}{2}(-1)$ ← *Multiply each side by $\dfrac{5}{2}$.*

$y < -\dfrac{5}{2}$, or -2.5

number line with -2.5 marked: $-3\ -2\ -1\ \ 0\ \ 1\ \ 2\ \ 3\ \ 4$ — y

TRY THIS Solve $-\dfrac{3}{2}y < -6$. Graph the solution on a number line.

EXERCISES

KEY SKILLS

What multiplier would you choose to solve each inequality? Will the inequality symbol be reversed?

1. $\frac{w}{4} \le -3$

2. $5 > \frac{t}{-3}$

3. $\frac{2}{-3} b \le -4$

4. $\frac{12}{5} \ge \frac{-7}{5} c$

PRACTICE

Solve each inequality. Graph the solution on a number line.

5. $\frac{r}{2} \le -1$

6. $\frac{t}{-3} \ge 4$

7. $-3 \le \frac{x}{-2}$

8. $-1 \le \frac{f}{7}$

9. $\frac{p}{6} > \frac{5}{-2}$

10. $\frac{n}{-21} > \frac{2}{-7}$

11. $\frac{5}{6} > \frac{1}{-3} u$

12. $-\frac{5}{2} \ge \frac{5}{2} c$

13. $\frac{n}{-2.91} \ge 0$

14. $0 < \frac{k}{14.3}$

15. $\frac{2x}{3} > 6$

16. $18 \le \frac{9x}{5}$

17. $\frac{3}{5} p > -9$

18. $-\frac{1}{8} x < 3$

19. $-\frac{2}{3} w < -6$

20. $\frac{5k}{2.5} > -3$

21. $-4 < \frac{-7k}{3.5}$

22. $\left(-\frac{5}{2}\right)\left(-\frac{4}{15}\right) \ge -\frac{2}{5} z$

23. $\left(\frac{5}{2}\right)\left(-\frac{6}{25}\right) s \ge -\frac{2}{3}$

24. $-\frac{2}{7} \le \left(\frac{1}{3}\right)\left(-\frac{9}{21}\right) a$

25. $\left(\frac{5}{2}\right)\left(-\frac{6}{25}\right) s \ge \left(-\frac{2}{3}\right)\left(-\frac{3}{5}\right)$

Show that each pair of inequalities has the same solution.

26. $\frac{x}{-3} \le \frac{2}{5};\ \frac{5}{2} x \ge -3$

27. $\frac{7}{3} > \frac{x}{4};\ \frac{3}{7} x < 4$

28. Critical Thinking Suppose that x is a variable, a and b are real numbers, and a is nonzero. Solve $\frac{x}{a} > b$ for x.

MIXED REVIEW

Find each quotient. (2.4 Skill B)

29. $\frac{-12}{12}$

30. $\frac{-300}{-5}$

31. $\frac{-12}{48}$

32. $\frac{7}{-3.5}$

33. $\left(\frac{7}{24}\right) \div \left(-\frac{49}{12}\right)$

34. $\left(-\frac{25}{24}\right) \div \left(-\frac{100}{96}\right)$

35. $\left(-\frac{3}{4}\right) \div \left(-\frac{4}{3}\right)$

36. $\left(\frac{3}{4}\right) \div \left(-\frac{15}{20}\right)$

37. $\left(-\frac{3}{4}\right) \div 4$

You may need to divide to solve an inequality.

Division Property of Inequality

If a and b are real numbers, c is positive, and $a < b$, then $\dfrac{a}{c} < \dfrac{b}{c}$.

If a and b are real numbers, c is negative, and $a < b$, then $\dfrac{a}{c} > \dfrac{b}{c}$.
Similar statements can be written for $a > b$, $a \le b$, and $a \ge b$.

Notice that you must reverse the inequality symbol if you divide both sides of an inequality by a negative number.

EXAMPLE

Solve each inequality. Graph the solution on a number line.

a. $4v < 6$ b. $-4x < 6$

▶ **Solution**

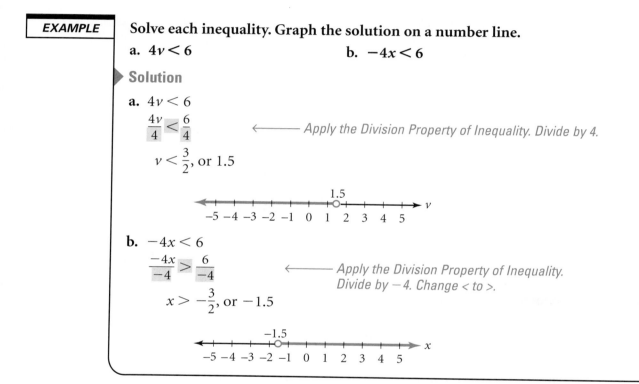

a. $4v < 6$

$\dfrac{4v}{4} < \dfrac{6}{4}$ ⟵—— *Apply the Division Property of Inequality. Divide by 4.*

$v < \dfrac{3}{2}$, or 1.5

b. $-4x < 6$

$\dfrac{-4x}{-4} > \dfrac{6}{-4}$ ⟵—— *Apply the Division Property of Inequality. Divide by −4. Change < to >.*

$x > -\dfrac{3}{2}$, or -1.5

TRY THIS

Solve each inequality. Graph the solution on a number line.   a. $4c \ge -10$ b. $-4c \ge -10$

The multiplication inequality $ax < b$ and the division inequality $\dfrac{x}{a} < b$ can both be solved using multiplication. Suppose $a > 0$.

$ax < b$ $\dfrac{x}{a} < b$

$\left(\dfrac{1}{a}\right)ax < \left(\dfrac{1}{a}\right)b$ ⟵—— *Multiply by $\dfrac{1}{a}$.* $a \cdot \dfrac{x}{a} < a \cdot b$ ⟵—— *Multiply by a.*

$\left(\dfrac{1}{a} \cdot a\right) \cdot x < \left(\dfrac{1}{a}\right)b$ $\left(a \cdot \dfrac{1}{a}\right) \cdot x < a \cdot b$ ⟵—— *Definition of division*

$x < \dfrac{b}{a}$ $x < ab$

EXERCISES

By what number would you choose to divide each side of the inequality?
Will the inequality symbol be reversed?

1. $4d \geq -12$ **2.** $-5r > 13$ **3.** $6t \leq 0$ **4.** $-7.5g < 3$

Solve each inequality. Graph the solution on a number line.

5. $2r > 8$ **6.** $-3t \geq 12$ **7.** $18 > -6x$ **8.** $-25 \geq -5d$

9. $12x < -1$ **10.** $-7 > 14t$ **11.** $-1.5d > 3$ **12.** $2.5h < -5$

13. $-1.5 \geq -3d$ **14.** $2.5 < -5h$ **15.** $\frac{1}{3}z \leq -2$ **16.** $-6 \leq \frac{1}{5}n$

You can multiply by a fraction to solve each inequality below. Identify
the fraction. Then use it to solve the inequality.

17. $\frac{2m}{-5} \leq 2$ **18.** $\frac{-3g}{7} > -9$ **19.** $-\frac{7g}{2} > -2$

Tell whether the solution contains only negative numbers, only positive
numbers, or both positive numbers and negative numbers.

20. $-3k > 5$ **21.** $4h \geq 12$ **22.** $-3c \geq -12$

If $-1 < x < 3$, then we say "x is between -1 and 3." If $-1 \leq x \leq 3$, we
say "x is between -1 and 3, **inclusive**," because -1 and 3 are *included* in
the possible values for x.

Write and graph an inequality for each description.

23. x is between 0 and 3.

24. y is between -2 and 4.

25. x is between 0 and 3, inclusive.

26. t is between 6 and 8, inclusive.

27. Critical Thinking   Suppose that a is between 3 and -5, inclusive.
Between what two numbers will the solution to $ax < 15$ be found?

Simplify by combining like terms. (2.6 Skill A)

28. $-3n + 6.3 - 4n$ **29.** $4j - 3j - 4j$ **30.** $7.2c + (-5.8c) + 1$

Simplify each expression. (2.6 Skill B)

31. $-\frac{7}{3}(9x + 15)$ **32.** $\frac{y - 4}{-2}$ **33.** $-\frac{-2x + 18}{2}$

An inequality is useful for solving problems that involve a *maximum* or a *minimum*.

EXAMPLE 1

Mrs. Blake wants to give each of her four children an equal amount of money. How much money should each child receive if she will distribute a minimum of $300?

▸ **Solution**

1. Let *d* represent the amount of money that each child will receive. *Minimum* implies that she will distribute an amount greater than or equal to $300.

Write an inequality.

2. Write and solve an inequality.

$$4d \geq 300$$
$$\frac{4d}{4} \geq \frac{300}{4} \quad \longleftarrow \text{ Divide each side by 4.}$$
$$d \geq 75$$

Each child will receive a minimum of $75.

TRY THIS

Mrs. James plans to drive at 55 miles per hour and cover a maximum of 440 miles in one day. How much time will she spend driving that day?

Example 2 involves a maximum *and* a minimum, so you will need to write and solve *two* inequalities.

EXAMPLE 2

Donny wants to give his mom and 4 grandparents presents that all cost the same amount. How much money should he save if each gift costs a minimum of $2.00 and a maximum of $5.00?

▸ **Solution**

1. Let *a* represent the amount in dollars that Donny should save.
   minimum: $2.00 → Each gift costs $2.00 or more.
   maximum: $5.00 → Each gift costs $5.00 or less.

Look for key words.

2. Write and solve two inequalities.

$$\frac{a}{5} \geq 2 \qquad\qquad \frac{a}{5} \leq 5$$
$$5\left(\frac{a}{5}\right) \geq 5(2) \qquad 5\left(\frac{a}{5}\right) \leq 5(5) \quad \longleftarrow \text{ Multiply each side of each inequality by 5.}$$
$$a \geq 10 \qquad\qquad a \leq 25$$

Donny should save at least $10 but not more than $25.

TRY THIS

Mr. Kim wants to put a minimum of 5 flowers and a maximum of 11 flowers in each of 6 vases. How many flowers must he buy?

EXERCISES

KEY SKILLS

Identify the minimum and maximum in each problem.

1. Each canning jar will contain between 15 ounces and 17 ounces, inclusive.

2. A certain truck driver may not drive more than 10 hours on a given day.

PRACTICE

Solve each problem.

3. How much time will a motorist spend driving if he drives between 500 miles and 600 miles at 50 miles per hour?

4. How much money must be budgeted if each of 6 school clubs receives an equal amount that is at least $2400 but no more than $3600?

5. What is the range of positive numbers, n, such that n divided by 5 is less than 32?

6. What is the range of negative numbers, n, such that the product of n and 7 is more than -63?

7. **a.** What is the range of positive real numbers, n, such that the product of n and 11 is less than 121?
   b. What are the positive integers, n, such that the product of n and 11 is less than 121?

8. **Critical Thinking**   When five objects with the same weight are put on a scale, the scale reads 1.5 kilograms. If the scale's reading is as much as 0.001 kilogram too little or too much, between what two numbers is the actual weight of one of the objects?

MIXED REVIEW APPLICATIONS

Find the domain and range of each function. (3.2 Skill C)

9. the weight, w, of n identical textbooks if each book weighs 2.3 pounds

10. the volume of water, v, in a tank that initially contains 800 gallons and drains at the rate of 20 gallons per minute for t minutes

11. the number of light bulbs, n, if there are c cartons that each contain 12 packages of 2 bulbs

Solving Simple One-Step Inequalities

SKILL A *Choosing an operation to solve an inequality in one step*

When you solve an inequality, it is important to think about which property of inequality you will use. Remember that you can rewrite inequalities so that the variable is on the left. For example, you can rewrite $5 < x$ as $x > 5$.

EXAMPLE 1 **Choose a method for solving $8 \leq t + 4.5$. Then solve.**

▶ **Solution**

$$8 \leq t + 4.5$$
$$t + 4.5 \geq 8 \qquad \longleftarrow \text{ Rewrite the inequality so that } t \text{ is on the left.}$$
$$t + 4.5 - 4.5 \geq 8 - 4.5 \qquad \longleftarrow \text{ Apply the Subtraction Property of Inequality.}$$
$$t \geq 3.5$$

TRY THIS Choose a method for solving $-6 < c + 1.5$. Then solve.

EXAMPLE 2 **Choose a method for solving $25 > \frac{5}{2}d$. Then solve.**

▶ **Solution**

$$25 > \frac{5}{2}d$$
$$\frac{5}{2}d < 25 \qquad \longleftarrow \text{ Rewrite the inequality so that } d \text{ is on the left.}$$
$$\left(\frac{2}{5}\right)\frac{5}{2}d < \left(\frac{2}{5}\right)25 \qquad \longleftarrow \text{ Apply the Multiplication Property of Inequality.}$$
$$d < 10$$

TRY THIS Choose a method for solving $\frac{1}{10} > \frac{2}{5}z$. Then solve.

EXAMPLE 3 **Choose a solution method. Then solve.**

    a. $-t \geq 5$ **b.** $0 < 6 - s$

▶ **Solution**

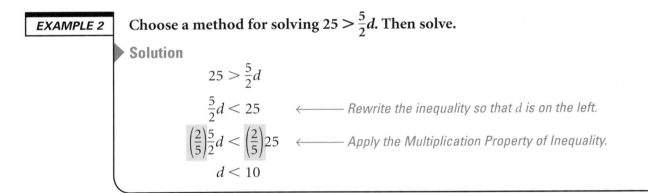

TRY THIS Choose a solution method. Then solve.   **a.** $-j > 0$ **b.** $0 \geq 4 - x$

EXERCISES

What step would you take to solve each inequality? Do not solve.

1. $x - 6 < -3$

2. $x + 3.5 > 8.2$

3. $5x \geq -40$

4. $\frac{x}{7} > 2$

PRACTICE

For Exercises 5–22:
   a. **What step would you take to solve each inequality?**
   b. **Will the inequality symbol be reversed?**
   c. **Solve the inequality.**

5. $t - \frac{3}{4} \geq \frac{1}{4}$

6. $j - 2.3 \leq -6$

7. $\frac{w}{-3} > 2.8$

8. $-3p < 3$

9. $-y < 0$

10. $-r \geq 9$

11. $j + 1.5 \leq -2$

12. $q + 2.6 < 6.6$

13. $z - 6 \leq -12.5$

14. $12.5 > 2.7 + w$

15. $-12 < \frac{v}{-5}$

16. $-\frac{3}{11} < \frac{a}{-22}$

17. $\frac{-3d}{11} > -3$

18. $2 > \frac{-13f}{2}$

19. $0 < 13 - h$

20. $-d > 4$

21. $5 - z > 0$

22. $-g \leq -2$

Find the solution that is common to both inequalities. Graph the common solution on a number line.

23. $x - 5 > 1$ and $x + 6 < 13$

24. $3w \geq -3$ and $3w \leq 21$

25. **Critical Thinking**
   a. Find all real numbers that are solutions to all of the following inequalities: $x - 5 > -1$ and $x - 5 > -2$ and $x - 5 > -3$ and $x - 5 > -4$.
   b. Graph the common solution on a number line.

MIXED REVIEW

Solve each equation. (2.8 Skill B)

26. $5 - 3x = 23$

27. $5a - 3a = 56$

28. $-13 + 3x = 4x - 17$

29. $4(x + 3) = 3x$

30. $-5(10 - z) = 4z$

31. $11 - 4x = -3x - 1$

32. $3k - 2k - 6 = -7$

33. $-3(n - 3n) = 54$

34. $1 - 3x = -3x + x + 11$

EXAMPLE 1

Jeff's score on this week's test is 10 points higher than his score on last week's test. His score this week is less than 95 points. What are Jeff's possible scores on last week's test?

▶ **Solution**

1. Let *t* represent Jeff's score on last week's test.
    Then $t + 10$ represents his score on this week's test.

Write an inequality.

2. Write and solve an inequality.
$$t + 10 < 95$$
$$t + 10 - 10 < 95 - 10 \quad \longleftarrow \text{ Apply the Subtraction Property of Inequality.}$$
$$t < 85$$

3. Check: $85 + 10 \overset{?}{=} 95$ ✔
        A score of less than 85 plus 10 points will be less than 95.

Jeff's score on last week's test was less than 85 points.

TRY THIS

Marcy is driving on the highway. She notices that if she increases her speed by 15 miles per hour, her speed will still be under the speed limit of 65 miles per hour. How fast could Marcy be driving?

EXAMPLE 2

In the school election, the number of eighth graders who voted was one-half of the number of seventh graders who voted. If fewer than 220 eighth graders voted, how many seventh graders could have voted?

▶ **Solution**

1. Let *n* represent the number of seventh graders who voted.
    Then $\frac{1}{2}n$ is the number of eighth graders who voted.

Write an inequality.

2. Write and solve an inequality.
$$\frac{1}{2}n < 220$$
$$(2)\frac{1}{2}n < (2)220 \quad \longleftarrow \text{ Apply the Multiplication Property of Inequality.}$$
$$n < 440$$

3. Check: $\frac{1}{2}(440) \overset{?}{=} 220$ ✔

        If fewer than 220 eighth graders voted, then the number of
        seventh graders must be less than 440.

Fewer than 440 seventh graders voted.

TRY THIS

The Hanes family bought a radio and a television set. The price of the television was four times the price of the radio. The television's price was less than $300. How much could the radio have cost?

EXERCISES

KEY SKILLS

Match each situation with its corresponding algebraic inequality.

1. The amount I have now and my next $5.00 allowance will give me at least $11.00.

2. I need a certain minimum amount to buy 5 CD's that cost at least $11.00 each.

3. I would have this amount if my four brothers and I split more than $11 equally.

a. $\frac{x}{5} \geq 11$

b. $5x > 11$

c. $x + 5 \geq 11$

PRACTICE

Solve each problem.

4. The difference of a number and 5 is less than -12. What are possible values for the number?

5. Three more than a number is greater than 23. What are possible values for the number?

6. Tina's softball glove cost $30 more than her softball bat. The price of the glove was less than $42. What are the possible prices of the bat?

7. The Martin family bought a new washer and dryer. The dryer cost $350, and the total cost was less than $600. What are the possible prices of the washer?

8. In a poll, the number of women surveyed was one-fourth of the number of men surveyed. If fewer than 100 women were surveyed, how many men could have been surveyed?

9. If the Smith family's monthly electric bill is less than $115, what is the range of their electric cost for one year?

MID-CHAPTER REVIEW

Graph each compound inequality on a number line. (4.1 Skill C)

10. $-2 \leq x < 3$

11. $0 < n$ or $n \leq -4$

Solve each inequality. Check your solution. (4.2 Skills A and B, 4.3 Skills A and B)

12. $-3.6 > x - 5$

13. $d + 4.3 \leq 1.1$

14. $-12.9 > h + 12.9$

15. $t - 3.5 \geq -3.6$

16. $4 > -6y$

17. $\frac{-9x}{17} > 1$

18. $-5g \geq 1$

19. $-2 \geq \frac{-x}{7}$

20. The amount of rainfall in March was 7.3 inches greater than the amount of rainfall in April. More than 9.9 inches of rain fell in March. How many inches of rain could have fallen in April? (4.4 Skill B)

4.5 Solving Multistep Inequalities

LESSON

SKILL A ▶ *Solving an inequality in multiple steps*

You may need to use more than one property to solve an inequality.

EXAMPLE 1   Solve $4j + 2 \geq -18$.

▶ **Solution**

$$4j + 2 \geq -18$$
$$4j \geq -20 \quad \longleftarrow \text{ Subtract 2 from each side.}$$
$$j \geq -5 \quad \longleftarrow \text{ Divide each side by 4.}$$

TRY THIS   Solve each inequality.   **a.** $-3t + 5 \leq -10$ **b.** $0 \geq 5 - 4v$

EXAMPLE 2   Solve $3a > 5a - 1$.

▶ **Solution**

$$3a > 5a - 1$$
$$-2a > -1 \quad \longleftarrow \text{ Subtract } 5a \text{ from each side.}$$
$$a < \frac{1}{2} \quad \longleftarrow \text{ Divide each side by } -2. \text{ Change } > \text{ to } <.$$

TRY THIS   Solve each inequality.   **a.** $-2w \leq 6w + 16$ **b.** $-3z + 2 < -7z$

EXAMPLE 3   Solve $2r + 5 < 3r - 1$. Check your solution.

▶ **Solution**

$$2r + 5 < 3r - 1$$
$$2r - 3r + 5 < 3r - 3r - 1 \quad \longleftarrow \text{ Apply the Subtraction Property of Inequality.}$$
$$-r + 5 - 5 < -1 - 5 \quad \longleftarrow \text{ Apply the Subtraction Property of Inequality.}$$
$$-r < -6 \quad \longleftarrow \text{ Multiply by } -1 \text{ and change } < \text{ to } >.$$
$$r > 6$$

Check: First substitute 6 into the related equation.

$$2(6) + 5 \overset{?}{=} 3(6) - 1$$
$$17 = 17 \ ✔$$

To check the direction of the inequality symbol, choose a number larger than 6 and substitute it into the original inequality.

$$r = 8 \quad \rightarrow \quad 2(8) + 5 < 3(8) - 1$$
$$21 < 23$$

The inequality is true, so the direction of the symbol in $r > 6$ is correct. ✔

TRY THIS   Solve and graph $2 - 6z > 6 - 5z$. Check your solution.

EXERCISES

**Write the inequality that results when you perform each operation.
What is the next step you would take to solve the inequality?**

1. $3n - 5 < -8$; add 5 to each side.

2. $4 > 2t + 8$; subtract 8 from each side.

3. $\frac{2x}{3} \leq 6$; multiply each side by 3.

4. $4x + 1 \geq 17$; subtract 1 from each side.

PRACTICE

Solve each inequality.

5. $-4r < -6r + 2$

6. $5z < -7z - 36$

7. $-a < -7a + 36$

8. $5z + 11 < -6z$

9. $-5y + 30 < 20$

10. $3f + 6 > 15$

11. $18 > 3g + 15$

12. $-2 > -2h + 2$

13. $0 > 4s - 8$

14. $4v + 20 \leq 0$

15. $-3u - 3 + 4u \leq 0$

16. $-4 > -4u + 4 + 5u$

17. $-8w - 8 + 9w > -4$

18. $-1 \leq -p + 1 + 2p$

19. $2q - 2 > 3q - 6$

20. $-10b - 1 > -11b + 3$

21. $-c - 1 \leq -2c + 2$

22. $\frac{1}{3}r - \frac{1}{3} < -\frac{2}{3}r$

Solve and graph each inequality. Check your solution.

23. $2x > 4$

24. $d - 4 > -1$

25. $3m - 2 \leq 7$

26. $-n + 2 \leq 3$

27. Critical Thinking For what value of n will the solution to $2(x - n) - 4 \geq 0$ be $x \geq 0$?

28. Critical Thinking Let a, b, and c be real numbers with $a \neq 0$. Solve $ax + b > c$ for x.

MIXED REVIEW

Graph each solution on a number line. (4.1 Skill B)

29. $x > -5$

30. $x \leq 5$

Graph each compound inequality. (4.1 Skill C)

31. $-2 < n < 4$

32. $n > 4$ and $n \leq 6$

33. $k < 4$ or $k \geq 6$

34. $m < 0$ or $m > 3$

EXAMPLE 1   **Solve and graph $2(x + 5) + 3 > 19$. Check your solution.**

▶ **Solution**

$$2(x + 5) + 3 > 19$$
$$2x + 10 + 3 > 19 \quad \longleftarrow \text{Apply the Distributive Property.}$$
$$2x + 13 > 19 \quad \longleftarrow \text{Combine like terms.}$$
$$2x > 6 \quad \longleftarrow \text{Apply the Subtraction Property of Inequality.}$$
$$x > 3 \quad \longleftarrow \text{Apply the Division Property of Inequality.}$$

Check: First substitute 3 into the related equation. The true statement confirms that the open circle is in the correct position. To test the direction of the ray, test points on either side of 3.

$$2(3 + 5) + 3 \overset{?}{=} 19$$
$$16 + 3 \overset{?}{=} 19$$
$$19 = 19 \ ✔$$

$$x = 0 \quad \rightarrow \quad 2(0 + 5) + 3 = 13 \qquad 13 > 19 \ ✗$$
$$x = 4 \quad \rightarrow \quad 2(4 + 5) + 3 = 21 \qquad 21 > 19 \ ✔$$

Because $13 > 19$ is not a true statement, 0 is not a solution. Because $21 > 19$ is true, 4 is a solution and it should be included in the graph.

TRY THIS   Solve $3(p - 6) + 3 < -9$. Check your solution.

All of the inequalities that you have studied so far have had some real numbers as their solution. But there are two other solution possibilities for an inequality: all real numbers or no solution.

• If you arrive at an equivalent inequality that is always true, the original inequality is satisfied by all real numbers.

• If you arrive at an equivalent inequality that is always false, the original inequality is not satisfied by any real number; it has no solution.

EXAMPLE 2   **Solve each inequality.**
  a. $3(b + 2) - 7 \geq 2b - 5 + b$ **b. $6h + 5 + 4h < 5(2h - 1)$**

▶ **Solution**

  a. Simplify each side of the inequality.

$$3(b + 2) - 7 \geq 2b - 5 + b$$
$$3b - 1 \geq 3b - 5$$
$$-1 \geq -5 \ ✔$$

  This true statement indicates that the solution is all real numbers.

  b. Simplify each side of the inequality.

$$6h + 5 + 4h < 5(2h - 1)$$
$$10h + 5 < 10h - 5$$
$$5 < -5 \ ✗$$

  This false statement indicates that there are no solutions to the original inequality.

TRY THIS   Solve each inequality.

  a. $-3(n - 1) - 5 > 2 - 3n$ **b. $3 - (4t + 2) \leq 4 - 2t + 2(1 - t)$**

EXERCISES

KEY SKILLS

Simplify each side of the inequalities below. Do not solve.

1. $5x - 4x > -7 + 3$

2. $6x - 7 - x < 3 - 2$

3. $3(x + 2) \leq 5 + 1$

4. $2(7 - x) + 1 \geq 4 - 6$

PRACTICE

Solve each inequality. Check your solution.

5. $-2(c - 5) - 7 > 10$

6. $7 - 2(n - 5) \leq -5$

7. $0 \leq -5(q + 2) - 4$

8. $4 \leq 2(w - 2) + 4$

9. $3 + 2(d - 1) > 3(d - 3) + 2$

10. $5 - 4(d - 1) < -5(d - 3) + 2$

11. $5(n + 2) - 5 \leq 3(n - 1) + 4$

12. $-(m + 2) - 2 \leq -2(m - 1) + 1$

13. $\frac{1}{2}(x + 1) + \frac{5}{2}(x - 1) > -3$

14. $\frac{2}{3}(a + 2) < \frac{7}{3}(a + 2) + 5$

15. $\frac{1}{8}(t + 2) + \frac{7}{8}(t + 2) + 5t < 0$

16. $\frac{1}{5}(5z + 2) > \frac{1}{5}(z - 2) + \frac{1}{5}(2z - 1)$

In Exercises 17–22, does the inequality have no solution or all real numbers as its solution?

17. $5(c + 1) \leq 5(c - 1)$

18. $-3(t - 2) \geq -3t$

19. $2(y + 9) \geq 2(y - 9)$

20. $-3(v + 1) > -3(v - 1) + 1$

21. $-3(v + 1) \leq -3(v - 1) + 1$

22. $-3(z + 9) \geq -2z - z$

23. Solve $\frac{6}{11}(y - 1) + \frac{4}{11}(y - 1) + \frac{1}{11}(y - 1) > 0$.

24. Solve $-\frac{1}{5}(z + 1) - \frac{2}{5}(z + 1) - \frac{2}{5}(z + 1) > 0$.

MIXED REVIEW

Show that each statement is either always true or never true. (2.9 Skill C)

25. $3(4x + 4) - 11 = 12x + 2$

26. $3a + 4a - 11 + a = -9 + 8a - 2$

27. $3(t + t) - t = 5(t + 1)$

28. $d(3 + 2) - 4d - 1 = d - 1$

Making a table is usually helpful in solving mixture problems.

EXAMPLE 1

A chemical technician has two solutions, A and B, that are to be mixed to make a third solution, C. Use the data shown at right. To the nearest milliliter, how much of each solution is needed to make 500 milliliters of solution C?

Solution	Sugar
A	10%
B	25%
C	no more than 20%

▶ **Solution**

Make a table.

1. Let a represent the amount of solution A. Then the amount of Solution B is $500 - a$.

2. Make a table.

Solution A + Solution B = Solution C

Amount of Solution	a	$500 - a$	500
Amount of Sugar	$0.10a$	$0.25(500 - a)$	$\leq 0.20(500)$

3. Write and solve an inequality.

$$0.10a + 0.25(500 - a) \leq 0.20(500)$$
$$a \geq 166.67$$

The chemist needs at least 167 milliliters of solution A and no more than $500 - 167$, or 333 milliliters, of solution B.

TRY THIS Rework Example 1 if solution C is to be no more than 15% sugar.

Sometimes you can adapt a formula to help solve an inequality.

EXAMPLE 2

Mrs. James has $1200 in a bank account that earns 6% simple interest annually. How much should she deposit now so that she will have at least $2800 in one year?

▶ **Solution**

Use a formula.

1. The formula below gives amount, A, in terms of principal, p, the annual interest rate as a decimal, r, and time, t, in years.

$$A = p(1 + rt), \text{ or } A = p + prt$$

Let d represent the amount to add to the $1200 already in the account.

2. Write and solve the inequality $p + prt \geq A$ using $p = 1200 + d$, $r = 0.06$, and $t = 1$.

$$(1200 + d) + (1200 + d)(0.06)(1) \geq 2800$$
$$d \geq 1441.51$$

Mrs. James should deposit at least $1441.51.

TRY THIS Rework Example 2 if the goal is at least $3000, the rate of interest is 5%, and there is $2000 in the account.

EXERCISES

KEY SKILLS

1. Refer to Example 1. Use the data below to set up an inequality.
   Do not solve.
   solution A: 20% sugar   solution B: 30% sugar
   solution C: 800 milliliters at no more than 25% sugar

2. Refer to Example 2. Use the data below to set up an inequality. Do not
   solve.
   current amount: $2400   annual rate: 4%   savings goal: $4000

PRACTICE

3. Use the data below to rework Example 1.
   solution A: 10% sugar   solution B: 4% sugar
   solution C: 500 milliliters at no more than 6% sugar

4. Use an inequality to show that more of solution A than solution B can
   be used to make solution C, given the data below.
   solution A: 20% sugar   solution B: 40% sugar
   solution C: 450 milliliters at no more than 30% sugar

5. Use an inequality to show that it is impossible to make solution C
   given the data below.
   solution A: 20% sugar   solution B: 40% sugar
   solution C: 800 milliliters at no more than 15% sugar

6. Use an inequality to show that Mrs. James will need to double her
   current amount of $1500 earning interest at 5% per year to reach a
   goal of $3150 in one year.

7. Write a mixture problem similar to Example 1 that could be solved by
   this inequality: $0.15a + 0.30(600 - a) \le 0.20(600)$.

8. Write a problem similar to Example 2 that could be solved by this
   inequality: $(1500 + d) + (1500 + d)(0.06) \ge 2000$.

9. **Critical Thinking**   Use the table in Example 1 as a model to construct
   a table that illustrates Example 2.

MIXED REVIEW APPLICATIONS

Use an inequality to solve each problem. (4.2 Skill C)

10. Angela has saved $112.45 and wants to have a total of at least $180 by
    the end of the month. At least how much more money must she save?

11. What positive integers can you decrease by 9 and get an integer no
    more than -2?

4.6 LESSON
Solving Compound Inequalities

Recall from Lesson 4.1 Skill C that a compound inequality may be a pair of inequalities joined by *and*. This type of compound inequality is called a **conjunction**. The solution to the conjunction $x > 1$ *and* $x < 7$ is graphed below.

$$x > 1 \qquad and \qquad x < 7$$

-2 -1 0 1 2 3 4 5 6 7 8 → x

EXAMPLE 1   **Solve $2x - 1 < 5$ *and* $2x + 9 \geq 7$. Graph the solution on a number line.**

▶ **Solution**

$$2x - 1 < 5 \qquad and \qquad 2x + 9 \geq 7$$
$$2x - 1 + 1 < 5 + 1 \qquad\qquad 2x + 9 - 9 \geq 7 - 9$$
$$2x < 6 \qquad\qquad 2x \geq -2$$
$$x < 3 \qquad and \qquad x \geq -1$$

$$-1 \leq x < 3$$

-4 -3 -2 -1 0 1 2 3 4 5 6 → x

TRY THIS   Solve $3b - 1 \leq 11$ *and* $5b > -25$. Graph the solution on a number line.

When a conjunction is written in shortened form, such as $a \leq x \leq b$, the pair can be solved simultaneously as shown below.

EXAMPLE 2   **Solve $-5 \leq -2a + 5 \leq 10$. Graph the solution on a number line.**

▶ **Solution**

Perform each operation on all three expressions simultaneously.

$$-5 \leq -2a + 5 \leq 10$$
$$-5 - 5 \leq -2a + 5 - 5 \leq 10 - 5 \qquad \longleftarrow \text{Subtract 5 from each of the three expressions.}$$
$$-10 \leq -2a \leq 5$$
$$\frac{-10}{-2} \geq \frac{-2a}{-2} \geq \frac{5}{-2} \qquad \longleftarrow \text{Divide each of the three expressions by } -2 \text{ and change } \leq \text{ to } \geq.$$
$$5 \geq a \geq -\frac{5}{2}$$
$$-\frac{5}{2} \leq a \leq 5 \qquad \longleftarrow$$

Compound inequalities are normally written with the least value on the left.

$-\frac{5}{2}$

-4 -3 -2 -1 0 1 2 3 4 5 6 → a

TRY THIS   Solve $-9 < 2(d - 5) < 1$. Graph the solution on a number line.

EXERCISES

KEY SKILLS

Write each pair of inequalities in shortened form as in Example 2.

1. $x > -7$ and $x \leq 8$

2. $-3n \geq 2$ and $-3n < 10$

3. $2(d + 1) \leq 3$ and $2(d + 1) > -2$

4. $-3a - 5 \geq 0$ and $-3a - 5 < 10$

PRACTICE

Solve each conjunction. Graph the solution on a number line.

5. $3x + 6 \geq 0$ and $x - 5 < 0$

6. $-2t - 2 \geq 0$ and $t + 5 > 3$

7. $g + 1 < 0$ and $-2(g + 3) \leq 0$

8. $p - 3 \geq 0$ and $-5(p - 5) > -15$

9. $2(n + 3) + 2 \geq 0$ and $-2(n + 3) - 2 \geq -2$

10. $-(w - 1) + 1 > 0$ and $-2(w + 1) - 3 \leq 1$

11. $13 > 3(b - 1) + 10$ and $0 \geq -(b - 4) - 5$

12. $-12 \geq -2(x + 4) + 1$ and $2(x + 2) - 8 \leq 3$

Solve each compound inequality. Graph the solution on a number line.

13. $-4 \leq 2d + 3 < 0$

14. $0 < 2d - 5 < 3$

15. $3 < -3(w - 1) \leq 12$

16. $-6 \leq -2(a + 3) \leq 8$

17. $-1 \leq -2(b - 1) + 3 \leq 7$

18. $-9 \leq -4(z + 4) - 1 \leq 7$

19. $0 < 3 + 3(k - 3) \leq 7$

20. $1 < 2(k - 1) - (k - 2) \leq 7$

21. $-2 < 3(q - 2) - 2(q - 3) \leq 2$

22. Show that $2(p - 3) + 3 < 3 - 2(p + 3)$ *and* $-4 + 2(p + 2) > 8 - (p + 1)$ has no solution.

23. Show that $-3 - 4(s + 3) > 0$ *and* $4 - 3(s - 3) < 10$ has no solution.

24. **Critical Thinking** Find a so that the pair of inequalities below has exactly one number as its solution.
$$2(x + a) + 4 \geq 0 \text{ and } -3(x - a) - 5 \geq -2$$

MIXED REVIEW

Write a linear equation in two variables to represent the data in each table. Predict the 50th term. (3.8 Skill A)

25.

x	1	2	3	4	5
y	-30	-19	-8	3	14

26.

n	1	2	3	4	5
y	24	12	0	-12	-24

Represent each compound inequality on a number line. (4.1 Skill C)

27. $x < -2$ or $x \geq 3$

28. $y \leq 0$ or $y > 2$

29. $s \leq -2$ or $s \geq -2$

30. $z < -2$ or $z > -2$

A pair of inequalities joined by *or* is called a **disjunction**. The solution to the disjunction $x \leq 0$ *or* $x > 4$ is shown on the number line below.

$$x \leq 0 \quad or \quad x > 4$$

Notice that the graph is a pair of nonintersecting and opposite rays.

EXAMPLE 1 **Solve $2(x + 3) < 5$ *or* $5x - 7 > 8$. Graph the solution on a number line.**

▶ **Solution**

Solve each inequality separately.

$$
\begin{array}{lcl}
2(x + 3) < 5 & or & 5x - 7 > 8 \\
2x + 6 < 5 & or & 5x > 15 \\
x < -0.5 & or & x > 3
\end{array}
$$

$$x < -0.5 \text{ or } x > 3$$

TRY THIS Solve $3(t + 1) \leq 3$ *or* $2t + 4 > 8$. Graph the solution on a number line.

Just as with simple inequalities, it is possible for a compound inequality to have no solution or a solution of all real numbers. The compound inequality $x < 0$ *and* $x > 5$ has no solution because x cannot be both less than zero and greater than 5.

Example 2 shows a compound inequality whose solution is all real numbers.

EXAMPLE 2 **Solve $-6y + 5 > 23$ *or* $12 \geq -3y$. Graph the solution on a number line.**

▶ **Solution**

Solve each inequality separately.

$$
\begin{array}{lcl}
-6y + 5 > 23 & or & 12 \geq -3y \\
-6y > 18 & & -4 \leq y \\
y < -3 & or & y \geq -4
\end{array}
$$

$$y < -3 \qquad\qquad y \geq -4$$

The solution to $-6y > 18$ *or* $12 \geq -3y$ is all real numbers.

TRY THIS Solve $2a > -6$ *or* $0 \leq -5a$. Graph the solution on a number line.

EXERCISES

Use mental math to decide whether the solution is a pair of non-intersecting opposite rays, the entire number line, or the entire number line with one point missing.

1. $x < 0 \text{ or } x > 1$

2. $h < 2 \text{ or } h > 1$

3. $a < -5.5 \text{ or } a > 2.3$

4. $z < -3 \text{ or } z > -3$

PRACTICE

Solve each disjunction. Graph the solution on a number line.

5. $2z - 5 > 1 \text{ or } -3z \geq 9$

6. $2c + 1 < 1 \text{ or } c - 3 > -2$

7. $2d - 1 \geq -4 \text{ or } -2d - 5 > -2$

8. $-2v - 1 < 0 \text{ or } 3v + 1 < 7$

9. $2(f - 1) \geq 0 \text{ or } -2(f + 3) > 3$

10. $-2(w - 1) < 0 \text{ or } 3(w + 3) < 6$

11. $2(g - 4) - 1 \geq 0 \text{ or } -3(g + 1) + 5 > 3$

12. $-(y + 2) - 1 < 0 \text{ or } 2(y - 1) + 1 > 5$

13. $3 - 2(n - 1) > 0 \text{ or } 7 + 3(n + 1) > 0$

14. $11 - 3(r - 2) < 0 \text{ or } 2 - 3(r + 2) > 1$

Show that the solution to each compound inequality is a single ray.

Sample: $3x > -6 \text{ or } -4x + 6 < 2$ Solution: $x > -2 \text{ or } x > 1$

$$-4 \ -3 \ -2 \ -1 \ \ 0 \ \ 1 \ \ 2 \ \ 3 \ \ 4$$

15. $2(x + 1) \geq 0 \text{ or } -6(x - 1) \leq 3$

16. $-(p + 2) < 0 \text{ or } 2p - 1 > 5$

17. $1 - (n + 2) > 1 \text{ or } 2 - (n + 3) > 0$

18. $1 - (r - 3) < 1 \text{ or } 2 - (r + 1) < 5$

19. Show that, for all values of a, the solution to $-3(x + a) > 0$ or $-2(x - a) > 0$ is a single ray.

20. Critical Thinking Show that, for all values of a, the solution to $-3(x + a) > 0$ or $2(x + a) > 0$ is the number line with one point missing.

MIXED REVIEW

Find the opposite and absolute value of each number. (2.1 Skill C)

21. -12.9

22. 111

23. $-(-2.7)$

24. $-(31.3)$

Solve each inequality. Graph the solutions on a number line. (4.1 Skill A)

25. $x - 2 > 2$

26. $d - 6 < -1$

27. $0 \geq a - 2.05$

28. $-1 > d - 1\frac{3}{4}$

You can use compound inequalities to solve real-world problems which involve a maximum or a minimum of a quantity.

EXAMPLE 1

Mary and Dean want to plant a square garden enclosed by a fence. They have 200 feet of fencing available, and they want the sides of their garden to be at least 10 feet in length. What side lengths are possible?

▶ **Solution**

Make a diagram.

1. Recall that the four sides of a square are equal in length. If s represents the length of one side and P represents perimeter, then $P = 4s$.

2. Make a drawing to represent the situation.

3. Identify the maximum and minimum values.
   maximum length of fence is 200 feet: $P \leq 200$, or $4s \leq 200$
   minimum length of each side is 10 feet: $s \geq 10$, so $4s \geq 4(10)$

Write an inequality.

4. Write and solve a compound inequality.

   minimum length ⟶ $4(10) \leq 4s \leq 200$ ⟵ *maximum length*
   $$10 \leq s \leq 50$$

   The side length can be any real number between 10 and 50, inclusive.

5. **Check:** Sides 10 feet or longer satisfy their minimum requirement.

   $4(50) = 200$, so 50 feet per side is the maximum length of fence available. ✔

TRY THIS   Rework Example 1 for an equilateral triangle and 540 feet of fencing. Each side of the triangle must be at least 30 feet long.

EXAMPLE 2

Depending on the cost of cashews, the ratio of cashews to peanuts in a nut mix will be either $1\frac{1}{2}$ to 1 or 2 to 1. How many pounds of peanuts should be used to make at least 5 pounds of the mix?

▶ **Solution**

1. Let p represent the number of pounds of peanuts needed. Then the number of pounds of cashews needed is:

   $1\frac{1}{2}p$ at the ratio $1\frac{1}{2}$ to 1   or $2p$ at the ratio 2 to 1

Write an inequality.

2. Write and solve a compound inequality.

   $$p + 1\frac{1}{2}p \geq 5 \quad or \quad p + 2p \geq 5$$
   $$p \geq 2 \quad or \quad p \geq \frac{5}{3}, \text{ or } 1\frac{2}{3}$$

   At least 2 pounds of peanuts should be used if the ratio is $1\frac{1}{2}$ to 1 and at least $1\frac{2}{3}$ pounds of peanuts should be used if the ratio is 2 to 1.

TRY THIS   Rework Example 2 if the ratio of cashews to peanuts will be either $2\frac{1}{2}$ to 1 or 3 to 1.

EXERCISES

Write an algebraic expression for each situation.

1. There are more than 3 times as many apples as oranges.

2. The ratio of science books to math books is at least 3 to 1.

PRACTICE

Solve each problem.

3. The sum of two consecutive integers is at least 51 but not more than 199. What might the integers be?

4. A triangular garden is shown at right. At least 490 feet of fencing but no more than 790 feet of fencing is to be used. What could be the length of the shortest side of the garden?

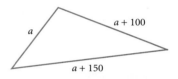

5. A motorist has driven 240 miles. If she is traveling at 50 miles per hour, how much more time will it take her to drive at least 480 miles but not more than 560 miles?

6. An engineer opens a valve to drain a tank that contains 1800 gallons of a mixture. If the mixture drains at the rate of 24 gallons per minute, how long will it take for the tank to be one-half to three-quarters empty, inclusive?

7. A party plannner wants to make a nut mix of peanuts and cashews. The ratio of cashews to peanuts will be either $1\frac{1}{2}$ to 1 or 2 to 1. How many pounds of peanuts should be used to make at least 9 pounds of the mix?

8. **Critical Thinking** Jaime has taken 3 tests. His score on the second test was 12 points higher than on the first test. The third score was 85. His average for the three tests was a B (80 to 89, inclusive). What is the possible range of his grade on the first test?

MIXED REVIEW APPLICATIONS

Use a direct-variation equation to solve each problem. (3.3 Skill B)

9. The distance that an object can be slid along the floor varies directly with the force applied. If 50 pounds of force move a box 4 feet, how far will the box move if 60 pounds of force are applied?

10. The distance that an object stretches a spring is directly proportional to the weight of the object. If a 25-pound object stretches a spring 0.01 inch, how heavy must an object be to stretch the spring 0.03 inch?

Solving Absolute-Value Equations and Inequalities

LESSON

SKILL A *Solving absolute-value equations*

Recall from Lesson 2.1 Skill C that the absolute value of a number is the distance between the number and zero on a number line. Equations and inequalities involving absolute value are solved using compound sentences.

distance from 0 is 3 units

$-4\ -3\ -2\ -1\ \ 0\ \ 1\ \ 2\ \ 3\ \ 4$

$|x| = 3$ means $x = 3$ *or* $x = -3$

To solve an absolute-value equation such as $|x| = 3$, write it as two simpler equations joined by *or*. Then solve these two equations.

Solving Absolute-Value Equations

If $|x| = a$ and $a > 0$, then $x = a$ or $x = -a$.

EXAMPLE 1   **Solve.**   **a.** $|x - 5| = 3$ **b.** $|-3t| = 18$

▶ **Solution**

a. If $|x - 5| = 3$, then $x - 5 = 3$ *or* $x - 5 = -3$. Thus, $x = 2$ *or* $x = 8$.

b. If $|-3t| = 18$, then $-3t = 18$ *or* $-3t = -18$. Thus, $t = -6$ *or* $t = 6$.

TRY THIS   Solve.   **a.** $|x + 4| = 3.5$ **b.** $|-2t| = 10$

EXAMPLE 2   **Solve $|3r - 1| = 5$.**

▶ **Solution**

If $|3r - 1| = 5$, then $3r - 1 = 5$ *or* $3r - 1 = -5$. Thus, $r = 2$ *or* $r = -\dfrac{4}{3}$.

TRY THIS   Solve $|-3x - 5| = 7$.

Be sure to isolate the absolute-value expression on one side.

EXAMPLE 3   **Solve $2|y - 1| - 2 = 4$.**

▶ **Solution**

$$2|y - 1| - 2 = 4$$
$$2|y - 1| = 6 \qquad \longleftarrow \text{Apply the Addition Property of Equality.}$$
$$|y - 1| = 3 \qquad \longleftarrow \text{Apply the Division Property of Equality.}$$
$$y - 1 = 3 \quad or \quad y - 1 = -3$$
$$y = 4 \quad or \quad y = -2$$

TRY THIS   Solve $2|z + 1| + 5 = 7$.

EXERCISES

KEY SKILLS

Use mental math to solve each equation.

1. $|w| = 3.6$

2. $|n| = 6$

3. $|z| = 0$

Write each equation in the form $|x| = a$. Do not solve.

4. $|x| - 5 = 13$

5. $2|x| = 16$

6. $3|x| + 5 = 8$

PRACTICE

Solve each equation. Check your solution.

7. $|x - 3| = 5$

8. $|w + 4| = 11$

9. $|2d| = 6$

10. $|3g| = 15$

11. $-|x + 2| = -2$

12. $-|k + 3| = -13$

13. $-5|h| = -25$

14. $-7|n| = -63$

15. $|3m - 5| = 11$

16. $|5x + 7| = 13$

17. $2|c + 1| - 7 = 10$

18. $3|z + 1| + 5 = 7$

19. $4|z - 4| - 3 = 7$

20. $3|y + 5| - 1 = 9$

21. $-7 = -2|k - 3| - 1$

22. $-11 = -4|h - 3| - 1$

23. $-4 = -|3z - 1| - 1$

24. $-11 = -2|2x - 3| - 1$

25. Solve $2|4(x - 2) - 3(x - 1)| = x$.

26. Show that $2|x + 3| = x$ has no solutions.

27. Find n such that $2|x| - 5 = n$ has exactly one solution.

28. Find m such that $3|x| - 4 = m$ has -2 and 2 as solutions.

29. For what values of n will $7|x| - 11 = n$ have no solutions?

30. Let a, b, and c be real numbers with $a \neq 0$ and $c \geq 0$.
    Solve $|ax + b| = c$ for x.

31. **Critical Thinking**   What does $|x - 5| = 3$ mean in terms of distance
    on a number line?

MIXED REVIEW

**Solve each inequality and check your solution. Use mental math where
possible.** (4.4 Skill A)

32. $x + (-11) > -22$

33. $y - 5.9 \geq -11$

34. $7.5 \leq 9.1 + z$

35. $-3p > 18$

36. $4n > -1$

37. $-13 \geq -3y$

Solve each inequality. Check your solution. (4.5 Skill A)

38. $2x - 3x \geq 4x + 1$

39. $5 - 3y < 3 - 5y$

40. $-4(t - 3) \leq t$

An absolute-value inequality can take one of two forms shown at right. (Similar inequalities can be written using $\leq$ and $\geq$.) Use the following rules to write absolute-value inequalities as compound inequalities:

$$|x| < a \qquad |x| > a$$

Solving Absolute-Value Inequalities

If $|x| < a$ and $a > 0$, then $x > -a$ and $x < a$. That is, $-a < x < a$.

If $|x| > a$ and $a > 0$, then $x < -a$ or $x > a$.

If $|x| < 2$, then $-2 < x < 2$.

$-2 < x$ and $x < 2$

If $|x| > 2$, then $x < -2$ or $x > 2$.

$x < -2$ or $x > 2$

EXAMPLE 1   Solve. Graph the solutions on a number line.   **a.** $|z - 3| \leq 3$ **b.** $|z - 3| \geq 3$

▶ **Solution**

a. If $|z - 3| \leq 3$, then $-3 \leq z - 3 \leq 3$.
Therefore, $0 \leq z \leq 6$.

b. If $|z - 3| \geq 3$, then $z - 3 \leq -3$ or $z - 3 \geq 3$. Therefore, $z \leq 0$ or $z \geq 6$.

TRY THIS   Solve. Graph the solutions on a number line.   **a.** $|d + 2| < 3$ **b.** $|d + 2| > 3$

EXAMPLE 2   Solve $|2(x - 5)| \leq 2$.

▶ **Solution**

> You can save two steps by not distributing the 2 over $(x - 5)$.

$$-2 \leq 2(x - 5) \leq 2$$
$$-1 \leq x - 5 \leq 1 \qquad \longleftarrow \text{Apply the Division Property of Inequality. Divide by 2.}$$
$$4 \leq x \leq 6 \qquad \longleftarrow \text{Apply the Addition Property of Inequality. Add 5.}$$

TRY THIS   Solve $|2(n + 3)| < 8$.

EXAMPLE 3   Solve $|12 - 3d| > 3$.

▶ **Solution**

$$\begin{array}{lll} 12 - 3d < -3 & or & 12 - 3d > 3 \\ -3d < -15 & or & -3d > -9 \\ d > 5 & or & d < 3 \end{array}$$

TRY THIS   Solve $|3k + 2| \geq 5$.

EXERCISES

Write the next step in solving each absolute-value inequality. Do not solve.

1. $|x| \leq 5$

2. $|x| \geq 7$

3. $|3y| < 9$

4. $|5z| > 12$

5. $|3c - 5| > 11$

6. $|2t + 1| \leq 3$

PRACTICE

Solve each inequality. Graph the solution on a number line.

7. $|d| \leq 1$

8. $|2x| \leq 4$

9. $|d| \geq 1$

10. $|2x| \geq 4$

11. $|c - 2| < 3$

12. $|z + 3| < 4$

13. $|2s - 1| \geq 5$

14. $|2t + 3| > 5$

15. $9 > |3(v + 1)|$

16. $6 \geq |4(y + 2)|$

17. $2 \leq |3(b - 2)|$

18. $3 \leq |-2(z + 2)|$

19. $|3(f + 1) - 2| > 0$

20. $-|3(f + 1) - 2| > 0$

21. $2|3(g + 2) - 2| \leq 8$

22. $5|-2(n + 1) - 4| \leq 10$

23. $-|-3(m - 2) - 3| > -9$

24. $-|1 - 2(t - 2)| \leq -1$

25. Find r so that the solution to $|3x + 4| > r$ is the entire number line.

26. Show there is no nonzero number r such that the endpoints of the solution to $|x + 4| > r$ are opposites of one another.

27. Solve $|2x - 1| < |x|$.

28. Solve the conjunction at right. $|2x - 1| \leq 11$ *and* $|3x| > 15$.

29. **Critical Thinking**   What does $|x| > 3$ mean in terms of distance on the number line?

30. **Critical Thinking**   What does $|x + 2| < 3$ mean in terms of distance on the number line?

31. **Critical Thinking**   Let a, b, and c be real numbers with a and $c > 0$. Prove that the solution to $|ax + b| \leq c$ is $\dfrac{-b - c}{a} \leq x \leq \dfrac{c - b}{a}$.

MIXED REVIEW

Is the given statement sometimes, always, or never true? Justify your response. (2.9 Skill C)

32. If x is any real number, then $-x + 4x - 1 = 3x - 1$.

33. If x is any real number, then $-(x - 4) = 3x - 1$.

34. If r is any real number, then $-(r - 4r) = 3r - 1$.

Tolerance describes the acceptable range that a measurement may differ from its ideal value. The examples show how to use absolute-value inequalities to solve problems involving tolerance.

EXAMPLE 1

The ideal weight of a can of vegetables is 16 ounces. The tolerance is 0.01 ounce. This means that a can is acceptable if it is within 0.01 ounce of 16 ounces and unacceptable otherwise. What are the acceptable weights and the unacceptable weights?

▶ **Solution**

1. Let w represent the actual weight of one can.

Make a diagram.

2. Make a drawing to represent the situation.

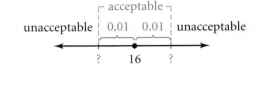

3. acceptable: $|w - 16| \leq 0.01$

$$16.00 - 0.01 \leq w \leq 16.00 + 0.01$$
$$15.99 \leq w \leq 16.01$$

unacceptable: $|w - 16| > 0.01$

$$w < 16.00 - 0.01 \text{ or } w > 16.00 + 0.01$$
$$w < 15.99 \text{ or } w > 16.01$$

The acceptable weights are between 15.99 and 16.01 ounces, inclusive. The unacceptable weights are less than 15.99 ounces and greater than 16.01 ounces. Another restriction on the variable is that weight, w, has to be greater than or equal to zero.

TRY THIS   Rework Example 2 if the ideal weight is 12.5 ounces and the tolerance is 0.02 ounce.

EXAMPLE 2

At a watch factory, a consultant found that the weekly cost of producing w watches is $10,000 + 2.5w$ dollars. If the company's ideal weekly cost is $30,000 with a tolerance of $2000, how many watches can they produce?

▶ **Solution**

$$|\text{weekly cost} - 30,000| \leq 2000$$

Write an inequality.

$\|(10,000 + 2.5w) - 30,000\| \leq 2000$	⟵——— *Write an absolute-value inequality.*
$\|2.5w - 20,000\| \leq 2000$	⟵——— *Simplify within the absolute value.*
$-2000 \leq 2.5w - 20,000 \leq 2000$	⟵——— *Recognize the form $\|x\| \leq a$ and solve $-a \leq x \leq a$.*
$18,000 \leq 2.5w \leq 22,000$	⟵——— *Apply the Addition Property of Inequality.*
$7200 \leq w \leq 8800$	⟵——— *Apply the Division Property of Inequality.*

The factory can produce between 7200 and 8800 watches and stay within the cost limits.

TRY THIS   Rework Example 2 if the ideal weekly cost is $25,000 and the tolerance is $3000.

EXERCISES

KEY SKILLS

Let *w* represent a weight. Write an absolute-value inequality to represent the acceptable weights for each ideal weight and tolerance.

1. ideal weight = 5 pounds; tolerance = 0.2 pound

2. ideal weight = 14 ounces; tolerance = 0.01 ounce

PRACTICE

Use absolute value to express the acceptable ranges and the unacceptable ranges.

3. plants whose height may vary at most 2 inches from the ideal height of 18 inches

4. a machine part whose length may vary at most 0.002 centimeter from the ideal length of 32 centimeters

5. glass whose thickness may vary at most 0.0025 centimeter from the ideal thickness of 1.2 centimeters

Solve each problem.

6. In Example 2, find the number of watches that can be produced if the weekly cost of producing *w* watches is $12{,}000 + 2.5w$, the ideal cost is $25,000, and the tolerance is $3000.

7. Margo's doctor tells her that her ideal weight is $3h - 56$, where *h* is Margo's height in inches. The tolerance is 9 pounds. The doctor then says that Margo's weight of 115 pounds is within this acceptable range. How tall can Margo be?

MIXED REVIEW APPLICATIONS

Solve each problem. (2.8 Skill C)

8. A chemist wants to separate 640 milliliters of solution into two containers in such a way that one container holds 200 milliliters less than twice the first container. How much should be put into each container?

9. How much money should be put into bank accounts paying 5% and 6% simple interest if $5000 is invested and the investor wants to earn $290 in one year?

4.8 LESSON

Deductive Reasoning With Inequalities

SKILL A *Showing that an inequality is always, sometimes, or never true*

To determine whether an inequality is true over the domain of all real numbers, simplify it. If the simplification is an inequality that is always true, then the original inequality is also always true.

EXAMPLE 1

Is $2(x + 5) > 2x + 9$ always, sometimes, or never true?

▶ **Solution**

$$2(x + 5) > 2x + 9$$
$$2x + 10 > 2x + 9 \quad \longleftarrow \text{ Apply the Distributive Property.}$$
$$2x + 1 > 2x \quad \longleftarrow \text{ Subtract 9 from each side.}$$
$$1 > 0 \quad \longleftarrow \text{ Subtract 2x from each side.}$$

Recall from Lesson 4.5 that if you arrive at an equivalent inequality that is always true, then the original inequality is true for all real numbers. $1 > 0$ is always true. Therefore, $2(x + 5) > 2x + 9$ is *always true.*

TRY THIS   Show whether $3(x - 4) > 3(x - 5)$ is always, sometimes, or never true.

If an inequality simplifies to a numerical inequality that is never true, then the original statement is never true. For example, $2x - 5 + 3x > 5x$ simplifies to $-5 > 0$, which is not true. Thus, $2x - 5 + 3x > 5x$ is *never true.*

EXAMPLE 2

Is $r > -r$ always, sometimes, or never true?

▶ **Solution**

$$r > -r$$
$$2r > 0 \quad \longleftarrow \text{ Add r to each side.}$$
$$r > 0 \quad \longleftarrow \text{ Divide each side by 2.}$$

The result shows that the inequality is true only when r is positive. When $r \leq 0$, the inequality is not true. Thus, the statement is only *sometimes true.*

TRY THIS   Show whether $r > 2r$ is always, sometimes, or never true.

In simplifying inequalities, *be careful of division by a variable.* Whenever you divide both sides of an inequality by a variable, you have to consider three cases: $r > 0$, $r < 0$ and $r = 0$.

In Example 2, if each side of $r > -r$ is divided by r, the result seems to be $1 > -1$ (true). However, if r is negative, the inequality sign must be reversed. Then the result is $1 < -1$ (false). And if $r = 0$, the division cannot be performed at all. So, $r > -r$ is true only when r is positive.

EXERCISES

1. What can you conclude about an inequality that simplifies to $1 > -1$?

2. What can you conclude about an inequality that simplifies to $1 < -1$?

3. What can you conclude about an inequality that simplifies to $x > 0$?

PRACTICE

Show whether each statement is always, sometimes, or never true.

4. $3 - (x + 1) > 1 - x$

5. $3x < 3(x + 1)$

6. $-2(x + 1) < -2x$

7. $\frac{1}{2}r < r$

8. $4x - 3 < 2(2x - 2)$

9. $4(x - 3x) > 9 - 8x$

10. $-2(x + 1) > -2x$

11. $3z - 5 + 4z \geq z + 6z - 5$

12. $-2(3z - 2z) > z$

13. $9 - 2(z + 2) < -2z$

14. $500x > 250x$

15. $500x > 250x^2$

16. $-(x + 1) \geq x + 1$

17. $-(x + 1) \leq x - 1$

Critical Thinking Show whether each statement is always, sometimes or never true.

18. $x^2 > x$

19. $x(-x) \leq 0$

20. $x^2 + 1 \geq 1$

21. $x^2 \geq 2x$

22. $\frac{x}{5} \geq \frac{5}{x}$

23. $\frac{x}{3} \leq \frac{3}{x}$

24. **Critical Thinking** Is the following statement always, sometimes, or never true? Justify your answer.
   The reciprocal of a number is greater than the opposite of the number.

MIXED REVIEW

Use intercepts to graph each equation. (3.5 Skill A)

25. $-4x - 7y = 28$

26. $x - y = 5.5$

27. $3x - 7y = 21$

Use the slope and the y-intercept to graph each equation. (3.5 Skill A)

28. $y = \frac{-2}{5}x$

29. $y = \frac{5}{2}x - 3.5$

30. $y = -x - 4$

1. Graph -3 and 5 on a number line and write two inequalities that describe their relationship.

2. Represent the following statement in symbols.
   The cost of a textbook, c, is more than $25.

3. Graph the solutions to $a \leq 1$ *or* $a > 2.5$ on a number line.

4. Graph the solutions to $t \geq -3$ *and* $t \leq -1.5$ on a number line.

5. Solve $a - 3.5 > 1$. Graph the solution on a number line.

6. Find and graph the solution to $x - 1 > 0$ *and* $x - 2 < 4$.

7. Solve $n + 2.5 \leq 6$. Graph the solution on a number line.

8. Find and graph the solution to $x + 1 > 3$ *and* $x + 2 < 8$.

9. Kathy wants to buy at least 2.5 pounds of hamburger. The scale shows 2.1 pounds. How much more should be put on the scale so that Kathy has at least 2.5 pounds of hamburger?

Solve. Graph each solution on a separate number line.

10. $\dfrac{k}{3} \geq -1$

11. $\dfrac{3}{5}d < 6$

12. $\dfrac{p}{-2} \geq -1.5$

13. $4g \geq 20$

14. $-4r \leq 20$

15. $-7w > -14$

16. The perimeter of a square playground is at least 225 feet. What is the length of each side?

Solve each inequality.

17. $7 \geq s + 2.5$

18. $-x > 5$

19. $2.5x \leq 2$

20. Ron wants to send roses to his wife, Sheila. Roses cost $1.50 each, and Ron has $25. What are the possible numbers of roses that Ron can send to Sheila?

Solve. Graph each solution on a separate number line.

21. $3b + 4 \leq 16 - b$

22. $4a + 4 > 3a + 1$

23. $7 > 5s - 3$

24. $5c - 3c > 10$

25. $7h - 6 < 6h$

26. $-7w > -14 - 5w$

Solve. Graph each solution on a separate number line.

27. $3(x - 1) - 2 \leq 19$

28. $4p + 4(p - 1) > 12$

29. $4r + 2(r + 1) < 3(r + 5)$

30. $-3(d - 1) + 5 > -2(d + 1)$

31. $-2(w + 1) - 3(w + 3) \geq 5 - (w + 6)$

32. Mr. Kent has \$1500 in a bank account that pays 5% interest annually. How much should he deposit now so that he will have at least \$1800 in one year?

Solve. Graph each solution on a separate number line.

33. $4(w + 1) < 3(w + 3) < 5 - (w + 6)$

34. $3a + 5 \leq 32$ *and* $4a - 3 \geq 21$

35. $-x - 2 \leq 6$ *or* $4x + 9 \leq 1$

36. $2(w - 1) \geq 3w + 1$ *or* $3(w - 2) \leq 5 - (w + 6)$

37. $3a + 5 \leq 23$ *or* $4a - 3 \geq 21$

38. $3x - 2 \geq 6$ *and* $4x + 7 \leq -3$

39. In a party mix, the ratio of peanuts to walnuts is 2 to 1. How many pounds of walnuts are needed to make at least 6.6 pounds of the mix but not more than 7.8 pounds of the mix?

Solve each absolute-value equation. Check your solution.

40. $|3x + 4| = 12$ **41.** $|-2d + 1| - 3 = 15$ **42.** $4 + |4x| = 16$

Solve. Graph each solution on a separate number line.

43. $|3(x + 1)| \geq 1$ **44.** $2|x + 1| - 2 < 4$ **45.** $5 - 2|3x| > -1$

46. The ideal weight of a can of oil is 32 ounces with a tolerance of 0.02 ounce. Find the acceptable weights and the unacceptable weights.

47. Is $2r > r$ always, sometimes, or never true?

Solving Systems of Equations and Inequalities

▶ **What You Already Know**

In earlier chapters, you learned how to solve an equation for a specified variable, graph the equation, and find how pairs of lines represented by linear equations in two variables are related.

▶ **What You Will Learn**

In Chapter 5, you will be given a pair of linear equations in two variables, called a *system of equations*. The task will be to learn how to find any ordered pair of real numbers that satisfies both equations.

In particular, you will learn how to solve a system of linear equations in two variables by

• using graphs,

• using the substitution method, and

• using the elimination method.

Your study will also include ways to find out whether a system of equations has a solution at all or has infinitely many solutions.

Finally, you will have the opportunity to explore linear inequalities in two variables. You will see that the graph of such an inequality is a region in the coordinate plane rather than a line. How to find the solution to a pair of such inequalities graphically will also be explored.

VOCABULARY	
boundary line	linear inequality in two variables
consistent	solution of a system
dependent	solution region
elimination method	substitution
inconsistent	substitution method
independent	system of linear equations
intersection	system of linear inequalities in two
linear inequality in standard form	variables

The diagram below shows how mathematical skills and mathematical reasoning are interrelated with the skills and concepts in Chapter 5. Notice that in this chapter you will learn how to solve systems of equations and linear inequalities in two variables.

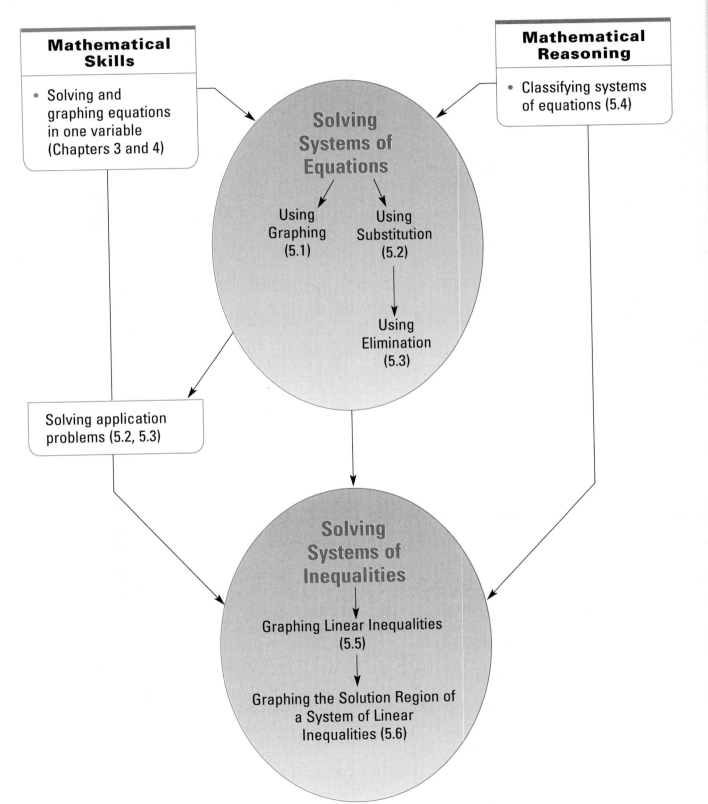

CHAPTER 5

Mathematical Skills

- Solving and graphing equations in one variable (Chapters 3 and 4)

Mathematical Reasoning

- Classifying systems of equations (5.4)

Solving Systems of Equations

Using Graphing (5.1)

Using Substitution (5.2)

Using Elimination (5.3)

Solving application problems (5.2, 5.3)

Solving Systems of Inequalities

Graphing Linear Inequalities (5.5)

Graphing the Solution Region of a System of Linear Inequalities (5.6)

Solving Systems of Equations by Graphing

LESSON 5.1

SKILL A *Using a graph to solve a system of linear equations*

A **system of linear equations** is a set of two or more equations with the same variables. The **solution of a system** in x and y is any ordered pair (x, y) that satisfies each of the equations in the system.

The solution of a system of equations is the *intersection* of the graphs of the equations.

EXAMPLE 1

Use a graph to solve the system of equations at right.
Verify that your solution lies on both lines.

$$\begin{cases} x + y = 5 \\ y = 2x - 1 \end{cases}$$

▶ **Solution**

Graph both equations on the same coordinate plane.
• Graph $x + y = 5$ using the intercepts: $(5, 0)$ and $(0, 5)$.
• Graph $y = 2x - 1$ using the slope and y-intercept.
Locate the point where the lines intersect. From the graph, the
solution appears to be $(2, 3)$.

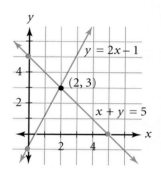

Check: To be sure that $(2, 3)$ is the solution, substitute 2 for x
and 3 for y into each equation.

$$\begin{array}{ll} x + y = 5 & y = 2x - 1 \\ 2 + 3 \overset{?}{=} 5 \checkmark & 3 \overset{?}{=} 2(2) - 1 \checkmark \end{array}$$

Because $(2, 3)$ satisfies both equations, it lies on both lines.

TRY THIS

Use a graph to solve the system of equations at right.
Verify that your solution lies on both lines.

$$\begin{cases} x + y = 3 \\ y = x - 7 \end{cases}$$

EXAMPLE 2

Use a graph to solve the system of equations at right.
Check your solution.

$$\begin{cases} x + y = 4 \\ x + y = 2 \end{cases}$$

▶ **Solution**

Graph $x + y = 4$ using the intercepts: $(4, 0)$ and $(0, 4)$.
Graph $x + y = 2$ using the intercepts: $(2, 0)$ and $(0, 2)$.
The lines appear to be parallel. Because parallel lines never
intersect, the system has no solution.

Check: To be sure that the lines are parallel, rewrite each
equation in slope-intercept form and then compare the slopes.
$$x + y = 4 \rightarrow y = -x + 4 \qquad x + y = 2 \rightarrow y = -x + 2$$
The slopes, -1 and -1, are equal, so the lines are parallel. ✔

TRY THIS

Use a graph to solve the system of equations at right.
Check your solution.

$$\begin{cases} x + y = 4 \\ x + y = 3 \end{cases}$$

EXERCISES

KEY SKILLS

Compare slopes to determine whether each system of equations has a solution or not. Do not solve.

1. $\begin{cases} y = 3x - 2 \\ y = 3x - 1 \end{cases}$

2. $\begin{cases} 2y = 10x - 4 \\ y = 5x - 1 \end{cases}$

3. $\begin{cases} 4x + 2y = 8 \\ 4x + 3y = -2 \end{cases}$

PRACTICE

Use a graph to solve each system of equations. If the system has no solution, write *none*.

4. $\begin{cases} y = -x - 1 \\ y = 5 - x \end{cases}$

5. $\begin{cases} y = 6x - 2 \\ y = 1 + 6x \end{cases}$

6. $\begin{cases} y = -x - 2 \\ y = x + 6 \end{cases}$

7. $\begin{cases} y = -x - 5 \\ y = x + 1 \end{cases}$

8. $\begin{cases} y = x + 11 \\ y + x = 3 \end{cases}$

9. $\begin{cases} y = x + 5 \\ y + x = 3 \end{cases}$

10. $\begin{cases} x - y = 1 \\ x - y = 3 \end{cases}$

11. $\begin{cases} -2x + y = 1 \\ -2x + y = -2 \end{cases}$

12. $\begin{cases} y = \frac{2}{3}x - 1 \\ -2x + 3y = 6 \end{cases}$

13. $\begin{cases} y = -2x + 10 \\ y = -\frac{1}{3}x + 5 \end{cases}$

14. $\begin{cases} y = \frac{1}{3}x - 3 \\ y = \frac{-5}{3}x - 9 \end{cases}$

15. $\begin{cases} y = -\frac{1}{2}x - \frac{1}{2} \\ y = \frac{1}{2}x - \frac{1}{2} \end{cases}$

16. $\begin{cases} y = \frac{1}{4}x - 4 \\ y = -\frac{1}{2}x + 2 \end{cases}$

17. $\begin{cases} 2x - y = -3 \\ 4x - 2y = -5 \end{cases}$

18. $\begin{cases} 5y = x - 10 \\ x - y = 6 \end{cases}$

Use the slopes and the y-intercepts to show that each system has a solution. Then graph the system to verify your conclusion.

19. $\begin{cases} y = \frac{4}{3}x - 2 \\ y = 2 \end{cases}$

20. $\begin{cases} y = -\frac{3}{5}x + 3 \\ y = 2x + 1 \end{cases}$

21. $\begin{cases} -x + 5y = 5 \\ 2x + y = 4 \end{cases}$

MIXED REVIEW

Simplify each expression. (2.6 Skill B)

22. $\frac{2}{3}(3x - 6) - 2$

23. $2 - \frac{4}{5}(x + 5)$

24. $-2(5 + 4n) - 4$

25. $-3(3y - 1) + 1$

26. $-0.5(3 - 5k) + 5$

27. $-2.5(1 - 2x) + 5$

Solving Systems of Equations by Substitution

SKILL A *Using the substitution method with one variable already isolated*

A second method for solving systems of equations is *substitution*. This method is quicker and more accurate than the graphing method. It is especially practical when one variable has a coefficient of 1.

The Substitution Method

Step 1. Solve one equation for x (or y).

Step 2. Substitute the expression from Step **1** into the other equation. The result is an equation in only one variable.

Step 3. Solve for y (or x).

Step 4. Take the value of y (or x) found in Step **3** and substitute it into one of the original equations. Then solve for the other variable.

Step 5. The ordered pair of values from Steps **3** and **4** is the solution. If the system has no solution, a contradictory statement will result in either Step **3** or **4**.

EXAMPLE   Solve $\begin{cases} y = 3x + 8 \\ x + 2y = 9 \end{cases}$ using the substitution method.

▶ **Solution**

1. The first equation is already solved for y, so Step **1** is already done.

2. In $x + 2y = 9$, substitute $3x + 8$ for y.

$$x + 2y = 9$$
$$x + 2(3x + 8) = 9$$

3. Solve $x + 2(3x + 8) = 9$ for x.

$$x + 2(3x + 8) = 9$$
$$x + 6x + 16 = 9$$
$$7x = -7$$
$$x = -1$$

4. Substitute -1 for x in $y = 3x + 8$ and then solve for y.

$$y = 3x + 8$$
$$y = 3(-1) + 8 = 5$$

5. The solution to the system is $(-1, 5)$.

Check: Check the values of x and y in the original equations.

$$y \overset{?}{=} 3x + 8 \qquad\qquad x + 2y = 9$$
$$5 \overset{?}{=} 3(-1) + 8 \qquad -1 + 2(5) \overset{?}{=} 9$$
$$5 \overset{?}{=} 5 ✔ \qquad\qquad 9 \overset{?}{=} 9 ✔$$

TRY THIS   Solve using the substitution method.   **a.** $\begin{cases} y = x - 9 \\ 4x + 3y = 22 \end{cases}$ **b.** $\begin{cases} x + y = -9 \\ 2y = 3x + 2 \end{cases}$

EXERCISES

KEY SKILLS

In each system below, one of the equations is solved for y. Replace that expression for y in the other equation and write the equation in x that results. Do not solve.

1. $\begin{cases} -2x + 3y = 1 \\ y = -3x + 1 \end{cases}$

2. $\begin{cases} 2x + 3y = 6 \\ y = 2x - 5 \end{cases}$

3. $\begin{cases} -\frac{1}{5}x + 3 = y \\ -3x + y = -5 \end{cases}$

PRACTICE

Use the substitution method to solve each system. Check your answers.

4. $\begin{cases} y = -5x - 1 \\ y = -1 \end{cases}$

5. $\begin{cases} y = -5x + 1 \\ y = -4 \end{cases}$

6. $\begin{cases} y = -2x + 4 \\ x + 3y = -3 \end{cases}$

7. $\begin{cases} y = -x + 4 \\ 2x + y = -5 \end{cases}$

8. $\begin{cases} 2x + y = 5 \\ x = 2y - 4 \end{cases}$

9. $\begin{cases} y = -\frac{3}{7}x + 2 \\ y = \frac{2}{7}x - 3 \end{cases}$

10. $\begin{cases} -2x + y = -11 \\ y = \frac{2}{5}x - 3 \end{cases}$

11. $\begin{cases} \frac{1}{5}x + 2y = 8 \\ y = x - 7 \end{cases}$

12. $\begin{cases} \frac{3}{4}x + 5y = 4 \\ y = 2x - \frac{7}{2} \end{cases}$

13. $\begin{cases} y = -3x + \frac{23}{7} \\ y = \frac{2}{7}x \end{cases}$

14. $\begin{cases} y = -3x - 7 \\ 5x - 2y = 20 \end{cases}$

15. $\begin{cases} 3 + 4y = x \\ -2x - 3y = 6 \end{cases}$

16. Let a and b be fixed real numbers. For what values of a and b will the solution to the system $\begin{cases} x + ay = 5 \\ x = by + 10 \end{cases}$ be $(2, -3)$?

17. Let a and b be fixed real numbers. Solve $\begin{cases} x = a + y \\ x + y = b \end{cases}$ for x and y.

18. **Critical Thinking** Solve the system at right by substitution. $\begin{cases} x = z - 2 \\ y = z - 1 \\ z = x + y \end{cases}$

MIXED REVIEW

Solve each inequality. (4.5 Skill B)

19. $3d + 4(d + 1) \geq 4$

20. $-4t - (t + 3) < 0$

21. $3(g - 4) > -2(g - 1)$

22. $3z - 4(z + 2) \geq -3z$

23. $\frac{1}{3}(a - 2) - \frac{1}{2}(a + 1) \geq 0$

24. $4 - (w + 4) < -3(w - 1)$

25. $\frac{1}{3}(b - 2) > \frac{2}{5}(b + 2)$

26. $\frac{c - 2}{3} > \frac{c + 2}{-5}$

27. $\frac{-c}{3} \leq \frac{3(c + 2)}{-2}$

A system of equations such as $\begin{cases} 3u + 2v = 12 \\ u - 2v = 3 \end{cases}$ may be given with neither variable already isolated.

In a case such as this, you must first isolate one of the variables before you can use substitution. It is easiest to isolate a variable that has a coefficient of 1, such as u in the second equation above.

EXAMPLE

Solve $\begin{cases} 3u + 2v = 12 \\ u - 2v = 3 \end{cases}$. **Check your solution.**

▶ **Solution**

1. Write the second equation so that u is isolated.

$$\begin{cases} 3u + 2v = 12 \\ u - 2v = 3 \end{cases} \rightarrow \begin{cases} 3u + 2v = 12 \\ u = 2v + 3 \end{cases}$$

2. Apply the substitution method.

$$\begin{cases} 3u + 2v = 12 \\ u = 2v + 3 \end{cases} \rightarrow 3(2v + 3) + 2v = 12$$

3. Solve $3(2v + 3) + 2v = 12$.

$$3(2v + 3) + 2v = 12$$
$$8v + 9 = 12 \rightarrow v = \frac{3}{8}$$

4. If $v = \frac{3}{8}$, then $u = 2\left(\frac{3}{8}\right) + 3$, so $u = \frac{15}{4}$.

5. The solution (u, v) to the given system is $\left(\frac{15}{4}, \frac{3}{8}\right)$.

Check: Check the values of u and v in both equations.

$$3u + 2v = 12 \qquad\qquad u - 2v = 3$$
$$3\left(\frac{15}{4}\right) + 2\left(\frac{3}{8}\right) \overset{?}{=} 12 \qquad \frac{15}{4} - 2\left(\frac{3}{8}\right) \overset{?}{=} 3$$
$$\frac{90 + 6}{8} \overset{?}{=} 12 \ ✔ \qquad\qquad \frac{30 - 6}{8} \overset{?}{=} 3 \ ✔$$

TRY THIS

Solve each system. Check your solution.

a. $\begin{cases} a + 2b = -2 \\ 2a + 3b = 1 \end{cases}$ 　　　 **b.** $\begin{cases} 2m - 5n = 4 \\ 7 = 3m + n \end{cases}$

Note that it would have been difficult to solve the system in the Example by graphing. The exact solution would be hard to determine from a graph because the coordinates are not integers. Solving a system algebraically is better than graphing when you need an accurate solution.

EXERCISES

KEY SKILLS

Isolate one variable in one of the equations. Do not solve the system.

1. $\begin{cases} -2x + 3y = 5 \\ x + 5y = 2 \end{cases}$

2. $\begin{cases} -2a - b = -2 \\ 2a - 3b = 7 \end{cases}$

3. $\begin{cases} -2r - 3s = 2 \\ 7r - s = -3 \end{cases}$

PRACTICE

Solve each system using the substitution method. Check your solution.

4. $\begin{cases} x - y = 2 \\ x + y = -4 \end{cases}$

5. $\begin{cases} a + b = -2 \\ a - b = 4 \end{cases}$

6. $\begin{cases} 2x + y = -1 \\ 3x - y = 11 \end{cases}$

7. $\begin{cases} -u + 3v = 1 \\ 2u - v = 5 \end{cases}$

8. $\begin{cases} 4c + 3d = 12 \\ -2c + d = 1 \end{cases}$

9. $\begin{cases} -2y + 5z = 16 \\ y + 3z = 3 \end{cases}$

10. $\begin{cases} 2x + 2y = 12 \\ x - y = 2 \end{cases}$

11. $\begin{cases} x + 2y = 3 \\ 3x - \frac{2}{3}y = 1 \end{cases}$

12. $\begin{cases} 6x + 8y = 9 \\ 3x + 2y = \frac{3}{2} \end{cases}$

13. $\begin{cases} \frac{1}{2}r + \frac{1}{2}s = -3 \\ \frac{1}{3}r + \frac{1}{2}s = 4 \end{cases}$

14. $\begin{cases} \frac{3}{5}m + \frac{2}{3}n = -3 \\ \frac{3}{5}m - \frac{1}{3}n = 4 \end{cases}$

15. $\begin{cases} a + \frac{2}{3}b = -3 \\ \frac{1}{3}a + \frac{2}{3}b = 1 \end{cases}$

16. $\begin{cases} 2m + 3n = 2 \\ 2m + 5n = 1 \end{cases}$

17. $\begin{cases} 3c + 2d = 2 \\ -2c + 2d = 5 \end{cases}$

18. $\begin{cases} 3a + 2b = 2 \\ 2a = 2b + 1 \end{cases}$

MIXED REVIEW

Simplify each expression. (2.6 Skill A)

19. $3k - 4k + 5 - k$

20. $5 - n + 6 - 2n$

21. $z - 2z - 3z - 1 - 2 - 3$

Simplify each expression. (2.6 Skill B)

22. $-2(d + 5) - 4(d - 1)$

23. $-(n - 3) - (n + 1)$

24. $4(a - 3) - 4(a + 3)$

25. $\frac{12w + 3}{3}$

26. $\frac{-18b + 12}{-2}$

27. $\frac{24b + 50}{-2}$

When there are two unknowns in a real-world problem, you can solve it using a system of two equations.

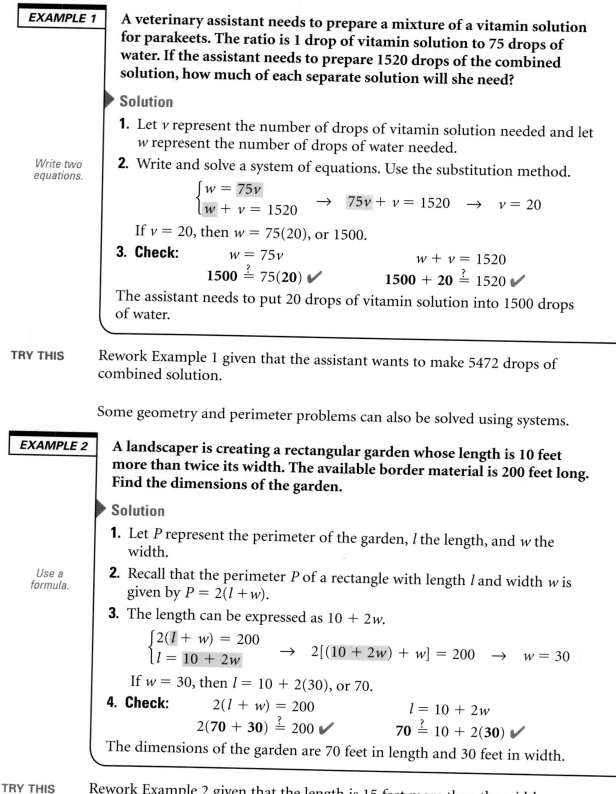

EXAMPLE 1

A veterinary assistant needs to prepare a mixture of a vitamin solution for parakeets. The ratio is 1 drop of vitamin solution to 75 drops of water. If the assistant needs to prepare 1520 drops of the combined solution, how much of each separate solution will she need?

▶ **Solution**

Write two equations.

1. Let v represent the number of drops of vitamin solution needed and let w represent the number of drops of water needed.

2. Write and solve a system of equations. Use the substitution method.

$$\begin{cases} w = 75v \\ w + v = 1520 \end{cases} \rightarrow 75v + v = 1520 \rightarrow v = 20$$

If $v = 20$, then $w = 75(20)$, or 1500.

3. **Check:**

$$w = 75v \qquad\qquad w + v = 1520$$
$$1500 \overset{?}{=} 75(20) ✔ \qquad 1500 + 20 \overset{?}{=} 1520 ✔$$

The assistant needs to put 20 drops of vitamin solution into 1500 drops of water.

TRY THIS

Rework Example 1 given that the assistant wants to make 5472 drops of combined solution.

Some geometry and perimeter problems can also be solved using systems.

EXAMPLE 2

A landscaper is creating a rectangular garden whose length is 10 feet more than twice its width. The available border material is 200 feet long. Find the dimensions of the garden.

▶ **Solution**

1. Let P represent the perimeter of the garden, l the length, and w the width.

Use a formula.

2. Recall that the perimeter P of a rectangle with length l and width w is given by $P = 2(l + w)$.

3. The length can be expressed as $10 + 2w$.

$$\begin{cases} 2(l + w) = 200 \\ l = 10 + 2w \end{cases} \rightarrow 2[(10 + 2w) + w] = 200 \rightarrow w = 30$$

If $w = 30$, then $l = 10 + 2(30)$, or 70.

4. **Check:**

$$2(l + w) = 200 \qquad\qquad l = 10 + 2w$$
$$2(70 + 30) \overset{?}{=} 200 ✔ \qquad 70 \overset{?}{=} 10 + 2(30) ✔$$

The dimensions of the garden are 70 feet in length and 30 feet in width.

TRY THIS

Rework Example 2 given that the length is 15 feet more than the width and there are 450 feet of border material available.

EXERCISES

Refer to Example 1. Write a system of equations to describe each situation.

1. The ratio is 1 drop of vitamin solution to 75 drops of water. The assistant wants a total of 1824 drops of the combined solution.

2. The ratio is 1 drop of vitamin solution to 82 drops of water. The assistant wants a total of 1520 drops of the combined solution.

PRACTICE

Use a system of equations to solve each problem. Check your solution.

3. Rework Example 1 given that the assistant wants to make 5396 drops of a solution that is made by adding 1 drop of vitamin solution to 70 drops of water.

4. Rework Example 2 that the width is 20 feet shorter than three times length and there are 1200 feet of border material available.

5. April sold 75 tickets to a school play and collected a total of $495. If adult tickets cost $8 each and child tickets cost $5 each, how many adult tickets and how many child tickets did she sell?

6. In the last basketball game, George scored a total of 16 points on 11 baskets. He made only 2-point field goals and 1-point free throws. How many field goals and how many free throws did George make?

7. If two solutions are mixed in the ratio of 1.5 to 1 and together they make 600 milliliters of a new solution, how many milliliters of each solution were used?

8. Two bags of fertilizer and 3 bags of peat moss together weigh 140 pounds. If one bag of fertilizer weighs the same as 2 bags of peat moss, how much does each bag weigh?

MIXED REVIEW APPLICATIONS

Solve each problem. Check your solution. (4.5 Skill C)

9. A chemist has two solutions, A and B, that are to be mixed to make a third solution, C, that is no more than 18% salt. Solution A is 12% salt and solution B is 20% salt. How much of each solution is needed to make 250 milliliters of solution C?

10. Mr. Lyons wants to add money to his account, which currently contains $8000, so that two years from now he will have at least $10,000 in the account. If the bank pays 5% simple interest annually, how much should he add?

Solving Systems of Equations by Elimination

LESSON 5.3

The system at right could be solved by substitution, but isolating one of the variables would involve several steps. Using the *elimination method* shortens the process.

$$\begin{cases} 2x + 3y = 4 \\ 3x - 3y = 11 \end{cases}$$

EXAMPLE 1 Solve $\begin{cases} 2x + 3y = 4 \\ 3x - 3y = 11 \end{cases}$ using the elimination method.

▶ **Solution**

Notice that the two coefficients of y are opposites.

1. Combine the two equations. The variable y is eliminated. Solve for x.

$$\begin{array}{r} 2x + 3y = 4 \\ 3x - 3y = 11 \\ \hline 5x + 0 \quad 15 \end{array} \rightarrow 5x = 15 \rightarrow x = 3$$

2. Replace x with 3 in either original equation. Solve for y.

$$3(3) - 3y = 11 \rightarrow y = -\frac{2}{3}$$

The solution to the system is $\left(3, -\frac{2}{3}\right)$.

TRY THIS Solve using elimination.

a. $\begin{cases} -2x + 2y = 2 \\ 5x - 2y = 4 \end{cases}$

b. $\begin{cases} 3x + 5y = 20 \\ -3x + 7y = 16 \end{cases}$

If opposite terms do not already exist within a system, you can apply the Multiplication Property of Equality to the original equation(s) to create them.

EXAMPLE 2 Solve $\begin{cases} 2x + 5y = 14 \\ 6x + 7y = 10 \end{cases}$ using the elimination method.

▶ **Solution**

1. Multiply each side of the first equation by -3.

$$\begin{cases} (-3)(2x + 5y) = (-3)14 \\ 6x + 7y = 10 \end{cases} \rightarrow \begin{cases} -6x - 15y = -42 \\ 6x + 7y = 10 \end{cases}$$

2. Combine the two equations so that x is eliminated.

$$\begin{array}{r} -6x - 15y = -42 \\ 6x + 7y = 10 \\ \hline 0 - 8y = -32 \end{array}$$

3. Solve for y.

$$y = 4$$

4. Replace y with 4 in either original equation. Solve for x.

$$2x + 5(4) = 14 \rightarrow x = -3$$

The solution to the system is $(-3, 4)$.

TRY THIS Solve using elimination.

a. $\begin{cases} -8x + 3y = 34 \\ 4x + 7y = -34 \end{cases}$

b. $\begin{cases} 3x + 4y = 10 \\ 5x + 12y = 38 \end{cases}$

EXERCISES

Explain how you would eliminate the specified variable. Do not solve the system.

1. $\begin{cases} -6x + 5y = 4 \\ 2x - 4y = 8 \end{cases}$; eliminate x

2. $\begin{cases} -3a + b = 4 \\ 2a + 3b = -3 \end{cases}$; eliminate b

3. $\begin{cases} 2r + 6s = 0 \\ 2r - 4s = -3 \end{cases}$; eliminate r

PRACTICE

Use elimination to solve each system of equations.

4. $\begin{cases} 2x - 5y = 1 \\ -2x - 3y = -9 \end{cases}$

5. $\begin{cases} 4m + 5n = 1 \\ -7m - 5n = 2 \end{cases}$

6. $\begin{cases} -4u + 3v = 7 \\ 4u + 3v = 23 \end{cases}$

7. $\begin{cases} 7p - 2q = 6 \\ 7p + 2q = 8 \end{cases}$

8. $\begin{cases} 2x + 6y = 0 \\ 2x - 4y = -3 \end{cases}$

9. $\begin{cases} 5c - 6d = 18 \\ 4c + 3d = 17 \end{cases}$

10. $\begin{cases} -3c - 8d = 4 \\ 6c - 2d = 1 \end{cases}$

11. $\begin{cases} -3c - 8d = 0 \\ 7c + 2d = 0 \end{cases}$

12. $\begin{cases} 5m - 8n = 0 \\ -10m + 7n = 0 \end{cases}$

13. $\begin{cases} \frac{1}{4}x + \frac{1}{3}y = 3 \\ \frac{1}{2}x - \frac{1}{6}y = 1 \end{cases}$

14. $\begin{cases} \frac{c}{3} + \frac{d}{3} = 5 \\ \frac{c}{2} - \frac{d}{9} = 2 \end{cases}$

15. $\begin{cases} \frac{c}{2} - \frac{d}{3} = 1 \\ \frac{c}{4} + \frac{d}{5} = 0 \end{cases}$

In Exercises 16–18, let a, b, and c represent real numbers. Solve each system for x and y.

16. $\begin{cases} 3ax - 3y = 1 \\ ax + 4y = 6 \end{cases}$, $a \neq 0$

17. $\begin{cases} 3x + 4by = 2 \\ 5x + by = 7 \end{cases}$, $b \neq 0$

18. $\begin{cases} 3x + 4y = c \\ 6x + 5y = c \end{cases}$

19. **Critical Thinking** Show that the system $\begin{cases} 3x + 2y = -c \\ 6x + 4y = 1 - 2c \end{cases}$

   has no solution for any value of c.

MIXED REVIEW

Evaluate each expression. (1.2 Skill C)

20. $ad - bc$, given $a = 3$, $b = -3$, $c = 5$, and $d = 6$

21. $ad - bc$, given $a = -4$, $b = 14$, $c = -2$, and $d = 7$

Graph each equation. (3.5 Skill B)

22. $-2x + 3y = 9$

23. $3x + 5y = -10$

24. $2x - 4y = 8$

To solve a system of equations, you may need to choose a multiplier for each equation in the system.

EXAMPLE 1   Solve $\begin{cases} 2a - 5b = -9 \\ -3a + 2b = 8 \end{cases}$ using the elimination method.

▶ **Solution**

1. Multiply each side of the first equation by 3.
   Multiply each side of the second equation by 2.

$$\begin{cases} 2a - 5b = -9 \\ -3a + 2b = 8 \end{cases} \rightarrow \begin{cases} 3(2a - 5b) = 3(-9) \\ 2(-3a + 2b) = 2(8) \end{cases} \rightarrow \begin{cases} 6a - 15b = -27 \\ -6a + 4b = 16 \end{cases}$$

2. Add the like terms of the two equations to eliminate a. Solve for b.
$$-11b = -11 \rightarrow b = 1$$

3. Replace b with 1 in either original equation. Solve for a.
   If $b = 1$, then $2a - 5(1) = -9$. So, $a = -2$.

4. **Check:** Check using the other original equation. $-3(-2) + 2(1) = 8$ ✔

The solution to the system is $(-2, 1)$.

TRY THIS   Solve $\begin{cases} 6r + 8s = 4 \\ 9r + 10s = 7 \end{cases}$ using the elimination method.

Before adding the like terms of the equations, be sure that the corresponding variables are aligned.

EXAMPLE 2   Solve $\begin{cases} 2c + 3d = -10 \\ -2d + 3c = -2 \end{cases}$ using the elimination method.

▶ **Solution**

1. Align c terms and d terms. $\begin{cases} 2c + 3d = -10 \\ -2d + 3c = -2 \end{cases} \rightarrow \begin{cases} 2c + 3d = -10 \\ 3c - 2d = -2 \end{cases}$

2. Choose appropriate multipliers.

$$\begin{cases} 2c + 3d = -10 \\ 3c - 2d = -2 \end{cases} \rightarrow \begin{cases} 3(2c + 3d) = 3(-10) \\ -2(3c - 2d) = -2(-2) \end{cases} \rightarrow \begin{cases} 6c + 9d = -30 \\ -6c + 4d = 4 \end{cases}$$

3. Combine the two equations so that c is eliminated. Solve for d.
$$13d = -26 \rightarrow d = -2$$

   Replace d with -2 in either original equation. Solve for c.
   If $d = -2$, then $2c + 3(-2) = -10$. So, $c = -2$.

4. **Check:** $-2(-2) + 3(-2) \overset{?}{=} -2$ ✔

The solution to the system is $(-2, -2)$.

TRY THIS   Solve $\begin{cases} 4y = 11 - 4x \\ 3x - 5 = -5y \end{cases}$ using the elimination method.

EXERCISES

KEY SKILLS

Explain how you would eliminate the specified variable. Do not solve the system.

1. $\begin{cases} -3x + 2y = 4 \\ 2x + 4y = -3 \end{cases}$; eliminate x
2. $\begin{cases} -3a + 5b = -2 \\ 5a + 3b = -1 \end{cases}$; eliminate b
3. $\begin{cases} -7r - 6s = 10 \\ 2r - 5s - 7 \end{cases}$; eliminate r

PRACTICE

Use the elimination method to solve each system of equations.

4. $\begin{cases} 2x - 3y = -10 \\ 3x + 2y = -2 \end{cases}$
5. $\begin{cases} -3x + 7y = -14 \\ -4x - 2y = 4 \end{cases}$
6. $\begin{cases} 4c - 2d = 0 \\ 3c - 3d = 0 \end{cases}$

7. $\begin{cases} 5a - 2b = 0 \\ -4a - 5b = 0 \end{cases}$
8. $\begin{cases} 4f - 7g = 1 \\ -3f + 5g = 0 \end{cases}$
9. $\begin{cases} x - 3y = 17 \\ 3x + 2y = 18 \end{cases}$

10. $\begin{cases} 4a - 3b = 7 \\ 5b - 4a = -1 \end{cases}$
11. $\begin{cases} -2r + 2s = 4 \\ 5s - 7r = 0 \end{cases}$
12. $\begin{cases} -2h + 2n = 17 \\ 3n + 3h = 58 \end{cases}$

13. $\begin{cases} 4m - 7n = 37 \\ 2n + 3m = 6 \end{cases}$
14. $\begin{cases} -5a = 18 - 4z \\ -3z + 2a = -16 \end{cases}$
15. $\begin{cases} 5q - 8p = 7 \\ 4p = 7q - 17 \end{cases}$

16. $\begin{cases} \dfrac{x}{3} - \dfrac{y}{4} = 1 \\ \dfrac{x}{2} + \dfrac{y}{3} = 1 \end{cases}$
17. $\begin{cases} \dfrac{a}{5} - \dfrac{c}{3} = 1 \\ \dfrac{a}{4} - \dfrac{c}{5} = 1 \end{cases}$
18. $\begin{cases} \dfrac{m}{7} + \dfrac{n}{2} = 1 \\ \dfrac{m}{6} + \dfrac{n}{3} = 1 \end{cases}$

19. **Critical Thinking** Let m and n represent fixed real numbers.

Solve $\begin{cases} 2x + my = 1 \\ 3x + ny = 1 \end{cases}$ for x and y. State restrictions on variables.

MIXED REVIEW

Are the given lines parallel? Justify your response. (3.8 Skill A)

20. $3x - 11y = 4$ and $-5x + 7y = 21$
21. $-4x - y = 14$ and $-8x - 2y = 0$

Are the given lines perpendicular? Justify your response. (3.8 Skill B)

22. $2.5x - y = 13$ and $-0.8x - 2y = 34$
23. $4x + 7y = -2$ and $7x + 4y = 3$

Making a table can be helpful when you need to set up a system of equations for a problem.

Also, it is often helpful to change the fractions or decimals in an equation to integers. To do this, multiply both sides of the equation by a common denominator.

EXAMPLE 1

A chemist has two solutions, A and B, that are to be mixed to make a third solution, C. The data is shown at right. How many whole milliliters of solutions A and B are needed to make solution C?

solution A: 10% sugar
solution B: 25% sugar
solution C: 500 mL of
20% sugar

▶ **Solution**

1. Let a and b represent the numbers of milliliters of solution A and B, respectively, that are needed. Organize the data in a table.

Make a table.

	A	+ B	= C
amount of solution	a	b	500
amount of sugar	0.10a	0.25b	(0.20)500

2. Write and solve a system of equations. Multiply the second equation by 20 so that it contains only integers. Multiply the first equation by -2.

$$\begin{cases} a + b = 500 \\ 0.10a + 0.25b = 0.20(500) \end{cases} \rightarrow \begin{cases} a + b = 500 \\ 2a + 5b = 2000 \end{cases} \rightarrow \begin{cases} -2a - 2b = -1000 \\ 2a + 5b = 2000 \end{cases}$$

Combine the equations to get $3b = 1000$. Thus, $b = 333\frac{1}{3}$, and then $a = 166\frac{2}{3}$.

The problem asks that the answers be rounded to whole milliliters. So, the chemist needs 167 mL of solution A and 333 mL of solution B.

TRY THIS

Rework Example 1 given that solution A is 12% sugar, solution B is 24% sugar, and 480 milliliters of solution C is 18% sugar.

EXAMPLE 2

Refer to Example 1. Can A and B be mixed to make 500 mL of a new solution, D, that is 5% sugar? Explain.

▶ **Solution**

The system is a modification of the system in Example 1.

$$\begin{cases} a + b = 500 \\ 0.10a + 0.25b = 0.05(500) \end{cases} \rightarrow \begin{matrix} a + b = 500 \\ 2a + 5b = 500 \end{matrix} \rightarrow 3b = -500 \rightarrow b = -\frac{500}{3}$$

Because a and b represent physical amounts, neither may be negative. So, the mixture cannot be made.

TRY THIS

Solution A is 10% sugar and solution B is 25% sugar. Can A and B be mixed to create 500 mL of a new solution, C, that is 2% sugar? Explain.

EXERCISES

KEY SKILLS

Refer to Example 1.

1. Represent the data below in a table.
   solution A: 18% sugar   solution B: 26% sugar
           solution C: 350 milliliters of 20% sugar

2. Write $0.18a + 0.26b = (0.20)(350)$ as an equation with only integer coefficients.

PRACTICE

Solve each problem using a system of equations.

3. Rework Example 1 with the data below.

   solution A: 5% sugar   solution B: 10% sugar

   solution C: 450 milliliters of 8% sugar

4. A shopkeeper wants to make 5 pounds of a candy mix from peppermint candy and lemon-lime candy. How many pounds of each candy should be put into the mix? Use the data below.
   peppermint: $0.75 per pound   lemon-lime: $0.50 per pound
                    mix: sells for $0.65 per pound

5. A motorist drives for 8 hours and travels 390 miles. For some of that time, the speed is 55 miles per hour and for the rest of that time the speed is 45 miles per hour. How long does the motorist drive at each speed?

6. Is it possible to make a 600 milliliter solution that is 32% salt from a solution that is 30% salt and another solution that is 25% salt? Justify your response.

MID-CHAPTER REVIEW

7. Use a graph to solve the system of equations at right. (5.1 Skill A) $\begin{cases} y = x - 5 \\ y + x = 3 \end{cases}$

Solve each system using the substitution method. (5.2 Skills A and B)

8. $\begin{cases} y = 2x \\ -3x + 2y = 2 \end{cases}$

9. $\begin{cases} x + y = -8 \\ -3x + 2y = 9 \end{cases}$

Solve each system using the elimination method. (5.3 Skills A and B)

10. $\begin{cases} x - 2y = -3 \\ 3x + 2y = 7 \end{cases}$

11. $\begin{cases} 2x + 3y = 5 \\ 3x + 2y = 5 \end{cases}$

5.4

Classifying Systems of Equations

SKILL A   *Classifying a system of equations as consistent or inconsistent*

When a system of equations has at least one solution, the system is called **consistent**. When a system has no solution, it is called **inconsistent**.

EXAMPLE 1   Is $\begin{cases} -3x - y = -2 \\ 7x + 2y = -10 \end{cases}$ a consistent or inconsistent system? Explain.

▶ **Solution**

You can solve this system by graphing, substitution, or elimination. Here we solve using the elimination method.

$$\begin{cases} (2)(-3x - y) = (2)(-2) \\ 7x + 2y = -10 \end{cases} \rightarrow \begin{cases} -6x - 2y = -4 \\ 7x + 2y = -10 \end{cases} \rightarrow x = -14$$

If $x = -14$, then $y = 44$. The system has a unique solution, $(-14, 44)$. So the lines intersect at one point, $(-14, 44)$, and the system is consistent.

TRY THIS   Is $\begin{cases} -16x + 4y = -8 \\ 12x - 3y = 6 \end{cases}$ a consistent or inconsistent system? Explain.

EXAMPLE 2   Is $\begin{cases} 5x + 10y = -2 \\ 2x + 4y = 10 \end{cases}$ a consistent or inconsistent system? Explain.

▶ **Solution**

Rewrite each equation in slope-intercept form.

$$\begin{cases} 5x + 10y = -2 \\ 2x + 4y = 10 \end{cases} \rightarrow \begin{cases} 10y = -2 - 5x \\ 4y = 10 - 2x \end{cases} \rightarrow \begin{cases} y = -\dfrac{1}{2}x - \dfrac{1}{5} \\ y = -\dfrac{1}{2}x + \dfrac{5}{2} \end{cases}$$

The slopes are equal but the y-intercepts, $-\dfrac{1}{5}$ and $\dfrac{5}{2}$, are different.

So, the lines must be parallel. Because parallel lines do not intersect, the system has no solution and it is inconsistent.

TRY THIS   Is $\begin{cases} -3x - y = -2 \\ 6x + 2y = 10 \end{cases}$ a consistent or inconsistent system? Explain.

EXERCISES

KEY SKILLS

Write the equations in each system in slope-intercept form.

1. $\begin{cases} x + y = 4 \\ x + 2y = 6 \end{cases}$

2. $\begin{cases} 3x - y = 2 \\ 6x - 2y = 8 \end{cases}$

3. $\begin{cases} -7x - 5y = -3 \\ y - 4x = -12 \end{cases}$

PRACTICE

Identify each system as consistent or inconsistent.

4. $\begin{cases} 3x - y = -5 \\ 9x - 3y = 1 \end{cases}$

5. $\begin{cases} 3u - v = 7 \\ -2u - 5v = 2 \end{cases}$

6. $\begin{cases} 6r - 2s = 6 \\ 2r + 5s = 0 \end{cases}$

7. $\begin{cases} 4u - 14v = -3 \\ -2u + 7v = 3 \end{cases}$

8. $\begin{cases} 4p - 4q = -5 \\ -4q + 7p = 3 \end{cases}$

9. $\begin{cases} -5g - 35h = 15 \\ -h + 7g = -12 \end{cases}$

Does each system have a unique solution? If it does, find it.

10. $\begin{cases} 2x + 3y = 6 \\ 2x - 3y = 6 \end{cases}$

11. $\begin{cases} 2x - y = -5 \\ 6x - 3y = 1 \end{cases}$

12. $\begin{cases} 2x - y = 3 \\ -2x + y = 2 \end{cases}$

13. $\begin{cases} -3m + 4n = 8 \\ -4m + 6n = 9 \end{cases}$

14. $\begin{cases} 5m - 5n = 8 \\ -4m + 4n = 9 \end{cases}$

15. $\begin{cases} 7p - 2q = -3 \\ -9p + 3q = 5 \end{cases}$

16. Consider $\begin{cases} ax - 2y = -2 \\ -5x + 3y = 7 \end{cases}$. Find a such that the system has no solution.

17. Consider $\begin{cases} 2x - 7y = 12 \\ -3x + dy = 11 \end{cases}$. Find d such that the system has exactly one solution.

18. **Critical Thinking** Let a, b, c, and d be real numbers.
    Solve $\begin{cases} ax + by = e \\ cx + dy = f \end{cases}$ for x. How does $ad - bc$ determine whether there
    is a solution for x?

MIXED REVIEW

Use intercepts to graph each equation. (3.5 Skill A)

19. $4x - 5y = 20$

20. $-x + 5y = 5$

21. $2x - 7y = 7$

Use the slope and y-intercept to graph each equation. (3.5 Skill B)

22. $y = -\frac{3}{4}x + 1$

23. $y = \frac{5}{4}x - 3$

24. $y = \frac{1}{5}x + 2.5$

In Skill A, you learned that a system of equations with at least one solution is called consistent. Consistent systems can be further classified as **independent** and **dependent**. An independent system has exactly one solution. A dependent system has infinitely many solutions. All of the equations in a dependent system simplify to the same equation. Therefore, the lines of the equations coincide at every point and are considered to be the same line.

EXAMPLE 1 Classify the system $\begin{cases} y = 4 - x \\ 3x - y = 8 \end{cases}$ as specifically as possible.

▶ **Solution**

Solve the system using the substitution method by substituting $4 - x$ for y in the second equation.

$$\begin{cases} y = 4 - x \\ 3x - y = 8 \end{cases} \rightarrow 3x - (4 - x) = 8 \rightarrow x = 3$$

If $x = 3$, then $y = 4 - 3$, or 1. The solution is $(3, 1)$. Because the solution is one point, $(3, 1)$, the system is consistent and independent.

TRY THIS Classify the system $\begin{cases} -4x + 4y = 4 \\ 4x - y = -4 \end{cases}$ as specifically as possible.

EXAMPLE 2 Classify the system $\begin{cases} 6u + 10v = -2 \\ 3u + 5v = -1 \end{cases}$ as specifically as possible.

▶ **Solution**

Multiply the second equation by -2 and then combine the equations.

Use the elimination method.

$$\begin{cases} 6u + 10v = -2 \\ (-2)(3u + 5v) = (-2)(-1) \end{cases} \rightarrow \begin{cases} 6u + 10v = -2 \\ -6u + -10v = 2 \end{cases} \rightarrow 0 = 0 \ ✔ \text{ True}$$

The result is a statement that is true for any values of u or v, so all of the solutions of one equation are also solutions of the other equation. The equations describe the same line, so the system is consistent and dependent.

TRY THIS Classify the system $\begin{cases} 7r + 3s = 4 \\ 14r + 6s = 10 \end{cases}$ as specifically as possible.

If the result had been a contradictory statement such as $1 = 0$, then the system would have no solution and thus would be inconsistent.

Type of system	Solutions	Slopes	y-intercepts	Graphs
Consistent and independent	one	different	either same or different	
Consistent and dependent	infinitely many	same	same	
Inconsistent and independent	none	same	different	

EXERCISES

KEY SKILLS

Graph a pair of equations that satisfies the given condition.

1. There is no solution.

2. There are infinitely many solutions.

PRACTICE

Classify each system as specifically as possible.

3. $\begin{cases} x = 7 - 2y \\ 3x - 2y = -11 \end{cases}$

4. $\begin{cases} x = 14 - 5y \\ 2x - y = -5 \end{cases}$

5. $\begin{cases} 2x - y = 3 \\ 5x - 2y = 10 \end{cases}$

6. $\begin{cases} 3x + y = -3 \\ 2x - y = -7 \end{cases}$

7. $\begin{cases} x + y = 2 \\ 2x + 2y = 4 \end{cases}$

8. $\begin{cases} y - 2x = 7 \\ 2y - 4x = 14 \end{cases}$

9. $\begin{cases} 4x - y = 19 \\ 2x + 3y = -1 \end{cases}$

10. $\begin{cases} 5x + 2y = -7 \\ x + 3y = 9 \end{cases}$

11. $\begin{cases} -3.5m - 3n = -2 \\ 7m + 6n = 5 \end{cases}$

12. $\begin{cases} -28m - 31n = -20 \\ 56m + 62n = 40 \end{cases}$

13. $\begin{cases} 24x + 31y = 50 \\ 72x + 93y = 150 \end{cases}$

14. $\begin{cases} 75a - 33b = -120 \\ 25a - 11b = -50 \end{cases}$

15. $\begin{cases} -3x + 2y = 6 \\ -x + \frac{2}{3}y = 2 \end{cases}$

16. $\begin{cases} \frac{5}{2}u + 5v = 0 \\ \frac{2}{3}u + \frac{4}{3}v = -\frac{5}{3} \end{cases}$

17. $\begin{cases} \frac{2x}{4} - \frac{2y}{7} = -1 \\ \frac{3x}{4} - \frac{3y}{7} = 4 \end{cases}$

Critical Thinking Consider $\begin{cases} 5x + by = e \\ 3x + 2y = 10 \end{cases}$. Find b and e such that the system has:

18. no solution.

19. infinitely many solutions.

Critical Thinking Consider $\begin{cases} 5x + 4y = -2 \\ cx + 2y = f \end{cases}$. Find c and f such that:

20. the graphs coincide.

21. the graphs are distinct and parallel.

MIXED REVIEW

Solve using the substitution method. (5.2 Skill B)

22. $\begin{cases} 2x - y = -7 \\ 3x + y = -3 \end{cases}$

23. $\begin{cases} y = \frac{2}{3}x - 4 \\ 5y - x = 1 \end{cases}$

Solve using the elimination method. (5.3 Skill B)

24. $\begin{cases} 7x - 5y = 10 \\ 3x - 2y = 6 \end{cases}$

25. $\begin{cases} 7r - 2s = 9 \\ 2r + 3s = -1 \end{cases}$

Graphing Linear Inequalities in Two Variables

SKILL A ▷ *Graphing linear inequalities*

When the equal symbol in a linear equation is replaced with an inequality symbol, a *linear inequality* is formed. A **linear inequality in two variables**, x and y, is any inequality that can be written in one of the forms below.

$$y < mx + b \qquad\qquad y \leq mx + b$$

$$y > mx + b \qquad\qquad y \geq mx + b$$

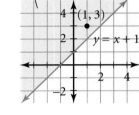

A solution to a linear inequality in two variables is any ordered pair that makes the inequality true. For example, $(1, 3)$ is a solution to $y \geq x + 1$ because $3 \geq 1 + 1$. The set of all possible solutions of a linear inequality is called the **solution region.**

Graphing a linear inequality in x and y

Step 1. Isolate y on one side of the inequality.

Step 2. Substitute an equal symbol ($=$) for the inequality symbol in the given inequality. Graph the resulting linear equation to form the **boundary line**. For $\leq$ or $\geq$, use a solid line. For $<$ or $>$, use a dashed line. The dashed line shows that the points on the line are not included in the solution region.

Step 3. Shade the region that contains the solutions of the inequality. For $y >$, shade above the line. For $y <$, shade below the line. (Note: To graph one-variable inequalities on a coordinate plane, $x < a$ or $x > a$, the boundary is a vertical line and the shading is on the left or right, respectively.)

EXAMPLE

Graph the solution region of $x - 2y < -6$.

▷ **Solution**

Recall that when you divide an inequality by a negative number you need to reverse the inequality sign.

1. $x - 2y < -6$ ⟵ Write y in terms of x.
   $-2y < -x - 6$

   $y > \dfrac{1}{2}x + 3$ ⟵ Divide each side by -2.

2. Graph $y = \dfrac{1}{2}x + 3$ with a dashed line.

3. Because $y > \dfrac{1}{2}x + 3$, the solution region should lie above the line. Shade this area.

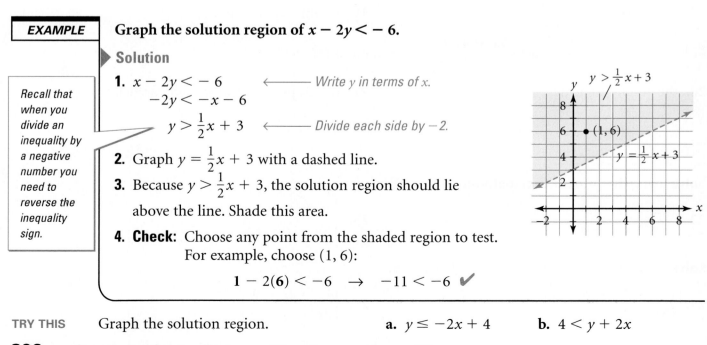

4. **Check:** Choose any point from the shaded region to test. For example, choose $(1, 6)$:

$$1 - 2(6) < -6 \quad \rightarrow \quad -11 < -6 \checkmark$$

TRY THIS   Graph the solution region.   **a.** $y \leq -2x + 4$ **b.** $4 < y + 2x$

206   Chapter 5   Solving Systems of Equations and Inequalities

EXERCISES

Write y in terms of x.

1. $5x + 2y \geq 10$

2. $4x - 3y < 9$

Is the given point above or below the graph of the given equation?

3. $A(0, 0)$; $3x - 2y = 3$

4. $B(1, -3)$; $-2x + 3y = 6$

PRACTICE

Graph the solution region of each inequality on a coordinate plane.

5. $x \geq 0$

6. $x > 4$

7. $y < 2$

8. $y \leq 0$

9. $x + y \geq -3$

10. $y - x \geq -5$

11. $3x + y < 0$

12. $-2x + y > 0$

13. $x + 3y > 2$

14. $x + 3y > -4$

15. $2x - y < 2$

16. $3x - y > 4$

17. $2x + 5y < 10$

18. $-5x + 4y \geq 12$

19. $2x + 4y \leq 0$

20. $3x - 4y \geq 0$

21. $-7x + 2y < 14$

22. $5x + 3y \geq 15$

Write a linear inequality in x and y that meets the given conditions.

23. The boundary contains $(2, 3)$ and $(5, 7)$ and the graph of the solution region contains $(3, 8)$.

24. The boundary contains $(-2, 3)$ and $(6, 9)$ and the graph of the solution region contains $(-2, -4)$ but does not contain the boundary.

Let $y > mx + b$, where m and b are real numbers and $m \neq 0$.
(*Hint:* If the boundary contains the origin, then $(0, 0)$ is a solution.)

25. Find m and b such that the boundary contains the origin and the solution region is the half-plane that contains $(1, 5)$.

26. Find m and b such that the boundary contains the origin and the solution region is the half-plane that contains $(-2, 7)$.

MIXED REVIEW

Solve each compound inequality. Graph the solution on a number line.
(4.6 Skill A)

27. $2x + 5 > 10 - 3x$ *and* $x - 3 < 9 - 3x$

28. $-2(a + 1) < -8$ *and* $-a + 4 \leq 16 - 3a$

Solve each compound inequality. Graph the solution on a number line.
(4.6 Skill B)

29. $y + 5 > 9$ *or* $3y + 4 < -8 - 3y$

30. $3(y + 5) \leq 9$ *or* $3(y - 4) \geq -3y$

When restrictions are placed on one or both variables, the graph of a solution will not necessarily be a shaded region as in Lesson 5.5 Skill A.

EXAMPLE 1

Given that x and y are whole numbers, graph all solutions to $x + 2y \leq 8$.

▶ **Solution**

Isolate y in the given inequality: $y \leq -\frac{1}{2}x + 4$.

Make an organized list.

Examine possible x-values.

If $x = 0$, then $y \leq 4$. So, $y = 0, 1, 2, 3$, or 4.

If $x = 1$, then $y \leq 3\frac{1}{2}$. So, $y = 0, 1, 2$, or 3.

If $x = 2$, then $y \leq 3$. So, $y = 0, 1, 2$, or 3.

If $x = 3$, then $y \leq 2\frac{1}{2}$. So, $y = 0, 1$, or 2.

⋮

If $x = 7$, then $y \leq 0.5$. So, $y = 0$.

If $x = 8$, then $y \leq 0$. So, $y = 0$.

If $x = 9$, $y \leq -\frac{1}{2}$. ✘ There are no whole numbers less than or equal to $-\frac{1}{2}$, so x cannot equal 9.

The graph is shown at right.

TRY THIS   Given that x and y are whole numbers, graph all solutions to $2x + y \leq 10$.

In many problems, there are natural restrictions on variables. For example, if variables represent distances or measurements, their values must be nonnegative. The graphs of many real-world equations are often restricted to Quadrant I because it is the only quadrant that contains nonnegative values for both variables.

EXAMPLE 2

Mickie and Michelle may use up to 120 feet of fencing to enclose a rectangular space. Draw a graph to represent the possible lengths and widths of the rectangular space.

▶ **Solution**

Use a formula.

1. The perimeter of a rectangle, P, is given by the formula $P = 2(l + w)$, where l represents length and w represents width.

2. Write and solve an inequality in l and w.
$$2(l + w) \leq 120, \text{ or } l + w \leq 60$$

3. Graph $l + w = 60$ using the intercepts: $(0, 60)$ and $(60, 0)$. Because the inequality is *less than or equal to*, shade *below* the line. Since length and width must be nonnegative, shade only in Quadrant I.

TRY THIS   Rework Example 2 given that Mickie and Michelle may use up to 150 feet of fencing.

EXERCISES

KEY SKILLS

Given that *x* and *y* are whole numbers, graph all solutions to each equation below.

1. $x + y \leq 5$ 2. $x + y < 7$ 3. $y < x - 3$

4. Sketch the triangular region in the plane determined by $A(0, 0)$, $B(4, 0)$, and $C(0, 6)$ as the vertices of the triangle.

PRACTICE

Graph the solution of each inequality in two variables.

5. Graphically represent all pairs of whole numbers whose sum is less than 6.

6. One whole number is twice another. Represent all such pairs of numbers whose sum is less than or equal to 12.

7. Suppose that you have 5 blue chips and 5 red chips. Graphically represent the possibilities for a collection of blue chips and red chips if the collection contains less than 7 chips altogether.

8. Rework Example 2 given that 180 feet of fencing is available.

9. Graphically represent all lengths and widths of rectangles whose perimeter is less than 140 units. For which rectangles will the length be twice the width?

10. Graphically represent all pairs of whole numbers that are each greater than or equal to 4 and whose sum is less than 12.

11. Use a graph to show there are no pairs of whole numbers that are each greater than or equal to 5 and whose sum is less than 9.

MIXED REVIEW APPLICATIONS

How many milliliters of solution A and of solution B are needed to make solution C? (5.3 Skill C)

12. solution A: 20% salt
    solution B: 25% salt
    solution C: 200 milliliters of 24% salt

13. solution A: 5% salt
    solution B: 15% salt
    solution C: 300 milliliters of 8% salt

Graphing Systems of Linear Inequalities in Two Variables

LESSON

Graphing systems of linear inequalities

A **system of linear inequalities in two variables** is a set of two or more linear inequalities in those variables. A solution to a system of such inequalities is any ordered pair that makes all of the inequalities in the system true. The graph of the solution of a system is the intersection or common region of the combined graphs of the inequalities.

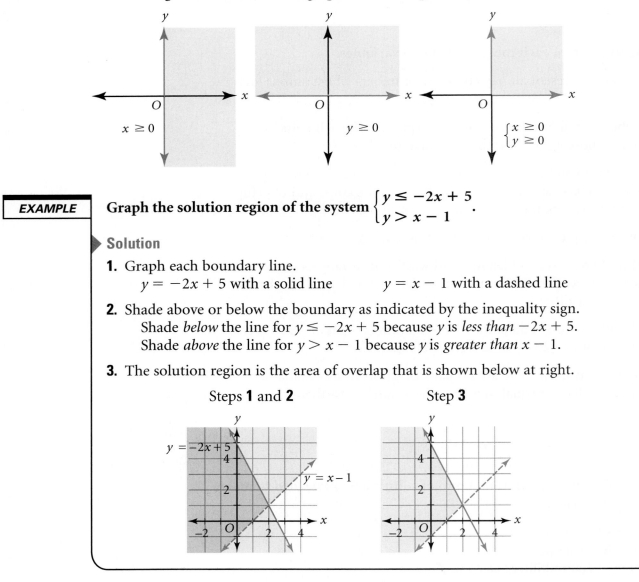

EXAMPLE **Graph the solution region of the system** $\begin{cases} y \le -2x + 5 \\ y > x - 1 \end{cases}$.

▶ **Solution**

1. Graph each boundary line.
$y = -2x + 5$ with a solid line $y = x - 1$ with a dashed line

2. Shade above or below the boundary as indicated by the inequality sign.
Shade *below* the line for $y \le -2x + 5$ because y is *less than* $-2x + 5$.
Shade *above* the line for $y > x - 1$ because y is *greater than* $x - 1$.

3. The solution region is the area of overlap that is shown below at right.

Steps **1** and **2**   Step **3**

TRY THIS Graph the solution region of each system. **a.** $\begin{cases} y \ge -\frac{1}{2}x + 1 \\ y \ge -2x - 1 \end{cases}$ **b.** $\begin{cases} y < -\frac{1}{2}x + 1 \\ y > -2x - 1 \end{cases}$

EXERCISES

Shade the described region.

1. Below the graph of $x + y = 8$ and above the graph of $x + y = 5$

2. Above the graph of $y = \frac{2}{3}x$ and below the graph of $y = 4$

3. What part of the coordinate plane is determined by $\begin{cases} x < 0 \\ y > 0 \end{cases}$?

PRACTICE

Graph the solution region of each system of inequalities.

4. $\begin{cases} y \le 2x \\ x \ge 0 \end{cases}$

5. $\begin{cases} y < -x \\ x \ge 0 \end{cases}$

6. $\begin{cases} y < 2 \\ y > 4x - 1 \end{cases}$

7. $\begin{cases} y \le -3x + 2 \\ y < -2 \end{cases}$

8. $\begin{cases} y \le 2x - 3 \\ x > 0 \end{cases}$

9. $\begin{cases} y > -2x + 1 \\ x < 0 \end{cases}$

10. $\begin{cases} y < x + 2 \\ y \ge x - 2 \end{cases}$

11. $\begin{cases} y < 2x + 1 \\ y \ge 2x - 3 \end{cases}$

12. $\begin{cases} y > -2x + 1 \\ y \le 3x - 1 \end{cases}$

13. $\begin{cases} y \le 3x - 1 \\ y > -2x - 3 \end{cases}$

14. $\begin{cases} y > -2x + 1 \\ y > 3x - 2 \end{cases}$

15. $\begin{cases} y < -2x + 1 \\ y > -2x - 2 \end{cases}$

16. $\begin{cases} y \ge -3x \\ y < 2x + 3 \\ x \ge 0 \\ y \ge 0 \end{cases}$

17. $\begin{cases} y > 2x - 1 \\ y < 3x + 2 \\ x \ge 0 \\ y \ge 0 \end{cases}$

18. $\begin{cases} y > -x + 3 \\ y \le x + 2 \\ x \ge 0 \\ y \ge 0 \end{cases}$

19. **Critical Thinking** Let $y > m_1 x + b_1$ and $y < m_2 x + b_2$, where m_1, b_1, m_2, and b_2 are real numbers. Under what conditions placed on m_1, b_1, m_2, and b_2 will the solution region be an infinite strip with parallel boundaries?

MIXED REVIEW

Classify each system as specifically as possible. (5.4 Skill B)

20. $\begin{cases} -9c + 3d = 10 \\ 12c - 4d = 1 \end{cases}$

21. $\begin{cases} 7p + 7q = -3 \\ 6p + 6q = 5 \end{cases}$

22. $\begin{cases} -3m - 2n = 2 \\ 12m + 8n = -8 \end{cases}$

A **linear inequality in standard form** is any inequality of one of the forms below. In the inequalities below, A and B cannot both be 0.

$$Ax + By < C \qquad Ax + By \leq C \qquad Ax + B > C \qquad Ax + By \geq C$$

If you are given a system of inequalities in standard form, they will need to be rewritten in slope-intercept form before graphing.

EXAMPLE 1   **Graph the solution region of** $\begin{cases} 2x - 5y \leq 0 \\ x + 4y \leq -8 \end{cases}$.

▶ **Solution**

Rewrite each inequality so that y is isolated.

$$\begin{cases} 2x - 5y \leq 0 \\ x + 4y \leq -8 \end{cases} \rightarrow \begin{cases} y \geq \frac{2}{5}x \\ y \leq -\frac{1}{4}x - 2 \end{cases}$$

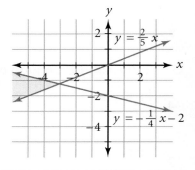

Graph each boundary line.
The solution is the region on or above the graph of $y = \frac{2}{5}x$

and on or below that of $y = -\frac{1}{4}x + 2$.

TRY THIS   Graph the solution region of $\begin{cases} x + y \geq 2 \\ -2x + y \geq -3 \end{cases}$.

A system of inequalities may have no solution or may have a line as its solution. If there is no solution to the system, you will discover this when you graph the system.

EXAMPLE 2   **Graph the solution region of** $\begin{cases} y \leq \frac{2}{5}x + 2 \\ -2x + 5y \geq 20 \end{cases}$.

▶ **Solution**

The inequality $-2x + 5y \geq 20$ can be rewritten as

$y \geq \frac{2}{5}x + 4$.

The graphs of the two inequalities are shown at right. Notice that there is no common region. This means that this system has no solution.

TRY THIS   Graph the solution region.   **a.** $\begin{cases} y \leq \frac{1}{3}x + 2 \\ 3y - 6 \geq x \end{cases}$ **b.** $\begin{cases} y < \frac{1}{3}x + 2 \\ 3y - 9 \geq x \end{cases}$

EXERCISES

KEY SKILLS

Write each system so that y is expressed in terms of x in both inequalities.

1. $\begin{cases} y \geq \frac{2}{5}x \\ 3x - 5y \leq 2 \end{cases}$

2. $\begin{cases} 2x + 7y < 0 \\ -2x - y < 3 \end{cases}$

3. $\begin{cases} -3x + 4y \geq 1 \\ -2x + 3y < 2 \end{cases}$

PRACTICE

Graph the solution region of the system of inequalities.

4. $\begin{cases} y < 2x \\ 2x - y < 3 \end{cases}$

5. $\begin{cases} y \leq 3x \\ 2x + y < 5 \end{cases}$

6. $\begin{cases} x + y \leq 5 \\ x + y > 3 \end{cases}$

7. $\begin{cases} x + y \leq 6 \\ x - y \geq 2 \end{cases}$

8. $\begin{cases} 2x + y \leq -2 \\ x - 2y < 4 \end{cases}$

9. $\begin{cases} x - 3y \geq -3 \\ 2x + y < 4 \end{cases}$

10. $\begin{cases} 3x - 5y < 15 \\ 2x + 3y \leq 6 \end{cases}$

11. $\begin{cases} -5x - 2y \leq 10 \\ 2x + 7y \leq 14 \end{cases}$

12. $\begin{cases} 4x - 5y < 20 \\ 2x - 7y > 14 \end{cases}$

13. Use algebraic reasoning to show that the graph of the solution of $\begin{cases} 2x + 3y \leq -1 \\ 2x + 3y \geq -1 \end{cases}$ is a line.

14. Use algebraic reasoning to show that the solution to the system at right is a single point.

$\begin{cases} y \geq x + 1 \\ y \leq x + 1 \\ y \geq -x + 1 \\ y \leq -x + 1 \end{cases}$

15. **Critical Thinking** How must a and b be related if $\begin{cases} y \geq -2x + a \\ y \leq -2x + b \end{cases}$ has a solution?

16. **Critical Thinking** Use algebraic reasoning to show that the graph of the solution to the system at right is a strip with parallel boundaries for all real numbers a.

$\begin{cases} y \geq -2x + a \\ y \leq -2x + (a + 1) \end{cases}$

MIXED REVIEW

Simplify. (previous courses)

17. 2^3

18. 2^5

19. $(-3)^2$

20. $(-5)^2$

21. $2^1 \times 2^3$

22. $3^2 \times 3^1$

23. $1^2 \times 1^2$

24. $10^2 \times 10^2$

25. $\frac{5^3}{5}$

26. $\frac{5^4}{5^2}$

27. $\frac{2^5}{2^2}$

28. $\frac{3^7}{3^7}$

EXAMPLE 1

Represent the pentagon-shaped region graphed at right as a system of inequalities.

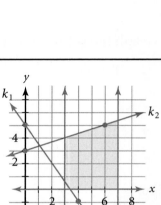

▶ **Solution**

The shaded region is partly determined by $x \geq 2$, $x \leq 8$, and $y \geq 0$. In addition, the shaded region is bounded by l_1 and l_2.

Using slope and y-intercept, the equations for l_1 and l_2 are:

$$l_1: y = -\frac{1}{3}x + 6 \qquad l_2: y = \frac{1}{2}x + 2$$

The system is $x \geq 2$, $x \leq 8$, $y \geq 0$, $y \leq -\frac{1}{3}x + 6$, and $y \leq \frac{1}{2}x + 2$.

TRY THIS   Represent this pentagon-shaped region as a system of inequalities.

EXAMPLE 2

A farmer has a total of 30 acres available for planting corn and soybeans. He wants to plant at least as much corn as soybeans. Represent the possible quantities of corn and soybeans in a graph.

▶ **Solution**

Look for key words.

1. Information about the total number of acres available indicates an inequality. The phrase "at least as much corn as soybeans" indicates a second inequality. The natural restriction is that the number of acres planted must be nonnegative.

Let c represent the number of acres of corn and let s represent the number of acres of soybeans to be planted.

Write inequalities.

2. Write and graph a system of inequalities.

$$\begin{cases} c + s \leq 30 \\ c \geq s \end{cases} \text{ and } \begin{cases} c \geq 0 \\ s \geq 0 \end{cases}$$

The graph is shown at right.

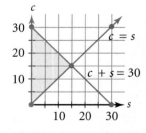

TRY THIS   A farmer has a total of 40 acres available for planting corn and soybeans. He wants to plant no more corn than soybeans. Write and graph a system of inequalities to represent this situation.

EXERCISES

KEY SKILLS

Refer to the diagram at right.

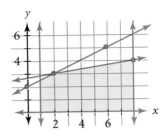

1. Write equations for the lines that contain the vertical sides and the horizontal side of the polygon.

2. Write equations for the slanted sides of the polygon.

PRACTICE

Write a system of linear inequalities to represent each polygon and its interior.

3.
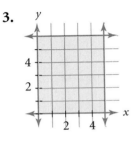

4.
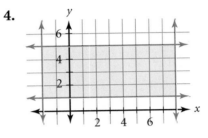

5.
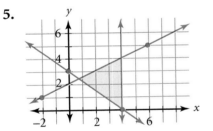

6.
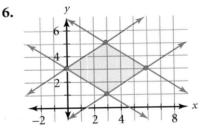

A farmer wants to plant up to 25 acres of corn and soybeans. Write a system of linear inequalities and graph the possible choices.

7. The acreage for soybeans is to be at least twice that of corn.

8. At least 5 acres of corn are to be planted. The acreage for soybeans is to be at least twice that of corn.

MIXED REVIEW APPLICATIONS

Solve each problem. (4.6 Skill C)

9. Jesse and Maria have 600 feet of fencing to enclose a square at least 20 feet on a side. What are the possible side lengths?

10. How many pounds of peanuts are needed to make between 5 pounds and 6 pounds of a mix if the ratio of cashews to peanuts is 2.5 to 1? Round quantities to the nearest tenth.

Use a graph to solve each system of equations.

1. $\begin{cases} 2x + 3y = 0 \\ x - y = 5 \end{cases}$

2. $\begin{cases} 3x + y = -1 \\ -2x - y = 5 \end{cases}$

3. $\begin{cases} -5x + 2y = -15 \\ y = \frac{2}{5}x + 3 \end{cases}$

4. $\begin{cases} x = -2y + 1 \\ y = -\frac{1}{2}x + \frac{3}{2} \end{cases}$

Solve each system using the substitution method.

5. $\begin{cases} 3m + 7 = n \\ m - 2n = 11 \end{cases}$

6. $\begin{cases} x = -3y + 2 \\ 3x + 2y = 6 \end{cases}$

7. $\begin{cases} y = -2z + 5 \\ 5 - 3z = y \end{cases}$

8. $\begin{cases} 2.5c + 1 = d \\ d = -\frac{3}{7}c + 1 \end{cases}$

9. $\begin{cases} j - 7k = 0 \\ -3j + 5k = -16 \end{cases}$

10. $\begin{cases} 2x - 3y = -2 \\ -2x + y = -6 \end{cases}$

11. $\begin{cases} \dfrac{a}{2} - \dfrac{b}{2} = 3 \\ \dfrac{a}{2} + \dfrac{b}{2} = 6 \end{cases}$

12. $\begin{cases} -1 = \dfrac{u}{5} - \dfrac{v}{2} \\ 0 = \dfrac{u}{5} + \dfrac{v}{3} \end{cases}$

Use a system of equations to solve each problem.

13. The ratio of milk to water for a certain recipe is 1 cup of milk to 4 cups of water. If the milk and water together make a 10 cup solution, how many cups of milk and how many cups of water are needed?

14. Find two whole numbers such that:
    a. one is twice the other, and
    b. five times the smaller minus two times the larger equals 7.

Solve each system using the elimination method.

15. $\begin{cases} 2g + 7h = 10 \\ g - 7h = 5 \end{cases}$

16. $\begin{cases} -3r - 3s = 8 \\ 3r + 5s = -6 \end{cases}$

17. $\begin{cases} 11y - 6z = 10 \\ 5y - 3z = 7 \end{cases}$

18. $\begin{cases} 2.5c + 5d = -1.5 \\ 5c - 6d = 5 \end{cases}$

Solve each system using the elimination method.

19. $\begin{cases} 5x + 7y = 1 \\ 3x + 4y = 1 \end{cases}$

20. $\begin{cases} 3x + 5y = -4 \\ 4x + 3y = 2 \end{cases}$

21. $\begin{cases} 5a + 4b = 0 \\ 3a - 6b = 4 \end{cases}$

22. $\begin{cases} -11j + 8k = 5 \\ 3j - 3k = -4 \end{cases}$

23. Is it possible to make a 720 milliliter salt solution that is 28% salt from one solution that is 20% salt and another solution that is 30% salt? Justify your response.

Solve, if possible.

24. $\begin{cases} 3x - 2y = 12 \\ x - 2y = 6 \end{cases}$

25. $\begin{cases} 3c + d = 3.5 \\ 9c + 3d = 7.5 \end{cases}$

26. $\begin{cases} 2c - 2d = 0.5 \\ 2c + 2d = -1.5 \end{cases}$

Classify each system as specifically as possible.

27. $\begin{cases} 3x + 2y = 6 \\ -6x - 4y = -0.5 \end{cases}$

28. $\begin{cases} 14m + 4n = -3 \\ 7m + 2n = -1.5 \end{cases}$

29. $\begin{cases} 9u - 3v = -1 \\ 27u - 9v = -3 \end{cases}$

30. $\begin{cases} b - 5b = 1 \\ 3b - 15b = -3 \end{cases}$

Graph the solution region of each linear inequality.

31. $y \geq \frac{3}{4}x - 4$

32. $y < -4$

33. $3x + 5y < 20$

34. Given that x and y are whole numbers, graph the solutions to $x + 3y \leq 9$.

Graph the solution region of each system of linear inequalities.

35. $\begin{cases} y < 5x - 3 \\ y > -3x - 3 \end{cases}$

36. $\begin{cases} y \geq 2 \\ y \leq -2x + 2 \end{cases}$

37. $\begin{cases} x < 3 \\ y > 2x - 3 \end{cases}$

38. $\begin{cases} 3x - y \geq 1 \\ 2x + 4y \leq 8 \end{cases}$

39. $\begin{cases} x - y < 0 \\ x + y > 0 \end{cases}$

40. $\begin{cases} 4x - y < 0 \\ 5x + 2y < 0 \end{cases}$

41. The perimeter of a rectangle may not exceed 20 inches and its length must be at least twice its width. Graph the possible lengths and widths of the rectangle.

CHAPTER 6

Operations With Polynomials

> ### What You Already Know

Adding, subtracting, multiplying, and dividing whole numbers are skills that you have learned and practiced for many years now. Your experience with whole numbers and operations on whole numbers will come into play when you study polynomials.

> ### What You Will Learn

To work successfully with polynomials, you first need to know about laws of exponents. The chapter begins with a study of them and their use in simplifying expressions involving exponents.

After studying laws of exponents but before getting into a study of polynomials, you will have the opportunity to extend your knowledge of functions. You can study simple polynomial functions called power functions and explore applications of them.

You then proceed to study vocabulary that is needed to differentiate one type of polynomial from another. However, a great deal of your work with polynomials will involve addition, subtraction, multiplication, and division of them. It is not unusual to learn how to perform operations on newly created mathematical objects.

VOCABULARY

base	leading coefficient	Exponents
binomial	leading term	Product Property of
congruent polygons	linear	Exponents
constant	monomial	quadratic
cubic	negative exponent	Quotient Property of
degree	perfect-square trinomial	Exponents
descending order	polynomial	scientific notation
difference of two	power function	term
squares	Power-of-a-Fraction	trinomial
evaluate	Property of	varies directly as the
exponent	Exponents	cube
exponential expression	Power-of-a-Power	varies directly as the
FOIL method	Property of	square
horizontal-addition	Exponents	vertical-addition format
format	Power-of-a-Product	zero exponent
	Property of	

The diagram below shows how mathematical skills and mathematical reasoning are interrelated with the skills and concepts in Chapter 6. Notice that this chapter involves operations with polynomials such as addition and multiplication.

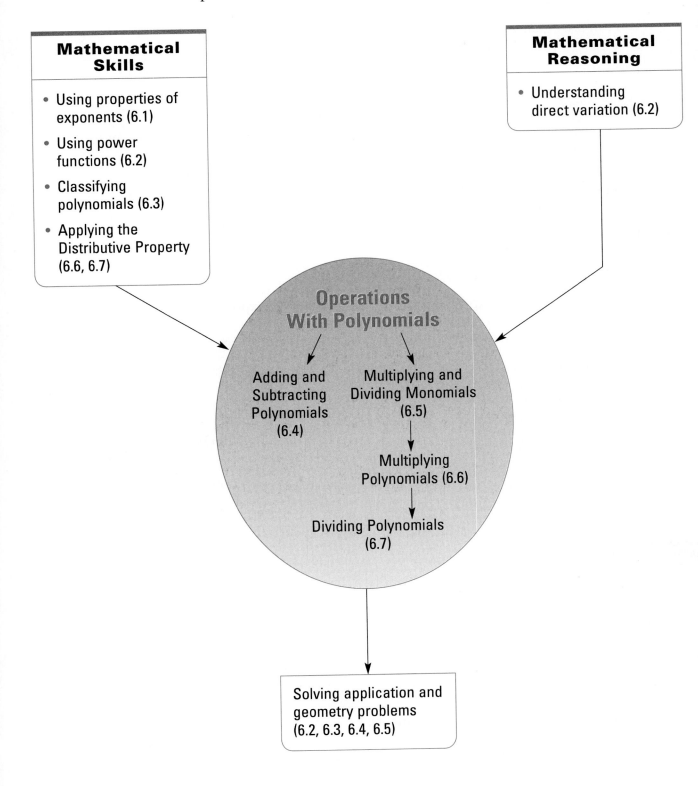

Mathematical Skills

- Using properties of exponents (6.1)
- Using power functions (6.2)
- Classifying polynomials (6.3)
- Applying the Distributive Property (6.6, 6.7)

Mathematical Reasoning

- Understanding direct variation (6.2)

Operations With Polynomials

Adding and Subtracting Polynomials (6.4)

Multiplying and Dividing Monomials (6.5)

Multiplying Polynomials (6.6)

Dividing Polynomials (6.7)

Solving application and geometry problems (6.2, 6.3, 6.4, 6.5)

Integer Exponents

SKILL A *Using properties of exponents for powers and products*

You can use *exponents* to express a repeated multiplication, such as $2 \cdot 2 \cdot 2 \cdot 2 \cdot 2$, in a shorter form.

Exponents

For any real number a and any positive integer m, $a^m = \overbrace{a \cdot a \cdot a \cdot \cdots \cdot a}^{m \text{ factors}}$.

a^m is called an **exponential expression**; a is the **base** of the expression, and m is the **exponent**.

Using this definition, you can write $2 \cdot 2 \cdot 2 \cdot 2 \cdot 2$ as 2^5, which is read *two to the fifth power* or *two raised to the fifth power*.

To multiply exponential expressions with the same base, add the exponents.

$$2^4 \cdot 2^3 = \overbrace{(2 \cdot 2 \cdot 2 \cdot 2)}^{4} \cdot \overbrace{(2 \cdot 2 \cdot 2)}^{3} = \overbrace{2 \cdot 2 \cdot 2 \cdot 2 \cdot 2 \cdot 2 \cdot 2}^{4 + 3 = 7} = 2^7$$

To raise an exponential expression to a power, multiply the exponents.

$$(3^2)^4 = (3^2)(3^2)(3^2)(3^2) = \overbrace{3 \cdot 3 \cdot 3 \cdot 3 \cdot 3 \cdot 3 \cdot 3 \cdot 3}^{2 \times 4 = 8} = 3^8$$

To raise a product to a power, distribute the exponent.

$$(4 \cdot 3)^2 = (4 \cdot 3)(4 \cdot 3) = (4 \cdot 4)(3 \cdot 3) = 4^2 \cdot 3^2$$

Let a and b be nonzero real numbers and let m and n be integers.

Product Property of Exponents $a^m a^n = a^{m+n}$

Power-of-a-Power Property of Exponents $(a^m)^n = a^{m \cdot n}$

Power-of-a-Product Property of Exponents $(ab)^m = a^m b^m$

EXAMPLE   **Write in the form a^b.**   **a.** $3^4 \cdot 3^2$ **b.** $(5^2)^6$ **c.** $2^3 \cdot 4^3$

Solution

a. $3^4 \cdot 3^2 = 3^{4+2} = 3^6$ ⟵ Apply the Product Property.
b. $(5^2)^6 = 5^{2 \cdot 6} = 5^{12}$ ⟵ Apply the Power-of-a-Power Property.
c. $2^3 \cdot 4^3 = (2 \cdot 4)^3 = 8^3$ ⟵ Apply the Power-of-a-Product Property.

TRY THIS   Write in the form a^b.   **a.** $10^3 \cdot 10^4$ **b.** $(3^2)^4$ **c.** $2^4 \cdot 3^4$

Be careful with powers of negative numbers. $(-3)^2 = (-3)(-3) = 9$, but $-3^2 = -(3^2) = -(3)(3) = -9$. The parentheses in $(-3)^2$ indicate that the exponent applies to the negative sign as well as to 3. However, in -3^2, the exponent applies only to 3.

EXERCISES

KEY SKILLS

Rewrite each expression as repeated multiplication.

1. 2^3 **2.** 3^4 **3.** 1^5 **4.** 10^3

Which property of exponents would you use to simplify each expression?

5. $5^3 \cdot 4^3$ **6.** $(4 \cdot 6)^3$ **7.** $5^2 \cdot 5^5$ **8.** $(7^2)^2$

PRACTICE

Write each expression in the form a^b.

9. $5^5 \cdot 5^4$ **10.** $2^2 \cdot 2^4$ **11.** $(3^2)^3$ **12.** $(2^3)^2$

13. $(10^7)^5$ **14.** $(10^6)^7$ **15.** $3^2 \cdot 3^2$ **16.** $1^3 \cdot 1^3$

17. $3^6 \cdot 9^6$ **18.** $3^9 \cdot 6^9$ **19.** $6^5 \cdot 6^4$ **20.** $5^7 \cdot 5^3$

21. $(3^2)^2$ **22.** $(7^3)^5$ **23.** $4^3 \cdot 4^2$ **24.** $5^2 \cdot 5^7$

25. $2^3 \cdot 2^5$ **26.** $3^2 \cdot 3^5$ **27.** $7^2 \cdot 2^2$ **28.** $3^4 \cdot 2^4$

29. $(-2)^3 \cdot (-3)^3$ **30.** $(-3)^3 \cdot (-3)^3$ **31.** $(-2)^4 \cdot 3^4$ **32.** $6^2 \cdot (-3)^2$

33. $(-3^4)^2$ **34.** $(-2^2)^5$ **35.** $(-2)^1(-2)^3(-2)^2$ **36.** $(-7)^2(-7)^1(-7)^5$

37. Explain why $(2x)^3$ is not the same as $2x^3$.

38. Can $3^5 \cdot 4^3$ be simplified using the properties of exponents? Explain your response.

39. **Critical Thinking** You have seen that $(-a)^2 \neq -a^2$. Is it also true that $(-a)^3 \neq -a^3$? Try this for several different powers. For what values of n does $(-a)^n = -a^n$?

MIXED REVIEW

Use the substitution method to solve each system of equations. (5.2 Skill A)

40. $\begin{cases} y = x + 3 \\ y = 2x - 4 \end{cases}$ **41.** $\begin{cases} x = y + 4 \\ 2x + 3y = 43 \end{cases}$ **42.** $\begin{cases} 4x + 2y = 20 \\ y = x - 2 \end{cases}$ **43.** $\begin{cases} x - y = 3 \\ 2x + 2y = 2 \end{cases}$

Use the elimination method to solve each system of equations. (5.3 Skill A)

44. $\begin{cases} 3x - y = 0 \\ 3x + 2y = -7 \end{cases}$ **45.** $\begin{cases} 2x + 2y = 3 \\ 5x - 2y = 0 \end{cases}$ **46.** $\begin{cases} x - 3y = 0 \\ 5x + 3y = 6 \end{cases}$ **47.** $\begin{cases} 2x - 2y = 10 \\ 4x + 3y = 12 \end{cases}$

You can use the definition of exponent to raise a fraction to a power. For example, $\left(\frac{1}{2}\right)^3 = \left(\frac{1}{2}\right)\left(\frac{1}{2}\right)\left(\frac{1}{2}\right) = \frac{1 \cdot 1 \cdot 1}{2 \cdot 2 \cdot 2} = \frac{1^3}{2^3} = \frac{1}{8}$. So, $\left(\frac{1}{2}\right)^3 = \frac{1^3}{2^3}$.

To divide exponential expressions, subtract the exponents. For example, $\frac{2^5}{2^3} = \frac{2 \cdot 2 \cdot 2 \cdot 2 \cdot 2}{2 \cdot 2 \cdot 2} = \frac{2 \cdot 2}{1} = 2^2 = 4$. So $\frac{2^5}{2^3} = 2^{5-3} = 2^2$.

Let a and b be nonzero real numbers and let m and n be integers.

Power-of-a-Fraction Property of Exponents $\quad \left(\frac{a}{b}\right)^m = \frac{a^m}{b^m}$

Quotient Property of Exponents $\quad \frac{a^m}{a^n} = a^{m-n}$

| **EXAMPLE 1** | Simplify. | a. $\left(\frac{3}{4}\right)^2$ | b. $\frac{4^8}{4^6}$ | c. $\frac{3^5 \cdot 3^2}{3^4}$ |

▶ **Solution**

a. $\left(\frac{3}{4}\right)^2 = \frac{3^2}{4^2} = \frac{9}{16}$

b. $\frac{4^8}{4^6} = 4^{8-6}$
$= 4^2 = 16$

c. $\frac{3^5 \cdot 3^2}{3^4} = \frac{3^{5+2}}{3^4} = \frac{3^7}{3^4}$
$= 3^{7-4} = 3^3 = 27$

TRY THIS Simplify. a. $\left(\frac{2}{3}\right)^3$ b. $\frac{3^8}{3^4}$ c. $\frac{5^6 \cdot 5^4}{5^6}$

Exponents can also be zero or negative. Examine the pattern below.

$6^2 = 6 \cdot 6 = \mathbf{36}$

$6^1 = \mathbf{6}$

$6^0 = \frac{6}{6} = \mathbf{1}$

$6^{-1} = \frac{1}{6}$

$6^{-2} = \frac{1}{6 \cdot 6} = \frac{1}{6^2} = \frac{1}{36}$

Let a be a nonzero real number and let n be an integer.

Zero Exponent $\quad a^0 = 1 \quad$ (0^0 does not exist.)

Negative Exponents $\quad a^{-n} = \frac{1}{a^n}$

| **EXAMPLE 2** | Simplify. | a. 54^0 | b. 5^{-2} | c. $\frac{3^4 \cdot 3^{-2}}{3^3}$ |

▶ **Solution**

a. $54^0 = 1$ b. $5^{-2} = \frac{1}{5^2} = \frac{1}{25}$ c. $\frac{3^4 \cdot 3^{-2}}{3^3} = \frac{3^{4+(-2)}}{3^3} = \frac{3^2}{3^3} = 3^{2-3} = 3^{-1} = \frac{1}{3}$

TRY THIS Simplify. a. 16^0 b. 10^{-2} c. $\frac{4^5 \cdot 4^{-2}}{4^6}$

EXERCISES

Which property or definition of exponents would you use to simplify each expression?

1. $\dfrac{3^5}{3^2}$

2. 8^{-12}

3. 7^0

4. $\left(\dfrac{9}{2}\right)^3$

5. $(6^2)^3$

6. $\dfrac{4^3}{4^2}$

7. $3^2 \cdot 3^5$

8. $(3 \cdot 5)^2$

Simplify each expression.

9. $\left(\dfrac{1}{3}\right)^2$

10. $\left(\dfrac{1}{8}\right)^2$

11. $\left(\dfrac{3}{5}\right)^3$

12. $\left(\dfrac{4}{3}\right)^3$

13. $\dfrac{3^5}{3^2}$

14. $\dfrac{5^6}{5^3}$

15. $\dfrac{2^3}{2}$

16. 10^{-1}

17. 2^{-3}

18. 8^{-2}

19. $\dfrac{3^6}{3^9}$

20. $\dfrac{2^5}{2^8}$

21. $\dfrac{3^2}{3^0}$

22. $\dfrac{2^0}{2^2}$

23. $\dfrac{3^2 \cdot 3^3}{3^5}$

24. $\dfrac{2^3 \cdot 2^3}{2^5}$

25. $\dfrac{4^6}{4^2 \cdot 4^3}$

26. $\dfrac{5^8}{5^3 \cdot 5^4}$

27. $\dfrac{7^3 \cdot 7^3}{7^8}$

28. $\dfrac{(-2)^2 \cdot (-2)^3}{(-2)^7}$

29. $\dfrac{(-5)^9}{(-5)^7 \cdot (-5)^2}$

30. $\dfrac{(-10)^3}{(-10) \cdot (-10)^2}$

31. $\left[(-1)^3 \, (-2)^3\right]^3$

32. $(3^{-1} \cdot 2^{-1})^2$

33. $\left[(-3)^{-2} \cdot (-3)^{-1}\right]^2$

34. $\left[(-4)^{-1} \cdot (-2)^{-1}\right]^3$

35. $\left(\dfrac{10^3 \cdot 10^2}{10^4 \cdot 10^2}\right)^2$

36. $\left(\dfrac{4^2 \cdot 4^2}{4^4 \cdot 4^3}\right)^2$

37. $\left(\dfrac{3^{-2} \cdot 3^{-2}}{2^4 \cdot 2^3}\right)^{-1}$

38. $\left(\dfrac{2^{-1} \cdot 3^{-1}}{5^4 \cdot 5^3}\right)^{-2}$

39. $\left(\dfrac{1^{-1} \cdot 1^{-1}}{2^{-1} \cdot 3^3}\right)^{-1}$

40. $\left(\dfrac{5^{-1} \cdot 5^{-1}}{2^4 \cdot 3^3}\right)\left(\dfrac{2^{-2} \cdot 3^{-1}}{5^2 \cdot 5^2}\right)^{-1}$

41. Critical Thinking How are the expressions a^n and a^{-n} related?

Find each product or quotient. Write your answer as a decimal. Use mental math where possible. (previous courses)

42. 10×235

43. $\dfrac{118}{100}$

44. 100×1389

45. $\dfrac{0.78}{10}$

46. 100×0.56

47. $\dfrac{2.57}{100}$

48. 1000×0.232

49. $\dfrac{18.3}{1000}$

A number is written in **scientific notation** when it has the form $a \times 10^n$, where n is an integer and $1 \le a < 10$. Scientific notation is a shortened way of writing very large or very small numbers. For example:

$$1{,}210{,}000{,}000{,}000{,}000{,}000{,}000{,}000{,}000{,}000{,}000{,}000 = 1.21 \times 10^{33}$$
$$0.00000000000000000000000000000368 = 3.68 \times 10^{-29}$$

To write a number in scientific notation, use the following procedure.

Scientific Notation

1. Move the decimal point so that the number is between 1 and 10 (excluding 10).

2. Multiply by a power of 10. To find the exponent of 10, count the number of places the decimal point must move to return to its original position. The exponent is negative if the decimal point must move to the left, and positive if it must move to the right.

EXAMPLE 1

Write each number in scientific notation.   **a.** 1364   **b.** 0.0258

▶ **Solution**

First move the decimal point so that the number is between 1 and 10. Then multiply by the correct power of 10.

a. $1364 = 1.364 \times 10^3$

> To return to its original position, the decimal point must move 3 places to the right.

b. $0.0258 = 2.58 \times 10^{-2}$

> To return to its original position, the decimal point must move 2 places to the left.

TRY THIS   Write each number in scientific notation.   **a.** 26.5   **b.** 0.0014

EXAMPLE 2

Find each product or quotient. Write the answer in scientific notation.

a. $(1.2 \times 10^3)(9.0 \times 10^4)$ **b.** $\dfrac{1.8 \times 10^3}{3.0 \times 10^5}$

▶ **Solution**

a. $(1.2 \times 10^3)(9.0 \times 10^4)$
$= (1.2 \times 9.0)(10^3 \times 10^4)$
$= (1.2 \times 9.0)(10^{3 + 4})$
$= 10.8 \times 10^7$
$= (1.08 \times 10^1) \times 10^7$
$= 1.08 \times 10^8$

b. $\dfrac{1.8 \times 10^3}{3.0 \times 10^5} = \left(\dfrac{1.8}{3.0}\right)\left(\dfrac{10^3}{10^5}\right)$
$= \left(\dfrac{1.8}{3.0}\right) \times 10^{3 - 5}$
$= 0.6 \times 10^{-2}$
$= (6 \times 10^{-1}) \times 10^{-2}$
$= 6 \times 10^{-3}$

TRY THIS   Find each product or quotient. Write the answer in scientific notation.

a. $(8.2 \times 10^2)(9.0 \times 10^2)$ **b.** $\dfrac{1.20 \times 10^6}{9.60 \times 10^2}$

EXERCISES

KEY SKILLS

Is each number in scientific notation? If not, write it in scientific notation.

1. 3.45×10^3

2. 34.5×10^3

3. 345

4. 0.345×10^{-2}

PRACTICE

Write each number in scientific notation.

5. 12.4

6. $15,334$

7. 0.012

8. 0.00025

9. $10,050$

10. $1,200,000$

11. 0.0165

12. 0.00335

Find each product or quotient. Write the answer in scientific notation.

13. $(1.5 \times 10^2)(2.0 \times 10^2)$

14. $(1.4 \times 10^3)(5.0 \times 10^3)$

15. $(6.5 \times 10^1)(3.0 \times 10^3)$

16. $(7.6 \times 10^4)(3.0 \times 10^1)$

17. $(5.4 \times 10^{-2})(6.0 \times 10^2)$

18. $(6.5 \times 10^{-3})(4.0 \times 10^{-2})$

19. $\dfrac{6.0 \times 10^4}{3.0 \times 10^2}$

20. $\dfrac{7.5 \times 10^5}{5.0 \times 10^4}$

21. $\dfrac{7.5 \times 10^3}{3.0 \times 10^5}$

22. $\dfrac{8.4 \times 10^4}{4.0 \times 10^7}$

23. $\dfrac{2.4 \times 10^2}{4.8 \times 10^3}$

24. $\dfrac{2.1 \times 10^2}{8.4 \times 10^5}$

25. **Critical Thinking**   Assume $1 \le a < 10$. When $(a \times 10^3)(5.0 \times 10^2)$ is written in scientific notation, the power of 10 is 5. What are the possible values of a?

26. **Critical Thinking**   Assume $1 \le a < 10$. When $(a \times 10^4) \div (5.0 \times 10^2)$ is written in scientific notation, the power of 10 is 2. What are the possible values of a?

MIXED REVIEW

Assume that y varies directly as x. (3.3 Skill B)

27. $y = 12$ when $x = 2.5$. Find y when $x = 6$.

28. $y = 3$ when $x = 4$. Find y when $x = 12.5$.

29. $y = 13$ when $x = 5$. Find y when $x = 0.5$.

30. $y = 2$ when $x = 7$. Find y when $x = 1$.

6.2

LESSON

The Power Functions $y = kx$, $y = kx^2$, and $y = kx^3$

SKILL A *Using power functions to solve direct-variation problems*

A **power function** is any function of the form $y = kx^n$, where k is nonzero and n is a positive integer.

Recall that if $y = kx$ and $k \neq 0$, then y varies directly as x and the constant of variation is k. This is a power function with $n = 1$.

If $y = kx^2$ and $k \neq 0$, then y **varies directly as the square** of x. The constant of variation is k. This is a power function with $n = 2$.

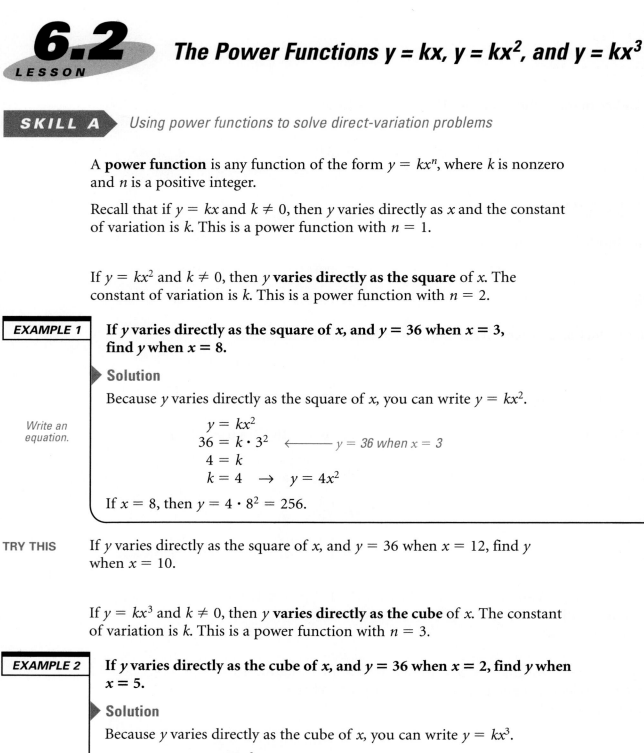

EXAMPLE 1

If y varies directly as the square of x, and $y = 36$ when $x = 3$, find y when $x = 8$.

▶ **Solution**

Because y varies directly as the square of x, you can write $y = kx^2$.

Write an equation.

$$y = kx^2$$
$$36 = k \cdot 3^2 \quad \longleftarrow \quad y = 36 \text{ when } x = 3$$
$$4 = k$$
$$k = 4 \quad \rightarrow \quad y = 4x^2$$

If $x = 8$, then $y = 4 \cdot 8^2 = 256$.

TRY THIS

If y varies directly as the square of x, and $y = 36$ when $x = 12$, find y when $x = 10$.

If $y = kx^3$ and $k \neq 0$, then y **varies directly as the cube** of x. The constant of variation is k. This is a power function with $n = 3$.

EXAMPLE 2

If y varies directly as the cube of x, and $y = 36$ when $x = 2$, find y when $x = 5$.

▶ **Solution**

Because y varies directly as the cube of x, you can write $y = kx^3$.

Write an equation.

$$y = kx^3$$
$$36 = k \cdot 2^3 \quad \longleftarrow \quad y = 36 \text{ when } x = 2$$
$$4.5 = k$$
$$k = 4.5 \quad \rightarrow \quad y = 4.5x^3$$

If $x = 5$, then $y = 4.5 \cdot 5^3 = 562.5$.

TRY THIS

If y varies directly as the cube of x, and $y = 1$ when $x = 2$, find y when $x = 12$.

EXERCISES

> **KEY SKILLS**

Let $x = 0, 1, 2, 3, 4,$ and 5. Make a table of ordered pairs for each function. Find the difference in successive y values. Is this difference constant?

1. $y = 3x$

2. $y = 3x^2$

3. $y = 3x^3$

> **PRACTICE**

In Exercises 4–9, find k and write an equation for y in terms of x. Then find the specified value of y.

4. y varies directly as the square of x. If $y = 18$ when $x = 3$, find y when $x = 5$.

5. y varies directly as the square of x. If $y = 40$ when $x = 4$, find y when $x = 8$.

6. y varies directly as the square of x. If $y = 36$ when $x = 3$, find y when $x = 4$.

7. y varies directly as the cube of x. If $y = 40$ when $x = 2$, find y when $x = 6$.

8. y varies directly as the cube of x. If $y = 94.5$ when $x = 3$, find y when $x = 10$.

9. y varies directly as the cube of x. If $y = 48$ when $x = 2$, find y when $x = 3$.

10. The length of one side of a square is x units.
   a. Write a direct-variation equation in the form $y = kx$ for the perimeter, P, of the square.
   b. Write a direct-variation equation in the form $y = kx^2$ for the area, A, of the square.
   c. Write a direct-variation equation in the form $y = kx^3$ for the volume, V, of a cube whose side length is x units.

11. Critical Thinking Suppose that y varies directly as the square of x and that (x_1, y_1) and (x_2, y_2) satisfy this direct-variation relationship. Find a relationship among x_1, x_2, y_1, and y_2 that does not include the constant of variation.

> **MIXED REVIEW**

Simplify. (2.6 Skill A)

12. $\frac{3}{4}a + 5 - \frac{1}{4}a$

13. $3s + 4s - 3s + 4$

14. $4 - (y + 2) - 3y + 1$

15. $3(u - 3) + 4(u - 3)$

16. $\frac{2}{3}t - 3 - \frac{1}{3}t + 2$

17. $0.6z + 4(0.2z - 2)$

The distance required for a moving car to stop varies directly as the square of the car's speed. Therefore, we can use an equation of the form $y = kx^2$ to represent this situation. The value of k will change for different types of cars, road conditions, etc.

EXAMPLE

A car traveling at 55 miles per hour had a stopping distance of 110 feet.

a. Write an equation that describes the car's stopping distance in terms of its speed.

b. If the speed is reduced by half, will the stopping distance also be reduced by half? Justify your response.

▷ **Solution**

Write an equation.

a. We know that stopping distance, d, varies directly as the square of the speed, s, so let $d = ks^2$ for some constant k.

$$110 = k(55)^2$$

$$k = \frac{110}{(55)^2} = \frac{2 \cdot 55}{55 \cdot 55} = \frac{2}{55}$$

$$k = \frac{2}{55} \quad \rightarrow \quad d = \frac{2}{55}s^2$$

b. One-half of the original speed, 55 mph, is 27.5 mph.

If $s = 27.5$, then $d = \frac{2}{55} \times (27.5)^2 = 27.5$.

When the speed is reduced by half to 27.5 mph, the stopping distance is reduced to 27.5 ft, which one-fourth of the original stopping distance of 110 ft. So, the stopping distance is reduced by more than one-half.

TRY THIS

A car traveling at 55 miles per hour had a stopping distance of 130 feet.

a. Find the stopping distance for a speed of 60 miles per hour.

b. Will the stopping distance be reduced by half if the speed is reduced by half? Justify your response.

You can show that when speed is reduced by half, stopping distance is always reduced by more than half. Let d_s represent the stopping distance at a speed of s miles per hour, and let $d_{\frac{1}{2}s}$ represent the stopping distance at a speed of $\frac{1}{2}s$ miles per hour.

$$\frac{d_{\frac{1}{2}s}}{d_s} = \frac{k\left(\frac{1}{2}s\right)^2}{ks^2} = \frac{\left(\frac{1}{2}s\right)^2}{s^2} = \frac{\left(\frac{1}{2}\right)^2 s^2}{s^2} = \frac{1}{4}$$

When the speed is reduced by half, the stopping distance is reduced to $\frac{1}{4}$ of the original stopping distance. Speed and stopping distance are clearly *not* linearly related.

EXERCISES

KEY SKILLS

1. Copy and complete the following table for the car in the Example. Round answers to the nearest tenth.

s	0	5	10	15	20	25	30
d	?	?	?	?	?	?	?

2. Are the differences in successive values of d constant?

PRACTICE

A car traveling at 50 miles per hour had a stopping distance of 135 feet. Find the stopping distance for each speed. Round to the nearest tenth.

3. 45 miles per hour

4. 65 miles per hour

5. 55 miles per hour

6. 100 miles per hour

7. **Critical Thinking** Let $d = ks^2$, where k is a nonzero constant. If d_s is the stopping distance for a speed of s, find and simplify an expression for $\frac{d_{rs}}{d_s}$, where $0 < r < 1$.

The vertical distance that a falling object travels varies directly as the square of time. Let d represent distance in feet, and let t represent time in seconds. The constant of variation is 16 feet per second squared. Find d for each value of t.

8. 2 seconds

9. 3 seconds

10. 5 seconds

11. 10 seconds

12. 15 seconds

13. 30 seconds

MIXED REVIEW APPLICATIONS

Use an inequality to solve each problem. (4.4 Skill B)

14. Lin spent $50 on a skirt and a sweater. The sweater cost less than $22.50. How much could the skirt have cost?

15. Video games cost $15 each. If Percy spent less than $120, how many video games could he have bought?

16. The Gibbons family spent more than $220 for a barbecue grill and patio furniture. The cost of the grill was $70 more than twice the cost of the patio furniture. How much could the patio furniture have cost?

6.3 LESSON Polynomials

SKILL A *Classifying polynomials by degree and number of terms*

Recall that a monomial in x is the product of a number and a whole-number power of x. For example, $-3x^2, \frac{2}{3}x^3, 2x, 10$ are each monomials. (Note that $2x = 2x^1$ and $10 = 10x^0$.)

A **polynomial in x** is a sum of one or more monomials in x. For example, $5x^2 - 4x + 7$ is a polynomial with three *terms*, $5x^2, -4x,$ and 7. A **term** is any monomial in a polynomial. The **degree** of a polynomial is the greatest exponent in any of its terms, which in $5x^2 - 4x + 7$ is 2. The term with the greatest exponent, $5x^2$, is called the **leading term**. The **leading coefficient** is the numerical part of the leading term, in this case 5.

A polynomial is written in **descending order** if the exponents decrease as you read the polynomial from left to right. For example, $3x^4 + 5x^2 + 2x + 5$ is written in descending order.

EXAMPLE 1 **Write $-3x + 5x^2 - 7x^3 + 5$ in descending order.**

▶ **Solution**

$-3x + 5x^2 - 7x^3 + 5 = -7x^3 + 5x^2 - 3x + 5$ ⟵ $-3x = -3x^1$ and $5 = 5x^0$

TRY THIS Write $13y^2 + 15y - 7y^3 + 5$ in descending order.

Polynomials have special names based on degree and number of terms.

Degree	Name	Example
0	constant	6
1	linear	$3x - 5$
2	quadratic	$-4x^2 + x + 1$
3	cubic	$9x^3 + x^2 - 5$

Terms	Name	Examples
1	monomial	$-5x, 2x^3, x^8, 4$
2	binomial	$3x - 1, x^2 + 6x$
3	trinomial	$x^2 - 5x + 1$

EXAMPLE 2 **Classify $-2a^3 + 2a^3 + 4a^2 - 3a - 5$ by degree and number of terms.**

▶ **Solution**

First simplify the polynomial.
$-2a^3 + 2a^3 + 4a^2 - 3a - 5$
$\qquad = (-2 + 2)a^3 + 4a^2 - 3a - 5$ ⟵ Combine like terms.
$\qquad = 4a^2 - 3a - 5$ ⟵ The greatest exponent is 2.

The degree is 2 and there are 3 terms. It is a quadratic trinomial.

TRY THIS Classify $5b^3 - 3.5b^3 - 3b + 3b^2 - b^2$ by degree and number of terms.

EXERCISES

Write the degree, number of terms, leading term, and leading coefficient
for each polynomial.

1. $r^3 + r^2 - 5$

2. $a^3 + a^2 - 2a + 1$

3. $3k^4 + k^3 - 2k^2 + k$

Write each polynomial in descending order.

4. $2x + 3x^2 - 1$

5. $5g - 7 + g^2$

6. $-2n + 1 - n^2$

7. $-m + 7 - 3m^2$

8. $3x^2 + 5x - 4 + 5x^3$

9. $3c^2 + 5c^2 - 4 + 5c^4$

10. $-2p + 6p - p^2 - p^3$

11. $y^2 + y + y + 5y^3 - y^2$

12. $-a^3 + 5a^2 - 4 + 5a^2 - a$

Classify each polynomial by degree and number of terms.

13. $2x^2 + 6x + x^3$

14. $-3s^2 + 6 - s^3 - 3s^4$

15. $q^2 + 6 - q^3 + 3q^4$

16. $c^2 + 7 - 2c^3 + 2c^2$

17. $-1 - 2z^3 + z^2 + 2z^3$

18. $-y^2 - 4y^3 + 2y^2 + 4y^3$

19. $5k^2 + 7k^3 - 2k^3 + 2k^2$

20. $5k + k^2 - 5k^2 + 2k$

21. $-3v + 6v^2 - 5v^2 + 3v$

22. Classify $2(3n - 2) - 3(2n^2 + 2n) + 6(n^2 - 2)$ by degree and number
of terms.

23. Every linear function can be written in the form $y = mx + b$, where
m and b are real numbers. Suppose that $m \neq 0$ and $b \neq 0$. Classify
$mx + b$ by degree and number of terms.

24. Let k be a nonzero real number. Classify kx^2 and kx^3 by degree and
number of terms.

25. Critical Thinking Find real numbers a and b such that the
polynomial $ax + 6x^3 - bx^3 + 3x + x - 5$ is a linear binomial.

26. Critical Thinking Find the missing polynomial in the equation
below.
$(\underline{\ ?\ }) - (2k^4 - 2k^2 - 3) = 4k^4 + k^3 + 2k^2 + 7$

Evaluate each formula for the given value(s) of the variables. (1.2 Skill C)

27. $S = 2r^2 + 20r$ given $r = 3$

28. $S = 2r^2 + 20r$ given $r = 4$

29. $W = 4.5x + 5y$ given $x = 6$ and $y = 6$

30. $z = \frac{2}{3}x + \frac{3}{7}y$ given $x = 3$ and $y = 7$

31. $t = x^2 + y^2$ given $x = \frac{1}{2}$ and $y = \frac{1}{3}$

32. $y = (x - 2)^2 + 2$ given $x = 4$

Polynomials are used in many geometry formulas. In the examples below, we will *evaluate* polynomials to answer questions about the surface area and volume of containers. To evaluate a polynomial, substitute the given value(s) for the variable into the expression and then simplify.

Note: You may find it helpful to use a calculator for the exercises in this section.

The **volume** of an object is the amount of space the object occupies. Volume is measured in *cubic units.*

EXAMPLE 1

The total volume of a cylinder whose height is 15 centimeters can be represented by the polynomial $15\pi r^2$, where r represents the radius. Find the volume of the container at right if the radius is 4 cm, 5 cm, and 6 cm. Use 3.14 to approximate π.

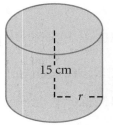
15 cm
r

▶ **Solution**

Evaluate $15\pi r^2$ for $r = 4$, 5, and 6.

$r = 4$: $(15)(3.14)(4)^2$ ⟵——— *Replace r with 4.*
$= 753.6$ cm^3
$r = 5$: $(15)(3.14)(5)^2$ ⟵——— *Replace r with 5.*
$= 1177.5$ cm^3
$r = 6$: $(15)(3.14)(6)^2$ ⟵——— *Replace r with 6.*
$= 1695.6$ cm^3

TRY THIS   Rework Example 1 if the radius is 1 cm, 2 cm, and 3 cm.

The **surface area** of an object is the total area of all the outer surfaces of the object. Surface area is measured in square units.

EXAMPLE 2

The total surface area of a cylinder whose height is 10 feet can be represented by the polynomial $2\pi r^2 + 20\pi r$. Find the surface area of the cylindrical tank at right if the radius is 5 ft, 6 ft, and 7 ft. Use 3.14 to approximate π.

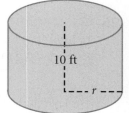
10 ft
r

▶ **Solution**

Evaluate $2\pi r^2 + 20\pi r$ for $r = 5$, 6, and 7.

$r = 5$: $(2)(3.14)(5)^2 + (20)(3.14)(5) = 471$ ft^2
$r = 6$: $(2)(3.14)(6)^2 + (20)(3.14)(6) = 602.88$ ft^2
$r = 7$: $(2)(3.14)(7)^2 + (20)(3.14)(7) = 747.32$ ft^2

TRY THIS   Rework Example 2 if the radius is 8 ft, 9 ft, and 10 ft.

EXERCISES

KEY SKILLS

Evaluate each polynomial for the given value of the variable.

1. $x^2 + 3x - 2; x = 3$

2. $2s^2 - 5s + 2; s = 4$

3. $v^3 + v^2 + v + 1; v = 3$

4. $2n^3 - 5n^2 - 5; n = -2$

PRACTICE

Refer to Example 1. Find the volume of the container for each radius. Use 3.14 to approximate π.

5. 7 cm

6. 8 cm

7. 8.5 cm

8. 9.5 cm

Refer to Example 2. Find the surface area of the tank for each radius. Use 3.14 to approximate π.

9. 1 foot

10. 2 feet

11. 3 feet

12. 3.5 feet

Sports balls come in many different sizes. Use the given polynomials to complete the table below. Use 3.14 to approximate π. Round answers to the nearest hundredth.

	Type of ball	Radius, r	Volume $\dfrac{4\pi r^3}{3}$	Surface Area $4\pi r^2$
13.	Golf ball	0.8 inches	?	?
14.	Tennis ball	1.3 inches	?	?
15.	Softball	1.75 inches	?	?
16.	Soccer ball	4.5 inches	?	?
17.	Basketball	4.8 inches	?	?

MID-CHAPTER REVIEW

Simplify each exponential expression. (6.1 Skill B)

18. $\dfrac{7^1 \cdot 7^2}{7^3}$

19. $[(-2)^2]^{-2}$

20. $\dfrac{(-3)^6 \cdot (-3)^3}{(-3)^{10}}$

21. $5^3 \times (5^2)^{-2}$

22. If y varies directly as the square of x, and $y = 11$ when $x = 2.5$, find y when $x = 15$. (6.2 Skill A)

Classify by degree and number of terms. (6.3 Skill A)

23. $-5x^3 + 2x + 5x^2 + 5x^3$

24. $3n^2 + n - 3n^2 + 1 + 5n^3$

6.4 LESSON

Adding and Subtracting Polynomials

Just as you can perform operations on integers, you can perform operations on polynomials.

The following statements say that the set of polynomials in the same variable(s) is *closed* under addition and under subtraction.

> **Closure of Polynomials Under Addition and Subtraction**
> The sum of two polynomials is a polynomial.
> The difference of two polynomials is a polynomial.

You can add polynomials by using the **vertical-addition format.** When you perform addition in this format, place like terms vertically above one another.

EXAMPLE 1

Use the vertical-addition format to find the sum
$(-2x^3 + 2.5x^2 - 5x + 3) + (5x^2 + 4x - 5)$.

▶ **Solution**

Align corresponding terms.
Add corresponding coefficients.

$$
\begin{array}{r}
-2x^3 + 2.5x^2 - 5x + 3 \\
+ \quad\quad\quad 5x^2 + 4x - 5 \\
\hline
-2x^3 \quad 7.5x^2 \; (-1)x - 2
\end{array}
$$

The sum is $-2x^3 + 7.5x^2 - x - 2$.

TRY THIS

Use the vertical-addition format to find $(2.6a^3 - 1.5a^2 - 5a) + (0.4a^3 - 5a + 3)$.

Example 2 shows how to perform addition using the Distributive Property and what is called the **horizontal-addition format.** When you use this format, you add by grouping and combining like terms.

EXAMPLE 2

Use the horizontal-addition format to find the sum in Example 1.

▶ **Solution**

$(-2x^3 + 2.5x^2 - 5x + 3) + (5x^2 + 4x - 5)$
$= (-2 + 0)x^3 + (2.5 + 5)x^2 + (-5 + 4)x + (3 - 5)$
$= -2x^3 + (7.5)x^2 + (-1)x + (-2)$
$= -2x^3 + 7.5x^2 - x - 2$

TRY THIS

Use the horizontal-addition format to find $(2.6a^3 - 1.5a^2 - 5a) + (0.4a^3 - 5a + 3)$.

EXERCISES

Add the polynomials.

1.
$$\begin{array}{r} -3x^2 - 5x + 1 \\ +\ \ 2x^2 \qquad - 5 \\ \hline \end{array}$$

2.
$$\begin{array}{r} a^3 - 3a^2 - a + 1 \\ +\ -4a^3 - 5a^2 \qquad - 7 \\ \hline \end{array}$$

3.
$$\begin{array}{r} -3y^3 \qquad - 5y + 1 \\ +\ \ 4y^3 - y^2 \qquad + 1 \\ \hline \end{array}$$

Find each sum.

4. $(2x^2 - 2x + 1) + (5x^2 - x - 2)$

5. $(-b^2 - 2b) + (5b^2 - b - 1)$

6. $(c^3 - 2c^2 - 5c + 3) + (c^2 + 2c - 1)$

7. $(z^3 + 3z^2) + (-3z^2 + z)$

8. $(9n^4 - n^2 - 7n) + (5n^3 - n^2 + 9)$

9. $(3k^4 - k^2 - 1) + (k^3 - k + 1)$

10. $(h^4 - h^2 + 1) + (-h^2 + 3h - 5)$

11. $(-b^5 - b^3 + b) + (b^5 + b^3 - 1)$

12. $(z^5 + z^3 - z^2 + 1) + (z^3 + z^2 + 3z - 1)$

13. $(5d^5 - d^3 + d^2 + 2d) + (d^5 + d^3 - 2d)$

14. $3x^3 - 4x^2 + 2x - 1,\ x^2 + 2x + 1,$ and $2x^4 - 3x^2 - 4x + 5$

15. $-5x^3 - 3x + 3,\ 2x^4 + 3x^2 + 3x + 2,$ and $-2x^4 - 4x^2 + 3$

16. Find a, b, and c such that the sum of $ax^2 + bx + c$ and $-2x^2 + 3x + 7$ equals 0.

17. Let a, b, c, k, m, and n represent real numbers. When is the sum $(ax^2 + bx + c) + (kx^2 + mx + n)$ only a constant term?

18. **Critical Thinking** Let a, b, r, and s represent real numbers, where $a \neq -r$ and $b \neq -s$. Show that the sum of $ax + b$ and $rx + s$ is a linear binomial.

Find each product or quotient. Write the answer in scientific notation.
(6.1 Skill C)

19. $(2.1 \times 10^{-1})(3.0 \times 10^2)$

20. $(1.5 \times 10^{-1})(4.0 \times 10^{-1})$

21. $(6.0 \times 10^2)(6.0 \times 10^3)$

22. $(2.5 \times 10^3)(2.5 \times 10^{-1})$

23. $\dfrac{9.0 \times 10^5}{2.0 \times 10^2}$

24. $\dfrac{7.5 \times 10^2}{3.0 \times 10^{-1}}$

25. $\dfrac{3.5 \times 10^3}{7.0 \times 10^4}$

Recall from Lesson 2.3 the Opposite of a Sum Property and the Opposite of a Difference Property.

$$-(a + b) = -a - b \qquad\qquad -(a - b) = -a + b$$

Notice that to find the opposite of a sum or the opposite of a difference, you take the opposite of *each term* in the expression. To find the opposite of a polynomial, you must find the opposite of each of its terms.

EXAMPLE 1 **Find the opposite of $-5x^3 + 3x^2 - 5x - 3$.**

▶ **Solution**

$$-(-5x^3 + 3x^2 - 5x - 3)$$
$$= -(-5x^3) - (3x^2) - (-5x) - (-3) \quad \longleftarrow \text{ Find the opposite of each term.}$$
$$= 5x^3 - 3x^2 + 5x + 3$$

TRY THIS Find the opposite of $-3a^3 - 3.5a^2 - 0.5a$.

Recall the definition of subtraction for real numbers: $a - b = a + (-b)$. In other words, subtracting is the same as adding the opposite. This is also true for polynomials. To subtract one polynomial from another, add the opposite.

Because subtraction is really addition of the opposite, you can use the vertical-addition format when you subtract. Subtraction in vertical format for $(-5x^3 + 2x^2 - x + 5) - (-5x^3 + 3x^2 - 5x - 3)$ is shown below.

$$\begin{array}{r} -5x^3 + 2x^2 - x + 5 \\ - {-5x^3 + 3x^2 - 5x - 3} \\ \hline \end{array} \longrightarrow \begin{array}{r} -5x^3 + 2x^2 - x + 5 \\ + 5x^3 - 3x^2 + 5x + 3 \\ \hline -x^2 + 4x + 8 \end{array}$$

The following example shows how to use the Distributive Property to subtract polynomials.

EXAMPLE 2 **Find $(-5x^3 + 2x^2 - x + 5) - (-5x^3 + 3x^2 - 5x - 3)$.**

▶ **Solution**

$$(-5x^3 + 2x^2 - x + 5) - (-5x^3 + 3x^2 - 5x - 3)$$
$$= (-5x^3 + 2x^2 - x + 5) + (-1)(-5x^3 + 3x^2 - 5x - 3) \quad \longleftarrow \begin{array}{l}\textit{Rewrite subtraction as}\\ \textit{addition of the opposite.}\end{array}$$
$$= (-5x^3 + 2x^2 - x + 5) + (5x^3 - 3x^2 + 5x + 3) \quad \longleftarrow \textit{Distribute } -1 \textit{ to each term.}$$
$$= (-5 + 5)x^3 + (2 - 3)x^2 + (-1 + 5)x + (5 + 3) \quad \longleftarrow \textit{Combine like terms.}$$
$$= -x^2 + 4x + 8$$

TRY THIS Find $(3n^3 - n - 1) - (3n^2 - 5n - 2)$.

EXERCISES

KEY SKILLS

Write the opposite of each expression.

1. $(a + b)$

2. $(a - b)$

3. $(a - b + c)$

Rewrite each subtraction as addition of the opposite.

4. $(2x^2 - x + 4) - (x^2 + x + 2)$

5. $(3z^3 + z^2 - z + 5) - (4z^3 + z^2 - z + 5)$

PRACTICE

Find the opposite of each polynomial.

6. $-3x^3 + 2x^2 - x + 7$

7. $6x^3 - 5x^2 - 3x + 1$

8. $9x^4 + 3x^3 + x^2 - 2$

9. $-7x^5 + 8x^4 - 3x + 5$

Find each difference.

10. $(4n^3 + n^2 - 2n + 1) - (n^2 - 3n + 1)$

11. $(4m^3 - 2m + 1) - (5m^3 + 4m + 6)$

12. $(6p^4 - 3p^3 + 5p + 2) - (5p^2 + p - 1)$

13. $(6b^4 - 3b^3 - 5b + 1) - (6b^4 + b^3 - 5b + 1)$

14. $(s^4 - 5s^3 + s^2) - (-s^4 + s^3 - s^2 + 3)$

15. $(a^4 - a^3 + a^2 - a) - (a^4 + a^3 - a^2 - a)$

16. $(2c^3 - 3c^4 + 5c + 1) - (5c + c^2 - 3)$

17. $(d^3 + 3d^4 + 5d + 2) - (2d^3 - 5d + d^2 - 1)$

18. Subtract $2a^2 + 3ab - b^2$ from $6a^2 - 5b^2$.

19. Subtract $3x^2 - 3xy - y^2$ from $9x^2 + 7y^2$.

20. Subtract $3x^4 + x^3 + x^2 - 5x - 2$ from $2x^4 - x^3 + 2x^2 - 3x + 2$.

21. Subtract $x^4 - x^3 - x^2 - x - 1$ from $x^4 - x^3 + x^2 - x + 1$.

22. Subtract $4a^3 + 2a^2 - a + 1$ from $-2a^3 + a^2 + 2a + 1$.

MIXED REVIEW

Find an equation in slope-intercept form for the line that contains each point and that is parallel to the given line. (3.9 Skill A)

23. $A(-3, -3); y = \frac{1}{2}x - 2$

24. $P(0, 5); 2x - 3y = 7$

25. $Q(2, 7); y = 4$

Find an equation in slope-intercept form for the line that contains each point and that is perpendicular to the given line. (3.9 Skill B)

26. $Z(3, 6); y = -\frac{3}{4}x - \frac{3}{2}$

27. $N(2, 0); 4x + 7y = 1$

28. $Q(-1, 4); x = -3$

EXAMPLE 1

Represent the perimeter of the geometric figure below as a polynomial. In the figure, the four gray rectangles are congruent.

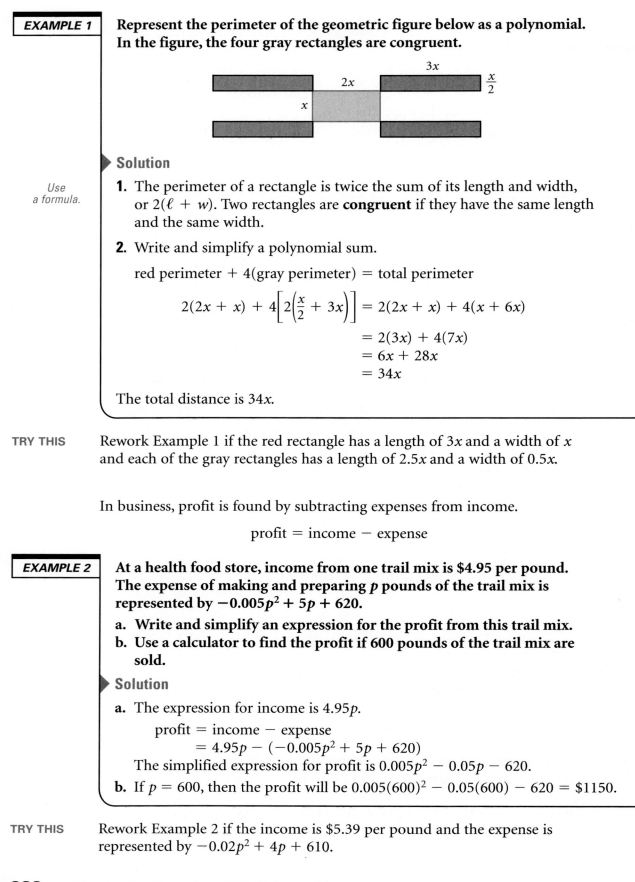

Use a formula.

▸ **Solution**

1. The perimeter of a rectangle is twice the sum of its length and width, or $2(\ell + w)$. Two rectangles are **congruent** if they have the same length and the same width.

2. Write and simplify a polynomial sum.

red perimeter + 4(gray perimeter) = total perimeter

$$2(2x + x) + 4\left[2\left(\frac{x}{2} + 3x\right)\right] = 2(2x + x) + 4(x + 6x)$$
$$= 2(3x) + 4(7x)$$
$$= 6x + 28x$$
$$= 34x$$

The total distance is $34x$.

TRY THIS

Rework Example 1 if the red rectangle has a length of $3x$ and a width of x and each of the gray rectangles has a length of $2.5x$ and a width of $0.5x$.

In business, profit is found by subtracting expenses from income.

profit = income − expense

EXAMPLE 2

At a health food store, income from one trail mix is \$4.95 per pound. The expense of making and preparing p pounds of the trail mix is represented by $-0.005p^2 + 5p + 620$.

a. **Write and simplify an expression for the profit from this trail mix.**
b. **Use a calculator to find the profit if 600 pounds of the trail mix are sold.**

▸ **Solution**

a. The expression for income is $4.95p$.

profit = income − expense
$$= 4.95p - (-0.005p^2 + 5p + 620)$$
The simplified expression for profit is $0.005p^2 - 0.05p - 620$.

b. If $p = 600$, then the profit will be $0.005(600)^2 - 0.05(600) - 620 = \$1150$.

TRY THIS

Rework Example 2 if the income is \$5.39 per pound and the expense is represented by $-0.02p^2 + 4p + 610$.

EXERCISES

KEY SKILLS

Refer to the geometric figure at right.
Rectangles of the same color are congruent.

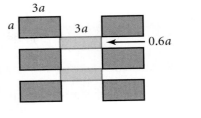

1. Write an expression for the perimeter of
   a. one red region. **b.** all the red regions.

2. Write an expression for the perimeter of
   a. one gray region. **b.** all the gray regions.

3. Write and simplify an expression for the entire perimeter.

PRACTICE

Write and simplify an expression for the total perimeter of each figure.
Rectangles of the same color are congruent.

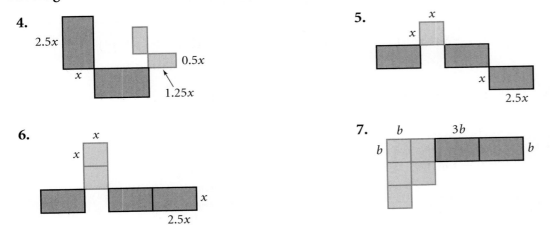

At a health food store, one trail mix sells for $5.50 per pound. The
polynomial $-0.005m^2 + 4m + 500$ represents the expense in dollars of
making p pounds of the mix. Use a calculator to find the profit or loss
on each sale.

8. 300 pounds   9. 400 pounds   10. 100 pounds   11. 150 pounds

MIXED REVIEW APPLICATIONS

Solve each problem. (6.2 Skills A and B)

12. If y varies directly as the square of x, and $y = 12$ when $x = 2$, find
    y when $x = 3$.

13. The bending of a beam varies directly as the mass of the load it
    supports. A beam is bent 20 mm by a mass of 40 kg. How much will
    the beam bend when supporting a mass of 100 kg?

Multiplying and Dividing Monomials

LESSON

Multiplying monomials

Recall from Lesson 6.1 the Properties of Exponents:

If a and b are nonzero and m and n are integers, then
$$a^m \cdot a^n = a^{m+n}, \ (a^m)^n = a^{m \cdot n}, \text{ and } (ab)^m = a^m b^m.$$

These properties allow you to multiply monomials.

EXAMPLE 1   **Multiply.**   **a.** $z^2 \cdot z^3$ **b.** $(2y)(-5y)$ **c.** $(-d^2)(-3d^3)$

▶ **Solution**

a. $z^2 \cdot z^3 = z^{2+3} = z^5$

b. $(2y)(-5y) = [(2)(-5)](y \cdot y) = -10y^2$

c. $(-d^2)(-3d^3) = [(-1)(-3)](d^2 \cdot d^3) = (3)(d^{2+3}) = 3d^5$

TRY THIS   Multiply.   **a.** $(-3h)(-h)$ **b.** $(2n^2)(4n)$ **c.** $(-4p^3)(-p^3)$

EXAMPLE 2   **Multiply.**   **a.** $(-2a^3)^3 \cdot a^2$ **b.** $(4x^2)\left(\frac{1}{3}x^4\right)(12x)$

▶ **Solution**

a. $(-2a^3)^3 \cdot a^2 = (-2)^3 \cdot a^{3 \cdot 3} \cdot a^2$

$\qquad = -8 \cdot a^9 \cdot a^2 = -8a^{11}$

> *Apply both the Power-of-a-Product and the Power-of-a-Power Properties.*

b. $(4x^2)\left(\frac{1}{3}x^4\right)(12x) = 4 \cdot \frac{1}{3} \cdot 12 \cdot (x^2 \cdot x^4 \cdot x^1)$

$\qquad = 16(x^{2+4+1}) = 16x^7$ ⟵ *Apply the Product Property of Exponents.*

TRY THIS   Multiply.   **a.** $\left(\frac{1}{4}t^3\right)^2(t^4)$ **b.** $(3a^2)\left(\frac{1}{3}a\right)(3a^0)$

EXAMPLE 3   **Multiply.**   **a.** $(2st^2)(3s^2t^4)$ **b.** $(3x^2y)(2x^4y^3)$

▶ **Solution**

a. $(2st^2)(3s^2t^4) = (2 \cdot 3)(s \cdot s^2)(t^2 \cdot t^4)$ ⟵ *Separate the variables.*

$\qquad = 6(s^{1+2})(t^{2+4})$ ⟵ *Apply the Product Property of Exponents.*

$\qquad = 6s^3t^6$

b. $(3x^2y)(2x^4y^3) = (3 \cdot 2)(x^2 \cdot x^4)(y \cdot y^3)$

$\qquad = 6(x^{2+4})(y^{1+3}) = 6x^6y^4$

TRY THIS   Multiply.   **a.** $(-3a^2b^2)\left(\frac{1}{3}a^3\right)$ **b.** $\left(\frac{1}{3}m^2n^3\right)(27mn^2)$

EXERCISES

Complete each simplification.

1. $[(-2)(-5)](x \cdot x^4)$

2. $[(6)(-6)](v^3 \cdot v^2)$

3. $(c^3 c^3)(d^2 d^4)$

4. $[(6)(1)](c^3 c^2)(d^2)$

Multiply.

5. $(a^2)(a)$

6. $(n^3)(n)$

7. $(m^3)(m^7)$

8. $(p^2)(p^2)$

9. $(3z^2)(10z^3)$

10. $(2a^2)(2a^2)$

11. $(-b^4)(-5b^3)$

12. $(-3q^{12})(-5q^3)$

13. $(4ab)^3$

14. $(3m^2 n^2)^2$

15. $(-2y^3 z^5)^3$

16. $(2r^2)(3r^4)(2r)$

17. $(-s^2)(3s^2)(-2s^2)$

18. $(-t)(-t^2)(-t^4)$

19. $(2t^2)(4t^4)(8t^8)$

20. $(7w^7)(5w^5)(w^3)$

21. $(-u^2)(-2.5u)(2u^4)$

22. $(4u^2 v^2)(5u^4 v^2)$

23. $(3a^3 b^5)(50a^2 b^3)$

24. $(-y^3 z^5)(-y^3 z^5)$

25. $(-2m^2 n^2)(-2m^4 n^4)$

26. $(7h^4 p^4)(7h^4 p^4)$

27. $(-3d^3 r^5)(-3d^5 r^3)$

28. $(2u^6 v^5)(3u^5 v^4)(4u^4 v^3)(5u^3 v^2)$

29. $(mn^2)(2m^2 n^3)(3m^3 n^4)(4m^4 n^5)$

30. $(-2a^3 b^4 c^3)(-2a^2 b^4 c^2)(-2a^5 b^3 c^5)$

31. $(3e^4 r^2 t^2)(5e^2 r^4 t^2)(-e^5 r^3 t^3)$

32. **Critical Thinking**   If $(a^3 b^3)(3a^m b^n)(a^{m+2} b^{n-2}) = 3a^7 b^9$, find m and n.

Find the slope of the line that contains each pair of points. If the line has no slope, write *none.* (3.4 Skill B)

33. $P\left(-1, \frac{1}{4}\right)$ and $Q\left(3, \frac{1}{4}\right)$

34. $A(3.5, 7)$ and $B(5, 8)$

35. $C(3, 0)$ and $D(3, -1)$

36. $M(3, 8)$ and $N(5, 12)$

37. $G(-1, 10)$ and $H(1, -10)$

38. $Y(-3, 4)$ and $Z(5, 1)$

You can use the Quotient Property of Exponents to simplify quotients containing monomials. In this section, you may assume that no denominator has a value of 0.

EXAMPLE 1

Divide. Write the answer with positive exponents.

a. $\dfrac{n^5}{n^3}$

b. $\dfrac{m^2}{m^5}$

c. $\dfrac{10r^2}{2r^6}$

▶ Solution

a. $\dfrac{n^5}{n^3} = n^{5-3} = n^2$

b. $\dfrac{m^2}{m^5} = m^{2-5} = m^{-3} = \dfrac{1}{m^3}$

c. $\dfrac{10r^2}{2r^6} = 5r^{2-6} = \dfrac{5}{r^4}$

TRY THIS

Divide. Write the answer with positive exponents. **a.** $\dfrac{b^5}{b^4}$ **b.** $\dfrac{d^6}{d^7}$ **c.** $\dfrac{-16y}{10y^2}$

You can use the Power-of-a-Product Property along with the Quotient Property to simplify expressions that contain both products and quotients.

EXAMPLE 2

Divide. Write the answer with positive exponents.

a. $\dfrac{(2a^2b)^2}{(4ab^2)^3}$

b. $\dfrac{(a^2b^3)^2}{(5a^6b^5)^2}$

▶ Solution

a. $\dfrac{(2a^2b)^2}{(4ab^2)^3} = \dfrac{4a^4b^2}{64a^3b^6}$ ⟵——— *Apply the Power-of-a-Product Property of Exponents.*

$= \dfrac{4}{64} \cdot \dfrac{a^4}{a^3} \cdot \dfrac{b^2}{b^6}$ ⟵——— *Separate the variables.*

$= \dfrac{4}{64} \cdot a^{4-3} \cdot b^{2-6}$ ⟵——— *Apply the Quotient Property of Exponents.*

$= \dfrac{1}{16} \cdot a^1 \cdot b^{-4}$

$= \dfrac{a}{16b^4}$ ⟵——— *Write with positive exponents.*

b. $\dfrac{(a^2b^3)^2}{(5a^6b^5)^2} = \dfrac{a^4b^6}{25a^{12}b^{10}}$ ⟵——— *Apply the Power-of-a-Product Property.*

$= \dfrac{1}{25} \cdot \dfrac{a^4}{a^{12}} \cdot \dfrac{b^6}{b^{10}}$ ⟵——— *Separate the variables.*

$= \dfrac{1}{25} \cdot a^{4-12} \cdot b^{6-10}$ ⟵——— *Apply the Quotient Property of Exponents.*

$= \dfrac{1}{25} \cdot a^{-8} \cdot b^{-4}$

$= \dfrac{1}{25a^8b^4}$ ⟵——— *Write with positive exponents.*

TRY THIS

Divide $\dfrac{(5u^2v^2)^3}{(5u^2v)^2}$. Write the answer with positive exponents.

EXERCISES

Rewrite each expression so that all exponents are positive.

1. a^{7-3}

2. b^{3-7}

3. m^{2-3}

4. v^{3-2}

PRACTICE

Divide. Write each answer with positive exponents.

5. $\dfrac{d^7}{d^2}$

6. $\dfrac{c^5}{c^6}$

7. $\dfrac{w^5}{w^7}$

8. $\dfrac{z^6}{z}$

9. $\dfrac{12a^5}{-3a^4}$

10. $\dfrac{-2f^4}{6f^3}$

11. $\dfrac{15g^4}{6g^5}$

12. $\dfrac{96q^7}{-12q^3}$

13. $\dfrac{(2s^2)^3}{s^2}$

14. $\dfrac{(-3t^2)^3}{(2t^2)^2}$

15. $\dfrac{(4a^3)^2}{(2a)^4}$

16. $\dfrac{(5n^3)^3}{(2n^2)^5}$

17. $\dfrac{24p^3q^2}{12p^3q^3}$

18. $\dfrac{12a^2d^4}{24a^3d^2}$

19. $\dfrac{y^3z^3}{5yz}$

20. $\dfrac{100y^3z}{25y^4z^3}$

21. $\dfrac{(2mn^3)^3}{(mn)^2}$

22. $\dfrac{(3a^3z)^4}{(9az)^3}$

23. $\dfrac{(2a^3b^2)^2}{(2a^4b^3)^2}$

24. $\dfrac{(2kp^2)^3}{(2k^4p)^2}$

Simplify the individual quotients. Then multiply.

25. $\left(\dfrac{a^3}{a^2}\right)\left(\dfrac{c^2}{c^3}\right)$

26. $\left(\dfrac{m^4}{m^3}\right)\left(\dfrac{n^5}{n^3}\right)$

27. $\left(\dfrac{-q^5}{-q^4}\right)\left(\dfrac{k^4}{k^3}\right)$

28. $\left(\dfrac{-d^3}{d^3}\right)\left(\dfrac{s^4}{s^3}\right)$

29. $\left(\dfrac{12x^3}{7x^3}\right)\left(\dfrac{21y^5}{15y^4}\right)$

30. $\left(\dfrac{-2a^7}{3a^3}\right)\left(\dfrac{27z^5}{8z^5}\right)$

31. $\left(\dfrac{-7b^7}{15b^5}\right)\left(\dfrac{5c^5}{-14c^3}\right)$

32. $\left(\dfrac{-49u^7}{11u^5}\right)\left(\dfrac{22v^7}{-7v^6}\right)$

33. $\left(\dfrac{16p^4}{9p^2}\right)^2\left(\dfrac{3a^3}{4a^2}\right)^2$

34. $\left(\dfrac{h^6r^3}{2h^2r^2}\right)^2\left(\dfrac{4h^3r^4}{r^3}\right)^2$

35. $\left(\dfrac{f^4b^3}{4f^3b^2}\right)^2\left(\dfrac{4f^3b^3}{b^2}\right)^2$

36. $\left(\dfrac{-9a^5b}{3a^2b}\right)^2\left(\dfrac{2a^2b^3}{ab^2}\right)^3$

MIXED REVIEW

Find the equation in point-slope form for the line with the given slope that contains the given point. (3.6 Skill A)

37. slope -2.5; $P(-3, -5)$

38. slope 6; $N(2.5, -3)$

39. slope $\dfrac{3}{7}$; $Z(0, 0)$

40. slope $-\dfrac{7}{2}$; $Q(2, 0)$

41. slope 0; $W(3, 5)$

42. slope $\dfrac{1}{5}$; $A(0, 3)$

Recall that the volume of a rectangular solid is given by the following equation.

$$\text{Volume} = \text{length} \times \text{width} \times \text{height}$$

EXAMPLE 1

A box's length is twice its width, and its height is three times its width. Write an expression for the total volume of 24 boxes.

▶ **Solution**

Use a formula.

Let w represent the width. Then length is $2w$ and height is $3w$.

Write and simplify an expression for volume.

$$24(w)(2w)(3w) = (24 \cdot 1 \cdot 2 \cdot 3)(w \cdot w \cdot w) = 144w^3$$

The total volume of 24 boxes is $144w^3$ cubic units.

TRY THIS

Find the total volume of 48 boxes whose dimensions are x, $5x$, and $10x$.

The volume and surface area of a cube with edge length l are given by the following equations.

$$\text{Volume} = l^3 \qquad \text{Surface area} = 6l^2$$

Recall from Lesson 3.3 that a ratio is the comparison of two quantities by division. In the next example, we will find the ratio of volume to surface area.

EXAMPLE 2

A cube has an edge length of $3d$ units. Write a ratio to compare the cube's volume to its surface area.

▶ **Solution**

Use a formula.

1. Write expressions for volume and surface area.

Volume: $(3d)^3$ Surface area: $6(3d)^2$

2. Write and simplify a ratio.

$$\frac{\text{Volume}}{\text{Surface area}} \rightarrow \frac{(3d)^3}{6(3d)^2} = \frac{27d^3}{54d^2}$$

$$= \frac{27}{54} \cdot d^{3-2} = \frac{1}{2} \cdot d = \frac{d}{2}$$

Therefore, the ratio of volume to surface area for this cube is $\frac{d}{2}$.

TRY THIS

A cube has an edge length of $4d$ units. Write a ratio to compare the cube's volume to its surface area.

Monomials can be found in several other geometry formulas, including some that you will study in later chapters as well as in future courses.

EXERCISES

Refer to the diagram at right.

1. Identify the length, width, and height.

2. Write an expression for the volume.

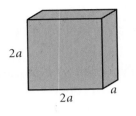

Solve each problem.

3. The dimensions of a rectangular box are x units, $3x$ units, and $5x$ units. Write an expression for the total volume of 36 boxes.

4. A cube has an edge length of $2x$ units. Write a ratio to compare the cube's volume to its surface area.

5. The sides of a square are x units long. The sides of a second square are kx units long, where k is a positive number. Write a ratio to compare the area of the second square to the area of the first square.

6. The area, A, of a circle whose radius is r units is $A = \pi r^2$. Let k be a positive number. Write a ratio to compare the area of a circle whose radius is kr with the area of a circle whose radius is r.

7. **Critical Thinking** The surface area, S, of a rectangular box with length l, width w, and height h is $S = 2(lw + lh + hw)$. Write a ratio to compare volume to surface area when $w = 2l$ and $h = 5l$.

8. Write a ratio to compare the total area of the red squares with the total perimeter of the figure below. (All red squares are congruent.)

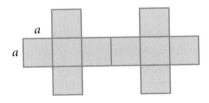

An automobile traveling at 55 miles per hour had a stopping distance of 125 feet. (6.2 Skill B)

9. Find the stopping distance if the speed is 60 miles per hour.

10. Write a ratio to compare the stopping distance for 60 miles per hour to the stopping distance for 30 miles per hour.

6.6 LESSON
Multiplying Polynomials

SKILL A *Multiplying a polynomial by a monomial*

To multiply a polynomial by a monomial, apply the Distributive Property. Then use the Product Property of Exponents to simplify the expression.

EXAMPLE 1 **Simplify $2m(m + 3)$.**

▶ **Solution**

$$2m(m + 3) = 2m(m) + 2m(3) \quad \longleftarrow \text{\textit{Apply the Distributive Property.}}$$
$$= 2m^2 + 6m$$

TRY THIS Simplify $2n^2(n - 4)$.

The Distributive Property can be extended to sums with more than two terms. In the examples below, 2 is distributed to three terms, and $3x$ is distributed to four terms.

$$2(x^2 + x + 3) \qquad\qquad 3x(x^3 + 2x^2 + 4x + 1)$$
$$2(x^2) + 2(x) + 2(3) \qquad 3x(x^3) + 3x(2x^2) + 3x(4x) + 3x(1)$$
$$2x^2 + 2x + 6 \qquad\qquad 3x^4 + 6x^3 + 12x^2 + 3x$$

Use this extended form of the Distributive Property to multiply a monomial by a polynomial with many terms.

EXAMPLE 2 **Simplify $2x^2(3x^3 + 5x^2 - 4)$.**

▶ **Solution**

$$2x^2(3x^3 + 5x^2 - 4)$$
$$= 2x^2(3x^3) + 2x^2(5x^2) + 2x^2(-4) \quad \longleftarrow \text{\textit{Apply the Distributive Property.}}$$
$$= 6x^5 + 10x^4 - 8x^2$$

TRY THIS Simplify $2x^3(5x^2 - x - 5)$.

EXAMPLE 3 **Simplify $2a^2b(3ab^2 - 4ab)$.**

▶ **Solution**

$$2a^2b(3ab^2 - 4ab) = 2a^2b(3ab^2) + 2a^2b(-4ab)$$
$$= 6(a^2 \cdot a)(b \cdot b^2) - 8(a^2 \cdot a)(b \cdot b) \quad \longleftarrow \text{\textit{Separate the variables.}}$$
$$= 6a^3b^3 - 8a^3b^2$$

TRY THIS Simplify $r^2s(3r^2s - r^3s)$.

EXERCISES

Write the sum that results when you apply the Distributive Property. Do not simplify further.

1. $3z^2(2z + 4)$

2. $3p^2(p^2 + 3p - 1)$

3. $uv(2uv^2 + 3u^2v)$

Complete each simplification.

4. $(5x^3)(-2x^3) + (5x^3)(2x^2)$

5. $(y^3)(-2y^4) + (y^3)(4y)$

6. $(5s^2)(-2s^4) + (5s^2)(4s)$

PRACTICE

Find each product.

7. $3x(4x + 5)$

8. $4y(5y - 2)$

9. $-2a^2(3a - 1)$

10. $3b^3(3b - 2)$

11. $c^3(10c^2 + 5c)$

12. $2n^3(-3n^2 - n)$

13. $q^3(3q^2 - q + 3)$

14. $r^3(2r^2 - 5r - 1)$

15. $w(2w^4 - 5w^3 - w^2)$

16. $-2d^2(d^4 - 3d^3 + 2d^2)$

17. $(2k^5 + 4k^3 - 2k^2)(4k^3)$

18. $(12h^4 + 4h^2 - 3h)(-h^2)$

19. $rs(rs^2 - rs)$

20. $m^2n(m^2n^2 + mn^2)$

21. $-2x^2y^2(3x^2y - 5x^2y^2)$

22. $-2a^4b^2(7a^3b + 4a^4b^3)$

23. $2u^2v^2(7u^3 + 4u^2v^2 - 4v^3)$

24. $u^3v^2(-3u^2 - 5u^3v^3 + v^2)$

25. Simplify $2m^2(3m^2 - m) + 3m(2m^3 + 2m^2)$.

26. Simplify $3z^2(2z^2 - z) - 3z(5z^3 + z^2)$.

27. Find a such that $2x^a(5x^{2a-3} + 2x^{2a+2}) = 10x^3 + 4x^8$.

28. Critical Thinking   Find m and n such that $x^my^n(x^ny^m + x^{n-2}y) = x^5y^5 + x^3y^3$.

29. Critical Thinking   Write $2x(a + b) + 3y(a + b)$ as a product of two binomials.

MIXED REVIEW

Graph each inequality in the coordinate plane. (5.5 Skill A)

30. $y \geq 4$

31. $x < -3$

32. $y > -\frac{2}{3}x + 1$

33. $y < \frac{5}{4}x - 1$

34. $y \leq \frac{1}{6}x$

35. $y \geq -3.5x$

You can use a vertical format to multiply polynomials.

EXAMPLE 1   **Multiply $(2r - 3)(3r + 2)$.**

▶ Solution

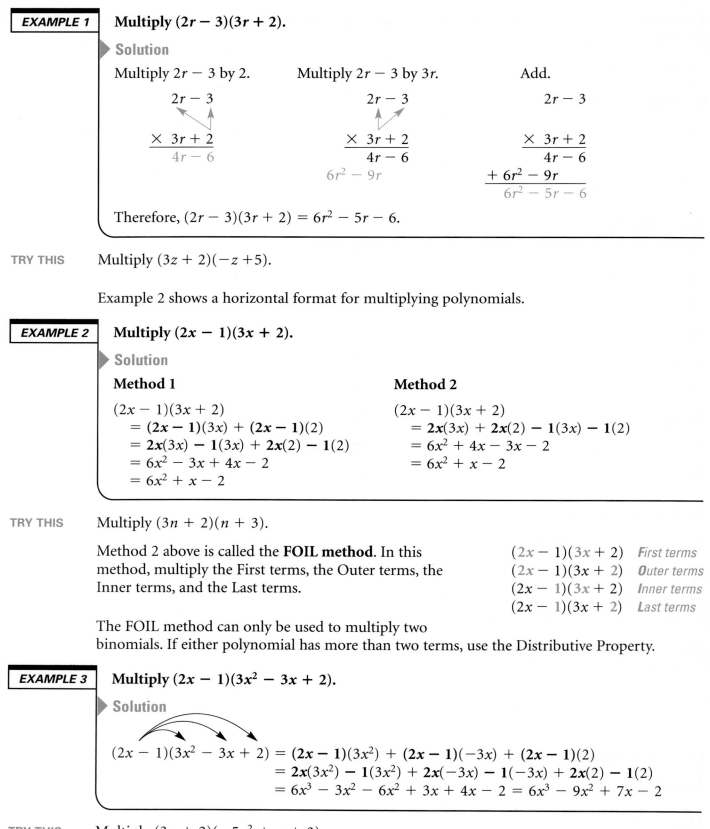

Multiply $2r - 3$ by 2.

$$
\begin{array}{r}
2r - 3 \\
\times\ 3r + 2 \\
\hline
4r - 6
\end{array}
$$

Multiply $2r - 3$ by $3r$.

$$
\begin{array}{r}
2r - 3 \\
\times\ 3r + 2 \\
\hline
4r - 6 \\
6r^2 - 9r
\end{array}
$$

Add.

$$
\begin{array}{r}
2r - 3 \\
\times\ 3r + 2 \\
\hline
4r - 6 \\
+\ 6r^2 - 9r \\
\hline
6r^2 - 5r - 6
\end{array}
$$

Therefore, $(2r - 3)(3r + 2) = 6r^2 - 5r - 6$.

TRY THIS   Multiply $(3z + 2)(-z + 5)$.

Example 2 shows a horizontal format for multiplying polynomials.

EXAMPLE 2   **Multiply $(2x - 1)(3x + 2)$.**

▶ Solution

Method 1

$(2x - 1)(3x + 2)$
$= (2x - 1)(3x) + (2x - 1)(2)$
$= 2x(3x) - 1(3x) + 2x(2) - 1(2)$
$= 6x^2 - 3x + 4x - 2$
$= 6x^2 + x - 2$

Method 2

$(2x - 1)(3x + 2)$
$= 2x(3x) + 2x(2) - 1(3x) - 1(2)$
$= 6x^2 + 4x - 3x - 2$
$= 6x^2 + x - 2$

TRY THIS   Multiply $(3n + 2)(n + 3)$.

Method 2 above is called the **FOIL method**. In this
method, multiply the First terms, the Outer terms, the
Inner terms, and the Last terms.

$(2x - 1)(3x + 2)$ *First terms*
$(2x - 1)(3x + 2)$ *Outer terms*
$(2x - 1)(3x + 2)$ *Inner terms*
$(2x - 1)(3x + 2)$ *Last terms*

The FOIL method can only be used to multiply two
binomials. If either polynomial has more than two terms, use the Distributive Property.

EXAMPLE 3   **Multiply $(2x - 1)(3x^2 - 3x + 2)$.**

▶ Solution

$(2x - 1)(3x^2 - 3x + 2) = (2x - 1)(3x^2) + (2x - 1)(-3x) + (2x - 1)(2)$
$= 2x(3x^2) - 1(3x^2) + 2x(-3x) - 1(-3x) + 2x(2) - 1(2)$
$= 6x^3 - 3x^2 - 6x^2 + 3x + 4x - 2 = 6x^3 - 9x^2 + 7x - 2$

TRY THIS   Multiply $(3n + 2)(-5n^2 + n + 3)$.

EXERCISES

Complete each multiplication.

1. $(x + 3)(x + 10) = (x + 3)(x) + (x + 3)(10)$

2. $(a - 2)(a - 7) = (a - 2)(a) + (a - 2)(-7)$

3. $(w + 3)(w^2 - 5w) = (w + 3)(w^2) - (w + 3)(5w)$

4. $(n - 1)(n^2 + 7n) = (n - 1)(n^2) + (n - 1)(7n)$

PRACTICE

Multiply.

5. $(x + 2)(x + 3)$

6. $(a + 2)(a + 3)$

7. $(b + 2)(b - 3)$

8. $(r + 1)(-r + 5)$

9. $(2z + 2)(z - 3)$

10. $(n + 2)(3n + 3)$

11. $(2m + 2)(4m + 3)$

12. $(3s + 1)(3s + 5)$

13. $(-v - 4)(v + 4)$

14. $(-k + 1)(-k + 7)$

15. $(3p - 3)(-2p + 5)$

16. $(-2s - 3)(-2s + 4)$

17. $(x + 1)(x^2 + x + 1)$

18. $(w + 2)(w^2 + 3w + 2)$

19. $(2r - 1)(r^2 - r + 1)$

20. $(a - 4)(2a^2 - a - 3)$

21. $(-c + 3)(-c^2 + c + 3)$

22. $(-d + 1)(-d^2 - 2d + 5)$

23. $(-5n + 4)(3n^2 + 2n - 7)$

24. $(3q - 5)(-4q^2 + 5q + 7)$

25. $(-5h + 5)(5h^2 + 5h + 5)$

26. $x(3x - 5)(x - 2)$

27. $(2z - 1)(z + 2)(z - 3)$

28. $(3x + 1)(2x - 3)(x - 1)$

29. $3ab(a + b)(a - b)$

30. $2mn(2n + m)(2m + n)$

31. $2c^2d(2d^2 - c)(2c^2 - d)$

32. Find a so that $(a + 2)(2a + 2) = (a - 2)(2a + 2)$.

33. Find r so that $(r - 3)(2r + 1) = (r - 2)(2r - 3)$.

34. Critical Thinking   Show that $(a + n)[a + (n + 1)] = a^2 + n^2 + 2an + a + n$.

MIXED REVIEW

Find each sum. (6.4 Skill A)

35. $(4c^2 + 2c - 2) + (-4c^2 - 2c + 3)$

36. $(n^3 + 5n - 1) + (-4n^2 + 2n)$

37. $(-n^2 + 7n) + (n - 3)$

38. $(t^2 + 7t + 2) + (t^2 - 7t - 2)$

Find each difference. (6.4 Skill B)

39. $(2d^3 - 7d^2 + 2) - (d^3 - 7d^2 - 1)$

40. $(-2m^3 - m^2 + m) - (m^3 - 7m^2 + 4m + 1)$

A **perfect-square trinomial** is the result of squaring a binomial. The **difference of two squares** is an expression of the form $a^2 - b^2$.

Special Products

Perfect-square trinomial: $a^2 + 2ab + b^2 = (a + b)^2$

Difference of two squares: $a^2 - b^2 = (a + b)(a - b)$

EXAMPLE 1 Write $(3x - 4)^2$ as a perfect-square trinomial.

▶ **Solution**

Rewrite $(3x - 4)^2$ as $[3x + (-4)]^2$.
Use the form for a perfect-square trinomial with $a = 3x$ and $b = -4$.
$$(3x - 4)^2 = (3x)^2 + 2(3x)(-4) + (-4)^2$$
$$= 9x^2 - 24x + 16$$

TRY THIS Write as a perfect-square trinomial.   **a.** $(-2x + 5)^2$ **b.** $(5d - 2)^2$

EXAMPLE 2 Multiply $(5st - 4)(5st + 4)$.

▶ **Solution**

Use the form for a difference of two squares with $a = 5st$ and $b = 4$.
$$(5st - 4)(5st + 4) = (5st)^2 - (4)^2$$
$$= 25s^2t^2 - 16$$

TRY THIS Multiply $(5c + 6d)(5c - 6d)$.

You can use a vertical format to find the special products given above.

$$\begin{array}{r} a + b \\ \times\, a + b \\ \hline ab + b^2 \\ +\, a^2 + ab \\ \hline a^2 + 2ab + b^2 \end{array} \qquad\qquad \begin{array}{r} a + b \\ \times\, a - b \\ \hline -ab - b^2 \\ +\, a^2 + ab \\ \hline a^2 \qquad - b^2 \end{array}$$

You can also use an area model to find special products.

Area: $(a + b)(a + b)$
Area: $a^2 + ab + ab + b^2$
$\quad = a^2 + 2ab + b^2$

Area: $a^2 - b^2$
Area: $(a - b)^2 + b(a - b) + b(a - b) = (a - b)(a - b) + 2b(a - b) = (a - b)(a - b + 2b) = (a - b)(a + b)$

EXERCISES

Identify a and b in each special product.

1. $(2d + 4)^2$ **2.** $(5z - 3)^2$ **3.** $(3k + 5)(3k - 5)$ **4.** $(-m + 3)(-m - 3)$

PRACTICE

Find each special product.

5. $(k + 1)^2$ **6.** $(v + 2)^2$ **7.** $(x - 2)^2$

8. $(g - 5)^2$ **9.** $(2h + 3)^2$ **10.** $(2h - 1)^2$

11. $(-2t - 3)^2$ **12.** $(-3x + 5)^2$ **13.** $(t + 1)(t - 1)$

14. $(z + 5)(z - 5)$ **15.** $(2z - 1)(2z + 1)$ **16.** $(3z - 2)(3z + 2)$

17. $(3 - d)(3 + d)$ **18.** $(5 + d)(5 - d)$ **19.** $(2d + 3)(2d - 3)$

20. $(7v - 5)(7v + 5)$ **21.** $(2u + 5v)^2$ **22.** $(3m - 2n)^2$

23. $(5r + 4s)(5r - 4s)$ **24.** $(2c - 3d)(2c + 3d)$ **25.** $(2rs + 3)^2$

26. $(3xy - 5)^2$ **27.** $(4pq + 1)(4pq - 1)$ **28.** $(3yz - 2)(3yz + 2)$

29. Use multiplication to show that $(a - b)^2 = a^2 - 2ab + b^2$.

30. Find m such that $(m - 3)^2 = m^2 - 3^2$.

31. Find n such that $(n + 3)^2 = n^2 + 3^2$.

32. How are m and n related if $[(m + n) + 2]^2 = (m + n)^2 + 2^2$?

33. Simplify $[(a + b) + c][(a + b) - c]$.

34. Use $(a + b)^3 = (a + b)(a + b)^2$ to write $(a + b)^3$ as a sum.

35. **Critical Thinking** Use mental math to find the product of 38 and 42. (*Hint:* Use the difference of two squares.)

MIXED REVIEW

Find each quotient. Write the answers with positive exponents. Use mental math where possible. (6.5 Skill B)

36. $\dfrac{4f^2}{2f}$ **37.** $\dfrac{21y^6}{3y^3}$ **38.** $\dfrac{5h^2}{25h^4}$ **39.** $-\dfrac{ab}{2a^2b^2}$

40. $\dfrac{(2w^2)^3}{2w^7}$ **41.** $\dfrac{(5mn)^2}{25m^3n}$ **42.** $\dfrac{6(uv)^4}{36(uv)^3}$ **43.** $\dfrac{(-2)^4a}{32ab^3}$

6.7 Dividing Polynomials

SKILL A   *Dividing a polynomial by a monomial*

To divide an expression by a monomial, use the following forms of the Distributive Property.

Division and the Distributive Property

If a, b, and c are real numbers and $c \neq 0$, then

$$\frac{a + b}{c} = \frac{a}{c} + \frac{b}{c} \quad \text{and} \quad \frac{a - b}{c} = \frac{a}{c} - \frac{b}{c}.$$

You can prove the first part of the statement above as shown below.

$$\frac{a + b}{c} = \frac{1}{c}(a + b) \qquad \longleftarrow \text{ Definition of division}$$

$$= \frac{1}{c} \cdot a + \frac{1}{c} \cdot b \qquad \longleftarrow \text{ Distributive Property}$$

$$= \frac{a}{c} + \frac{b}{c} \qquad \longleftarrow \text{ Definition of division}$$

EXAMPLE 1
   a. Divide $45x^3 - 25x^2$ by $5x$.   **b.** Divide $4a^3b^2 + 8a^2b$ by $2ab$.

▶ **Solution**

a.
$$\frac{45x^3 - 25x^2}{5x}$$

$$= \frac{45x^3}{5x} - \frac{25x^2}{5x}$$

$$= \frac{45}{5} \cdot \frac{x^3}{x} - \frac{25}{5} \cdot \frac{x^2}{x}$$

$$= 9x^2 - 5x$$

b.
$$\frac{4a^3b^2 + 8a^2b}{2ab}$$

$$= \frac{4a^3b^2}{2ab} + \frac{8a^2b}{2ab}$$

$$= \frac{4}{2} \cdot \frac{a^3}{a} \cdot \frac{b^2}{b} + \frac{8}{2} \cdot \frac{a^2}{a} \cdot \frac{b}{b}$$

$$= 2a^2b + 4a$$

TRY THIS
   a. Divide $32n^4 + 8n^2$ by $16n^2$.   **b.** Divide $4u^3v^3 - 4u^2v^2$ by $4u^2v^2$.

When you divide one polynomial by another, you may get an expression that is not a polynomial. The set of polynomials is not *closed* under division. For example, $\frac{x}{x^2} = x^{1-2} = x^{-1} = \frac{1}{x}$, which is *not* a polynomial.

EXAMPLE 2
   Divide $20x^2 - 15$ by $5x$.

▶ **Solution**

$$\frac{20x^2 - 15}{5x} = \frac{20x^2}{5x} - \frac{15}{5x} \qquad \longleftarrow \text{ Recall that } \frac{a - b}{c} = \frac{a}{c} - \frac{b}{c}.$$

$$= \frac{20}{5} \cdot \frac{x^2}{x} - \frac{15}{5} \cdot \frac{1}{x} = 4x - \frac{3}{x} \qquad \longleftarrow \text{ This is not a polynomial.}$$

TRY THIS
   Divide $30m^4 - 15m$ by $5m^2$.

EXERCISES

Complete each division.

1. $\dfrac{15n^2 + 6n}{3n} = \dfrac{15n^2}{3n} + \dfrac{6n}{3n}$

2. $\dfrac{5k^3 - 12k^2}{5k} = \dfrac{5k^3}{5k} - \dfrac{12k^2}{5k}$

3. $\dfrac{y^3z^3 - yz^2}{yz} = \dfrac{y^3z^3}{yz} - \dfrac{yz^2}{yz}$

4. $\dfrac{3p^3q^2 + 6p^2q^2}{2pq} = \dfrac{3p^3q^2}{2pq} + \dfrac{6p^2q^2}{2pq}$

PRACTICE

Find each quotient below. Is the quotient a polynomial?

5. $\dfrac{12a + 15}{3}$

6. $\dfrac{14y + 7}{7}$

7. $\dfrac{20x - 8}{-4}$

8. $\dfrac{-100d + 50}{25}$

9. $\dfrac{8c^2 - 8c}{2c}$

10. $\dfrac{-10r^2 + 2r}{-2r}$

11. $\dfrac{10s^3 - 20s^2}{4s}$

12. $\dfrac{6v^3 - 24v^2}{-6v}$

13. $\dfrac{a^3b^3 + a^2b^2}{ab}$

14. $\dfrac{c^3d^3 - 3cd}{cd}$

15. $\dfrac{7k^3p - 49k^2p^3}{7kp}$

16. $\dfrac{6r^2d + 27r^2d^2}{3rd}$

17. $\dfrac{s^2 - 1}{s}$

18. $\dfrac{2r^2 + 1}{r}$

19. $\dfrac{-12d^2 + 6}{3d}$

20. $\dfrac{-10q^2 + 2}{-q}$

21. $\dfrac{14z^3 - 28}{7z}$

22. $\dfrac{15c^3 - 27}{3c}$

23. $\dfrac{x^3 + x}{x^2}$

24. $\dfrac{y^4 - y}{y^2}$

25. $\dfrac{2z^3 - 6z}{z}$

26. When $24y + 36$ is divided by a, the quotient has only positive integer coefficients. What are the possible values of a?

27. **Critical Thinking** Use $\dfrac{a - b}{c} = \dfrac{a + (-b)}{c}$ to show that $\dfrac{a - b}{c} = \dfrac{a}{c} - \dfrac{b}{c}$.

MIXED REVIEW

Multiply. (6.5 Skill A)

28. $(-2a^2)(5a)(-2a^2)$

29. $(-3u)(-3u)(-3u^3)$

30. $(a^2b^2)(a^2b)(ab^2)$

Multiply. (6.6 Skill A)

31. $(2d^2)(3d^2 + 2d + 1)$

32. $(-v^2)(3v^3 - 2v^2 + 2v)$

33. $(a^2)(a^3 + a^2 + a + 1)$

To divide a polynomial by a binomial, you can use long division, as shown in Example 1.

EXAMPLE 1   **Divide $2x^2 + 5x - 3$ by $x + 3$.**

▶ **Solution**

1. Divide x into $2x^2$.

$$\begin{array}{r} 2x \\ x + 3 \overline{)\, 2x^2 + 5x - 3} \end{array}$$

2. Multiply $x + 3$ by $2x$.

$$\begin{array}{r} 2x \\ x + 3 \overline{)\, 2x^2 + 5x - 3} \\ 2x^2 + 6x \end{array}$$

3. Subtract.

$$\begin{array}{r} 2x \\ x + 3 \overline{)\, 2x^2 + 5x - 3} \\ -(2x^2 + 6x) \\ \hline -x \end{array}$$

4. Bring down -3.

$$\begin{array}{r} 2x \\ x + 3 \overline{)\, 2x^2 + 5x - 3} \\ 2x^2 + 6x \downarrow \\ \hline -x - 3 \end{array}$$

5. Divide x into $-x$.

$$\begin{array}{r} 2x - 1 \\ x + 3 \overline{)\, 2x^2 + 5x - 3} \\ 2x^2 + 6x \\ \hline -x - 3 \end{array}$$

6. Multiply $x + 3$ by -1. Then subtract.

$$\begin{array}{r} 2x - 1 \\ x + 3 \overline{)\, 2x^2 + 5x - 3} \\ 2x^2 + 6x \\ \hline -x - 3 \\ -(-x - 3) \\ \hline 0 \end{array}$$

The quotient is $2x - 1$.

TRY THIS   Divide $3y^2 - y - 2$ by $y - 1$.

The dividend must contain every possible power of the variable. If one is missing, include a coefficient of 0 for that term. For example, to divide $x^2 - 4$ by $x - 2$, rewrite $x^2 - 4$ as $x^2 + 0x - 4$, and then proceed as in Example 1. The quotient is $x + 2$.

$$x - 2 \overline{)\, x^2 - 4} \quad \rightarrow \quad \begin{array}{r} x + 2 \\ x - 2 \overline{)\, x^2 + 0x - 4} \\ -(x^2 - 2x) \downarrow \\ \hline 2x - 4 \\ -(2x - 4) \\ \hline 0 \end{array}$$

EXAMPLE 2   **The product of two polynomials is $a^2 + 6a - 16$. If one polynomial is $a - 2$, what is the other?**

▶ **Solution**

$$(a - 2)(\text{polynomial}) = a^2 + 6a - 16$$

Divide $a^2 + 6a - 16$ by $a - 2$ to find that the unknown polynomial is $a + 8$.

$$\begin{array}{r} a + 8 \\ a - 2 \overline{)\, a^2 + 6a - 16} \\ -(a^2 - 2a) \downarrow \\ \hline 8a - 16 \\ -(8a - 16) \\ \hline 0 \end{array}$$

TRY THIS   The product of two polynomials is $m^2 - 49$. If one polynomial is $m - 7$, what is the other?

EXERCISES

In Exercises 1–3, check the division by multiplying the divisor and the quotient. If the division is correct, your answer will be the dividend.

1.
$$x + 2\overline{)x^2 + 5x + 6}^{\,x + 3}$$

2.
$$x - 3\overline{)x^2 + 2x - 15}^{\,x + 5}$$

3.
$$x - 2\overline{)x^2 - 9x + 14}^{\,x - 7}$$

PRACTICE

Divide.

4. $\dfrac{z^2 + 3z - 10}{z - 2}$

5. $\dfrac{b^2 + 2b - 8}{b - 2}$

6. $\dfrac{c^2 + 8c + 7}{c + 7}$

7. $\dfrac{u^2 - 7u - 30}{u - 10}$

8. $\dfrac{n^2 - 5n - 36}{n - 9}$

9. $\dfrac{w^2 - w - 42}{w + 6}$

10. $\dfrac{3a^2 - 13a - 10}{a - 5}$

11. $\dfrac{3h^2 - 4h - 15}{h - 3}$

12. $\dfrac{2v^2 + 15v + 28}{v + 4}$

13. $4k^2 - 19k - 5$ by $k - 5$

14. $3c^2 + 17c - 56$ by $c + 8$

15. $5p^2 + 18p - 8$ by $p + 4$

16. $n^2 - 25$ by $n + 5$

17. $x^2 - 64$ by $x + 8$

18. $25p^2 - 9$ by $5p + 3$

Solve each problem.

19. The product of two polynomials is $100z^2 - 49$. If one polynomial is $10z - 7$, what is the other?

20. The product of two polynomials is $81n^2 - 121$. If one polynomial is $9n + 11$, what is the other?

21. The product of two polynomials is $6a^2 + ab - b^2$. If one polynomial is $2a + b$, what is the other?

22. Show that if $6x^2 + 11x - 35$ is divided by $3x - 5$, the quotient is $2x + 7$.

MIXED REVIEW

Solve each absolute-value inequality. Graph the solutions on a number line. (4.7 Skill B)

23. $3|a - 2| \geq 6$

24. $|6(t + 1) - 5t| \geq 0$

25. $|4t - (t + 1)| < 0$

26. $|-2u + 5| < 5$

27. $3 + 3|x| \leq 12$

28. $4|r + 5| > 8$

Write each expression in the form a^b.

1. $3^2 \cdot 3^5$

2. $\dfrac{2.5^4}{2.5^4}$

3. $6^{-2} \cdot 6 \cdot 6^0$

4. $\dfrac{(-7)^3}{(-7)^5}$

Simplify.

5. $\dfrac{(-2)^3(-5)^5}{(-2)^5(-5)^3}$

6. $[(3^{-2} \cdot 3^2)^2 \cdot 3^2]^2$

7. $\dfrac{3^3 \cdot 3^{-5}}{(-2)^{-5}(-2)^3}$

Write each number in scientific notation.

8. 0.000312

9. $18{,}560$

Write each product or quotient in scientific notation.

10. $(2.1 \times 10^2)(5.0 \times 10^3)$

11. $\dfrac{6.9 \times 10^3}{3.0 \times 10^5}$

12. If y varies directly as the square of x, and $y = 20$ when $x = 4$, find y when $x = 10$.

13. If y varies directly as the cube of x, and $y = 250$ when $x = 5$, find y when $x = 12$.

The distance, d, that a falling object travels downward varies directly as the square of elapsed time, t. The constant of variation is 16 feet per second squared. Find d for each value of t.

14. 4.5 seconds

15. 12 seconds

Classify each polynomial by degree and number of terms.

16. $3a^2 - 2a^3 + a^2 - a^2$

17. $t - 3 - 2t^3 + 2t^2 + 4$

The surface area, s, of a cylindrical container with a height of 6 feet can be represented by $s = 2\pi r^2 + 12\pi r$. Find the surface area of each cylindrical container with the given radius. Use 3.14 to approximate π.

18. 6.5 feet

19. 10 feet

Simplify.

20. $3g^5 + 2g^3 - g^2 + g^3 + g^2 + 3g - 1$

21. $x^3 + 5x^2 + 2x - 2 + 3x^2 + 5x + 2 + 2x^4 - 3x^2 - 7x$

Simplify.

22. $(2b^4 - 2b^3 + 3b^2) - (-2b^4 + b^3 - 2b^2 + 5)$

23. $(2v^4 - 5v^3 - v^2 + 5v + 3) - (3v^4 + v^3 + v^2 - 5v - 2)$

24. Write an expression for the perimeter of the geometric figure below. Rectangles of the same color are congruent.

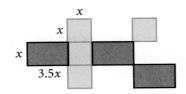

Simplify. Write each answer with positive exponents only.

25. $\left(\dfrac{3y^2}{6}\right)\left(\dfrac{4y^3}{5y^5}\right)(15y^2)$

26. $(3a^3b^2)(4a^5b)$

27. $\dfrac{(3x^2)^3}{(6x^2)^2}$

28. $\dfrac{(9m^2n^2)^2}{(3m^2n)^3}$

29. A rectangular box has dimensions $1.5c$, $3c$, and $9c$. Write an expression for the total volume of 36 of these boxes.

Find each product.

30. $3g^3(g^2 - 3g + 7)$

31. $2n^3p(np^2 + 6n^2p^2)$

32. $(3s + 5)(-2s + 11)$

33. $(-2t + 7)(2t^2 + 11t - 1)$

34. $(-5z + 7)(5z + 7)$

35. $(11a + 1)^2$

Find each quotient.

36. $\dfrac{5n^2 + 30n}{15n}$

37. $\dfrac{16y^5z^3 - 10y^2z^4}{2y^2z^3}$

38. $\dfrac{2b^3 + 2b^2 + 3}{2b}$

39. $\dfrac{2d^4 - 2d^2h + 5d^2h^2}{2d^2h}$

40. $(a^2 - 15a + 56) \div (a - 8)$

41. $(4q^2 - 44q + 121) \div (2q - 11)$

42. The product of two polynomials is $12n^2 + 5n - 72$. If one polynomial is $3n + 8$, what is the other?

CHAPTER

Factoring Polynomials

▸ **What You Already Know**

In Chapter 6, you learned how to multiply two binomial expressions and write the product of them in simplest form. In earlier mathematics courses, you learned what factoring a positive integer means and you learned how to do it.

▸ **What You Will Learn**

In Chapter 7, you are going to learn how to factor polynomials.

Studying how to factor polynomials involves many procedures.

- Recognize factoring patterns such as the difference of two squares and perfect-square trinomials.

- Factoring expressions of the form $ax^2 + bx + c$, where $a = 1$.

- Factoring expressions of the form $ax^2 + bx + c$, where $a \neq 1$.

- Factoring expressions by finding the greatest common factor.

- Factoring polynomial expressions by grouping.

Finally, you will learn how to apply your factoring skills to

- find the number of x-intercepts of the graph of a quadratic function, and

- find the number of real solutions to a quadratic equation.

VOCABULARY

axis of symmetry	prime factorization
common binomial factor	prime number
composite number	quadratic equation
cubic equation	quadratic function
difference of two squares	relatively prime
double root	root
factor	splitting the middle term
factored completely	standard form of a cubic equation
greatest common factor (GCF)	standard form of a quadratic equation
parabola	vertex
perfect-square trinomials	Zero-Product Property

The diagram below shows how mathematical skills and mathematical reasoning are interrelated with the skills and concepts in Chapter 7. Notice that the focus of this chapter is on factoring polynomials.

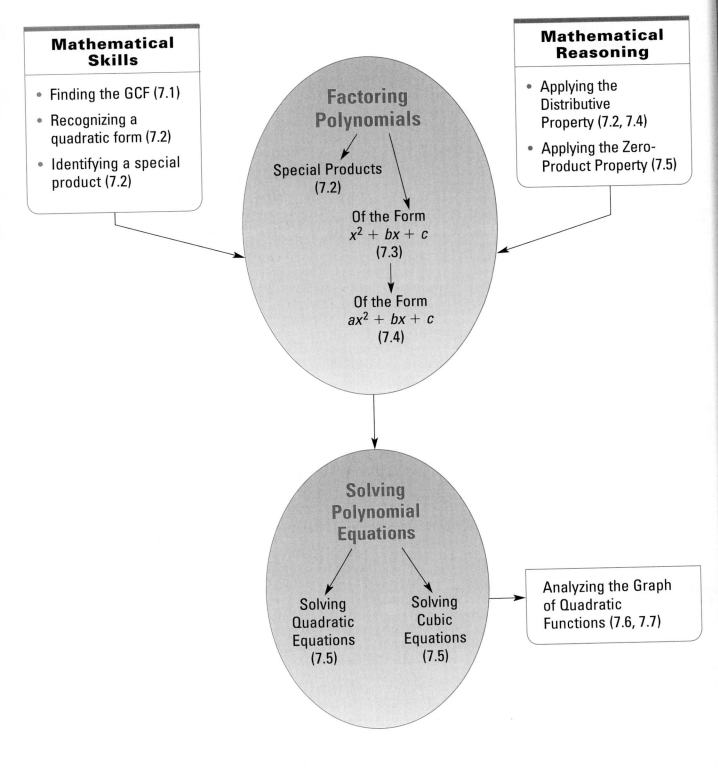

Mathematical Skills

- Finding the GCF (7.1)
- Recognizing a quadratic form (7.2)
- Identifying a special product (7.2)

Factoring Polynomials

Special Products (7.2)

Of the Form $x^2 + bx + c$ (7.3)

Of the Form $ax^2 + bx + c$ (7.4)

Mathematical Reasoning

- Applying the Distributive Property (7.2, 7.4)
- Applying the Zero-Product Property (7.5)

Solving Polynomial Equations

Solving Quadratic Equations (7.5)

Solving Cubic Equations (7.5)

Analyzing the Graph of Quadratic Functions (7.6, 7.7)

7.1 LESSON

Using the Greatest Common Factor to Factor an Expression

SKILL A ▸ *Factoring integers*

From your study of multiplication with real numbers, recall the following facts:

If $3 \times 4 = 12$, then

- 12 is the product of 4 and 3.
- 4 and 3 are *factors*, or *divisors*, of 12.
- The quotient of 12 divided by 4 is 3.
- The quotient of 12 divided by 3 is 4.

To **factor** a number is to write it as the product of two or more numbers, usually natural numbers. Factoring and division are closely related.

EXAMPLE 1   **Write three different factorizations of 16. Use natural numbers.**

▸ **Solution**

Divide 16 by several natural numbers to find divisors.

Make a table.

16 divided by	1	2	3	4	5	6	7	8
natural number?	16 ✔	8 ✔	no	4 ✔	no	no	no	2 ✔

Three factorizations of 16 are 1×16, 2×8, and 4×4.

TRY THIS   Write four different factorizations of 36. Use natural numbers as factors.

A **prime number** is any natural number greater than 1 whose only factors are 1 and itself. For example, 7 is prime because it has no factors besides 1 and 7. A number that has additional factors is called a **composite number.** An example of a composite number is 24; its factors are 1, 2, 3, 4, 6, 8, 12, and 24. Note that the number 1 is neither prime nor composite.

The **prime factorization** of a natural number is the factorization that contains only prime numbers or powers of prime numbers.

EXAMPLE 2   **Write the prime factorization of 300.**

▸ **Solution**

Make a diagram.

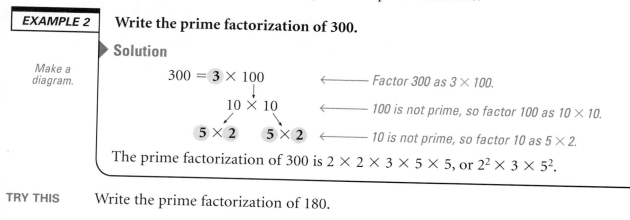

$$300 = \mathbf{3} \times 100 \qquad \longleftarrow \text{Factor 300 as } 3 \times 100.$$
$$10 \times 10 \qquad \longleftarrow \text{100 is not prime, so factor 100 as } 10 \times 10.$$
$$\mathbf{5} \times \mathbf{2} \quad \mathbf{5} \times \mathbf{2} \qquad \longleftarrow \text{10 is not prime, so factor 10 as } 5 \times 2.$$

The prime factorization of 300 is $2 \times 2 \times 3 \times 5 \times 5$, or $2^2 \times 3 \times 5^2$.

TRY THIS   Write the prime factorization of 180.

EXERCISES

KEY SKILLS

Identify each number as prime, composite, or neither.

1. 1   **2.** 100   **3.** 31   **4.** 75

PRACTICE

Write three different factorizations of each number. Use natural numbers as factors.

5. 24   **6.** 30   **7.** 40   **8.** 48

Write the prime factorization of each number, or write *prime*.

9. 24   **10.** 32   **11.** 19   **12.** 17

13. 25   **14.** 36   **15.** 72   **16.** 27

17. 100   **18.** 81   **19.** 51   **20.** 35

21. 125   **22.** 8   **23.** 64   **24.** 441

Use prime factorization to show that each number is a power of a single prime number.

25. 243   **26.** 343   **27.** 625   **28.** 1024

Two numbers are *relatively prime* if their only common factor is 1. Write the prime factorization of both numbers in each pair below. Then use the prime factorizations to show that the two numbers are relatively prime.

29. 12 and 13   **30.** 10 and 21   **31.** 18 and 25   **32.** 36 and 49

33. How many factors, other than 1, does 1024 have?

34. Critical Thinking   In the table in Example 1, why would it be necessary to use only the natural numbers up through 7 in the search for factor pairs?

MIXED REVIEW

Solve each equation. Check your solution. (2.5 Skill A)

35. $x - 7 = -10$ **36.** $z + 2.5 = -0.5$ **37.** $x - 6.4 = -10.5$

38. $x + 6.4 = -4.3$ **39.** $2 = a + 6.9$ **40.** $-22.6 = t - 100$

Solve each equation. Check your solution. (2.5 Skill B)

41. $(2 - 6)z = -18$ **42.** $20 = \dfrac{c}{-2 - 3}$ **43.** $\left(4 - \dfrac{1}{2}\right)z = -14$

The **greatest common factor,** or **GCF,** of two or more numbers is the greatest integer that is a common factor to all of those numbers.

EXAMPLE 1 | **Find the GCF of 36 and 54.**

▶ **Solution**

Write the prime factorization of each number. Choose the smallest power of each prime factor that appears in both numbers. The GCF will be the product of those prime factors.

$$36 = 2^2 \times 3^2 \qquad 54 = 2^1 \times 3^3$$

The GCF of 36 and 54 is $2^1 \times 3^2$, or 18.

TRY THIS   Find the GCF of 60 and 140.

The GCF of two or more monomials is the product of all the integer and variable factors that are common to those monomials. To find the GCF, first find the GCF of the coefficients. Then find the product of the smallest power of each variable factor that appears in all the monomials. The GCF is the product of these two results.

EXAMPLE 2 | **Find the GCF of $36c^3$ and $54c^2$.**

▶ **Solution**

From Example 1, the GCF of 36 and 54 is 18.
The factor c appears in both monomials; the smallest power of c that occurs is c^2.
Therefore, the GCF of $36c^3$ and $54c^2$ is $18c^2$.

TRY THIS   Find the GCF.   **a.** $18d^5$ and $108d$ **b.** $18d$ and 5

EXAMPLE 3 | **Find the GCF of a^3b and $5a^2b^2$.**

▶ **Solution**

GCF of 1 and 5 is 1.   GCF of a^3 and a^2 is a^2.   GCF of b and b^2 is b.
Therefore, the GCF of a^3b and $5a^2b^2$ is $1a^2b$, or a^2b.

TRY THIS   Find the GCF of $3m^3n^3$ and $9m^2n^2$.

EXAMPLE 4 | **Find the GCF of $3y^3z$, $15y^2z^2$, and $-45y^2z$.**

▶ **Solution**

GCF of 3, 15, and -45 is 3.   GCF of y^3 and y^2 is y^2.   GCF of z and z^2 is z.
Therefore, the GCF of $3y^3z$, $15y^2z^2$, and $-45y^2z$ is $3y^2z$.

TRY THIS   Find the GCF of $4mn^3$, $4m^2n^3$, and $16m^2n^2$.

EXERCISES

List the variables common to both monomials in each pair.

1. $3d^3z$ and $12z$

2. $2a^2n$ and $2mn$

3. a^3n and $2a^2n^2$

4. $3a^3$ and $2n^2$

PRACTICE

Find the GCF of each pair of numbers.

5. 6 and 9

6. 27 and 54

7. 14 and 8

8. 14 and 16

9. 16 and 32

10. 15 and 45

11. 14 and 51

12. 16 and 27

13. 100 and 2

14. 11 and 77

15. 63 and 81

16. 19 and 57

Find the GCF of each pair of monomials.

17. $3x^3$ and $6x^2$

18. $-4a^3$ and $8a^2$

19. $6c^3$ and $9c$

20. $6c^3$ and $15c$

21. $4t^3$ and 6

22. $-5z^3$ and 30

23. $14x^3$ and 21

24. $10y^3$ and 25

25. $2x^3y$ and $4x^2$

26. $-4b^3t$ and $4b^2$

27. $5d^2z$ and $12dz$

28. $3c^2r$ and $-15cr$

29. $-4t^2y^2$ and 6

30. $-10z^2y^3$ and 20

31. $21a^3b^3$ and $3a^3b^3$

32. $10a^2v^3$ and $10a^2v^2$

Find the GCF of each set of monomials.

33. $3a^2b$, $15a^2b^2$, and $15ab$

34. $4c^2d$, $6c^2d^2$, and $8cd$

35. $6r^2s$, $6rs^2$, and 11

36. $7u^3v$, $7uv^2$, and 1

37. $3x^3y$, $6xy^2$, and $8x^2y^2$

38. $5m^3n$, $5m^3n^2$, and $7m^2n$

39. p^3q, $-p^3q^2$, and $2p^2q$

40. $3b^3t$, $-3b^3t^2$, and 15

41. $-4b^9t$, $-8b^8t^7$, and 32

42. 1, x, x^2, x^3, x^4, and x^5

43. 1, $2x$, 2^2x, 2^3x, 2^4x, and 2^5x

MIXED REVIEW

Find an equation in point-slope form for the line that contains the given point and has the given slope. (3.6 Skill A)

44. $A(3, -4)$; slope: $\frac{3}{4}$

45. $B(-2, 4)$; slope: $-\frac{1}{3}$

46. $T(3, -1)$; slope: 0

Find the equation in slope-intercept form for the line that contains each pair of points. (3.7 Skill A)

47. $A(3, 4)$ and $B(5, -2)$

48. $K(-2, 4)$ and $L(5, 4)$

49. $P(-3, -1)$ and $Q(6, 8)$

To factor a polynomial means to write it as a product of two or more polynomials. The first step is to find the GCF of all terms.

EXAMPLE 1

Use the GCF to factor $6x^2 - 4x$.

▶ **Solution**

1. Find the GCF of all the terms. The GCF of $6x^2$ and $-4x$ is $2x$.

2. Write each term in $6x^2 - 4x$ as a product involving the GCF.
$$6x^2 = (2x)(3x) \qquad -4x = (2x)(-2)$$

3. Apply the Distributive Property.
$$6x^2 - 4x = (2x)(3x) + (2x)(-2)$$
$$= 2x(3x - 2)$$

TRY THIS Use the GCF to factor $-3v^2 + 15v$.

EXAMPLE 2

Use the GCF to factor $3p^3 + 12p^2 - 3p$.

▶ **Solution**

1. The GCF of $3p^3$, $12p^2$, and $-3p$ is $3p$.

2. Write each term in $3p^3 + 12p^2 - 3p$ as a product involving the GCF.
$$3p^3 = (3p)(p^2) \qquad 12p^2 = (3p)(4p) \qquad -3p = (3p)(-1)$$

3. Apply the Distributive Property.
$$3p^3 + 12p^3 - 3p = (3p)(p^2) + (3p)(4p) + (3p)(-1)$$
$$= 3p(p^2 + 4p - 1)$$

TRY THIS Use the GCF to factor $13t^4 + 26t^3 + 13t^2$.

EXAMPLE 3

Use the GCF to factor $2ab^2 + 6a^2b^2 + 9a^2b$.

▶ **Solution**

1. The GCF of $2ab^2$, $6a^2b^2$, and $9a^2b$ is ab.

2. Write each term in $2ab^2 + 6a^2b^2 + 9a^2b$ as a product involving the GCF.
$$2ab^2 = (ab)(2b) \qquad 6a^2b^2 = (ab)(6ab) \qquad 9a^2b = (ab)(9a)$$

3. Apply the Distributive Property.
$$2ab^2 + 6a^2b^2 + 9a^2b = (ab)(2b) + (ab)(6ab) + (ab)(9a)$$
$$= ab(2b + 6ab + 9a)$$

TRY THIS Use the GCF to factor $-3c^3d^2 + 6c^2d^2 + 9c^2d$.

EXERCISES

Use multiplication to verify that the factorization shown is correct.

1. $2x^2 + 6x = 2x(x + 3)$

2. $2x^3 - 10x = 2x(x^2 - 5)$

3. $3x^3 + 3x^2 = 3x^2(x + 1)$

4. $x^3 - 4x = x(x^2 - 4)$

PRACTICE

Use the GCF to factor each expression.

5. $7x + 14$

6. $5a - 10$

7. $10b^2 + 12b$

8. $-6y^2 + 24y$

9. $6z^2 - 9z$

10. $r^2 + r$

11. $12n^3 + 4n^2 - 4n$

12. $5m^3 + 5m^2 + 15m$

13. $5k^3 - 25k^2 - 75k$

14. $2n^3 + 17n^2 - 41n$

15. $2w^4 + 18w^3 - 18w^2$

16. $x^4 - 11x^3 - 7x^2$

17. $-12z^5 + 15z^4 - 18z^3$

18. $7v^5 + 49v^4 - 98v^2$

19. $11v^5 - 11v^4 + 22v^3$

20. $2ab^2 - 8a^2b^2 - 8ab$

21. $r^2s^2 - 8rs^2 + 8r^2s$

22. $3d^2n^2 - 8dn^2 + 8d^2n$

23. $3mn^2 + 10m^2n^2 - 7mn - 2n$

24. $5h^3p^3 + 10hp^2 - 15hp - 15h$

25. $7u^2v + 28uv^2 - 7uv^4 + 7uv$

26. $h^5p^3 + 10hp^2 - 15hp^4 - 15h^3$

27. $ab^2 + a^2b^2 - a^3b^3 + a^2b^5$

28. $-4sc^3 + 10s^2c^2 - 16sc^4 - 10sc^2$

29. $3(a + b)^3 + 3a(a + b)^2$

30. $3(x + y)^3 - 5x(x + y)^2$

31. $3(m + n)^3 - 5m(m + n)^2 + 2n(m + n)^2$

32. $(p + q)^3 - 5p(p + q)^2 + 2q(p + q)^2$

33. $(x^2y^3)^2 + (x^3y^2)^3$

MIXED REVIEW

Find each product. (6.6 Skill B)

34. $(2x + 4)(x - 5)$

35. $(2x - 5)(2x + 1)$

36. $(3r + 1)(3r + 2)$

37. $(x + 7)(2 + x)$

38. $(2 - 5x)(1 + 6x)$

39. $(3 - 2y)(3y - 2)$

Find each product. (6.6 Skill C)

40. $(2a + 4)^2$

41. $(2x - 3)(2x + 3)$

42. $(3y + 1)^2$

Factoring Special Polynomials

Factoring polynomials with special forms

Some expressions can be quickly factored if you can recognize special patterns. For example, perfect-square trinomials and differences of two squares are easily factorable.

Factoring Special Products

Perfect-square trinomial: $a^2 + 2ab + b^2 = (a + b)^2$

Difference of two squares: $a^2 - b^2 = (a + b)(a - b)$

The polynomial $49x^2 + 14x + 1$ is a perfect-square trinomial.

$$49x^2 + 14x + 1 = 7^2x^2 + (2)(7)x + 1^2$$
$$= (7x)^2 + 2(7x)(1) + 1^2 \quad \longleftarrow \text{ Power-of-a-Product Property}$$
$$= (7x + 1)^2$$

EXAMPLE 1   Factor.   a. $m^2 + 12m + 36$ b. $4n^2 - 20n + 25$

▶ **Solution**

a. $m^2 + 12m + 36 = m^2 + 2(m)(6) + 6^2$
$$= (m + 6)^2 \quad \longleftarrow a = m \text{ and } b = 6$$

b. $4n^2 - 20n + 25 = (2n)^2 + 2(2n)(-5) + (-5)^2 \quad \longleftarrow 4n^2 = (2n)^2 \text{ and } 25 = (-5)^2$
$$= [2n + (-5)]^2 \quad \longleftarrow a = 2n \text{ and } b = -5$$
$$= (2n - 5)^2$$

TRY THIS   Factor.   a. $d^2 - 18d + 81$ b. $9z^2 + 6z + 1$

The polynomial $25x^2 - 81$ is a difference of two squares.

$$25x^2 - 81 = 5^2x^2 - 9^2$$
$$= (5x)^2 - 9^2 \quad \longleftarrow \text{ Power-of-a-Product Property}$$
$$= (5x + 9)(5x - 9)$$

EXAMPLE 2   Factor.   a. $v^2 - 121$ b. $9x^2 - 25$

▶ **Solution**

a. $v^2 - 121 = v^2 - 11^2 \quad \longleftarrow 121 = 11^2$
$$= (v + 11)(v - 11) \quad \longleftarrow a = v \text{ and } b = 11$$

b. $9x^2 - 25 = (3x)^2 - 5^2 \quad \longleftarrow 9x^2 = (3x)^2 \text{ and } 25 = 5^2$
$$= (3x + 5)(3x - 5) \quad \longleftarrow a = 3x \text{ and } b = 5$$

TRY THIS   Factor.   a. $h^2 - 100$ b. $16x^2 - 49$

EXERCISES

KEY SKILLS

Each polynomial below is in the form of a perfect-square trinomial, $a^2 + 2ab + b^2$, or a difference of two squares, $a^2 - b^2$. Write the values of a and b. Do not factor.

1. $x^2 + 6x + 9$ **2.** $64x^2 - 16x + 1$

3. $k^2 - 9$ **4.** $25n^2 - 36$

PRACTICE

Factor.

5. $d^2 + 2d + 1$ **6.** $c^2 + 10c + 25$ **7.** $u^2 - 8u + 16$

8. $y^2 - 10y + 25$ **9.** $h^2 - 144$ **10.** $g^2 - 4$

11. $k^2 - 225$ **12.** $q^2 - 400$ **13.** $4y^2 + 20y + 25$

14. $16y^2 - 40y + 25$ **15.** $49m^2 + 14m + 1$ **16.** $16y^2 - 72y + 81$

17. $16x^2 - 121$ **18.** $64c^2 - 1$ **19.** $u^2v^2 + 2uv + 1$

20. $y^2z^2 - 8yz + 16$ **21.** $4m^2n^2 - 16mn + 16$ **22.** $3m^2n^2 + 6mn + 3$

23. $64c^2d^2 - 25$ **24.** $100k^2q^2 - 1$ **25.** $121s^2t^2 - 81$

26. $4m^2n^2 + 24mnxy + 36x^2y^2$ **27.** $25m^2n^2 - 49x^2y^2$

28. Suppose that a is an integer between 1 and 100 inclusive. Find all values of a such that $ak^2 - 49$ is the difference of two squares.

29. Critical Thinking Suppose that $25y^2 - 50y + 25 = (my + n)^2$. Find m and n.

30. Critical Thinking Suppose that $36x^2y^2 + 12xy + 1 = (mxy + n)^2$. Find m and n.

MIXED REVIEW

Find each product. (6.6 Skill B)

31. $(3a - 10)(2a + 3)$ **32.** $(11t - 1)(2t - 1)$ **33.** $(-2n + 3)(2n + 5)$

Find each product. (6.6 Skill C)

34. $(3nt + 4)(3nt - 4)$ **35.** $(3t - 4c)^2$ **36.** $(6xy + 1)(6xy - 1)$

Write the prime factorization of each number. (7.1 Skill A)

37. 154   **38.** 104   **39.** 231   **40.** 112

A polynomial may have a common factor that is a binomial instead of a monomial.

EXAMPLE 1 | **Factor $3(a + b) + g(a + b)$.**

▶ **Solution**

Notice that $3(a + b)$ and $g(a + b)$ each contain the **common binomial factor** $(a + b)$. Use the Distributive Property to write

$$3(a + b) + g(a + b) = (3 + g)(a + b).$$

common factor

$$3(a + b) + g(a + b)$$

TRY THIS | Factor $3(2a - b) - y(2a - b)$.

You may need to group terms before you can find a common binomial factor.

EXAMPLE 2 | **Factor $ax + 2a + bx + 2b$ by grouping.**

▶ **Solution**

Group the terms that have a common number or variable as a factor. Treat $ax + 2a$ as one expression, and treat $bx + 2b$ as another expression.

$$\begin{aligned}
ax + 2a + bx + 2b &= (ax + 2a) + (bx + 2b) &&\longleftarrow \text{ Group terms.}\\
&= a(x + 2) + b(x + 2) &&\longleftarrow \text{ Factor each group.}\\
&= (a + b)(x + 2) &&\longleftarrow \text{ Apply the Distributive Property.}
\end{aligned}$$

TRY THIS | Factor $9m + 15n + 3my + 5ny$ by grouping.

When factoring by grouping, be careful when the terms involve subtraction. Notice that

$$x^3 + 3x^2 - ax - 3a = (x^3 + 3x^2) - (ax + 3a) \quad ✔$$

and *not* $x^3 + 3x^2 - ax - 3a = (x^3 + 3x^2) - (ax - 3a).$ ✘

You may find it helpful to rewrite each subtraction as addition of the opposite.

EXAMPLE 3 | **Factor $x^3 + 3x^2 - ax - 3a$ by grouping.**

▶ **Solution**

$$\begin{aligned}
x^3 + 3x^2 - ax - 3a &= (x^3 + 3x^2) - (ax + 3a) &&\longleftarrow \begin{array}{l}\text{Group terms.}\\ \text{Apply the Multiplicative Property of } -1.\end{array}\\
&= x^2(x + 3) - a(x + 3) &&\longleftarrow \text{ Factor each group.}\\
&= (x^2 - a)(x + 3) &&\longleftarrow \text{ Apply the Distributive Property.}
\end{aligned}$$

TRY THIS | Factor $d^2 - 8d + bd - 8b$ by grouping.

EXERCISES

What is the common binomial factor in each polynomial?

1. $2x(3x - 7) - 5(3x - 7)$

2. $x^2(x + 2) + 4(x + 2)$

3. $2(x - y) - g(x - y)$

4. $5(p + q) + 3g(p + q)$

PRACTICE

Factor.

5. $2(2p - 5) + h(2p - 5)$

6. $9(2n + 1) - s(2n + 1)$

7. $-2(m + 7) + t(m + 7)$

8. $-(a - 9) + 3z(a - 9)$

9. $-(v - 3t) - w(v - 3t)$

10. $-(-u + s) - w(-u + s)$

11. $cx + cy + 3x + 3y$

12. $4a + 4b + ta + tb$

13. $x^2 - 10x + xy - 10y$

14. $x^2 + x + xw + w$

15. $y^2 - 3y + yd - 3d$

16. $t^2 - 9t + 3t - 27$

17. $9cn + 12cm + 15dn + 20dm$

18. $4mna + 4mnb + 5na + 5nb$

19. $15a - 10b + 21at - 14tb$

20. $6x - 2y + 12xz - 4yz$

21. $2x^2 - 4x + xy - 2y$

22. $7y^2 - 14y + by - 2b$

23. $4ab - 2c - 8a + bc$

24. $12xy - z - 4x + 3yz$

25. $9x + 6 - 6ax - 4a$

26. $20a + 12 - 25ax - 15x$

27. Find a such that $3(2y - a) + n(2y + 5)$ can be factored. Then factor.

28. Find a such that $3(2y - a) + n[2y + 2(a + 1)]$ can be factored. Then factor.

29. If $4z(3x - 7) + 3y(ax + b) = (4z + 3y)(3x - 7)$, what are a and b?

MIXED REVIEW

Graph the solution to each inequality. (5.5 Skill A)

30. $y \geq -2.5x + 1$

31. $y < -3x + 5$

32. $y \leq 5x - 1.5$

Graph the solution to each system of inequalities. (5.6 Skill A)

33. $\begin{cases} y < 2x \\ y \geq -2x - 3 \end{cases}$

34. $\begin{cases} y > 2x - 1 \\ y \leq -2.5x + 3 \end{cases}$

35. $\begin{cases} y < 1.5x + 1 \\ y < -1.5x + 1 \end{cases}$

Factoring $x^2 + bx + c$

SKILL A *Factoring $x^2 + bx + c$ when c is positive*

Suppose that you want to factor $x^2 + 5x + 6$.

- Because $x^2 + 5x + 6$ has no common monomial factor other than 1, you cannot factor out a monomial.
- Because $x^2 + 5x + 6$ is not a perfect-square trinomial or a difference of squares, you cannot write a special product.

The following fact will help you factor quadratic trinomials like $x^2 + 5x + 6$.

Factoring $x^2 + bx + c$

For $x^2 + bx + c$, if there are two real numbers r and s such that $c = rs$ and $b = r + s$, then $x^2 + bx + c = (x + r)(x + s)$.

In other words, to factor $x^2 + bx + c$, look for factor pairs, two numbers whose product is c. Choose the pair whose sum is b.

EXAMPLE 1

Make an organized list.

Factor $x^2 + 5x + 6$.

▶ **Solution**

1. $c = 6$: write factor pairs of 6. Remember to include negative factors.
$$1 \times 6 \qquad\qquad 2 \times 3 \qquad\qquad\qquad -1 \times -6 \qquad\qquad -2 \times -3$$

2. $b = 5$: choose the pair whose sum is 5.
$$\cancel{1 + 6 = 7} \qquad 2 + 3 = 5 ✔ \qquad \cancel{-1 + -6 = -7} \qquad \cancel{-2 + -3 = -5}$$

3. Write the product using 2 and 3. Thus, $x^2 + 5x + 6 = (x + 2)(x + 3)$.

TRY THIS   Factor $a^2 + 9a + 20$.

EXAMPLE 2

Make an organized list.

Factor $y^2 - 10y + 24$.

▶ **Solution**

1. $c = 24$: write factor pairs of 24.
$$\begin{array}{llll} 1 \times 24 & -1 \times -24 & 4 \times 6 & -4 \times -6 \\ 2 \times 12 & -2 \times -12 & 3 \times 8 & -3 \times -8 \end{array}$$

2. $b = -10$: choose the pair whose sum is -10.
$$\cancel{1 + 24 = 25} \qquad \cancel{-1 + (-24) = -25} \qquad \cancel{4 + 6 = 10} \qquad -4 + (-6) = -10 ✔$$

3. Write the product using -4 and -6. $[y + (-4)][y + (-6)] = (y - 4)(y - 6)$

Thus, $y^2 - 10y + 24 = (y - 4)(y - 6)$.

TRY THIS   Factor $n^2 - 13n + 36$.

EXERCISES

Given each quadratic trinomial and list of factor pairs, which pair can you use to write the factorization? Write the factorization.

1. $m^2 + 12m + 20$
   $1 \times 20, -1 \times -20,$
   $2 \times 10, -2 \times -10,$
   $4 \times 5, -4 \times -5$

2. $n^2 - 9n + 20$
   $1 \times 20, -1 \times -20,$
   $2 \times 10, -2 \times -10,$
   $4 \times 5, -4 \times -5$

3. $k^2 - 21k + 20$
   $1 \times 20, -1 \times -20,$
   $2 \times 10, -2 \times -10,$
   $4 \times 5, -4 \times -5$

4. $z^2 + 11z + 18$
   $1 \times 18, -1 \times -18,$
   $2 \times 9, -2 \times -9,$
   $3 \times 6, -3 \times -6$

5. $d^2 - 12d + 35$
   $1 \times 35, 5 \times 7,$
   $-1 \times -35, -5 \times -7$

6. $h^2 - 15h + 36$
   $2 \times 18, -2 \times -18,$
   $3 \times 12, -3 \times -12,$
   $4 \times 9, -4 \times -9$

Verify each factorization by multiplying.

7. $x^2 + 12x + 35 = (x + 5)(x + 7)$

8. $z^2 - 12z + 35 = (z - 5)(z - 7)$

Factor.

9. $m^2 + 5m + 4$

10. $x^2 + 7x + 12$

11. $z^2 + 8z + 15$

12. $g^2 + 6g + 5$

13. $k^2 - 11k + 30$

14. $m^2 - 4m + 3$

15. $m^2 - 8m + 7$

16. $m^2 - 12m + 11$

17. $x^2 + 13x + 40$

18. $x^2 + 19x + 48$

19. $y^2 + 26y + 48$

20. $w^2 + 15w + 56$

21. $t^2 + 23t + 42$

22. $s^2 - 16s + 63$

23. $t^2 - 23t + 42$

24. $m^2 - 20m + 64$

25. $p^2 - 50p + 400$

26. $m^2 - 30m + 144$

27. Factor $a^2b^2 + 10ab + 21$.

28. **Critical Thinking** In $x^2 + bx + c$, if c is positive, what do you know about the factors of c?

29. **Critical Thinking** In $x^2 + bx + c$, if c is positive and bx is positive, what do you know about the factors of c? What do you know if bx is negative?

Find the GCF of each pair of monomials. (7.1 Skill B)

30. $3a^3$ and a

31. $8v^3$ and $8v^2$

32. $3d^2$ and $9d$

33. $5n^2$ and $15n^3$

34. $3m^3$ and m^2

35. $4k^2$ and $8k^3$

36. z^5 and $9z$

37. $6h^4$ and $15h$

Use the GCF to factor each expression. (7.1 Skill C)

38. $n^2 - 2n$

39. $3p^3 - 15p^2 + 3p$

40. $2z^3 - 6z^2$

In Skill A, you learned how to factor expressions of the form $x^2 + bx + c$ when c is positive. You may be also able to factor such expressions when c is negative.

EXAMPLE 1

Factor $x^2 + x - 20$.

▶ **Solution**

Make an organized list.

1. $c = -20$: write factor pairs of -20.

$$-1 \times 20 \qquad -2 \times 10 \qquad -4 \times 5$$
$$1 \times -20 \qquad 2 \times -10 \qquad 4 \times -5$$

2. $b = 1$: find the pair whose sum is 1.
$$-4 + 5 = 1$$

3. Write the product using -4 and 5.
$$[x + (-4)](x + 5)$$

Thus, $x^2 + x - 20 = (x - 4)(x + 5)$.

TRY THIS   Factor $n^2 + 3n - 40$.

EXAMPLE 2

Factor $z^2 - 4z - 12$.

▶ **Solution**

Make an organized list.

1. $c = -12$: write factor pairs of -12.

$$-1 \times 12 \qquad -2 \times 6 \qquad -3 \times 4$$
$$1 \times -12 \qquad 2 \times -6 \qquad 3 \times -4$$

2. $b = -4$: find the pair whose sum is -4.
$$2 + (-6) = -4$$

3. Write the product using 2 and -6.
$$(z + 2)[z + (-6)]$$

Thus, $z^2 - 4z - 12 = (z + 2)(z - 6)$.

TRY THIS   Factor $n^2 - 3n - 40$.

Not every polynomial of the form $x^2 + bx + c$ is factorable.

EXAMPLE 3

Show that $v^2 + 3v - 1$ cannot be factored.

▶ **Solution**

$c = -1$: the only factors of -1 are 1 and -1.
$b = 3$: because $-1 + 1 \neq 3$, $v^2 + 3v - 1$ cannot be factored.

TRY THIS   Show that $q^2 + 5q - 8$ cannot be factored.

EXERCISES

KEY SKILLS

Each expression is in the form $ax^2 + bx + c$. Identify b and c, including their signs, in the following quadratic trinomials.

1. $x^2 - 7x - 18$

2. $x^2 + 5x - 24$

3. $x^2 - 2x + 35$

4. $t^2 - 3t - 1$

PRACTICE

Factor.

5. $x^2 - 6x - 7$

6. $y^2 + 10y - 11$

7. $v^2 + 4v - 5$

8. $n^2 - 4n - 5$

9. $k^2 - k - 2$

10. $a^2 + 6a - 7$

11. $w^2 - w - 42$

12. $p^2 - 2p - 63$

13. $d^2 + 2d - 63$

14. $b^2 + 31b - 32$

15. $h^2 - 17h - 38$

16. $x^2 - 10x - 96$

17. $z^2 + 10z - 96$

18. $s^2 - 10s - 75$

19. $t^2 + 4t - 45$

20. $d^2 + 9d - 36$

21. $m^2 - 42m - 43$

22. $s^2 + 11s - 80$

Show that the given polynomial cannot be factored.

23. $x^2 - 9x + 7$

24. $a^2 + 8a + 5$

25. $b^2 + 3b + 1$

26. $q^2 - 6q - 3$

27. $k^2 - 5k - 3$

28. $z^2 + 2z - 10$

29. Given $x^2 + bx - c$, what do you know about the signs of the factors of $-c$?

30. **Critical Thinking** Let r and s represent real numbers. Show that if $x^2 + bx + c = (x + r)(x + s)$, then $c = rs$ and $b = r + s$.

MIXED REVIEW

Factor. (7.2 Skill A)

31. $n^2 + 22n + 121$

32. $k^2 - 26k + 169$

33. $4a^2 + 20a + 25$

34. $36z^2 - 36$

35. $9y^2 - b^2$

36. $u^2v^2 - 9$

Factor. (7.2 Skill B)

37. $2(z + 1) + 3z(z + 1)$

38. $3(a + 2) - 2a(a + 2)$

39. $-(n - 1) + 3n(n - 1)$

40. $x^2y + x^2 - 3y - 3$

41. $x^2 - 20x - yx + 20y$

42. $4wx^2 + 5w - 4x^2 - 5$

7.4 LESSON

Factoring $ax^2 + bx + c$

Factoring $ax^2 + bx + c$ when c is positive

When the x^2-term in a quadratic trinomial has a coefficient other than 1, the task of factoring is a little more involved. However, the process uses what you have learned about factoring by grouping.

EXAMPLE 1 | Factor $2x^2 + 7x + 3$.

▶ **Solution**

1. Identify a, b, and c. $a = 2, b = 7, c = 3$
2. Find the product ac. $(2)(3) = 6$
3. Find factors of ac that add to b. $6 \times 1 = 6$ $6 + 1 = 7$
4. Rewrite the original equation, rewriting bx as the sum of the factors you found in Step **2.** (The order of the factors does not matter.) $2x^2 + 6x + 1x + 3$
5. Now apply factoring by grouping. Using parentheses, group the four terms into two groups of two terms. $(2x^2 + 6x) + (x + 3)$
6. Factor each group. (Factor out 1 if it is the GCF.) The expressions remaining in parentheses will be equal. $2x(x + 3) + 1(x + 3)$
7. Apply the Distributive Property. $(2x + 1)(x + 3)$

TRY THIS   Factor $2n^2 + 11n + 5$.

Because you rewrite bx as two terms in Step 4, this method is sometimes called *splitting the middle term*.

Remember to group carefully when terms involve subtraction.

EXAMPLE 2 | Factor $12y^2 - 7y + 1$.

▶ **Solution**

1. $a = 12, b = -7, c = 1$
2. $ac = (12)(1) = 12$
3. Factors of 12 that add to -7 are -3 and -4.
4. Rewrite the middle term: $12y^2 - 3y - 4y + 1$
5. Make two groups: $(12y^2 - 3y) - (4y - 1)$ *Remember that* $-4y + 1 = -(4y - 1)$.
6. Factor each group: $3y(4y - 1) - 1(4y - 1)$
7. Apply the Distributive Property: $(3y - 1)(4y - 1)$

TRY THIS   Factor $8a^2 - 14a + 3$.

EXERCISES

For each quadratic trinomial, identify *a*, *b*, and *c*. Then list the possible factor pairs of *ac* and find the pair whose sum is *b*.

1. $4x^2 + 9x + 2$

2. $8x^2 + 27x + 9$

3. $2x^2 + 8x + 8$

4. $3x^2 - 10x + 8$

PRACTICE

In Exercises 5–12, *ac* = 120. List the factor pairs of 120 and their sums. Then, for each expression, rewrite the middle term as two terms whose coefficients multiply to 120.

Sample: $3x^2 + 23x + 40$ Solution: $8x + 15x$

5. $8x^2 + 26x + 15$

6. $5x^2 + 34x + 24$

7. $40x^2 + 43x + 3$

8. $15y^2 + 22y + 8$

9. $2y^2 - 23y + 60$

10. $120x^2 - 121x + 1$

11. $3v^2 - 29v + 40$

12. $4v^2 - 62v + 30$

Factor.

13. $2a^2 + 7a + 3$

14. $3n^2 + 7n + 2$

15. $3k^2 + 5k + 2$

16. $5a^2 + 7a + 2$

17. $3h^2 + 8h + 4$

18. $7h^2 + 8h + 1$

19. $5z^2 + 17z + 6$

20. $3v^2 + 11v + 10$

21. $4v^2 + 7v + 3$

22. $8p^2 - 14p + 5$

23. $7x^2 - 15x + 2$

24. $8c^2 - 6c + 1$

25. $8c^2 - 11c + 3$

26. $12a^2 - 8a + 1$

27. $-9m^2 - 3m + 2$

28. $-6x^2 - 11x + 2$

29. $-2x^2 - 4x + 6$

30. $-8b^2 - 7b + 1$

31. $4x^2y^2 + 16xy + 15$

32. $4a^2b^2 + 16abmn + 15m^2n^2$

MIXED REVIEW

Solve each system of equations. (5.3 Skills A and B)

33. $\begin{cases} 5a - 2b = 0 \\ 2a + 2b = 3 \end{cases}$

34. $\begin{cases} 7m - 2t = 1 \\ 6m - 2t = 0 \end{cases}$

35. $\begin{cases} -2s - 2t = 1 \\ 5s - 6t = 1 \end{cases}$

36. $\begin{cases} \dfrac{x}{2} + \dfrac{y}{3} = 0 \\ \dfrac{x}{3} - \dfrac{y}{5} = 0 \end{cases}$

37. $\begin{cases} \dfrac{a}{2} - \dfrac{b}{3} = 1 \\ \dfrac{a}{3} + \dfrac{b}{2} = -1 \end{cases}$

38. $\begin{cases} \dfrac{m}{2} - \dfrac{n}{5} = 2 \\ \dfrac{m}{2} - \dfrac{n}{2} = 1 \end{cases}$

Factoring $ax^2 + bx + c$ *when c is negative*

To factor $ax^2 + bx + c$ when c is negative, follow the same procedure that you learned in Skill A. However, be careful when grouping terms that involve subtraction.

EXAMPLE 1

Factor $5b^2 + 4b - 1$.

▶ **Solution**

1. $a = 5$, $b = 4$, $c = -1$
2. $ac = (5)(-1) = -5$
3. Factors of -5 that add to 4 are 5 and -1.
4. Rewrite the middle term: $5b^2 + 5b - 1b - 1$
5. Make two groups: $(5b^2 + 5b) - (1b + 1)$
6. Factor each group: $5b(b + 1) - 1(b + 1)$
7. Apply the Distributive Property: $(5b - 1)(b + 1)$

TRY THIS   Factor $3k^2 + 4k - 4$.

EXAMPLE 2

Factor $10n^2 - 11n - 6$.

▶ **Solution**

1. $a = 10$, $b = -11$, $c = -6$
2. $ac = (10)(-6) = -60$
3. Factors of -60 whose sum is -11 are 4 and -15.
4. Rewrite the middle term: $10n^2 + 4n - 15n - 6$
5. Make two groups: $(10n^2 + 4n) - (15n + 6)$
6. Factor each group: $2n(5n + 2) - 3(5n + 2)$
7. Apply the Distributive Property: $(2n - 3)(5n + 2)$

TRY THIS   Factor $6m^2 - 7m - 49$.

Not all expressions of the form $ax^2 + bx - c$ can be factored.

EXAMPLE 3

Factor $-10n^2 + 21n - 5$.

▶ **Solution**

1. $a = -10$, $b = 21$, $c = -5$
2. $ac = (-10)(-5) = 50$
3. There are no factors of 50 whose sum is 21. Therefore, $-10n^2 + 21n - 5$ cannot be factored.

TRY THIS   Factor $-3x^2 + 15x - 2$.

EXERCISES

KEY SKILLS

Complete each grouping by filling in each blank with $(+)$ or $(-)$.

1. $8z^2 - 12z + 2z - 3 = (8z^2 - 12z) + (2z \underline{} 3)$

2. $15a^2 - 20a + 9a - 12 = (15a^2 - 20a) + (9a \underline{} 12)$

3. $3n^2 + 6n - 2n - 4 = (3n^2 + 6n) - (2n \underline{} 4)$

4. $20m^2 + 4m - 10m - 7 = (20m^2 + 4m) - (10m \underline{} 7)$

PRACTICE

Factor.

5. $3v^2 + 20v - 7$

6. $2b^2 + 3b - 5$

7. $3d^2 + 2d - 1$

8. $7p^2 + 12p - 4$

9. $11x^2 + 9x - 2$

10. $7y^2 + 4y - 3$

11. $3h^2 + h - 2$

12. $3g^2 - g - 2$

13. $3c^2 - 7c - 6$

14. $2z^2 - 7z - 15$

15. $2m^2 - 5m - 7$

16. $10a^2 + 21a - 10$

17. $2x^2 + 11x - 90$

18. $6x^2 - x - 5$

19. $6x^2 - 17x - 3$

20. $8x^2 - 4x - 4$

Show that the given polynomial cannot be factored.

21. $10x^2 + 21x + 6$

22. $6a^2 + a - 3$

23. $3x^2 + 3x - 4$

24. $9y^2 - y + 2$

25. **Critical Thinking** Suppose that a is a positive number. Find the value(s) of a so that $3x^2 + ax + 6$ can be factored as the product of two binomials. How does your answer change if a is a negative number?

MIXED REVIEW

Identify each system as consistent or inconsistent. Justify your response. (5.4 Skill A)

26. $\begin{cases} 4x + 2y = 0 \\ -6x - 5y = 1 \end{cases}$

27. $\begin{cases} 4m - 2n = 5 \\ -12m + 6n = -2 \end{cases}$

28. $\begin{cases} -3s - 5t = -7 \\ -2s + 6t = -7 \end{cases}$

Classify each system as specifically as possible. (5.4 Skill B)

29. $\begin{cases} 4c + 2d = 0 \\ -6c - 3d = 1 \end{cases}$

30. $\begin{cases} 7u + 2v = 1 \\ 14u + 4v = 2 \end{cases}$

31. $\begin{cases} -7m + 5n = 1 \\ 7m - 5n = 5 \end{cases}$

Simplifying before factoring a polynomial

The first step in factoring a polynomial is to check whether there is a GCF of all the terms. If so, factor it out before trying other factoring strategies.

Sometimes the GCF will be a number.

EXAMPLE 1   **Factor each polynomial.**   **a.** $-t^2 - 5t + 14$ **b.** $6x^2 + 10x + 4$

▶ **Solution**

a. $-t^2 - 5t + 14 = -(t^2 + 5t - 14)$ ⟵——— *Factor −1 from each term.*
$\qquad\qquad\qquad = -(t + 7)(t - 2)$ ⟵——— *Factor $t^2 + 5t - 14$.*
b. $6x^2 + 10x + 4 = 2(3x^2 + 5x + 2)$ ⟵——— *Factor 2 from each term.*
$\qquad\qquad\qquad = 2(3x + 2)(x + 1)$ ⟵——— *Factor $3x^2 + 5x + 2$*

TRY THIS   Factor each polynomial.   **a.** $-a^2 + 7a - 12$ **b.** $6n^2 + 15n + 9$

In the Example above, you were able to continue factoring after factoring out the GCF. A polynomial is **factored completely** when it is written as a product that cannot be factored any further. Whenever you are factoring a polynomial, be sure to factor completely.

The GCF can also be a monomial.

EXAMPLE 2   **Factor each polynomial.**   **a.** $12c^3 - 75c$ **b.** $5d^3 + 50d^2 + 125d$

▶ **Solution**

a. $12c^3 - 75c = 3c(4c^2 - 25)$ ⟵——— *Factor 3c from each term.*
$\qquad\qquad\quad = 3c(2c + 5)(2c - 5)$ ⟵——— *Factor the difference of two squares.*
b. $5d^3 + 50d^2 + 125d$
$\qquad\qquad\quad = 5d(d^2 + 10d + 25)$ ⟵——— *Factor 5d from each term.*
$\qquad\qquad\quad = 5d(d + 5)^2$ ⟵——— *Factor a perfect-square trinomial.*

TRY THIS   Factor each polynomial.   **a.** $27g^3 - 48g$ **b.** $4d^3 + 16d^2 + 16d$

EXAMPLE 3   **Factor $2z^3 + 6z^2 - 8z$.**

▶ **Solution**

$2z^3 + 6z^2 - 8z = 2z(z^2 + 3z - 4)$ ⟵——— *Factor 2z from each term.*
$\qquad\qquad\qquad = 2z(z + 4)(z - 1)$ ⟵——— *Factor $z^2 + 3z - 4$.*

TRY THIS   Factor $9r^3 - 24r^2 + 12r$.

EXERCISES

Is each factorization complete? Justify your response.

1. $5x^3 - 5x^2 = 5x(x^2 - x)$

2. $28a^2 + 29a - 35 = (4a + 7)(7a - 5)$

3. $3t^3 - 6t^2 - 45t = 3t(t - 5)(t + 3)$

4. $-2x^3 - 14x^2 - 20x = -2x(x^2 + 7x + 10)$

PRACTICE

Factor completely.

5. $-x^2 + x + 6$

6. $-n^2 + 4n + 5$

7. $-u^2 + 7u - 12$

8. $-c^2 - c + 20$

9. $-z^2 - 4z + 21$

10. $-a^2 + 7a - 10$

11. $3v^2 + 18v + 15$

12. $3x^2 + 15x - 42$

13. $7x^2 + 42x + 35$

14. $2x^2 + 2x - 40$

15. $-3t^2 + 33t - 90$

16. $-4x^2 + 10x + 6$

17. $-30b^2 - 35b + 25$

18. $-42g^2 - 49g - 14$

19. $63m^2 - 42m + 7$

20. $45q^2 + 30q + 5$

21. $14ab + 7b + 21$

22. $9c + 6b - 15$

23. $125q^2 - 45$

24. $7y^3 - 63y$

25. $3a^3 + 30a^2 - 33a$

26. $20n^3 + 20n^2 - 15n$

27. $2x^3 + 12x^2 - 110x$

28. $16x^2y - 8xy^2 + 12xy$

29. Find n such that $12x^3 + 6x^2 - 4nx^2 - 2nx = 2x(3x - 4)(2x + 1)$.

MID-CHAPTER REVIEW

Use the GCF to factor each expression. (7.1 Skill C)

30. $5y^4 + 20y^2$

31. $15x^3v^2 + 20xv^2$

32. $15a^2b^2 + 20a^2b - 3ab$

Factor. (7.2 Skill A)

33. $25x^2y^2 - 49$

34. $m^2n^2 + 14mn + 49$

35. $a^2b^2 - x^2y^2$

Factor. (7.3 Skills A and B)

36. $x^2 - 35x + 34$

37. $a^2 - 14a - 15$

38. $y^2 - 17y - 60$

Factor. (7.4 Skills A and B)

39. $3x^2 + 5x + 2$

40. $2n^2 + 9n + 10$

41. $5k^2 + 13k - 6$

7.5 Solving Polynomial Equations by Factoring

SKILL A *Solving quadratic equations by factoring*

A **quadratic equation** in x is any equation that can be written in the form $ax^2 + bx + c = 0$, where $a \neq 0$. This is called the **standard form of a quadratic equation**. The value of the variable in a standard form equation is called the *solution,* or the **root,** or the equation.

The **Multiplication Property of 0** states that if $a = 0$ or $b = 0$, then $ab = 0$. The *Zero-Product Property* enables you to conclude that if a product equals 0, then at least one of the factors of the product must equal 0.

Zero-Product Property

If a and b are real numbers and $ab = 0$, then $a = 0$ or $b = 0$.

Example 1 shows how to solve a quadratic equation once the quadratic expression has been factored into a product of two linear factors.

EXAMPLE 1 Solve $(4x + 5)(3x - 2) = 0$.

▶ **Solution**

> Apply the Zero-Product Property.

$$(4x + 5)(3x - 2) = 0 \quad \longleftarrow \textit{The quadratic expression has already been factored.}$$
$$4x + 5 = 0 \text{ or } 3x - 2 = 0 \quad \longleftarrow \textit{If (4x + 5)(3x - 2) = 0, then (4x + 5) = 0 or (3x - 2) = 0.}$$
$$4x = -5 \qquad\qquad 3x = 2$$
$$x = -\frac{5}{4} \qquad\qquad x = \frac{2}{3} \quad \longleftarrow \textit{Solve each equation for x.}$$

TRY THIS Solve $(3d + 5)(d - 2) = 0$.

Example 2 shows how to solve a quadratic equation when the quadratic expression is not given to you in factored form.

EXAMPLE 2 Solve $3t^2 - 8t + 5 = 0$ by factoring.

▶ **Solution**

$$3t^2 - 8t + 5 = 0$$
$$(3t - 5)(t - 1) = 0 \quad \longleftarrow \textit{Factor } 3t^2 - 8t + 5.$$
$$3t - 5 = 0 \text{ or } t - 1 = 0 \quad \longleftarrow \textit{Apply the Zero-Product Property.}$$
$$t = \frac{5}{3} \qquad\qquad t = 1 \quad \longleftarrow \textit{Solve each equation for t.}$$

TRY THIS Solve $6t^2 + t - 15 = 0$ by factoring.

EXERCISES

KEY SKILLS

Solve each equation.

1. $x + 3 = 0$ or $x - 5 = 0$

2. $3z + 1 = 0$ or $2z - 3 = 0$

3. $5n + 3 = 0$ or $2n + 7 = 0$

4. $-2b + 5 = 0$ or $2b - 9 = 0$

PRACTICE

Solve each quadratic equation.

5. $(x + 5)(x + 2) = 0$

6. $(x - 7)(x + 3) = 0$

7. $(x - 7)(x - 5) = 0$

8. $(3d + 2)(7d - 1) = 0$

9. $(-5n - 1)(3n - 9) = 0$

10. $(6p + 5)(7p + 3) = 0$

11. $t^2 + 2t - 15 = 0$

12. $a^2 - 10a + 21 = 0$

13. $y^2 + 13y + 40 = 0$

14. $v^2 - 14v + 45 = 0$

15. $2s^2 + s - 15 = 0$

16. $3k^2 - 4k + 1 = 0$

17. $x^2 + 7x - 44 = 0$

18. $y^2 + 7y + 10 = 0$

19. $5y^2 - 46y + 9 = 0$

20. $2g^2 - 15g + 7 = 0$

21. $9x^2 - 12x + 4 = 0$

22. $4a^2 + 23a - 6 = 0$

23. $2x^2 - 5x + 3 = 0$

24. $3m^2 + 13m - 10 = 0$

25. $2a^2 - 7a - 4 = 0$

26. $3k^2 - 7k - 6 = 0$

27. Let m and n be real numbers and let $6x^2 - 2mx - 3nx + mn = 0$. Solve for x.

28. **Critical Thinking** A quadratic equation has 3 and -11 as its solutions. Work backwards to find the equation.

29. Solve $(x^2 - 16)(x^2 + 10x + 25) = 0$.

MIXED REVIEW

Find each product. (6.6 Skill A)

30. $2x^2(x^2 - 2x + 1)$

31. $3a^2(2a^2 - 2a - 2)$

32. $n^2(n^3 - 2n^2 - 5)$

33. $-5m^3(m^3 + m^2 + m)$

34. $k(k^3 + k^2 + k + 1)$

35. $y^3(y^3 - 2y^2 + 2y - 1)$

36. $-3b^5(-b^3 + 2b^2 + 3)$

37. $8d(d^7 + 9d^5 + d^3)$

In Skill A, all of the quadratic equations were written in standard form, $ax^2 + bx + c = 0$. If an equation is not given in standard form, rewrite the equation in standard form. Then solve it by factoring.

EXAMPLE 1 **Solve by $8v^2 = 10v + 3$ by factoring.**

▶ **Solution**

$$8v^2 = 10v + 3$$
$$8v^2 - 10v - 3 = 0 \qquad \longleftarrow \text{First write in standard form.}$$
$$(4v + 1)(2v - 3) = 0 \qquad \longleftarrow \text{Factor.}$$
$$4v + 1 = 0 \quad \text{or} \quad 2v - 3 = 0 \qquad \longleftarrow \text{Apply the Zero-Product Property.}$$
$$v = -\frac{1}{4} \qquad\qquad v = \frac{3}{2}$$

TRY THIS Solve $6t^2 + t = 15$ by factoring.

EXAMPLE 2 **Solve $(z + 1)(z + 2) = 30$ by factoring.**

▶ **Solution**

To apply the Zero-Product-Property, the equation must be in the form $ab = 0$.
$$(z + 1)(z + 2) = 30$$
$$z^2 + 3z + 2 = 30 \qquad \longleftarrow \text{Multiply.}$$
$$z^2 + 3z - 28 = 0 \qquad \longleftarrow \text{Write in standard form.}$$
$$(z - 4)(z + 7) = 0 \qquad \longleftarrow \text{Factor.}$$
$$z - 4 = 0 \quad \text{or} \quad z + 7 = 0 \qquad \longleftarrow \text{Apply the Zero-Product Property.}$$
$$z = 4 \qquad\qquad z = -7$$

TRY THIS Solve $(b - 3)(b - 2) = 42$ by factoring.

Example 3 shows how you can make solving a quadratic equation easier if you divide both sides by the numerical GCF of all terms before factoring.

EXAMPLE 3 **Solve by factoring.** **a. $2r^2 - 18 = 0$** **b. $-3s^2 + 12s - 12 = 0$**

▶ **Solution**

a.
$$2r^2 - 18 = 0$$
$$r^2 - 9 = 0 \qquad \longleftarrow \text{Divide each side by 2, the GCF of } 2r^2 \text{ and 18.}$$
$$(r + 3)(r - 3) = 0 \qquad \longleftarrow \text{Factor.}$$
$$r = -3 \quad \text{or} \quad r = 3 \qquad \longleftarrow \text{The solutions are 3 and } -3.$$
b.
$$-3s^2 + 12s - 12 = 0$$
$$s^2 - 4s + 4 = 0 \qquad \longleftarrow \text{Divide each side by } -3.$$
$$(s - 2)(s - 2) = 0 \qquad \longleftarrow \text{Factor.}$$
$$s = 2$$

Because $(s - 2)$ appears twice as a factor, 2 is called a **double root**.

TRY THIS Solve by factoring. **a. $-9c^2 + 36 = 0$** **b. $10s^2 + 20s + 10 = 0$**

EXERCISES

Write each quadratic equation in standard form.

1. $6x^2 = 7x + 5$

2. $r^2 + 5r + 6 = 42$

3. $4n^2 + 9 = 12n$

4. $6 = 22d - 12d^2$

5. $15t^2 - 1 = 2t$

6. $-12w^2 + 12 = 24w$

PRACTICE

Use factoring to solve.

7. $x^2 + 4x = 21$

8. $t^2 - 3t = 40$

9. $q^2 = -14q - 33$

10. $12a + a^2 = -35$

11. $w + w^2 = 42$

12. $39 + v^2 = 16v$

13. $4b + 15b^2 = 3$

14. $-d + 21d^2 = 2$

15. $9 - 15y = 14y^2$

16. $(x + 5)(x - 2) = 18$

17. $(z - 3)(z - 7) = 32$

18. $(a - 4)(a - 7) = -2$

19. $(2x - 6)(x - 7) = 24$

20. $(c + 1)(3c + 2) = 44$

21. $(3x + 5)(2x - 2) = 22$

22. $(4x - 1)(2x - 3) = 3$

23. $(7x + 2)(2x + 3) = -5$

24. $(7x - 3)(3x - 7) = -11$

25. $12x^2 + 12x + 3 = 0$

26. $12a^2 + 36a + 27 = 0$

27. $12x^2 - 3 = 0$

28. $16t^2 - 36 = 0$

29. $-16v^2 - 80v - 100 = 0$

30. $28c^2 - 84c + 63 = 0$

31. $100h^2 = 160h - 64$

32. $-144 + 144y = 36y^2$

33. $20 - 60x = -45x^2$

34. $30a^2 = 18 + 33a$

35. Solve $(x - 5)^2 = 36$ by writing the equation as the difference of two squares and then factoring.

36. **Critical Thinking**   Let $a = 0, 1, 2, 3, 4, 5,$ or 6. For which values of a will $(x - 1)(x - 2) = a$ have integer solutions? What are the solutions?

MIXED REVIEW

Let r represent any real number. Is the given statement always, sometimes, or never true? Justify your response. (2.9 Skill A)

37. $r^2 = 7r$

38. $2(r - 5) + r = 2r - 2(r + 5)$

39. $2(r - 5) + 3(r - 5) = 5(r - 5) + 1$

40. $(r - 1)^2 = 2(r - 1)$

A **cubic equation** in x is any equation that can be written in the form $ax^3 + bx^2 + cx + d = 0$, where $a \neq 0$. This is called the **standard form of a cubic equation**. Using an extension of the Zero-Product Property, you can solve many cubic equations.

> If a, b, and c represent real numbers and $abc = 0$, then $a = 0$, $b = 0$, or $c = 0$.

For example, if $x(x - 2)(3x + 4) = 0$, you can write the following.

$$x = 0 \quad \text{or} \quad x - 2 = 0 \quad \text{or} \quad 3x + 4 = 0$$

Thus, the solutions are 0, 2, and $-\dfrac{4}{3}$.

EXAMPLE 1 Solve $2n^3 + 8n^2 - 42n = 0$.

▶ **Solution**

$$2n^3 + 8n^2 - 42n = 0$$
$$2n(n^2 + 4n - 21) = 0 \qquad \longleftarrow \text{Factor the GCF, } 2n, \text{ from each term.}$$
$$2n(n + 7)(n - 3) = 0 \qquad \longleftarrow \text{Factor } n^2 + 4n - 21.$$
$$2n = 0 \quad \text{or} \quad (n + 7) = 0 \quad \text{or} \quad (n - 3) = 0 \qquad \longleftarrow \text{Apply the Zero-Product Property.}$$
$$n = 0 \qquad\qquad n = -7 \qquad\qquad n = 3 \qquad \longleftarrow \text{Solve each equation for } n.$$

The solutions are 0, -7, and 3.

TRY THIS Solve $10k^3 - 13k^2 + 4k = 0$.

EXAMPLE 2 Solve $m^3 + 22m^2 + 121m = 0$.

▶ **Solution**

$$m^3 + 22m^2 + 121m = 0$$
$$m(m^2 + 22m + 121) = 0 \qquad \longleftarrow \text{Factor the GCF, } m, \text{ from each term.}$$
$$m(m + 11)(m + 11) = 0 \qquad \longleftarrow \text{Factor } m^2 + 22m + 121.$$
$$m = 0 \quad \text{or} \quad m + 11 = 0 \quad \text{or} \quad m + 11 = 0 \qquad \longleftarrow \text{Apply the Zero-Product Property.}$$
$$m = 0 \qquad\qquad m = -11 \qquad\qquad m = -11 \qquad \longleftarrow \text{Solve each equation for } m.$$

The solutions are 0 and -11.

TRY THIS Solve $z^3 - 14z^2 + 49z = 0$.

The steps for solving a polynomial equation by factoring are:
Step 1. Write the equation in standard form.
Step 2. Factor the GCF, if one exists, from each term in the equation.
Step 3. Factor the polynomial.
Step 4. Apply the Zero-Product Property and set each factor equal to zero.
Step 5. Solve for the variable.
Step 6. Check your solution(s) in the original equation.

EXERCISES

Use the Zero-Product Property to write the solutions to each equation.

1. $(x - 1)(x - 2)(x - 3) = 0$

2. $a(a + 2)(a - 2) = 0$

3. $n(2n + 1)(3n - 2) = 0$

4. $x^2(x - 1) = 0$

5. $y(y - 3)(y - 3) = 0$

6. $(x - a)(x - b)(x - c) = 0$

Use factoring to solve. Check your solutions.

7. $4x^3 - 3x^2 = 0$

8. $z^3 + 4z^2 = 0$

9. $7v^3 - 14v^2 = 0$

10. $2x^3 + 7x^2 = 0$

11. $k^3 - 4k^2 = 0$

12. $5b^3 + 10b^2 = 0$

13. $3y^3 + 9y^2 = 0$

14. $12g^3 - 144g^2 = 0$

15. $7s^3 - 5s^2 = 0$

16. $b^3 - 4b^2 + 3b = 0$

17. $t^3 + 11t^2 + 18t = 0$

18. $s^3 - 4s^2 + 3s = 0$

19. $x^3 - 2x^2 + x = 0$

20. $v^3 - 5v^2 + 4v = 0$

21. $7c^3 - 21c^2 = 0$

22. $6b^3 - 19b^2 + 15b = 0$

23. $4m^3 - 4m^2 - 35m = 0$

24. $6a^3 + 23a^2 + 21a = 0$

25. $2x^3 - 8x = 0$

26. $3h^3 - 27h = 0$

27. $25n^3 - 40n^2 + 16n = 0$

28. $36z^3 + 60z^2 + 25z = 0$

29. $2d^3 - 98d = 0$

30. $49y^3 + 14y^2 + y = 0$

31. Solve $(b^2 - 16)(b^2 - 9)(b^2 - 1) = 0$.

32. Solve $(n^2 - 25)(n^2 - 16)(n^2 + 2n + 1) = 0$.

33. Solve $(z^2 - 1)(z^2 - 4)(z^2 - 9)(z^2 - 16) = 0$.

34. Solve $(x^2 - 2xy + y^2)(x^2 + 2xy + y^2) = 0$.

35. Critical Thinking Justify each step in the following proof.

    1. $abc = a(bc)$
    2. If $abc = 0$, then $a(bc) = 0$.
    3. If $a(bc) = 0$, then $a = 0$ or $bc = 0$.
    4. If $a = 0$ or $bc = 0$, then $a = 0$ or $b = 0$ or $c = 0$.

Let r represent any real number. Is the given inequality sometimes, always, or never true? Justify your response. (4.8 Skill A)

36. $r \geq 20r$

37. $2(r - 3) - (r - 3) \geq r - 3$

38. $2(r + 5) - 3r \leq 9 - r$

39. $(r - 1)^2 \leq r^2$

7.6 LESSON
Quadratic Functions and Their Graphs

SKILL A   *Graphing a quadratic function*

A **quadratic function** is any function of the form $y = ax^2 + bx + c$, where $a \neq 0$.

The Graph of a Quadratic Function

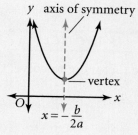

- The graph of $y = ax^2 + bx + c$ $(a \neq 0)$ has a U shape called a **parabola**. The lowest or highest point is called the **vertex**.
- The U shape is symmetric. It has an **axis of symmetry** whose equation is $x = -\dfrac{b}{2a}$. The value of $-\dfrac{b}{2a}$ is also the x-coordinate of the vertex.
- If $a > 0$, the graph opens upward and has a minimum point.
- If $a < 0$, the graph opens downward and has a maximum point.

EXAMPLE

Find the coordinates of the vertex and an equation for the axis of symmetry. Then graph the function.

a. $y = x^2 - 4x + 3$

b. $y = -x^2 + 4$

▶ **Solution**

a. $a = 1$, $b = -4$, $c = 3$

$$-\frac{b}{2a} \rightarrow -\frac{-4}{2(1)} = 2$$

axis of symmetry: $x = 2$
If $x = 2$, then $y = 2^2 - 4(2) + 3 = -1$.
vertex: $(2, -1)$

b. $a = -1$, $b = 0$, $c = 4$

$$-\frac{b}{2a} \rightarrow -\frac{0}{2(-1)} = 0$$

axis of symmetry: $x = 0$
If $x = 0$, then $y = -0^2 + 4 = 4$.
vertex: $(0, 4)$

Make a table.

x	-1	0	1	2	3	4	5
y	8	3	0	-1	0	3	8

x	-3	-2	-1	0	1	2	3
y	-5	0	3	4	3	0	-5

The graph opens down, and the vertex is the maximum point.

The graph opens up, and the vertex is the minimum point.

TRY THIS

Rework the Example using the following equations:

a. $y = x^2 + 4x + 3$

b. $y = -x^2 + 2$

EXERCISES

Find the coordinates of the vertex and write an equation for the axis of symmetry.

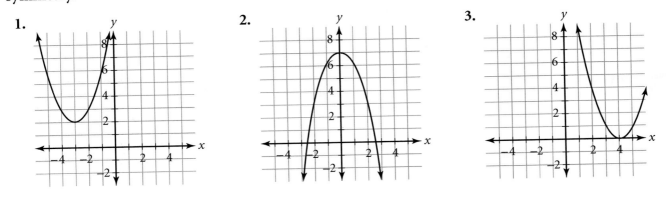

1.

2.

3.

Using $y = ax^2 + bx + c$, identify a, b, and c in each of the functions below.

4. $y = 2x^2 + 3x - 6$

5. $y = -x^2 + 2x + 8$

6. $y = x^2 - 5$

Each equation represents a quadratic function. Multiply and write each product in the form $ax^2 + bx + c$. Then identify a, b, and c.

7. $y = x(x - 3)$

8. $y = (2x - 1)(x - 2)$

9. $y = (x + 3)(x - 3)$

10. $y = (x - 1)(x + 1)$

11. $y = (x + 2)^2$

12. $y = -(x - 1)^2$

Does the graph of each function open upward or downward?

13. $y = x^2 + 3x - 2$

14. $y = -x^2 + 4x - 3$

15. $y = x^2 - 2x$

For Exercises 16–24:
  a. Find the coordinates of the vertex and write an equation for the axis of symmetry.
  b. Identify the vertex as the maximum or the minimum point.
  c. Then make a table of ordered pairs and use it to graph each function.

16. $y = -x^2 - 2x + 3$

17. $y = x^2 - 9$

18. $y = -x^2 - 1$

19. $y = x^2 + x - 6$

20. $y = x^2 - 3x + 2$

21. $y = x^2 - 1$

22. $y = x^2 + 3x - 2$

23. $y = -x^2 + 4x - 3$

24. $y = x^2 - 2x$

Find the intercepts of the graph of each equation. Graph the line. (3.5 Skill A)

25. $y = -3.5$

26. $x = 3.5$

27. $x + y = 6$

28. $x - y = 6$

29. $2x + 3y = 9$

30. $2x - y = 10$

31. $y = -\frac{2}{7}x + 1$

32. $y = \frac{2}{5}x + 3$

Recall that an *x*-intercept of a graph is the *x*-coordinate of any point where the graph crosses the *x*-axis. The graph of a quadratic equation, a parabola, may have one of three possibilities: no *x*-intercepts, exactly one *x*-intercept, or two distinct *x*-intercepts. The values of *x* at the *x*-intercepts are the solutions, or roots, of the equation.

One way to find the *x*-intercepts of a quadratic function is to graph the function and see where it crosses the *x*-axis.

EXAMPLE 1 Use a graph to count the *x*-intercepts of the graph of $y = x^2 + 2$.

▶ **Solution**

Use the coordinates of the vertex and a table of ordered pairs to graph $y = x^2 + 2$.

x-coordinate of vertex: $\dfrac{-b}{2a} = \dfrac{-0}{2(1)} = 0$

Make a table.

x	-2	-1	0	1	2
y	6	3	2	3	6

The graph never crosses the *x*-axis, so it has no *x*-intercepts.

TRY THIS Use a graph to count the *x*-intercepts of $y = x^2 + x + 1$.

You can also use an algebraic approach to find the *x*-intercepts. To find the *x*-intercepts of the graph of $y = ax^2 + bx + c$, set $ax^2 + bx + c$ equal to 0, because $y = 0$ at the *x*-axis. Then solve for *x*. The solution gives the *x*-intercept.

EXAMPLE 2 Find and count the *x*-intercepts of the graph of $y = 2x^2 - 3x - 9$.

▶ **Solution**

$$2x^2 - 3x - 9 = 0 \quad \longleftarrow \text{Set } 2x^2 - 3x - 9 \text{ equal to 0.}$$
$$(2x + 3)(x - 3) = 0 \quad \longleftarrow \text{Factor.}$$
$$2x + 3 = 0 \quad \text{or} \quad x - 3 = 0 \quad \longleftarrow \text{Apply the Zero-Product Property.}$$
$$x = -\frac{3}{2} \quad \text{or} \quad x = 3 \quad \longleftarrow \text{Solve each equation for } x.$$

There are two *x*-intercepts of the graph, $-\dfrac{3}{2}$ and 3.

TRY THIS Find and count the *x*-intercepts of the graph of $y = 5x^2 - 26x + 5$.

If you graph $y = x^2 + 2x + 1$ or use an algebraic approach to find the *x*-intercepts, you will find exactly one *x*-intercept, -1. If the graph of a quadratic function has exactly one *x*-intercept, then the *x*-intercept is a *double root* of the equation.

$$x^2 + 2x + 1 = 0$$
$$(x + 1)^2 = 0$$
$$x = -1$$

There is only one *x*-intercept, -1.

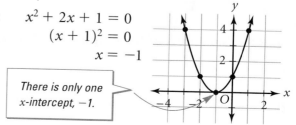

EXERCISES

How many *x*-intercepts does each graph have?

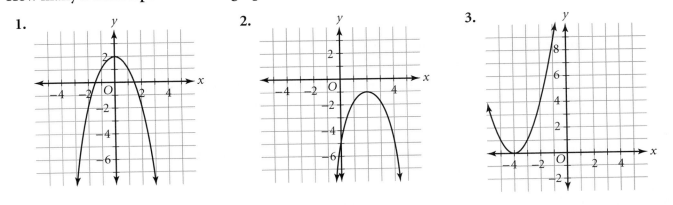

1. 2. 3.

Use a graph to count the *x*-intercepts.

4. $y = x^2 + 4x + 4$

5. $y = -x^2 + 1$

6. $y = x^2 + x + 2$

Find and count the *x*-intercepts of the graph of each equation.

7. $y = x^2 + 7x + 6$

8. $y = -x^2 + 4x - 4$

9. $y = x^2 + 3x + 2$

10. $y = 4x^2 + 4x - 3$

11. $y = -x^2 + 9$

12. $y = x^2 + 6x + 9$

Let *r* be a real number. Find the values of *r* such that the graph has the specified number of *x*-intercepts.

13. $y = x^2 + r$
    no *x*-intercepts

14. $y = x^2 + 2x + r$
    one *x*-intercept

15. $y = x^2 + rx$
    two *x*-intercepts

16. **Critical Thinking** Suppose that the vertex of the graph of $y = ax^2 + bx + c$ is above the *x*-axis and $a > 0$. How many *x*-intercepts does the graph have? Justify your response.

Find and simplify each sum. (6.4 Skill A)

17. $-3v^3 - 2v^2 + 1$ and $5v^3 - v^2 - 2$

18. $5.2c^2 - 2c$ and $-5.2c^2 - 3c - 3$

19. $2t^3 - 2t^2 - 5$ and $5t^2 - 2t - 3$

20. $-5n^2 - 2n + 1$ and $5n^2 - n - 1$

Find and simplify each product. (6.6 Skill B, C)

21. $2(a - 4)(a + 5)$

22. $(2r + 3)(3r + 7)$

23. $-(3d + 2)(5d + 1)$

24. $3(4g + 1)(4g - 1)$

25. $(7y - 2)(7y - 2)$

26. $-2(3w + 5)(2w - 3)$

Analyzing the Graph of a Quadratic Function

SKILL A   *Finding the range of a quadratic function*

The domain of a quadratic function is the set of all real numbers. The range of the function can be represented by an inequality. When you find the range of a quadratic function, draw a quick sketch of the graph to help you.

Recall that if $a > 0$, the graph opens upward and the y-coordinate of the vertex is the minimum value of y.

EXAMPLE 1   **Find the range of the function represented by $y = 3x^2 - 6x - 1$.**

▶ **Solution**

Find the coordinates of the vertex, V.

Use $\frac{-b}{2a}$ to find the x-coordinate.

x-coordinate: $-\dfrac{-6}{2(3)} = 1$ ←——— *$a = 3$ and $b = -6$*

y-coordinate: $3(1)^2 - 6(1) - 1 = -4$

Because the graph opens upward, all values of y are *at least* equal to the y-coordinate of the vertex.

Draw a quick sketch. a is positive, so the graph opens upward.

Therefore, the range is $y \geq -4$.

V $(1, -4)$

TRY THIS   Find the range of the function represented by $y = 2x^2 + 5x$.

Recall that if $a < 0$, the graph opens downward and the y-coordinate of the vertex is the maximum value of y.

EXAMPLE 2   **Find the range of the function represented by $y = -5x^2 - 10x + 3$.**

▶ **Solution**

Find the coordinates of the vertex, V.

Use $\frac{-b}{2a}$ to find the x-coordinate.

x-coordinate: $-\dfrac{-10}{2(-5)} = -1$ ←——— *$a = -5$ and $b = -10$*

y-coordinate: $-5(-1)^2 - 10(-1) + 3 = 8$

Draw a quick sketch. a is negative, so the graph opens downward.

Because the graph opens downward, all values of y are *at most* equal to the y-coordinate of the vertex.

V $(-1, 8)$

Therefore, the range is $y \leq 8$.

TRY THIS   Find the range of the function represented by $y = -x^2 + 7x - 1$.

EXERCISES

KEY SKILLS

Find the range of each quadratic function graphed below.

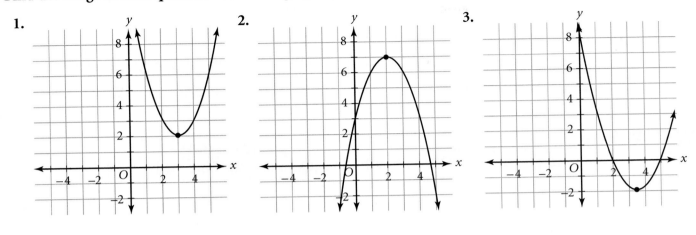

1.
2.
3.

PRACTICE

Find the range of each quadratic function.

4. $y = x^2 - 4x + 3$

5. $y = -x^2 - 6x + 2$

6. $y = -x^2 - 5x + 2$

7. $y = x^2 - 3x + 1$

8. $y = -2x^2 - 5x + 3$

9. $y = 2x^2 + 7x + 3$

10. $y = -3x^2 - 9x + 7$

11. $y = 5x^2 - 10x + 4$

12. $y = \frac{1}{2}x^2 - 3x - 7$

13. $y = \frac{1}{2}x^2 + 5x - 1$

14. $y = \frac{3}{4}x^2 + x + 3$

15. $y = -\frac{3}{4}x^2 - 3x$

16. $y = x(x - 1)$

17. $y = x(x + 2)$

18. $y = (x + 2)(x - 1)$

19. $y = -(x - 5)(x - 3)$

20. $y = (2x + 5)(x + 1)$

21. $y = (2x + 7)(3x - 5)$

In Exercises 22–24, use the given range to determine if the graph of the quadratic function opens up or down.

22. The range is $y \geq -2$.

23. The range is $y \leq 3$.

24. The range is $y \leq 0$.

25. Given the domain $-2 \leq x \leq 2$, what is the range of $y = x^2$?

MIXED REVIEW

Simplify each expression. (6.5 Skill A)

26. $(2n^2)(3n^4)$

27. $(3n^3)(2n^2)(-2n^2)$

28. $(z^3)(z^2)^2$

29. $(k^2)(k^2)^2$

Simplify each expression. (6.5 Skill B)

30. $\dfrac{2(x^2y^2)^2}{(xy)^2}$

31. $\dfrac{(3a^2b)^2}{(9a^2b)^2}$

32. $\dfrac{(2c^2d^2)^2}{(c^4d^3)}$

33. $\dfrac{(2gh^2)^2}{(4g^2h^2)^2}$

Just as you can write an equation for a line given sufficient information, you can write an equation for a quadratic function given sufficient information about its graph.

Using Intercepts to Write a Quadratic Function

If a parabola has x-intercepts r and s, then for some nonzero real number a, $y = a(x - r)(x - s)$ represents the parabola.

EXAMPLE 1

The graph of a quadratic function has x-intercepts -2 and 3. The y-intercept of the graph is -12. Write an equation for the function.

▶ **Solution**

1. Because the x-intercepts are -2 and 3, the function has the form $y = a[x - (-2)](x - 3)$ for some nonzero real number a.

2. To find a, use the coordinates of the y-intercept, $(0, -12)$. Substitute 0 for x and -12 for y and then solve the equation for a.

$$a(x + 2)(x - 3) = y$$
$$a(0 + 2)(0 - 3) = -12$$
$$a(-6) = -12$$
$$a = 2$$

3. Write the function and simplify it.
$$y = 2(x + 2)(x - 3) = 2x^2 - 2x - 12$$

An equation for the function is $y = 2x^2 - 2x - 12$.

TRY THIS   The graph of a quadratic function has x-intercepts 3 and 4. The y-intercept of the graph is 12. Write an equation for the function.

EXAMPLE 2

The graph of a quadratic function has x-intercepts -4 and 5. The point $(2, -36)$ is on the graph. Write an equation for the function.

▶ **Solution**

1. Since the x-intercepts are -4 and 5, the function has the form $y = a[x - (-4)](x - 5)$ for some nonzero real number a.

2. To find a, use $(2, -36)$. Substitute 2 for x and -36 for y and then solve the equation for a.

$$a(x + 4)(x - 5) = y$$
$$a(2 + 4)(2 - 5) = -36$$
$$a(-18) = -36$$
$$a = 2$$

3. Write the function and simplify it.
$$y = 2(x + 4)(x - 5) = 2x^2 - 2x - 40$$

An equation for the function is $y = 2x^2 - 2x - 40$.

TRY THIS   The graph of a quadratic function has x-intercepts -4 and 4. The point $(3, -14)$ is on the graph. Write an equation for the function.

EXERCISES

KEY SKILLS

Use the given x-intercept(s) to write a quadratic function in the form $y = a(x - r)(x - s)$.

1. x-intercepts: 2 and -2

2. x-intercepts: -5 and 2

3. x-intercepts: -1 and 3

4. x-intercepts: 0 and 4

5. x-intercept: 1

6. x-intercept: -3

PRACTICE

Use the given information to write a quadratic function in the form $y = ax^2 + bx + c$.

7. x-intercepts: -2 and 8; y-intercept 32

8. x-intercepts: 1 and 7; y-intercept 5

9. x-intercepts: -3 and 3; vertex $(0, 5)$

10. x-intercepts: -5 and 5; vertex $(0, -75)$

11. x-intercepts: -8 and 1
The point $(3, 44)$ is on the graph.

12. x-intercepts: 3 and 4
The point $(-1, -5)$ is on the graph.

Write an equation in the form $y = ax^2 + bx + c$ to represent each parabola.

13.

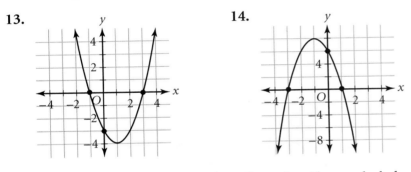

14.

15.

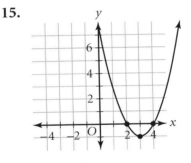

16. Let r be a nonzero real number. Show that if a parabola has x-intercepts r and $-r$, then the parabola is represented by $y = ax^2 - ar^2$ for some nonzero real number a.

17. Show that if a parabola has exactly one x-intercept r, then the parabola is represented by $y = a(x - r)^2$.

18. **Critical Thinking**   Explain why the graph of any function represented by $y = ax^2 + b$, where $a > 0$ and $b > 0$, has no x-intercepts.

MIXED REVIEW

Use factoring to solve each equation. (7.5 Skill A)

19. $3a^2 - 75 = 0$

20. $6x^2 + 3x - 9 = 0$

21. $4x^2 - 8x + 3 = 0$

Use factoring to solve each equation. (7.5 Skill B)

22. $3a^2 = 4a$

23. $6c^2 = 132 - 61c$

24. $-35n^2 = 74n + 35$

25. $(2n + 1)(2n + 3) = 99$

26. $(3p - 5)(5p - 3) = 7$

27. $(4a + 3)(7a + 11) = 33$

You can apply your knowledge of quadratic functions to real-world problems.

EXAMPLE 1

A soccer ball is kicked into the air. The function $y = -16x^2 + 32x$ describes its height in feet, y, after x seconds. Draw a graph of the function and use it to answer the questions below.

  a. What is the maximum height reached by the ball?
  b. How long does it take for the ball to reach its maximum height?
  c. How long does it take for the ball to return to the ground?

▶ **Solution**

  a. Because a, -16, is negative, the graph is a parabola that opens downward, and the y-coordinate of the vertex represents the maximum height, 16 feet.

  b. The x-coordinate of the vertex represents the time (in seconds) when the ball reaches its maximum height. The vertex is $(1, 16)$, which means that after 1 second, the ball reaches its maximum height of 16 feet.

  c. The y-coordinate, which represents the height, increases until it reaches the vertex and then decreases until it reaches 0 at $x = 2$. Therefore, the ball returns to the ground after 2 seconds.

TRY THIS   Rework Example 1 for the function $y = -16x^2 + 64x$.

EXAMPLE 2

The profit p a company can make on the sale of x calendars is given by the function $p = 0.05x(1200 - x)$. How many calendars need to be sold to achieve the maximum profit? What is the maximum profit?

▶ **Solution**

  1. Rewrite the function in standard form.
$$p = 0.05x(1200 - x) \quad \rightarrow \quad p = -0.05x^2 + 60x$$

  2. Because a, -0.05, is negative, the graph opens downward and the vertex contains the maximum value of p, the profit.

  3. Find the coordinates of the vertex.

x-coordinate: $-\dfrac{b}{2a} \quad \rightarrow \quad -\dfrac{60}{2(-0.05)} = \mathbf{600}$

p-coordinate: $p = -0.05(\mathbf{600})^2 + 60(\mathbf{600}) = 18{,}000$

The maximum profit of $18,000 is reached when 600 calendars are sold.

TRY THIS   Rework Example 2 for the profit function $p = 0.04x(1500 - x)$.

EXERCISES

Determine whether the graph of the function has a maximum or a minimum value.

1. $y = -5x^2 - 9x + 1$

2. $y = 2x^2 + 3x - 1$

3. $y = 9x^2 - 3x - 7$

4. $y = -x^2 + x$

PRACTICE

Refer to Example 1.

5. What is the axis of symmetry of the graph?

6. Does it make sense for the graph to extend beyond the first quadrant? Why or why not?

7. What do the x-intercepts of the graph represent?

In Exercises 8–15, each function represents the profit p made from the sale of x units of a product. How many units need to be sold to achieve the maximum profit? What is the maximum profit?

8. $p = 0.02x(4800 - x)$

9. $p = 0.08x(8000 - x)$

10. $p = 0.03x(4900 - x)$

11. $p = 0.05x(6300 - 9x)$

12. $p = 0.01x(2700 - 6x)$

13. $p = 0.04x(9600 - 12x)$

14. $p = 0.03x(2700 - 9x)$

15. $p = 0.06x(3600 - 4x)$

16. **Critical Thinking** Suppose that x units are sold and that profit p is given by $p = 0.04x(1500 - x)$. Show that maximum profit is reached when x is the average of the x-intercepts of the graph of the function.

MIXED REVIEW APPLICATIONS

Write an expression for the total volume of 36 boxes with each box having length, width, and height as specified. Do not simplify. (6.5 Skill C)

17. length $2.5a$ units, width $2a$ units, and height $4a$ units

18. length $3.5n$ units, width $2.5n$ units, and height $6n$ units

19. length $1.5z$ units, width $1.5z$ units, and height $2.5z$ units

20. length $\frac{4m}{5}$ units, width $\frac{5m}{16}$ units, and height $\frac{16m}{25}$ units

Write the prime factorization of each number.

1. 500   2. 1080   3. 31   4. 2401

Find the GCF of each set of monomials.

5. $5x^3$ and $35x^4$ 6. $6n^2d$ and $32n^4d^2$ 7. $5y^2z^2$, $18y^2$, and $60y^4z^2$

Use the GCF to factor each polynomial.

8. $3s^2 - 4s^3$

9. $14z^4 + 4z^2$

10. $7a^3 + 14a^2 - 7a$

11. $10n^4 - 14n^3 - 7n^2$

12. $3qp^4 - 12q^2p^2$

13. $a^5b^2 + 10ab^3 - ab^2 - 15b^2$

Factor each polynomial.

14. $25d^2 - 120d + 144$

15. $144z^2 - 25$

16. $4b^2 - 44bd + 121d^2$

17. $16z^2 - 81y^2$

18. $4a^2b^2 + 4abc + c^2$

19. $49m^2n^2 - 16p^2$

20. $4a^4 + 12a^2 + 9$

21. $9hj + 6h - 12kj - 8k$

22. $k^2 + 16k + 39$

23. $y^2 - 18y + 56$

24. $a^2 + 19a + 84$

Factor. If the expression cannot be factored using integers, write *not factorable* and justify your response.

25. $m^2 + 5m - 84$ 26. $c^2 + 13c - 8$ 27. $d^2 + 13d - 30$

Factor.

28. $15z^2 + 52z + 32$

29. $15a^2 + 67a + 44$

30. $14m^2 + 57m + 55$

31. $18b^2 + 49b - 49$

32. $33p^2 - 37p + 10$

33. $12w^2 - 53w + 55$

34. $-u^2 + 12u - 35$

35. $3u^2 - 36u + 81$

36. $5g^2 - 10g - 120$

37. $25x^3 - 121x$

38. $49k^3 - 140k^2 + 100k$

39. $42a^3 + 81a^2 + 30a$

Use factoring to solve each equation.

40. $14b^2 - 39b + 27 = 0$

41. $36n^2 - 36n + 5 = 0$

42. $25q^2 - 75q + 26 = 0$

43. $21d^2 - 50d - 99 = 0$

44. $8z^2 = 63z + 81$

45. $(3s - 10)(s - 1) = 40$

46. $7r^2 - 63 = 0$

47. $5v^2 - 10v + 5 = 0$

48. $3s^3 - 7s^2 = 0$

49. $n^3 - 7n^2 + 6n = 0$

50. $p^3 + 18p^2 + 81p = 0$

51. $15n^3 + 57n^2 - 12n = 0$

Find the coordinates of the vertex and an equation for the axis of symmetry. Is the vertex the maximum or minimum point on the graph?

52. $y = 3x^2 + 4x + 4$

53. $y = 2x^2 - 4x + 1$

54. $y = 0.5x^2 - 2x + 0.2$

55. $y = -1.5x^2 + 5x - 3$

Use a graph to determine if the graph of the function has zero, one, or two x-intercepts.

56. $y = -x^2 - 3$

57. $y = x^2 - 6x + 9$

Find and count the x-intercepts of the graph of each function.

58. $y = 4x^2 - 20x + 21$

59. $y = -x^2 - 2x - 1$

Find the range of each quadratic function.

60. $y = 2.5x^2 - 2x - 3$

61. $y = -4x^2 - 7x + 5$

62. $y = -x^2 - 4$

63. $y = 3x^2 - 9$

64. The graph of a quadratic function has x-intercepts -5 and 6. The point $(1, 60)$ is on the graph. Write an equation for the function.

65. The graph of a quadratic function has x-intercepts -6 and -2. The point $(0, 24)$ is on the graph. Write an equation for the function.

66. The profit a company can make on the same of x units of goods is represented by the function $p = 0.04x(1600 - 2x)$. How many units need to be sold to achieve the maximum profit possible? What is the maximum profit?

Quadratic Functions and Equations

What You Already Know

By now, you have gained much experience in solving linear equations. In particular, the four arithmetic operations of addition, subtraction, multiplication, and division of real numbers is key to your success here. Now it is time to extend those skills and learn new ones that will help you solve quadratic equations, equations of the form $ax^2 + bx + c$, where a, b, and c are real numbers and $a \neq 0$.

What You Will Learn

First, you will learn about taking the square root of a number. This operation is quite different from the four operations mentioned above. Then you will see how to solve certain quadratic equations by using the four arithmetic operations along with taking the square root of a number.

Special and new equation-solving skills presented to you include

• factoring and applying the Zero-Product Property,

• completing the square, and

• applying the quadratic formula.

Your study of quadratic equations will lead you to

• write a solution as a deductive argument,

• use inductive reasoning to represent patterns as quadratic functions, and

• solve real-world problems involving the accelerated motion of an object under the force of gravity.

VOCABULARY

completing the square	Pythagorean Theorem
discriminant	quadratic formula
imaginary unit	quadratic pattern
perfect square	Quotient Property of Square Roots
principal square root	second differences
Product Property of Square Roots	simplest radical form
pure imaginary number	square root

The diagram below shows how mathematical skills and mathematical reasoning are interrelated with the skills and concepts in Chapter 8. Notice that this chapter involves solving quadratic equations using different methods.

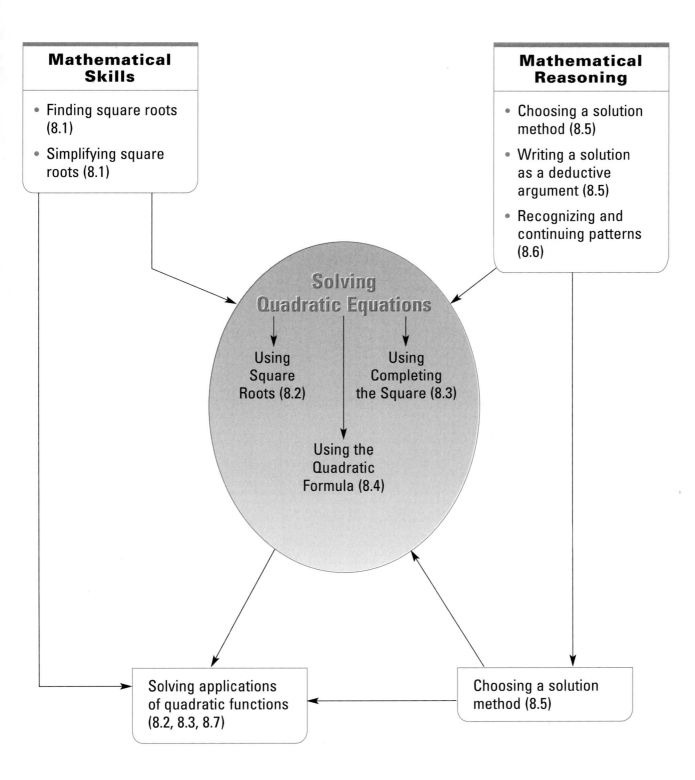

Mathematical Skills

- Finding square roots (8.1)
- Simplifying square roots (8.1)

Mathematical Reasoning

- Choosing a solution method (8.5)
- Writing a solution as a deductive argument (8.5)
- Recognizing and continuing patterns (8.6)

Solving Quadratic Equations

Using Square Roots (8.2)

Using Completing the Square (8.3)

Using the Quadratic Formula (8.4)

Solving applications of quadratic functions (8.2, 8.3, 8.7)

Choosing a solution method (8.5)

Square Roots and the Equation $x^2 = k$

Finding and approximating square roots

The symbol for the *principal,* or positive, *square root,* $\sqrt{}$, is called the *radical sign.*

> **Definition of Square Root**
>
> If $k \geq 0$, then $\sqrt{k}$ represents the positive square root of k and $-\sqrt{k}$ represents the negative square root of k. The square roots of k have the following property: $\sqrt{k} \cdot \sqrt{k} = k$ $(-\sqrt{k}) \cdot (-\sqrt{k}) = k$

EXAMPLE 1   **Find the square roots of each number.**   **a.** 100   **b.** $\dfrac{9}{4}$

▶ **Solution**

a. $(10)(10) = 100$ and $(-10)(-10) = 100$, so the square roots of 100 are ± 10.

b. $\left(\dfrac{3}{2}\right)\left(\dfrac{3}{2}\right) = \left(\dfrac{9}{4}\right)$ and $\left(-\dfrac{3}{2}\right)\left(-\dfrac{3}{2}\right) = \left(\dfrac{9}{4}\right)$, the square roots of $\dfrac{9}{4}$ are $\pm\dfrac{3}{2}$.

> $\pm\dfrac{3}{2}$ is a shortened way of writing "$\dfrac{3}{2}$ or $-\dfrac{3}{2}$."

TRY THIS   Find the square roots of each number.   **a.** 81   **b.** $\dfrac{16}{25}$

Rational numbers such as 100 and $\dfrac{9}{4}$, whose square roots are rational numbers, are called **perfect squares**. Not every rational number is a perfect square.

EXAMPLE 2   **Find two consecutive integers between which $\sqrt{58}$ lies.**

▶ **Solution**

$7 \times 7 = 49$ too small $8 \times 8 = 64$ too large

Thus, $\sqrt{58}$ is between 7 and 8. Using a calculator, $\sqrt{58} \approx 7.62$.

> The approximately equal symbol, $\approx$, indicates an inexact or rounded answer.

TRY THIS   Repeat Example 2 with $-\sqrt{77}$.

You can use the following fact to solve equations of the form $x^2 = k$.

If $x^2 = k$ and $k > 0$, then $x = \sqrt{k}$ or $x = -\sqrt{k}$, abbreviated $x = \pm\sqrt{k}$. This is referred to as "taking the square root of each side of an equation."

EXAMPLE 3   **Solve. Round the roots to the nearest hundredth.**   **a.** $x^2 = 49$ **b.** $x^2 = 53$

▶ **Solution**

a. $x^2 = 49 \rightarrow x = \pm\sqrt{49}$ **b.** $x^2 = 53$
$x = 7$ or $x = -7$ $x = \pm\sqrt{53} \approx \pm 7.28$

TRY THIS   Solve. Round the roots to the nearest hundredth.   **a.** $x^2 = 81$ **b.** $x^2 = 63$

EXERCISES

KEY SKILLS

Find the square roots of each number.

1. 121

2. 4

3. 900

4. 169

PRACTICE

Find the square roots of each number. If the square roots are not rational numbers, then approximate the roots to the nearest hundredth.

5. 144

6. 50

7. 1

8. 56

9. 0.25

10. 2

11. $\dfrac{121}{64}$

12. 5

Find two consecutive integers between which each square root lies. Use a calculator to approximate the root to the nearest hundredth.

13. $\sqrt{68}$

14. $-\sqrt{160}$

15. $\sqrt{33}$

16. $-\sqrt{601}$

List all integers that are perfect squares that lie between each given pair of numbers.

17. 1 and 9, inclusive

18. 10 and 20, inclusive

19. 40 and 70, inclusive

20. 80 and 100, inclusive

21. 200 and 400, inclusive

22. 600 and 1000, inclusive

Solve each equation. Round irrational roots to the nearest hundredth.

23. $x^2 = 144$

24. $x^2 = 14$

25. $x^2 = 42$

26. $x^2 = 400$

27. Critical Thinking   Show that there are no perfect squares between 961 and 1024.

28. Critical Thinking   Let a be a real number. Show that the solutions to $x^2 = a^2$ are $-a$ and a by factoring $x^2 - a^2 = 0$.

29. Critical Thinking   Show that the solutions to $x^2 - 5x + 4 = 0$ are perfect squares.

MIXED REVIEW

Write the prime factorization of each number. (7.1 Skill A)

30. 196

31. 725

32. 375

33. 2000

Simplify. (6.1 Skill A)

34. $1^4 \cdot 2^3$

35. $7^2 \cdot 7^2$

36. $3^5 \cdot 3$

37. $3^2 \cdot 5^0$

Using properties of square roots, you can simplify a radical expression. A square-root expression is in **simplest radical form** when all of the following are true.

    1. There are no perfect squares under the radical sign.
    2. There are no fractions under the radical sign.
    3. There are no radical expressions in the denominator.

One property of square roots that can help simplify expressions involving square roots is the Product Property of Square Roots.

Product Property of Square Roots

If a and b represent positive real numbers, then $\sqrt{ab} = \sqrt{a} \cdot \sqrt{b}$.

When you simplify a square root, you may need to take the square root of a perfect square. Use the following fact.

    If x is a real number, then $\sqrt{x^2} = |x|$. If $x > 0$, then $\sqrt{x^2} = x$.

For example, $\sqrt{4^2} = 4$ and $\sqrt{(-3)^2} = |-3| = 3$.

EXAMPLE 1   **Write $\sqrt{12}$ in simplest radical form.**

▶ **Solution**

$$\sqrt{12} = \sqrt{4 \cdot 3} = \sqrt{2^2 \cdot 3} = \sqrt{2^2} \cdot \sqrt{3} = 2\sqrt{3}$$ ←—— *Apply the Product Property of Square Roots.*

In simplest radical form, $\sqrt{12} = 2\sqrt{3}$.

TRY THIS   Write $\sqrt{72}$ in simplest radical form.

Another property of square roots that you can use to simplify a square root is the Quotient Property of Square Roots.

Quotient Property of Square Roots

If a and b represent positive real numbers, then $\sqrt{\dfrac{a}{b}} = \dfrac{\sqrt{a}}{\sqrt{b}}$.

EXAMPLE 2   **Write $\sqrt{\dfrac{5}{16}}$ in simplest radical form.**

▶ **Solution**

$$\sqrt{\frac{5}{16}} = \frac{\sqrt{5}}{\sqrt{16}} = \frac{\sqrt{5}}{\sqrt{4^2}} = \frac{\sqrt{5}}{4}$$ ←—— *Apply the Quotient Property of Square Roots.*

In simplest radical form, $\sqrt{\dfrac{5}{16}} = \dfrac{\sqrt{5}}{4}$.

TRY THIS   Write $\sqrt{\dfrac{20}{49}}$ in simplest radical form.

EXERCISES

Complete each simplification.

1. $\sqrt{288} = \sqrt{12^2 \cdot 2}$

2. $\sqrt{75} = \sqrt{5^2 \cdot 3}$

3. $\sqrt{98} = \sqrt{7^2 \cdot 2}$

4. $\sqrt{\dfrac{5}{36}} = \dfrac{\sqrt{5}}{\sqrt{6^2}}$

5. $\sqrt{\dfrac{7}{64}} = \dfrac{\sqrt{7}}{\sqrt{8^2}}$

6. $\sqrt{\dfrac{1}{25}} = \dfrac{\sqrt{1}}{\sqrt{5^2}}$

PRACTICE

Write each expression in simplest radical form.

7. $\sqrt{108}$

8. $\sqrt{245}$

9. $\sqrt{252}$

10. $\sqrt{192}$

11. $\sqrt{\dfrac{3}{25}}$

12. $\sqrt{\dfrac{7}{144}}$

13. $\sqrt{\dfrac{2}{49}}$

14. $\sqrt{\dfrac{5}{81}}$

15. $\sqrt{405}$

16. $\sqrt{363}$

17. $\sqrt{1200}$

18. $\sqrt{675}$

19. $\sqrt{\dfrac{72}{49}}$

20. $\sqrt{\dfrac{125}{81}}$

21. $\sqrt{468}$

22. $\sqrt{828}$

23. $\sqrt{\dfrac{343}{25}}$

24. $\sqrt{\dfrac{567}{16}}$

25. $\sqrt{3564}$

26. $\sqrt{2352}$

Find and simplify each product.

27. $\sqrt{\dfrac{15}{16}} \cdot \sqrt{\dfrac{32}{25}}$

28. $\sqrt{\dfrac{7}{9}} \cdot \sqrt{\dfrac{27}{49}}$

29. $\sqrt{\dfrac{2}{25}} \cdot \sqrt{\dfrac{2}{9}} \cdot \sqrt{\dfrac{1}{16}}$

30. $\sqrt{\dfrac{3}{16}} \cdot \sqrt{\dfrac{4}{25}} \cdot \sqrt{\dfrac{5}{36}}$

31. **Critical Thinking**   Let a and b represent positive real numbers. Prove that $\sqrt{\dfrac{a}{b}} \cdot \sqrt{\dfrac{a}{b}} = \dfrac{a}{b}$.

MIXED REVIEW

Simplify. (6.5 Skill A)

32. $(-3r)(-4r)$

33. $s^3(-5s^3)$

34. $2g^2(3g^2)(-3g)$

35. $-h^2(h^2)(-h^2)$

36. $(2r^2t^3)^2$

37. $(-2a^2b^4)^3$

38. $(4ns^3)(6n^2s^3)$

39. $(-a^2z^4)(-a^2z^2)$

Simplify. (6.5 Skill B)

40. $\dfrac{24n^3}{n^2}$

41. $\dfrac{26z^5}{39z^2}$

42. $\dfrac{(4d^3m^2)^5}{(16d^5m^3)^3}$

43. $\dfrac{(5a^3b^3)^3}{(25a^3b^4)^3}$

Radical expressions can be represented in an equivalent form using fractional exponents. The following definition shows the form for $\sqrt{a}$.

Definition of the Exponent $\frac{1}{2}$

If a is positive, then $a^{\frac{1}{2}}$ is defined as $\sqrt{a}$.

The following is an explanation of why this definition works mathematically.

By the definition of square root, any number whose square is a is represented as $\sqrt{a}$. And, by the Product Property of Exponents, $\left(a^{\frac{1}{2}}\right)\left(a^{\frac{1}{2}}\right) = a^1 = a$.

Therefore, because $\left(a^{\frac{1}{2}}\right)^2 = a$, $a^{\frac{1}{2}} = \sqrt{a}$. Some numerical examples are:

$$4^{\frac{1}{2}} = \sqrt{4} = 2 \qquad 9^{\frac{1}{2}} = \sqrt{9} = 3 \qquad \left(\frac{9}{4}\right)^{\frac{1}{2}} = \sqrt{\frac{9}{4}} = \frac{3}{2}$$

All of the laws of exponents you learned in Chapter 6 are true when the exponent is $\frac{1}{2}$ and the bases are positive.

EXAMPLE

Suppose $a > 0$. Using laws of exponents, prove that $\sqrt{4a} = 2\sqrt{a}$.

▶ **Solution**

$$\sqrt{4a} = (4a)^{\frac{1}{2}} \qquad \longleftarrow \textit{Definition of the exponent } \frac{1}{2}$$

$$= (4)^{\frac{1}{2}}a^{\frac{1}{2}} \qquad \longleftarrow \textit{Power-of-a-Product Law of Exponents}$$

$$= (2^2)^{\frac{1}{2}}a^{\frac{1}{2}} \qquad \longleftarrow \textit{Definition of positive integer exponent}$$

$$= 2^{\left(2 \times \frac{1}{2}\right)}a^{\frac{1}{2}} \qquad \longleftarrow \textit{Power-of-a-Power Law of Exponents}$$

$$= 2^1 a^{\frac{1}{2}} \qquad \longleftarrow \textit{Multiplicative inverse}$$

$$= 2\sqrt{a} \qquad \longleftarrow \textit{Definition of the exponents 1 and } \frac{1}{2}$$

Thus, $\sqrt{4a} = 2\sqrt{a}$.

TRY THIS

Suppose $a > 0$. Using laws of exponents, prove that $\sqrt{9a} = 3\sqrt{a}$.

You prove the Product Property of Square Roots as shown below.
For $a \geq 0$ and $b \geq 0$,

$$\sqrt{ab} = (ab)^{\frac{1}{2}} \qquad \longleftarrow \textit{Definition of the exponent } \frac{1}{2}$$

$$= a^{\frac{1}{2}}b^{\frac{1}{2}} \qquad \longleftarrow \textit{Power-of-a-Product Law of Exponents}$$

$$= \sqrt{a}\sqrt{b} \qquad \longleftarrow \textit{Definition of the exponent } \frac{1}{2}$$

EXERCISES

KEY SKILLS

Evaluate each expression.

1. $49^{\frac{1}{2}}$

2. $121^{\frac{1}{2}}$

3. $144^{\frac{1}{2}}$

4. $400^{\frac{1}{2}}$

PRACTICE

The proof below states that if a and b are positive numbers, then $\left(\sqrt{ab}\right)^2 = ab$. In Exercises 5–8, give a reason for each step in the proof.

$$
\begin{array}{ccccccccc}
& & \text{Step 1} & & \text{Step 2} & & \text{Step 3} & & \text{Step 4} \\
\left(\sqrt{ab}\right)^2 & = & \left((ab)^{\frac{1}{2}}\right)^2 & = & (ab)^{\frac{1}{2} \cdot 2} & = & (ab)^1 & = & ab
\end{array}
$$

5. Step 1   **6.** Step 2   **7.** Step 3   **8.** Step 4

The proof below states that if a is positive and n is an integer, then $\sqrt{a^n} = \left(\sqrt{a}\right)^n$. In Exercises 9–13, give a reason for each step in the proof.

$$
\begin{array}{ccccccccccc}
& & \text{Step 1} & & \text{Step 2} & & \text{Step 3} & & \text{Step 4} & & \text{Step 5} \\
\sqrt{a^n} & = & (a^n)^{\frac{1}{2}} & = & a^{n \cdot \frac{1}{2}} & = & a^{\frac{1}{2} \cdot n} & = & \left(a^{\frac{1}{2}}\right)^n & = & \left(\sqrt{a}\right)^n
\end{array}
$$

9. Step 1   **10.** Step 2   **11.** Step 3   **12.** Step 4   **13.** Step 5

Use the properties of exponents with the exponent $\frac{1}{2}$ to prove each statement.

14. If a is a positive real number, then $\sqrt{25a} = 5\sqrt{a}$.

15. If a and b are positive real numbers, then $\sqrt{\frac{a}{b}} = \frac{\sqrt{a}}{\sqrt{b}}$.

16. If a is a positive real number, then $\frac{\sqrt{4a^2}}{\sqrt{9a^2}} = \frac{2}{3}$.

17. If a and b are positive real numbers, then $\frac{\sqrt{a^4b^4}}{\sqrt{a^2b^2}} = ab$.

MIXED REVIEW

Solve each system. (5.3 Skills A and B)

18. $\begin{cases} 2x + 3y = 6 \\ 2x - 3y = 6 \end{cases}$

19. $\begin{cases} 2x + y = 5 \\ 4x + y = 6 \end{cases}$

20. $\begin{cases} 5x + 2y = -7 \\ x + 3y = 9 \end{cases}$

21. $\begin{cases} -3a + 11b = 0 \\ 4a - 11b = 3 \end{cases}$

22. $\begin{cases} 5a + 6b = 4 \\ 4a + 5b = 0 \end{cases}$

23. $\begin{cases} -7x - 6y = 14 \\ 8x + 7y = 5 \end{cases}$

8.2 LESSON

Solving Equations of the Form $ax^2 + c = 0$

In Lesson 8.1, you learned that the solutions to an equation of the form $x^2 = k$ are $x = \pm\sqrt{k}$. In this lesson, you will learn how to solve equations of the form $ax^2 + c = 0$ by first isolating x^2 on one side of the equation and *then* taking the square root of each side.

EXAMPLE 1 | **Solve $2d^2 - 32 = 0$.**

▶ Solution

$$2d^2 - 32 = 0$$
$$2d^2 = 32$$
$$d^2 = 16$$
$$d = \pm\sqrt{16} \quad \longleftarrow \text{ Take the square root of each side.}$$
$$d = \pm 4$$

The solutions to $2d^2 - 32 = 0$ are 4 and -4.

TRY THIS Solve $10d^2 - 1000 = 0$.

EXAMPLE 2 | **Solve $3t^2 - 54 = 0$. Write the solutions in simplest radical form.**

▶ Solution

$$3t^2 - 54 = 0$$
$$3t^2 = 54$$
$$t^2 = 18$$
$$t = \pm\sqrt{18} \quad \longleftarrow \text{ Take the square root of each side.}$$
$$t = \pm\sqrt{9 \cdot 2} = \pm 3\sqrt{2} \quad \longleftarrow \text{ Simplify.}$$

The solutions to $3t^2 - 54 = 0$ are $3\sqrt{2}$ and $-3\sqrt{2}$.

TRY THIS Solve $4t^2 - 96 = 0$. Write the solutions in simplest radical form.

EXAMPLE 3 | **Solve $4z^2 - 109 = 91$. Write the solutions in simplest radical form.**

▶ Solution

$$4z^2 - 109 = 91$$
$$4z^2 = 200$$
$$z^2 = 50$$
$$z = \pm\sqrt{50} \quad \longleftarrow \text{ Take the square root of each side.}$$
$$z = \pm\sqrt{25 \cdot 2} = \pm 5\sqrt{2} \quad \longleftarrow \text{ Simplify.}$$

The solutions to $4z^2 - 109 = 91$ are $5\sqrt{2}$ and $-5\sqrt{2}$.

TRY THIS Solve $3z^2 - 83 = -2$. Write the solutions in simplest radical form.

EXERCISES

Write each equation in the form $x^2 = k$.

1. $7t^2 - 28 = 0$

2. $-2x^2 + 50 = 0$

3. $25z^2 - 96 = 4$

4. $\frac{1}{2}h^2 - 2 = 0$

PRACTICE

Solve. Write the solutions in simplest radical form.

5. $x^2 - 1 = 0$

6. $y^2 - 49 = 0$

7. $2n^2 - 32 = 0$

8. $3m^2 - 27 = 0$

9. $-11s^2 + 44 = 0$

10. $-3v^2 + 75 = 0$

11. $-\frac{1}{10}z^2 + 10 = 0$

12. $-\frac{1}{2}m^2 + 2 = 0$

13. $7d^2 - 21 = 0$

14. $-2y^2 + 4 = 0$

15. $5q^2 - 35 = 0$

16. $3k^2 - 6 = 0$

17. $6t^2 - 72 = 0$

18. $-3p^2 + 72 = 0$

19. $8v^2 - 96 = 0$

20. $3h^2 - 60 = 0$

21. $3u^2 - 2 = 4$

22. $3u^2 - 14 = 4$

23. $-4u^2 + 36 = 4$

24. $-2s^2 + 113 = 13$

25. $3n^2 - 104 = -8$

26. $5z^2 - 21 = 39$

27. $-7r^2 + 63 = 7$

28. $-8w^2 + 107 = 11$

29. For what values of a will $av^2 - 144 = 0$ have one solution between 3 and 4 and the other solution between -4 and -3?

30. For what values of a will $av^2 - 144 = 0$ have one solution between 1 and 2 and the other solution between -2 and -1?

31. For what values of c will $3u^2 + c = 0$ have one solution between 3 and 4 and the other solution between -4 and -3?

32. Critical Thinking Let $ax^2 + c = 0$ and let a and c represent real numbers with $a \neq 0$.
  a. Write a formula that gives x in terms of a and c.
  b. Under what conditions will your formula give two solutions for x? one solution for x? no real solution for x?

MIXED REVIEW

Solve each equation for the specified variable. All variables represent real numbers. Assume that there is no division by 0. (2.7 Skill B)

33. $\frac{m}{n}x + a = y$; m

34. $y(x - c) = z$; x

35. $yz + 4 = xz - 4$; z

36. $x(y + b) = 2xy$; y

37. $a(x + 3) = b(x - 2)$; x

38. $x(y - n) = 2ay$; y

Some geometry problems can be solved using quadratic equations.

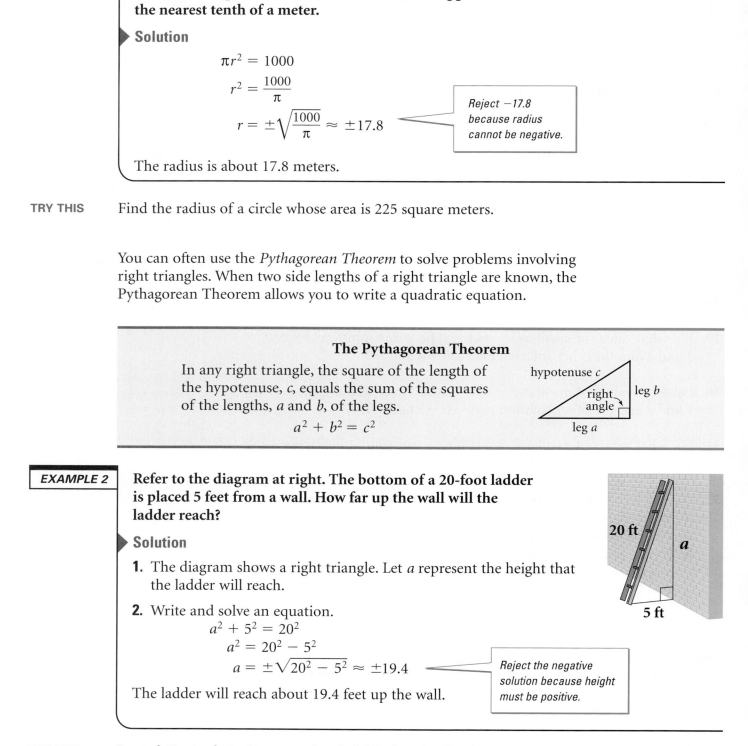

EXAMPLE 1 **Use the formula area = π · (radius)² to find the radius of a circle whose area is 1000 square meters. Use a calculator to approximate the radius to the nearest tenth of a meter.**

▶ **Solution**

$$\pi r^2 = 1000$$
$$r^2 = \frac{1000}{\pi}$$
$$r = \pm\sqrt{\frac{1000}{\pi}} \approx \pm 17.8$$

Reject −17.8 because radius cannot be negative.

The radius is about 17.8 meters.

TRY THIS Find the radius of a circle whose area is 225 square meters.

You can often use the *Pythagorean Theorem* to solve problems involving right triangles. When two side lengths of a right triangle are known, the Pythagorean Theorem allows you to write a quadratic equation.

The Pythagorean Theorem

In any right triangle, the square of the length of the hypotenuse, *c*, equals the sum of the squares of the lengths, *a* and *b*, of the legs.

$$a^2 + b^2 = c^2$$

hypotenuse *c*
right angle
leg *b*
leg *a*

EXAMPLE 2 **Refer to the diagram at right. The bottom of a 20-foot ladder is placed 5 feet from a wall. How far up the wall will the ladder reach?**

▶ **Solution**

1. The diagram shows a right triangle. Let *a* represent the height that the ladder will reach.

2. Write and solve an equation.
$$a^2 + 5^2 = 20^2$$
$$a^2 = 20^2 - 5^2$$
$$a = \pm\sqrt{20^2 - 5^2} \approx \pm 19.4$$

Reject the negative solution because height must be positive.

The ladder will reach about 19.4 feet up the wall.

20 ft *a*
5 ft

TRY THIS Rework Example 2 given a 28-foot ladder placed 7 feet from a wall.

EXERCISES

KEY SKILLS

Write an equation that you can use to solve each problem. Identify the variable. Do not solve.

1. What is the radius of a circle whose area is 1200 square feet?

2. If the hypotenuse of a triangle is 55 units and the length of one leg is 35 units, what is the length of the other leg?

PRACTICE

Use the given area to find the radius of each circle. Round answers to the nearest tenth of a unit.

3. area: 500 square feet

4. area: 620 square meters

5. area: 314 square inches

6. area: 1240 square meters

Find the length of the other leg of each right triangle. Round answers to the nearest tenth of a unit.

7. hypotenuse: 24 units
   one leg: 18 units

8. hypotenuse: 35 feet
   one leg: 30 feet

9. hypotenuse: 500 inches
   one leg: 250 inches

10. hypotenuse: 625 yards
    one leg: 62.5 yards

Refer to rectangle *ABCD* at right. (The notation *AD* represents the length of the line segment with endpoints *A* and *D*.)

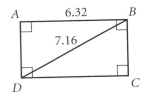

11. **a.** Write an expression for *AD*.
    b. Approximate *AD* to the nearest hundredth of a unit.

MIXED REVIEW APPLICATIONS

Write a simplified expression for the total perimeter of each figure. Lightly shaded rectangles are congruent and black rectangles are congruent. (6.4 Skill C)

12.
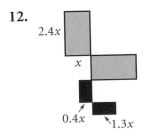

13.
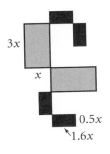

14.
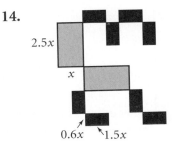

There are no real numbers that make $x^2 + 1 = 0$ true. This is because there is no real number whose square is negative. However, if you define the square root of a negative number, then you create a new number system in which the equation does have solutions.

$$x^2 + 1 = 0$$
$$x^2 = -1$$
$$x = \pm\sqrt{-1}$$

Definition of i and Purely Imaginary Numbers

The **imaginary unit** i is defined as $\sqrt{-1}$. Thus, $i^2 = -1$.

A **purely imaginary number** is defined as $\sqrt{-r}$, or $i\sqrt{r}$, where $r > 0$.

To simplify $\sqrt{-r}$ write $\sqrt{-r}$ as $i\sqrt{r}$.

EXAMPLE 1 Simplify each expression. **a.** $\sqrt{-81}$ **b.** $\sqrt{-75}$

▶ Solution

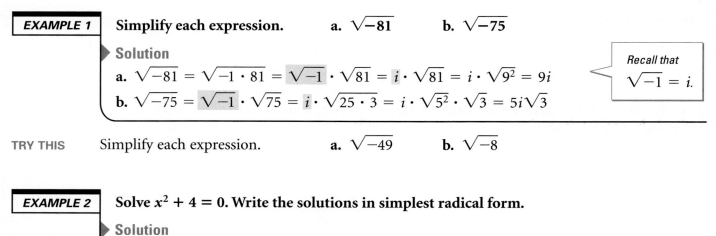

a. $\sqrt{-81} = \sqrt{-1 \cdot 81} = \sqrt{-1} \cdot \sqrt{81} = i \cdot \sqrt{81} = i \cdot \sqrt{9^2} = 9i$

b. $\sqrt{-75} = \sqrt{-1} \cdot \sqrt{75} = i \cdot \sqrt{25 \cdot 3} = i \cdot \sqrt{5^2} \cdot \sqrt{3} = 5i\sqrt{3}$

> *Recall that*
> $\sqrt{-1} = i$.

TRY THIS Simplify each expression. **a.** $\sqrt{-49}$ **b.** $\sqrt{-8}$

EXAMPLE 2 Solve $x^2 + 4 = 0$. Write the solutions in simplest radical form.

▶ Solution

$$x^2 + 4 = 0$$
$$x^2 = -4$$
$$x = \pm\sqrt{-4} \quad \longleftarrow \text{\textit{Take the square root of each side.}}$$
$$x = \pm 2i \quad \longleftarrow \sqrt{-4} = i\sqrt{4} = 2i$$

TRY THIS Solve $a^2 + 25 = 0$. Write the solutions in simplest radical form.

EXAMPLE 3 Solve $3z^2 + 144 = 0$. Write the solutions in simplest radical form.

▶ Solution

$$3z^2 + 144 = 0$$
$$z^2 = -48$$
$$z = \pm\sqrt{-48} \quad \longleftarrow \text{\textit{Take the square root of each side.}}$$
$$z = \pm i\sqrt{48} \quad \longleftarrow \sqrt{-48} = i\sqrt{48}$$
$$z = \pm i\sqrt{16 \cdot 3} \quad \longleftarrow \text{\textit{Factor 48 as 16 · 3 because 16 is the largest perfect square that is a factor of 48.}}$$
$$z = \pm 4i\sqrt{3} \quad \longleftarrow \sqrt{16 \cdot 3} = \sqrt{16}\sqrt{3} = 4\sqrt{3}$$

TRY THIS Solve $2z^2 + 250 = 0$. Write the solutions in simplest radical form.

EXERCISES

Identify each number as real or as purely imaginary.

1. $\sqrt{13}$

2. $\sqrt{-13}$

3. $i\sqrt{5}$

4. $\sqrt{-(-5)}$

Simplify each expression. Write the answers in simplest radical form.

5. $\sqrt{-16}$

6. $\sqrt{-25}$

7. $\sqrt{-100}$

8. $\sqrt{-9}$

9. $\sqrt{-32}$

10. $\sqrt{-48}$

11. $\sqrt{-200}$

12. $\sqrt{-45}$

Solve each equation. Write the solutions in simplest radical form.

13. $x^2 + 9 = 0$

14. $v^2 + 121 = 0$

15. $n^2 + 49 = 0$

16. $t^2 + 169 = 0$

17. $2m^2 + 8 = 0$

18. $3w^2 + 27 = 0$

19. $5p^2 + 80 = 0$

20. $4k^2 + 100 = 0$

21. $-3m^2 - 36 = 0$

22. $-5c^2 - 40 = 0$

23. $-2d^2 - 90 = 0$

24. $-6u^2 - 48 = 0$

25. $-2y^2 - 144 = 0$

26. $-11x^2 - 88 = 0$

27. $-4h^2 - 96 = 0$

28. $-8k^2 - 320 = 0$

29. $2x^2 - 8 = -12$

30. $3y^2 - 1 = -22$

31. $5z^2 + 8 = -17$

32. $7b^2 + 8 = -27$

33. $2(x^2 - 8) = -22$

34. $3(x^2 - 2) = -33$

35. $-(x^2 + 8) = 4$

36. $-(x^2 + 5) = 22$

37. Critical Thinking   Using $i = \sqrt{-1}$ and $i^2 = -1$, find i^3, i^4, i^5, and i^6. What pattern do you see? What pattern do you see in the values of i^n, where n is a positive integer?

38. Critical Thinking   **a.** Simplify each sum below.
$i + i^2$ $i + i^2 + i^3$ $i + i^2 + i^3 + i^4$
$i + i^2 + i^3 + i^4 + i^5$ $i + i^2 + i^3 + i^4 + i^5 + i^6$
b. What pattern do you see in the sums $i + i^2 + \cdots + i^n$, where n is a natural number?

Simplify each square root. Write the result in simplest radical form.
(8.1 Skill B)

39. $\sqrt{150}$

40. $\sqrt{63}$

41. $\sqrt{360}$

42. $\sqrt{44}$

43. $\sqrt{\dfrac{400}{49}}$

44. $\sqrt{\dfrac{169}{121}}$

45. $\sqrt{\dfrac{252}{25}}$

46. $\sqrt{\dfrac{450}{121}}$

8.3 LESSON

Solving Quadratic Equations by Completing the Square

SKILL A ▸ *Solving equations of the form $a(x - r)^2 = s$*

In Lessons 8.1 and 8.2, you learned how to use properties of equality and square roots to solve different types of quadratic equations, such as equations of the forms $x^2 = a$ and $ax^2 + c = 0$. In this lesson, you will learn how to solve an equation of the form $a(x - r)^2 = s$, where $a \neq 0$.

EXAMPLE 1 Solve $7(x - 1)^2 = 28$.

▸ **Solution**

$$7(x - 1)^2 = 28$$
$$(x - 1)^2 = 4 \qquad \longleftarrow \text{Apply the Division Property of Equality.}$$
$$\sqrt{(x - 1)^2} = \pm\sqrt{4} \qquad \longleftarrow \text{Take the square root of each side.}$$
$$x - 1 = \pm 2$$
$$x = 1 \pm 2$$
$$x = 3 \text{ or } x = -1$$

TRY THIS Solve $-2(z + 5)^2 = -128$.

Example 2 shows solutions that are not rational.

EXAMPLE 2 Solve $\frac{1}{2}(v - 2)^2 = 6$.

▸ **Solution**

$$\frac{1}{2}(v - 2)^2 = 6$$
$$(v - 2)^2 = 12 \qquad \longleftarrow \text{Apply the Multiplication Property of Equality.}$$
$$v - 2 = \pm\sqrt{12} \qquad \longleftarrow \text{Take the square root of each side.}$$
$$v = 2 \pm \sqrt{12}$$
$$v = 2 \pm 2\sqrt{3} \qquad \longleftarrow \text{Simplify } \sqrt{12} \text{ to } 2\sqrt{3}.$$
$$v = 2 + 2\sqrt{3} \text{ or } v = 2 - 2\sqrt{3}$$

TRY THIS Solve $10(n + 9)^2 = 750$.

The following is a summary of how to solve an equation of the form $a(x - r)^2 = s$.

Summary of the Method for Solving $a(x - r)^2 = s$

1. Isolate the squared expression on one side of the equation.

2. Take the square root of each side of the equation.

3. Solve for the variable.

4. Simplify the solution(s), if necessary.

EXERCISES

Explain how the method summarized on the previous page can be applied to solve each equation. Do not solve.

1. $(x - 4)^2 = 49$

2. $(x + 1)^2 = 4$

3. $\frac{1}{3}(x - 3)^2 = 12$

PRACTICE

Solve each equation. Write the solutions in simplest radical form.

4. $(x - 4)^2 = 1$

5. $(x + 2)^2 = 9$

6. $(x + 3)^2 = 64$

7. $(x - 5)^2 = 100$

8. $(x + 7)^2 = 16$

9. $(x - 2)^2 = 81$

10. $(x + 1)^2 = 4$

11. $(x + 7)^2 = 9$

12. $2(d - 3)^2 = 48$

13. $-(m + 5)^2 = -75$

14. $-2(s - 7)^2 = -144$

15. $\frac{1}{3}(t + 4)^2 = 9$

16. $\frac{1}{2}(x - 8)^2 = 40$

17. $-3(n + 2)^2 = -60$

18. $0.5(k - 1)^2 = 49$

19. $\frac{3}{2}(p + 10)^2 = 42$

20. $(2y + 1)^2 = 36$

21. $(3n - 1)^2 = 49$

22. $(5t + 3)^2 = 25$

23. $(5c - 4)^2 = 1$

24. $2(2w - 3)^2 = 100$

Solve each equation by completing the square.

25. $x^2 + 4x + 4 = 36$

26. $n^2 - 10n + 25 = 100$

27. $9z^2 + 12z + 4 = 25$

28. $4s^2 + 12s + 9 = 1$

29. $25k^2 + 10k + 1 = 4$

30. $16c^2 + 56c + 49 = 81$

31. Let $a(x - r)^2 = s$, where $a > 0$ and $s > 0$. Give a justification for each step in the reasoning at right.
  a. Step 1
  b. Step 2
  c. Step 3
  d. Step 4

Step 1: $a(x - r)^2 = s$

Step 2: $(x - r)^2 = \frac{s}{a}$

Step 3: $(x - r) = \pm\sqrt{\frac{s}{a}}$

Step 4: $x = r \pm \sqrt{\frac{s}{a}}$

32. Write a formula for solving $a(bx - r)^2 = s$, where a and b are nonzero real numbers and $a \cdot s > 0$.

33. Critical Thinking   For what real-number value(s) of n will $(x - 4)^2 = n - 5$ have two distinct real solutions?

MIXED REVIEW

Multiply. (6.5 Skill A)

34. $(2d^2n^2)^2(3dn^3)^2$

35. $(3a^2b^4)^3(-3a^2b^3)^2$

36. $(-4x^3y^3)^3(x^3y^3)^3$

Completing the square is often a good method to use for solving a quadratic equation when the equation is not factorable using integers.

Completing the Square to Solve $ax^2 + bx + c = 0$

1. Write the equation so that the constant term is isolated on the right side. Divide each side of the equation by a.

2. Find the square of one-half the coefficient of x. Add that number to each side of the equation.

3. Factor the left side of the equation. The result should have the form $(x + r)^2$.

4. Take the square root of each side of the equation. Then solve for x and simplify the solutions.

EXAMPLE 1

Solve $x^2 + 6x + 5 = 0$ by completing the square.

▶ **Solution**

1. $\qquad x^2 + 6x = -5 \qquad \longleftarrow$ *Write variable terms on the left and the constant on the right.*

2. $x^2 + 6x + 3^2 = -5 + 9 \qquad \longleftarrow$ *Add the square of one-half the coefficient of x to each side.*

3. $\qquad (x + 3)^2 = 4 \qquad \longleftarrow$ *Write the left side of the equation as a squared binomial.*

4. $\qquad x + 3 = \pm\sqrt{4} \qquad \longleftarrow$ *Take the square root of each side of the equation.*

$\qquad\qquad x + 3 = \pm 2$

$x = -3 + 2 = -1 \quad$ or $\quad x = -3 - 2 = -5$

The solutions to $x^2 + 6x + 5 = 0$ are -1 and -5.

The coefficient of x is 6.

$\left(\frac{1}{2} \cdot 6\right)^2 = 3^2 = 9$

TRY THIS Solve $z^2 - 2z - 15 = 0$ by completing the square.

EXAMPLE 2

Solve $4v^2 + 8v - 3 = 0$ by completing the square.

▶ **Solution**

1. $\qquad 4v^2 + 8v = 3 \qquad \longleftarrow$ *Write variable terms on the left and the constant on the right.*

$\qquad v^2 + 2v = \dfrac{3}{4} \qquad \longleftarrow$ *Divide each side by the coefficient of v^2.*

2. $v^2 + 2v + 1^2 = \dfrac{3}{4} + 1 \qquad \longleftarrow$ *Add the square of one-half the coefficient of v to each side.*

3. $\qquad (v + 1)^2 = \dfrac{7}{4} \qquad \longleftarrow$ *Write the left side as a squared binomial.*

4. $\qquad v + 1 = \pm\dfrac{\sqrt{7}}{2} \qquad \longleftarrow$ *Take the square root of each side of the equation.*

The solutions to $4v^2 + 8v - 3 = 0$ are $-1 + \dfrac{\sqrt{7}}{2}$ and $-1 - \dfrac{\sqrt{7}}{2}$.

The coefficient of v is 2.

$\left(\frac{1}{2} \cdot 2\right)^2 = 1^2 = 1$

TRY THIS Solve $9y^2 + 18y + 4 = 0$ by completing the square.

EXERCISES

KEY SKILLS

Write the number that makes the given expression a perfect-square trinomial.

1. $s^2 + 2s +$ ___?___

2. $y^2 + 10y +$ ___?___

3. $n^2 - 12n +$ ___?___

4. $v^2 - 16v +$ ___?___

Complete the solution to each equation.

5. $s^2 + 2s + 1 = 3 + 1$
$(s + 1)^2 = 4$

6. $n^2 + 6n + 9 = 91 + 9$
$(n + 3)^2 = 100$

7. $n^2 - 12n + 36 = 13 + 36$
$(n - 6)^2 = 49$

PRACTICE

Solve each equation by completing the square. Write the solutions in simplest radical form where necessary.

8. $x^2 + 2x = 3$

9. $m^2 + 4m = 5$

10. $t^2 - 10t = 11$

11. $q^2 - 6q = 7$

12. $u^2 + 4u = 77$

13. $h^2 + 8h = 9$

14. $x^2 - 12x = 28$

15. $r^2 - 18r = -17$

16. $d^2 - 4d - 12 = 0$

17. $f^2 - 10f - 11 = 0$

18. $h^2 + 8h - 20 = 0$

19. $m^2 - 16m - 36 = 0$

20. $4y^2 - 8y - 3 = 0$

21. $9z^2 + 18z - 11 = 0$

22. $4r^2 + 16r + 5 = 0$

23. $3t^2 - 18t + 21 = 0$

24. $4z^2 - 20z + 1 = 0$

25. $4m^2 + 12m - 11 = 0$

26. $12q^2 + 12q - 9 = 0$

27. $9q^2 - 36q + 1 = 0$

28. Use the method of completing the square to find n such that $x^2 + 2x = n$ has two distinct real solutions.

29. Use the method of completing the square to find m such that $x^2 + 6x = m$ has no real solutions.

30. Critical Thinking Find and graph the ordered pairs (a, b) that make $x^2 + 2x = a + b$ have two distinct real solutions. (*Hint:* Complete the square.)

31. Critical Thinking Solve the equation $x^2 + 2x - y^2 - 2y = 0$ for y.

MIXED REVIEW

Evaluate each expression given $a = -2$, $b = 5$, and $c = 12$. (1.2 Skill C)

32. $\frac{3}{2}(a^2 + b^2 - c^2)$

33. $\frac{5}{2}(a^2 + b^2)^2 - c^2$

34. $\frac{1}{4}(a^2 + b^2 - c^2)^2$

35. $(2a + 4b - 5c)^2$

36. $b^2 - 4ac$

37. $a(ab^2 - 4ac)$

You can use the *completing the square* method to solve quadratic equations which model real-world problems.

EXAMPLE

Boats A and B leave the dock at the same time and each sails at a constant speed. Boat A sails due north and Boat B sails due east. After 2 hours, they are 30 miles apart. Boat B sails 7 miles per hour faster than Boat A. **Find the speed of each boat to the nearest tenth of a mile per hour. Use a calculator to approximate the value of the radical expression.**

▶ **Solution**

1. Organize the information in a table. Let x be the speed of Boat A.

Make a table.

	speed (miles per hour)	distance (miles)
Boat A	x	$2x$
Boat B	$x + 7$	$2(x + 7)$

Draw a sketch. Because the paths of the boats form a right triangle, you can use the Pythagorean Theorem to help you solve the problem.

Make a diagram.

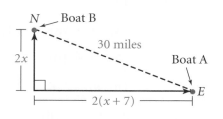

$$(\text{Boat A distance})^2 + (\text{Boat B distance})^2 = 30^2$$

Write an equation.

2. Write and solve an equation. Use the Pythagorean Theorem.

$$(2x)^2 + [2(x + 7)]^2 = 30^2$$

$$8x^2 + 56x + 196 = 900 \qquad \longleftarrow \text{ Simplify.}$$

$$x^2 + 7x = 88 \qquad \longleftarrow \text{ Move all constant terms to the right side and then divide by 8, the GCF.}$$

$$x^2 + 7x + \left(\frac{7}{2}\right)^2 = 88 + \frac{49}{4} \qquad \longleftarrow \text{ Complete the square; } \left(\frac{7}{2}\right)^2 = \frac{49}{4}.$$

$$\left(x + \frac{7}{2}\right)^2 = \frac{401}{4} \qquad \longleftarrow \text{ Factor the left side of the equation.}$$

$$x + \frac{7}{2} = \pm\sqrt{\frac{401}{4}} \qquad \longleftarrow \text{ Take the square root of each side.}$$

$$x = -\frac{7}{2} \pm \sqrt{\frac{401}{4}} \quad \rightarrow \quad x \approx 6.5 \text{ or } x \approx -13.5$$

Reject $-\frac{7}{2} - \sqrt{\frac{401}{4}}$ as a solution because speed cannot be negative.

The speed of Boat A is about 6.5 miles per hour. The speed of Boat B then is about $6.5 + 7$, or 13.5 miles per hour.

TRY THIS

Rework the Example given that the speed of Boat B is 1 mile per hour more than Boat A and the distance between them is 20 miles.

EXERCISES

Refer to the Example on the previous page.

1. Make a table to represent the information below. Boat B sails 5 miles per hour faster than Boat A. After 2 hours, the boats are 32 miles apart.

2. Use the Pythagorean Theorem to write an equation to represent the data in Exercise 1.

PRACTICE

Rework the Example on the previous page using the given modifications of the problem. Assume all other information is the same as in the Example.

3. The boats are 25 miles apart after 1 hour.

4. The boats are 45 miles apart after 2 hours.

5. The boats are 65 miles apart after 3 hours.

Boats A and B leave the dock at the same time and sail at a constant speed for two hours. Find the speed, s, of each boat in miles per hour.

6.

7.

8. **Critical Thinking** How far apart should points X and Z be so that $\triangle XYZ$ is a right triangle, $XZ = 2a - 4$, $YZ = a$, and $XY = 10$?

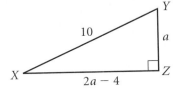

MIXED REVIEW APPLICATIONS

Represent the shaded region as a system of linear inequalities. (5.6 Skill C)

9.

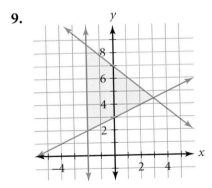

10.

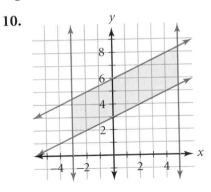

 **8.4**

LESSON

Solving Quadratic Equations by Using the Quadratic Formula

SKILL A *Using the quadratic formula to solve quadratic equations*

Recall from Lesson 7.5 that $ax^2 + bx + c = 0$ is the standard form of a quadratic equation. When a quadratic equation is written in standard form, you can apply the *quadratic formula* to find its solutions.

The Quadratic Formula

If $ax^2 + bx + c = 0$, where $a \neq 0$, then $x = \dfrac{-b \pm \sqrt{b^2 - 4ac}}{2a}$.

The proof of the quadratic formula is given in Skill C of this lesson.

EXAMPLE 1 **Solve $x^2 + 7x + 6 = 0$ using the quadratic formula.**

▶ **Solution**

$$x = \frac{-b \pm \sqrt{b^2 - 4ac}}{2a}$$

$$x = \frac{-7 \pm \sqrt{7^2 - 4(1)(6)}}{2(1)}$$ ⟵——— *Substitute $a = 1$, $b = 7$, and $c = 6$.*

$$x = \frac{-7 \pm \sqrt{25}}{2} = \frac{-7 \pm 5}{2}$$

$$x = -1 \quad \text{or} \quad x = -6$$

TRY THIS Solve $w^2 - 3w - 4 = 0$ using the quadratic formula.

EXAMPLE 2 **Solve $t^2 + 2t = 5$ using the quadratic formula.**

▶ **Solution**

$$t^2 + 2t = 5$$
$$t^2 + 2t - 5 = 0$$ ⟵——— *Write the equation in standard form.*

$$t = \frac{-2 \pm \sqrt{2^2 - 4(1)(-5)}}{2(1)}$$ ⟵——— *Apply the quadratic formula with $a = 1$, $b = 2$, and $c = -5$.*

$$t = \frac{-2 \pm \sqrt{24}}{2} = \frac{-2 \pm 2\sqrt{6}}{2}$$ ⟵——— *$\sqrt{24} = 2\sqrt{6}$*

$$t = -1 + \sqrt{6} \quad \text{or} \quad t = -1 - \sqrt{6}$$

TRY THIS Solve $a^2 - 3a = 5$ using the quadratic formula.

Another name for a solution to a polynomial equation is *root*. In Example 2, $-1 + \sqrt{6}$ and $-1 - \sqrt{6}$ are the roots of the equation $t^2 + 2t = 5$.

EXERCISES

Write each quadratic equation in standard form. Then identify a, b, and c.

1. $x^2 = 5x + 6$ **2.** $x^2 - 9x = -18$ **3.** $3x + 2x^2 = 20$ **4.** $-2 - 5x = 3x^2$

PRACTICE

Use the quadratic formula to solve each equation. Write the solutions in simplest radical form where necessary.

5. $x^2 + 11x + 30 = 0$ **6.** $y^2 - 15y + 56 = 0$ **7.** $w^2 + 15w + 56 = 0$

8. $v^2 + 13v + 42 = 0$ **9.** $z^2 - 8z = -7$ **10.** $m^2 + 27 = 12m$

11. $33 + 8x = x^2$ **12.** $45 - 4y = y^2$ **13.** $2k^2 = 11k - 15$

14. $6g^2 + 14 = 25g$ **15.** $6x^2 + 5 = 13x$ **16.** $7 + 10h^2 = 37h$

17. $y^2 = 6y - 2$ **18.** $m^2 = 2m + 6$ **19.** $p^2 = 8p - 11$

20. $44 + q^2 = 14q$ **21.** $x^2 = 4x - 2$ **22.** $z^2 = 4z - 1$

23. $4n^2 = 8n + 1$ **24.** $2t^2 - 6t = -1$ **25.** $3y^2 - 8y = 2$

Solve each equation for x in terms of a, b, and c. Then verify that you would arrive at the same expression for x if you solve using the quadratic formula.

26. $ax^2 + c = 0$ **27.** $ax^2 + bx = 0$ **28.** $x^2 + bx + c = 0$

29. For what value of b will $4x^2 + bx + 9 = 0$ have exactly one root? Use the quadratic formula to justify your response.

30. Use the quadratic formula to solve $2(v + 3)^2 - 4(v + 3) - 7 = 0$.

31. Use the quadratic formula to solve $6n^4 - 13n^2 + 5 = 0$.

32. Use the quadratic formula to show that $x^2 + 1 = 0$ has no real roots.

MIXED REVIEW

Find the coordinates of the vertex and an equation for the axis of symmetry. Identify the vertex as the maximum or the minimum point. Then graph the function. (7.6 Skill A)

33. $y = x^2 + 10x + 25$ **34.** $y = x^2 - 10x + 25$ **35.** $y = -x^2 + 4$

36. $y = -x^2 + 2x$ **37.** $y = x^2 - 4x$ **38.** $y = -\frac{1}{2}x^2 + x$

In the quadratic formula, the expression under the radical sign, $b^2 - 4ac$, is called the **discriminant** of $ax^2 + bx + c = 0$.

The Quadratic Formula and the Discriminant

If $ax^2 + bx + c = 0$, where $a \neq 0$, then the equation has

- two real and distinct roots if $b^2 - 4ac > 0$,
- one real root if $b^2 - 4ac = 0$, and
- no real roots if $b^2 - 4ac < 0$.

EXAMPLE 1 | **Find the number of real roots of $3x^2 - 5x - 12 = 0$.**

▶ **Solution**

Identify a, b, and c: $a = 3$, $b = -5$, and $c = -12$.

Evaluate the discriminant, $b^2 - 4ac$.

$$(-5)^2 - 4(3)(-12) = 169$$

Because the discriminant, 169, is positive, the equation has 2 real roots.

TRY THIS   Find the number of real roots of $2x^2 - 5x + 10 = 0$.

Recall from Lesson 7.6 that a parabola has either no x-intercept, one x-intercept, or two x-intercepts. The x-intercepts of the graph of $y = ax^2 + bx + c$ are the real roots of $ax^2 + bx + c = 0$.

EXAMPLE 2 | **Find the number of x-intercepts of the graph of $y = 3x^2 - 5x - 12$.**

▶ **Solution**

The graph of $y = 3x^2 - 5x - 12$ is a parabola. Evaluate the discriminant of $3x^2 - 5x - 12 = 0$. From Example 1, we know that the discriminant, 169, is positive.

Because the discriminant is positive, the equation has two real, distinct roots and thus, two x-intercepts.

TRY THIS   Find the number of x-intercepts of the graph of $y = x^2 + 14x + 49$.

A quadratic equation has no real roots if $b^2 - 4ac < 0$. For example, the discriminant of $x^2 + 7x + 20 = 0$ is $7^2 - 4(1)(20) = -31$. The graph of such an equation has no x-intercepts. It does not touch or cross the x-axis at any point.

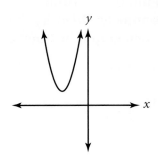

EXERCISES

Each expression is the discriminant of a quadratic equation. How many real roots does the equation have?

1. $(-6)^2 - 4(2)(5)$ **2.** $(-6)^2 - 4(2)(-5)$ **3.** $(8)^2 - 4(4)(4)$ **4.** $(-1)^2 - 4(1)(-3)$

PRACTICE

Use the discriminant to find the number of real roots of each quadratic equation.

5. $x^2 + 5x - 24 = 0$ **6.** $2z^2 - z - 3 = 0$ **7.** $n^2 + 6n + 11 = 0$

8. $k^2 + 10k + 28 = 0$ **9.** $4y^2 - 12y + 9 = 0$ **10.** $9d^2 - 42d + 49 = 0$

11. $25t^2 - 49 = 0$ **12.** $25s^2 + 49 = 0$ **13.** $2w^2 = -10w - 15$

14. $3h^2 + 2h = -2$ **15.** $9v^2 + 16 = 24v$ **16.** $p^2 + 1 = p$

Use the discriminant to find the number of x-intercepts of each quadratic function.

17. $y = x^2 - 12x + 36$ **18.** $y = 4x^2 - 16x + 19$ **19.** $y = 3x^2 - 8x + 7$

20. $y = 9x^2 - 42x + 49$ **21.** $y = 3x^2 - 6x + 2$ **22.** $y = 2x^2 + 6x + 1$

23. $y = 7x^2 + 1$ **24.** $y = -x^2 - 11$ **25.** $y = 6x^2 - 5x + 1$

26. $y = 6x^2 - 5x$ **27.** $y = x^2 + 18x + 70$ **28.** $y = x^2 + 18x + 68$

Write a quadratic equation that has the given number of real roots.

29. no real roots   **30.** one real root   **31.** two real roots

32. For what values of a will $ax^2 - 5x + 1 = 0$ have two distinct real roots?

33. For what values of c will $5x^2 - 7x + c = 0$ have no real roots?

34. For what values of b will $5x^2 - bx + 6 = 0$ have exactly one real root?

MIXED REVIEW

Solve each equation. Write the solution as a logical argument. Summarize the problem and its solution as a conditional statement. (2.9 Skill B)

35. $\frac{1}{2}(a + 3) = 5$ **36.** $3x = -5x + 4$ **37.** $-4t + 5 + 5t = 9$

38. $\frac{2}{3}(n + 3) = \frac{1}{3}n$ **39.** $4m + 5m = m + 2$ **40.** $\frac{2y - 2}{3} = 4$

Recall from Lesson 2.9 that you can write the solution to an equation as a logical argument. For example, given the linear equation $ax + b = 0$, you can solve for x to write a formula for the solution. Examine the logical argument of this solution below.

$$ax + b = 0, a \neq 0 \quad \text{Given}$$
$$ax = -b \qquad \text{Subtraction Property of Equality}$$
$$x = -\frac{b}{a} \qquad \text{Division Property of Equality}$$

From the proof above, if $ax + b = 0$ and $a \neq 0$, then a formula for the solution is $x = \dfrac{-b}{a}$.

Similarly, given the quadratic equation $ax^2 + bx + c = 0$ and $a \neq 0$, you can solve for x to write a formula for the solution, $x = \dfrac{-b \pm \sqrt{b^2 - 4ac}}{2a}$.
The statements for the proof of this formula are shown at right.

The proof begins with the hypothesis "$ax^2 + bx + c = 0$ and $a \neq 0$" and ends with the conclusion "$x = \dfrac{-b \pm \sqrt{b^2 - 4ac}}{2a}$." The reasoning and the sequence of statements from the hypothesis to the conclusion can be justified with properties of real numbers, properties of equality, and other facts.

You will be asked to use the properties to justify some of the steps as part of the exercises on the next page.

1. $ax^2 + bx + c = 0, a \neq 0$

2. $ax^2 + bx = -c$

3. $x^2 + \dfrac{b}{a}x = -\dfrac{c}{a}$

4. $x^2 + \dfrac{b}{a}x + \left(\dfrac{b}{2a}\right)^2 = -\dfrac{c}{a} + \left(\dfrac{b}{2a}\right)^2$

5. $\left(x + \dfrac{b}{2a}\right)^2 = -\dfrac{c}{a} + \dfrac{b^2}{4a^2}$

6. $x + \dfrac{b}{2a} = \pm\sqrt{\dfrac{-c}{a} + \dfrac{b^2}{4a^2}}$

7. $x = -\dfrac{b}{2a} \pm \sqrt{\dfrac{-c}{a} + \dfrac{b^2}{4a^2}}$

8. $x = -\dfrac{b}{2a} \pm \sqrt{\dfrac{-c}{a} \cdot \dfrac{4a}{4a} + \dfrac{b^2}{4a^2}}$

9. $x = -\dfrac{b}{2a} \pm \sqrt{\dfrac{-4ac}{4a^2} + \dfrac{b^2}{4a^2}}$

10. $x = -\dfrac{b}{2a} \pm \sqrt{\dfrac{b^2 - 4ac}{4a^2}}$

11. $x = -\dfrac{b}{2a} \pm \sqrt{\dfrac{b^2 - 4ac}{(2a)^2}}$

12. $x = -\dfrac{b}{2a} \pm \dfrac{\sqrt{b^2 - 4ac}}{\sqrt{(2a)^2}}$

13. $x = -\dfrac{b}{2a} \pm \dfrac{\sqrt{b^2 - 4ac}}{2a}$

14. $x = \dfrac{-b \pm \sqrt{b^2 - 4ac}}{2a}$

EXERCISES

KEY SKILLS

Give a justification for each step at right.

1. Step 1

2. Step 2

Step 1: If $ax + b = 0$, then $ax = -b$.

Step 2: If $ax = -b$ and $a \neq 0$, then $x = -\dfrac{b}{a}$.

PRACTICE

Refer to the proof of the quadratic formula on the previous page. Give a justification for each specified step of the proof.

3. Step 2

4. Step 3

5. Step 4

6. Step 5

7. Step 6

8. Step 7

9. Step 8

10. Step 9

11. Step 12

The reasoning at right provides a derivation for a formula that can be used to solve equations of the form $ax^2 + b = rx^2 + t$, where $a \neq r$. Give a justification for each step.

Step 1: $ax^2 + b = rx^2 + t,\ a \neq r$

Step 2: $ax^2 = rx^2 + t - b$

Step 3: $ax^2 - rx^2 = t - b$

Step 4: $(a - r)x^2 = t - b$

Step 5: $x^2 = \dfrac{t - b}{a - r}$

Step 6: $x = \pm\sqrt{\dfrac{t - b}{a - r}}$

12. Step 1

13. Step 2

14. Step 3

15. Step 4

16. Step 5

17. Step 6

18. Derive a formula to find the solutions of $ax^2 + bx = rx^2 + tx$, where $a \neq r$. (*Hint:* Use factoring.)

MID-CHAPTER REVIEW

Solve each equation.

19. $x^2 = 25$ (8.1 Skill A)

20. $3a^2 + 46 = 100$ (8.2 Skill A)

21. $2z^2 + 48 = 0$ (8.2 Skill C)

22. $(2n + 3)^2 = 50$ (8.3 Skill A)

23. $4v^2 + 8v = 41$ (8.3 Skill B)

24. $4t^2 - 16t + 13 = 0$ (8.4 Skill A)

Solve. Round the answers to the nearest tenth. (8.2 Skill B)

25. The bottom of a 32-foot ladder is placed 8 feet from a wall. How far up the wall will the ladder reach?

26. The bottom of a 16-foot ladder is placed 5 feet from a wall. How far up the wall will the ladder reach?

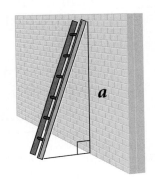

8.5 LESSON

Solving Quadratic Equations and Logical Reasoning

SKILL A   *Choosing a method for solving a quadratic equation*

You have learned many different methods for solving a quadratic equation:

- Factoring and the Zero-Product Property.
- Completing the square.

- Taking the square root of each side of an equation.
- Using the quadratic formula.

The method you choose should depend on the equation itself.

EXAMPLE 1   **Identify the method you might choose to solve $7x^2 = 175$. Justify your choice.**

▶ **Solution**

By dividing each side of the equation by 7 (Division Property of Equality), the result will be an equation that can be solved in one step by taking the square root of each side of the equation.

TRY THIS   Explain why the method in Example 1 is also a good choice for solving $-3x^2 = -48$.

EXAMPLE 2   **Identify the method you might choose to solve $(2x + 1)(3x - 7) = 0$. Justify your choice.**

▶ **Solution**

Because the left side of the equation is given in factored form and the right side is 0, apply the Zero-Product Property. Then solve for x in the two linear equations that result.

TRY THIS   Explain why the choice in Example 2 will not immediately apply to solving $(2x + 1)(3x - 7) = 3$.

If the coefficients a, b, and c of a quadratic equation are not integers, it is often easiest to use the quadratic formula to solve the equation.

EXAMPLE 3   **Identify the method you might choose to solve $2.5a^2 - 5.4a + 1 = 0$. Justify your choice.**

▶ **Solution**

Because the coefficients are decimals, both factoring and completing the square would be very difficult.

Therefore, identify a, b, and c, and then apply the quadratic formula.

TRY THIS   Explain why using the quadratic formula to solve $7x^2 = 175$ is not the quickest method.

EXERCISES

What solution method is used in each exercise?

1. $x^2 + 5x + 6 = 0$
   $(x + 2)(x + 3) = 0$
   $x + 2 = 0$ or $x + 3 = 0$
   $x = -2$ or $x = -3$

2. $3s^2 - 5s + 1 = 0$
   $s = \dfrac{-(-5) \pm \sqrt{(-5)^2 - 4(3)(1)}}{2(3)}$
   $s = \dfrac{5 \pm \sqrt{13}}{6}$

PRACTICE

Is the stated solution method suitable? Justify your response.

3. $2a^2 = 32$
   Take a square root.

4. $z^2 - 2.4z - 1.1 = 0$
   Use the quadratic formula.

5. $(v - 2)(v - 1.5) = 0$
   Apply the Zero-Product Property.

6. $3(n - 2)^2 = 27$
   Take a square root.

Identify the method you might choose to solve each quadratic equation. Justify your choice, but do not solve.

7. $0.5n^2 + 2.4n - 1 = 0$

8. $(2t + 3)(3t + 2) = 0$

9. $2h^2 - 98 = 0$

10. $-3(h + 3)^2 + 30 = 0$

11. $n^2 + 2n + 1 = 4$

12. $x^2 + 6x + 9 = 11$

13. $(-p + 5)(7p + 11) = 0$

14. $11y^2 = 13y - 37$

15. $0.5n^2 + 2.4n - 1 = 0$

Summarize a method for finding the number of x-intercepts.

16. $0 = x^2 + 2$

17. $0 = (2x + 3)(x + 11)$

18. $0 = 2.7x^2 - 2.4x - 1$

19. $0 = x^2 - 2$

20. $0 = (2x - 1)^2$

21. $0 = 1.3x^2 + 2x + 1.8$

22. How would you find b such that $y = 0.5x^2 + bx + 1$ has exactly one x-intercept?

23. **Critical Thinking** How would you find the real values of n such that $y = x^2 + nx + n + 1$ has one x-intercept?

MIXED REVIEW

Express the pattern shown in each table as a function with x as the independent variable. Predict the 60th term that would appear in the table. (3.8 Skill A)

24.

x	1	2	3	4	5
y	-0.5	1.5	3.5	5.5	7.5

25.

x	1	2	3	4	5
y	4	1	-2	-5	-8

After deciding on a solution strategy, you can use *deductive reasoning* to write the steps in a solution as a logical argument. Recall from Lesson 2.9 that deductive reasoning is the process of starting with a statement or a hypothesis and following a sequence of logical steps to reach a conclusion.

In Lesson 2.9, you wrote solutions to linear equations as deductive arguments. You can also write solutions to quadratic equations as deductive arguments. The examples below show a two-column format for such deductive arguments.

EXAMPLE 1   **Write the process for solving $x^2 - 15x + 56 = 0$ as a deductive argument in a two-column format.**

▶ **Solution**

Statement	Reason
1. $x^2 - 15x + 56 = 0$	**1.** Given
2. $(x - 7)(x - 8) = 0$	**2.** Distributive Property (factoring)
3. $x - 7 = 0$ or $x - 8 = 0$	**3.** Zero-Product Property
4. $x = 7$ or $x = 8$	**4.** Addition Property of Equality

Thus, if $x^2 - 15x + 56 = 0$, then $x = 7$ or $x = 8$.

TRY THIS   Write the process for solving $x^2 + 10x + 21 = 0$ as a deductive argument in a two-column format.

In many solutions, you can use the following fact from Lesson 8.1 as a justification in a logical argument.

$$\text{If } x^2 = a \text{ and } a > 0, \text{ then } x = \sqrt{a} \text{ or } x = -\sqrt{a}.$$

It is the basis for taking the square root of each side of an equation.

EXAMPLE 2   **Write the process for solving $x^2 + 6x = 7$ as a deductive argument in a two-column format.**

▶ **Solution**

Statement	Reason
1. $x^2 + 6x = 7$	**1.** Given
2. $x^2 + 6x + 9 = 7 + 9$	**2.** Addition Property of Equality
3. $(x + 3)^2 = 16$	**3.** Distributive Property (factoring)
4. $x + 3 = \pm 4$	**4.** Take the square root of each side.
5. $x = -3 \pm 4$	**5.** Subtraction Property of Equality
6. $x = -7$ or $x = 1$	**6.** Simplify.

Thus, if $x^2 + 6x = 7$, then $x = -7$ or $x = 1$.

TRY THIS   Write the process for solving $x^2 + 8x = 20$ as a deductive argument in a two-column format.

EXERCISES

KEY SKILLS

Give a reason for each step in the solution.

1. | Statements | | Reasons |
    |---|---|---|
    | $x^2 - 6x = 0$ | | Given |
    | $x(x - 6) = 0$ | **a.** | ? |
    | $x = 0$ or $x - 6 = 0$ | **b.** | ? |
    | $x = 0$ or $x = 6$ | **c.** | ? |

2. | Statements | | Reasons |
    |---|---|---|
    | $-x^2 + 8x = -9$ | | Given |
    | $x^2 - 8x = 9$ | **a.** | ? |
    | $x^2 - 8x - 9 = 0$ | **b.** | ? |
    | $(x - 9)(x + 1) = 0$ | **c.** | ? |
    | $x - 9 = 0$ or $x + 1 = 0$ | **d.** | ? |
    | $x = 9$ or $x = -1$ | **e.** | ? |

PRACTICE

The steps below are not listed in logical order. Write the steps in the solution of the given equation in logical order. Also give a reason for each step.

3. $x^2 - 6x + 5 = 32$ Given
    $x^2 - 6x + 9 = 36$
    $x - 3 = \pm 6$
    $x^2 - 6x + 5 + 4 = 32 + 4$
    $x = 9$ or $x = -3$
    $(x - 3)^2 = 36$

4. $6m^2 = 7m + 5$ Given
    $(2m + 1)(3m - 5) = 0$
    $m = -\dfrac{1}{2}$ or $m = \dfrac{5}{3}$
    $6m^2 - 7m - 5 = 0$
    $2m = -1$ or $3m = 5$
    $2m + 1 = 0$ or $3m - 5 = 0$

Write the process for solving each equation as a deductive argument in a two-column format.

5. $n^2 - 5n = 0$

6. $4(t + 5)^2 = 32$

7. $8b^2 + 2b - 3 = 0$

8. $d^2 - 12d = -36$

9. $4r^2 - 9^2 = 0$

10. $4(z - 5)^2 + 7 = 32$

Use a deductive argument to show that each statement is true.

11. The x-intercepts of $y = x^2 - 7x + 12$ are positive.

12. The x-intercepts of $y = x^2 - 3x - 18$ have opposite signs.

MIXED REVIEW

Solve each system of equations. (5.3 Skill B)

13. $\begin{cases} 3a - 2b = 1 \\ -2a + 7 = 0 \end{cases}$

14. $\begin{cases} -5r - 3s = 7 \\ -7r + 7s = 1 \end{cases}$

15. $\begin{cases} 2.5m - 3n = 0 \\ 3m + 2n = 2 \end{cases}$

16. $\begin{cases} 5x - 3y = 16 \\ 3x + 2y = 2 \end{cases}$

17. $\begin{cases} -7b - 3c = 5 \\ -2b + 2c = 30 \end{cases}$

18. $\begin{cases} 10u - 5v = 0 \\ 3u + 2v = 10 \end{cases}$

8.6 Quadratic Patterns

LESSON

SKILL A *Recognizing and continuing a quadratic pattern*

Recall from Lesson 3.9 that a table of ordered pairs (x, y) shows a linear pattern if the differences between successive x-terms are constant and the differences between successive y-terms are also constant. The table below represents a linear pattern.

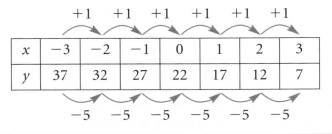

x	-3	-2	-1	0	1	2	3
y	37	32	27	22	17	12	7

Test for a Quadratic Pattern in a Table of Ordered Pairs

A table of ordered pairs (x, y) shows a quadratic pattern if the differences between successive x-values are constant and the differences between successive y-values form a linear pattern.

EXAMPLE

Does the table at right represent a quadratic pattern? Continue the table for $x = 4, 5,$ and 6.

x	-3	-2	-1	0	1	2	3
y	17	7	1	-1	1	7	17

▶ **Solution**

First find the differences between successive x- and y-values.

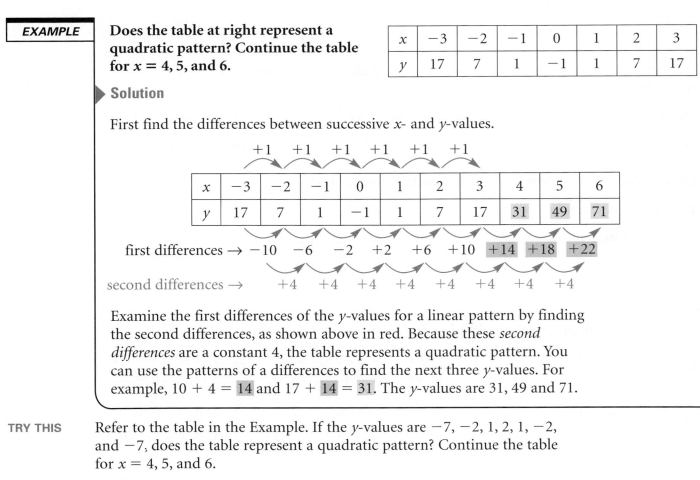

Examine the first differences of the y-values for a linear pattern by finding the second differences, as shown above in red. Because these *second differences* are a constant 4, the table represents a quadratic pattern. You can use the patterns of a differences to find the next three y-values. For example, $10 + 4 = 14$ and $17 + 14 = 31$. The y-values are 31, 49 and 71.

TRY THIS Refer to the table in the Example. If the y-values are $-7, -2, 1, 2, 1, -2$, and -7, does the table represent a quadratic pattern? Continue the table for $x = 4, 5,$ and 6.

EXERCISES

Write the successive differences of the *y*-values. Are the differences constant?

1.

x	−3	−2	−1	0	1	2	3
y	−7	−2	3	8	13	18	23

2.

x	−3	−2	−1	0	1	2	3
y	27	14	5	0	−1	2	9

Describe the pattern in each table as linear, quadratic, or neither.

3.

x	−3	−2	−1	0	1	2	3
y	4	1	0	1	4	9	16

4.

x	−3	−2	−1	0	1	2	3
y	4	0	−2	−2	0	4	10

5.

x	−3	−2	−1	0	1	2	3
y	48	22	6	0	4	18	42

6.

x	−3	−2	−1	0	1	2	3
y	24	22	20	18	16	14	12

7.

x	−3	−2	−1	0	1	2	3
y	−15	−10	−5	0	5	10	15

8.

x	−3	−2	−1	0	1	2	3
y	−23	−17	−11	−5	1	7	13

Describe the pattern in each list as linear or quadratic. Then give the next three ordered pairs for each list.

9. $(1, 1), (2, 4), (3, 9), (4, 16), (5, 25)$

10. $(−3, −1), (−2, −6), (−1, −9), (0, −10)$

11. $(−2, −14), (−1, −5), (0, 0), (1, 1), (2, −2)$

12. $(−2, −18), (−1, −8), (0, 0), (1, 6), (2, 10)$

13. Critical Thinking Consider $y = ax^2 + bx + c$, where a, b, and c are fixed real numbers and $a \neq 0$.
  a. Write an expression for y when x is replaced by n.
  b. Show that when x is replaced by $n + 1$, then
  $y = an^2 + 2an + bn + a + b + c$.
  c. Show that the expression in part **b** minus the expression in part **a** equals $(2a)n + (a + b)$, which is a linear expression in n.

Write a function to predict the number of objects, *t*, in the *n*th diagram. Then predict the number of objects in the 40th diagram. (3.9 Skill B)

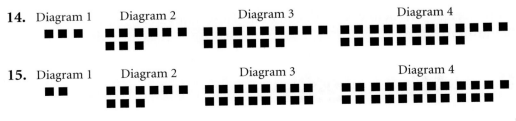

14. Diagram 1   Diagram 2   Diagram 3   Diagram 4

15. Diagram 1   Diagram 2   Diagram 3   Diagram 4

In Lesson 3.9, you learned how to represent a linear pattern shown geometrically with an equation. You can also write a quadratic function to represent a quadratic pattern shown geometrically.

EXAMPLE

a. **Represent the dot pattern below with an equation.**
b. **How many dots will be in the 50th group of dots?**

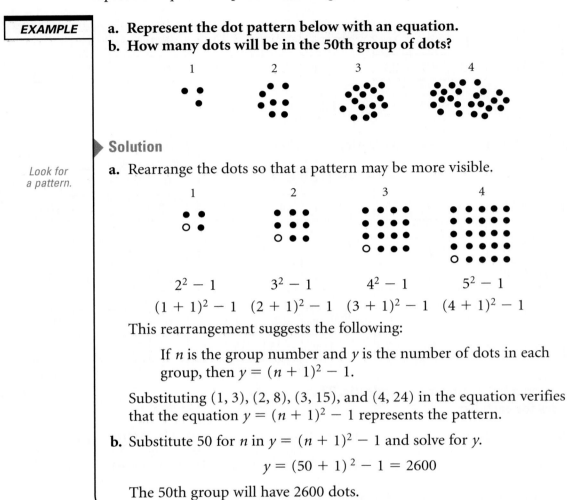

Look for a pattern.

▶ **Solution**

a. Rearrange the dots so that a pattern may be more visible.

$$2^2 - 1 \qquad 3^2 - 1 \qquad 4^2 - 1 \qquad 5^2 - 1$$
$$(1 + 1)^2 - 1 \quad (2 + 1)^2 - 1 \quad (3 + 1)^2 - 1 \quad (4 + 1)^2 - 1$$

This rearrangement suggests the following:

If n is the group number and y is the number of dots in each group, then $y = (n + 1)^2 - 1$.

Substituting $(1, 3)$, $(2, 8)$, $(3, 15)$, and $(4, 24)$ in the equation verifies that the equation $y = (n + 1)^2 - 1$ represents the pattern.

b. Substitute 50 for n in $y = (n + 1)^2 - 1$ and solve for y.

$$y = (50 + 1)^2 - 1 = 2600$$

The 50th group will have 2600 dots.

You can also solve the example above by counting the dots in each group and then representing the counts in a table.

n	1	2	3	4
y	3	8	15	24
y	$4 - 1$	$9 - 1$	$16 - 1$	$25 - 1$
y	$2^2 - 1$	$3^2 - 1$	$4^2 - 1$	$5^2 - 1$

From the table, $y = (n + 1)^2 - 1$.

TRY THIS   Represent the dot pattern shown below with an equation. How many dots will be in the 50th group?

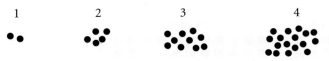

EXERCISES

Refer to the pattern below.

Group 1   Group 2   Group 3   Group 4

1. Make a table that shows the group number, n, and the number of squares in each group, y, for the pattern above.

2. Use guess-and-check to find a and b such that $y = 2(n - a)^2 + b$ represents the table you made in Exercise 1.

PRACTICE

Represent each pattern below with an equation. Then predict how many objects will appear in the 60th group of objects.

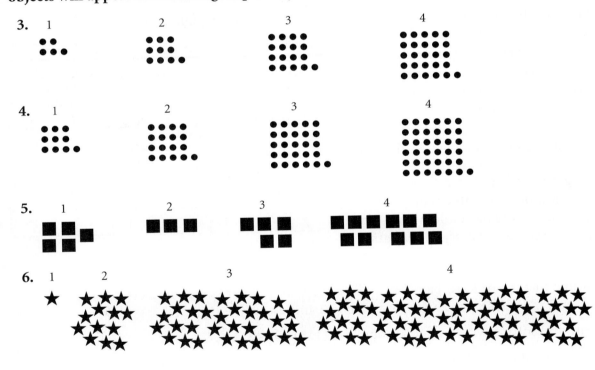

3. 1   2   3   4

4. 1   2   3   4

5. 1   2   3   4

6. 1   2   3   4

MIXED REVIEW

Solve each equation. (7.5 Skill A)

7. $(a + 4)(a + 7) = 0$

8. $(2n - 1)(n + 5) = 0$

9. $(7z + 1)(3z - 4) = 0$

10. $4t^2 - 12t + 9 = 0$

11. $4y^2 + 4y + 1 = 0$

12. $16d^2 - 56d + 49 = 0$

13. $9r^2 + 9r - 28 = 0$

14. $3z^2 - 10z - 77 = 0$

15. $16c^2 + 56c + 49 = 0$

16. $2h^2 + 11h - 90 = 0$

17. $3m^2 - 19m - 110 = 0$

18. $25g^2 - 5g - 2 = 0$

Using a system of equations in finding a quadratic pattern

If a table of ordered pairs can be represented by the equation $y = ax^2 + bx + c$, then you may be able to use a system of equations to find a, b, and c.

EXAMPLE 1

The table at right shows a quadratic pattern. Write a quadratic equation to represent the pattern.

x	-3	-2	-1	0	1	2	3
y	-7	-2	1	2	1	-2	-7

▶ **Solution**

Let $y = ax^2 + bx + c$ for some numbers a, b, and c.
From the table, we see that if $x = 0$, then $y = 2$.

$$2 = a(0)^2 + b(0) + c \;\rightarrow\; 2 = c \;\rightarrow\; y = ax^2 + bx + 2$$

Write and solve a system of equations. Use $(1, 1)$ and $(-1, 1)$.

$$\begin{array}{l} x = 1: \quad a(1)^2 + b(1) + 2 = 1 \\ x = -1: \; a(-1)^2 + b(-1) + 2 = 1 \end{array} \rightarrow \begin{cases} a + b = -1 \\ a - b = -1 \end{cases} \rightarrow \begin{cases} a = -1 \\ b = 0 \end{cases}$$

Substitute the values for a and b into $y = ax^2 + bx + 2$. An equation that represents the table of ordered pairs is $y = -x^2 + 2$.

TRY THIS

The table at right shows a quadratic pattern. Write a quadratic equation to represent the pattern.

x	-3	-2	-1	0	1	2	3
y	10	3	0	1	6	15	28

Recall that the graph of a quadratic function is a parabola.

EXAMPLE 2

The graphs of the ordered pairs at right lie along a parabola. Find an equation for the parabola. Verify that the ordered pairs satisfy the equation.

x	-3	-2	-1	0	1	2	3
y	29	14	5	2	5	14	29

▶ **Solution**

Assume that the ordered pairs satisfy $y = ax^2 + bx + c$ for some numbers a, b, and c. Because $y = 2$ when $x = 0$, $c = 2$ and $y = ax^2 + bx + 2$.

Write and solve a system of equations. Use $(1, 5)$ and $(2, 14)$.

$$\begin{array}{l} x = 1: \; a(1)^2 + b(1) + 2 = 5 \\ x = 2: \; a(2)^2 + b(2) + 2 = 14 \end{array} \rightarrow \begin{cases} a + b = 3 \\ 4a + 2b = 12 \end{cases} \rightarrow \begin{cases} a = 3 \\ b = 0 \end{cases}$$

Therefore, $y = 3x^2 + 2$.

Verify that $(-3, 29)$, $(-2, 14)$, $(-1, 5)$, and $(3, 29)$ satisfy $y = 3x^2 + 2$.

$$\begin{array}{ll} x = -3: \; 3(-3)^2 + 2 = \mathbf{29} ✔ & x = -1: \; 3(-1)^2 + 2 = \mathbf{5} ✔ \\ x = -2: \; 3(-2)^2 + 2 = \mathbf{14} ✔ & x = 3: \quad 3(3)^2 + 2 = \mathbf{29} ✔ \end{array}$$

TRY THIS

The graphs of the ordered pairs at right lie along a parabola. Find an equation for the parabola. Verify that the ordered pairs satisfy the equation.

x	-3	-2	-1	0	1	2	3
y	15	5	-1	-3	-1	5	15

EXERCISES

Verify that each table represents a quadratic pattern.

1.

x	-4	-3	-2	-1	0	1	2
y	80	45	20	5	0	5	20

2.

x	-3	-2	-1	0	1	2	3
y	16	6	0	-2	0	6	16

3.

x	-2	-1	0	1	2	3	4
y	-6	0	2	0	-6	-16	-30

4.

x	-4	-3	-2	-1	0	1	2
y	-47	-26	-11	-2	1	-2	-11

Each table in Exercises 1–4 represents a quadratic pattern.
Write a quadratic equation to represent each table.

5. Exercise 1   **6.** Exercise 2   **7.** Exercise 3   **8.** Exercise 4

The graphs of each set of ordered pairs lie along a parabola. Find an equation
for the parabola. Verify that the ordered pairs satisfy the equation.

9. $y = x^2 - bx + 4$ for some real b

x	1	2	3	4	5	6
y	3	4	7	12	19	28

10. $y = -3x^2 + 7x + c$ for some real c

x	1	2	3	4	5	6
y	-1	-3	-11	-25	-45	-71

11. $y = ax^2 + x + c$ for some real a and c

x	1	2	3	4	5	6
y	-7	-12	-21	-34	-51	-72

12. $y = ax^2 + c$ for some real a and c

x	1	2	3	4	5	6
y	-7	2	17	38	65	98

13. Critical Thinking Find an equation of the form $y = ax^2 + c$,
for some real numbers a and c, that represents the graph at right.
Then find y for each value of x in the table below.

x	10	11	12	13	14	15	16
y	?	?	?	?	?	?	?

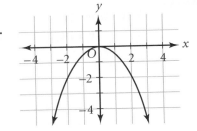

Solve each equation. Write the solutions in simplest radical form. (8.2 Skill A)

14. $4b^2 - 49 = 0$ **15.** $81v^2 - 4 = 0$ **16.** $9t^2 - 125 = 0$ **17.** $-5u^2 + 120 = 0$

18. $-9r^2 + 44 = -5$ **19.** $-3m^2 - 7 = -67$ **20.** $-\frac{1}{3}h^2 + 10 = -1$ **21.** $-\frac{1}{2}p^2 + 2 = -110$

8.7 Quadratic Functions and Acceleration

SKILL A > **APPLICATIONS** > *Using quadratic functions to solve problems involving falling objects that have no initial velocity*

Because of acceleration due to gravity, the speed of a falling object is not constant. The relationship between the distance that an object falls and the time that the object spends falling cannot be modeled by a linear equation.

Free-Falling Objects

If an object is dropped from an initial height of h_0 feet and is allowed to fall freely, then after t seconds it will fall $16t^2$ feet and will have height h given by $h = -16t^2 + h_0$.

EXAMPLE 1

A skydiver jumps from a plane at a height of 5200 feet above the ground. How far does the diver descend and what is the diver's height above the ground after 3 seconds, 4 seconds, and 5 seconds of free fall?

▶ **Solution**

Substitute 3, 4, and 5 for t into $16t^2$ and $-16t^2 + 5200$.

Make a table.

time	$t = 3$	$t = 4$	$t = 5$
distance	$16(3)^2 = 144$	$16(4)^2 = 256$	$16(5)^2 = 400$
height	$5200 - 144 = 5056$	$5200 - 256 = 4944$	$5200 - 400 = 4800$

After 3, 4, and 5 seconds, the diver descends 144 feet, 256 feet, and 400 feet; his height is 5056 feet, 4944 feet, and 4800 feet, respectively.

TRY THIS

Rework Example 1 given an initial height of 5400 feet and elapsed times of 2 seconds, 4 seconds, and 6 seconds.

EXAMPLE 2

After how many seconds of free fall will an object dropped from 1200 feet have a height of 120 feet above the ground? Use a calculator.

▶ **Solution**

$$h = -16t^2 + h_0$$
$$120 = -16t^2 + 1200 \qquad \longleftarrow \text{ Replace } h \text{ with 120 and } h_0 \text{ with 1200.}$$
$$-1080 = -16t^2$$
$$t = \pm\sqrt{\frac{1080}{16}} \qquad \longleftarrow \text{ Take the square root of each side.}$$
$$t \approx 8.2 \qquad \longleftarrow \text{ Reject the negative solution because time cannot be negative.}$$

The height will be 120 feet after about 8.2 seconds.

TRY THIS

After how many seconds of free fall will an object dropped from 2500 feet have a height of 500 feet above the ground?

EXERCISES

Find the exact positive value of *t*.

1. $125 = -16t^2 + 1300$ **2.** $140 = -16t^2 + 1500$ **3.** $160 = -16t^2 + 2800$

A skydiver jumps from a plane at the given height and falls freely. Find how far the diver descends and the diver's height for each elapsed time.

4. 5000 feet; after 4 seconds

5. 5600 feet; after 4 seconds

6. 5200 feet; after 3 seconds

7. 4800 feet; after 5 seconds

8. 6500 feet; after 4.5 seconds

9. 6300 feet; after 5.5 seconds

10. A skydiver jumps from a plane at a height of 6800 feet. How far does the diver descend and what is the diver's height above the ground after 2 seconds, 3 seconds, and 6 seconds of free fall?

To the nearest tenth of a second, find how long it takes for an object in free fall to have the specified height.

11. initial height: 5300 feet
specified height: 4000 feet

12. initial height: 6000 feet
specified height: 3000 feet

13. initial height: 4200 feet
specified height: 100 feet

14. initial height: 4800 feet
specified height: 1000 feet

15. initial height: 7500 feet
specified height: 0 feet

16. initial height: 100 feet
specified height: 10 feet

17. Critical Thinking   How long will it take an object in free fall to have a height that is one-half its initial height of *h* feet?

To the nearest tenth of a unit, approximate the unknown length of the third side in each right triangle. (8.2 Skill B)

18. hypotenuse: 100 meters
length of leg: 70 meters

19. length of leg: 60 feet
length of leg: 60 feet

20. hypotenuse: 2.5 inches
length of leg: 1.0 inches

21. length of leg: 12 yards
length of leg: 15 yards

22. length of leg: 120 feet
length of leg: 200 feet

23. hypotenuse: 300 meters
length of leg: 150 meters

In Skill A of this lesson, you learned how to solve problems involving objects that were dropped from a still position, or that had no *initial velocity*. Now you will learn how to solve problems involving objects that do have an initial velocity.

Accelerated Motion with Initial Velocity

After t seconds, an object propelled upward from an initial height of h_0 feet and that has an initial velocity of v_0 feet per second will have height h feet given by $h = -16t^2 + v_0 t + h_0$.

EXAMPLE 1

A ball thrown directly upward from the edge of a cliff 48 feet above sea level is thrown at an initial velocity of 96 feet per second. Find the ball's height above sea level after 2 seconds and after 3 seconds.

▶ **Solution**

The height function is $h = -16t^2 + 96t + 48$.

Make a table.

t	$t = 2$	$t = 3$
h	$-16(2)^2 + 96(2) + 48 = 176$	$-16(3)^2 + 96(3) + 48 = 192$

After 2 and 3 seconds, the ball's height above sea level is 176 feet and 192 feet, respectively.

TRY THIS

Rework Example 1 given an initial height of 50 feet and an initial velocity of 128 feet per second.

EXAMPLE 2

Refer to the problem in Example 1. After how many seconds will the ball have a height of 180 feet? Round answer(s) to the nearest hundredth.

▶ **Solution**

$$180 = -16t^2 + 96t + 48$$
$$0 = -16t^2 + 96t - 132 \qquad \longleftarrow \text{Write in standard form.}$$

Use a formula.

$$t = \frac{-96 \pm \sqrt{96^2 - 4(-16)(-132)}}{2(-16)} \qquad \longleftarrow \begin{array}{l}\text{Apply the quadratic formula} \\ \text{with } a = -16, b = 96, \text{ and} \\ c = -132.\end{array}$$
$$t \approx 2.13 \text{ or } 3.87$$

The height is 180 feet after about 2.13 and 3.87 seconds.

TRY THIS

Rework Example 2 given a desired height of 170 feet.

You can use the following fact to find a projectile's maximum height.

If $h = -16t^2 + v_0 t + h_0$ models height as a function of time t, then maximum height is achieved after $\frac{v_0}{32}$ seconds.

EXERCISES

KEY SKILLS

Write each equation in standard form. Then write the equation for *t* that results from applying the quadratic formula. Do not evaluate.

1. $-16t^2 + 96t + 60 = 180$

2. $-16t^2 + 96t + 100 = 200$

3. $-16t^2 + 96t = 180$

4. $-16t^2 + 96t + 500 = 1000$

PRACTICE

A ball thrown directly upward from the edge of a cliff. The height of the cliff is given in feet above sea level and the specified initial velocity is given in feet per second. Find the height of the ball after 3 seconds and after 5 seconds.

5. height 50 ft; velocity 96 ft/sec

6. height 60 ft; velocity 96 ft/sec

7. height 96 ft; velocity 128 ft/sec

8. height 10 ft; velocity 128 ft/sec

9. height 160 ft; velocity 48 ft/sec

10. height 250 ft; velocity 48 ft/sec

A ball thrown directly upward from the edge of a cliff. The cliff is 64 feet above sea level and the ball has an initial velocity of 96 feet per second. Find the amount of time it takes the ball to have the specified height above sea level.

11. 170 feet

12. 200 feet

13. 64 feet

14. 128 feet

15. 0 feet

16. 100 feet

17. A ball thrown directly upward from the edge of a cliff 48 feet above sea level has an initial velocity of 96 feet per second.
   a. How long will it take the ball to reach its maximum height?
   b. Find the maximum height.

MIXED REVIEW APPLICATIONS

Simplify. Write answers in scientific notation. (6.1 Skill C)

18. $(2.3 \times 10^2)(3.1 \times 10^3)$

19. $(5.0 \times 10^1)(2.5 \times 10^4)$

20. $(6.3 \times 10^2)(4.0 \times 10^2)$

21. $(9.0 \times 10^3)(9.0 \times 10^2)$

22. $\dfrac{18.6 \times 10^5}{6.0 \times 10^4}$

23. $\dfrac{2.25 \times 10^2}{1.5 \times 10^3}$

24. $\dfrac{1.69 \times 10^6}{1.3 \times 10^2}$

25. $\dfrac{2.4 \times 10^3}{4.8 \times 10^5}$

Find the square roots of each number. If the number is not a perfect square, approximate the roots to the nearest hundredth.

1. 225

2. 60

3. 0.49

4. 72

Write each square root in simplest radical form.

5. $\sqrt{98}$

6. $\sqrt{300}$

7. $\sqrt{\dfrac{2}{25}}$

8. $\sqrt{\dfrac{50}{81}}$

9. Let $a > 0$. Using laws of exponents, prove that $\sqrt{36a^2} = 6a$.

Solve. Write irrational solutions in simplest radical form.

10. $9n^2 - 1 = 0$

11. $4d^2 - 1 = 8$

12. $16y^2 - 5 = 15$

13. In a right triangle, the length of the hypotenuse is 110 feet and the length of one leg is 65 feet. To the nearest hundredth of a foot, find the length of the other leg.

Solve. Write the solutions in simplest radical form where necessary.

14. $p^2 + 5 = 0$

15. $2x^2 + 4 = 0$

16. $2z^2 - 3 = -11$

17. $(x + 2)^2 = 4$

18. $(k - 5)^2 = 27$

19. $3(z - 1)^2 = 36$

Solve each equation by completing the square. Write the solutions in simplest radical form where necessary.

20. $4g^2 + 4g - 3 = 0$

21. $z^2 - 4z - 46 = 0$

22. Boats A and B leave the dock at the same time and sail at a constant speed. Boat A sails due north and Boat B sails due east. After 2 hours, they are 26 miles apart. Boat B sails 7 miles per hour faster than Boat A. Find their speeds using completing the square.

Solve each equation using the quadratic formula. Write the solutions in simplest radical form where necessary.

23. $h^2 - 2h - 26 = 0$

24. $s^2 - 14s + 31 = 0$

25. $p^2 - 2p - 3 = 0$

26. $7z^2 + 24z - 55 = 0$

27. $2d^2 = 19d - 45$

28. $w^2 + 2w = 12$

Use the discriminant to find the number of real roots of each equation.

29. $3q^2 - q + 3 = 0$

30. $4n^2 = 8n - 4$

Use the discriminant to find the number of x-intercepts of the graph of each quadratic function.

31. $y = 3t^2 - t - 3$

32. $y = -9x^2 - 18x - 9$

33. $y = 3t^2 - t + 3$

34. $y = 16x^2 - 49$

35. Let a, b, and c represent real numbers with $a \neq 0$.

Prove that if $ax^2 + bx = -c$, then $a\left(x^2 + \dfrac{b}{a}x\right) = -c$.

36. Identify the solution method you might choose to solve $v^2 = 14v - 45$. Justify your choice but do not solve the equation.

37. Write the solution process to $2u^2 = -3u + 5$ as a deductive argument in a two-column format.

38. Continue the pattern in the table below for $x = 4, 5,$ and 6.

x	-3	-2	-1	0	1	2	3
y	30	16	6	0	-2	0	6

39. a. Represent the dot pattern below with an equation.

    b. How many dots will be in the 50th group of dots?

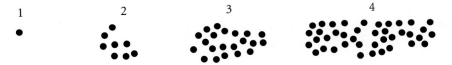

40. Write a quadratic function to represent the pattern in the table below.

x	-1	0	1	2	3	4	5
y	5	0	-1	2	9	20	35

41. A object is dropped from a height of 1375 feet. Approximate to the nearest tenth of a second how long it will take for the object to have a height above ground of 1000 feet. Use a calculator to approximate the value of the radical expressions.

42. A ball thrown directly upward from the edge of a cliff. The cliff is 100 feet above sea level and the ball has an initial velocity of 96 feet per second. After how many seconds will the ball have a height of 200 feet? Use a calculator to approximate the value of the radical expressions.

9

Rational Expressions, Equations, and Functions

▶ **What You Already Know**

In earlier mathematics courses, you learned how to add, subtract, multiply, and divide rational numbers. In Chapter 6, you learned how to add, subtract, multiply, and divide polynomials in one variable. In Chapter 9, you can bring these concepts and skills together to help you work with rational expressions and functions.

▶ **What You Will Learn**

First, you will learn what a rational expression is. In particular, you will learn how to find those real numbers for which the rational expression has meaning and those for which it does not.

A great deal of your time will be directed toward adding, subtracting, multiplying, and dividing rational expressions. Simplifying a rational expression will also be part of your study. You will also learn how to

- solve and use rational equations to solve problems involving electricity, ratios, and work,

- solve problems involving inverse variation and inverse variation as the square, problems such as those dealing with area and illumination, and

- use deductive reasoning to prove mathematical statements about proportions.

VOCABULARY	
complex fraction	least common denominator (LCD)
constant of variation	least common multiple (LCM)
converse	rational equation
Cross-Product Property of Proportions	rational expression
domain (of a rational expression)	rational function
excluded value	reciprocal function
extraneous solution	similar (geometric figures)
inverse variation	work rate
inverse-square variation	

The diagram below shows how mathematical skills and mathematical reasoning are interrelated with the skills and concepts in Chapter 9. Notice that the focus of this chapter involves operations with rational expressions such as multiplication and addition.

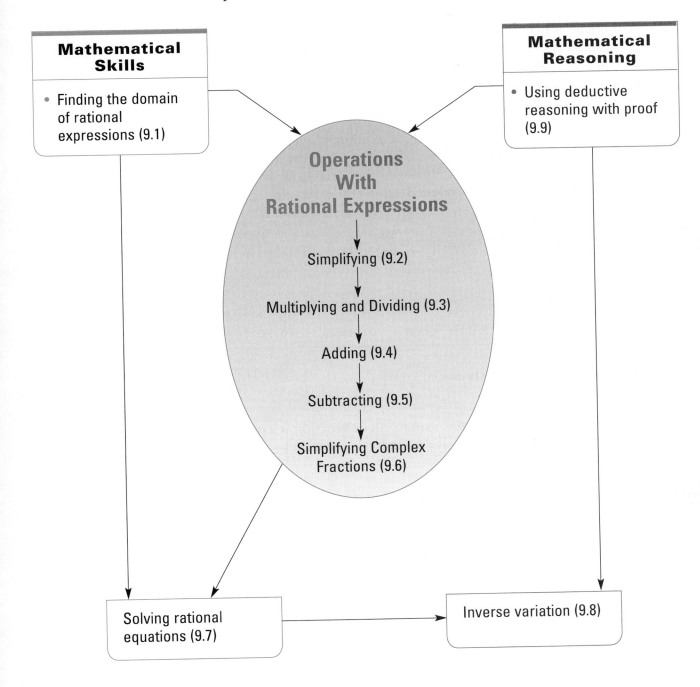

9.1

LESSON

Rational Expressions and Functions

SKILL A *Finding the domain of a rational expression or function*

A quotient of two polynomials is called a **rational expression.** Every polynomial is a rational expression since every polynomial can be written as a quotient whose denominator is 1. A **rational function** in x is a function defined by a rational expression in x.

rational expression: $\dfrac{3x-1}{2x-7}$ rational function: $y=\dfrac{3x-1}{2x-7}$

The **domain** of a rational expression or a rational function in x, unless otherwise stated, is the set of all real values of x for which the denominator is not equal to 0. A value of the variable for which the denominator equals 0 is an **excluded value** of the domain.

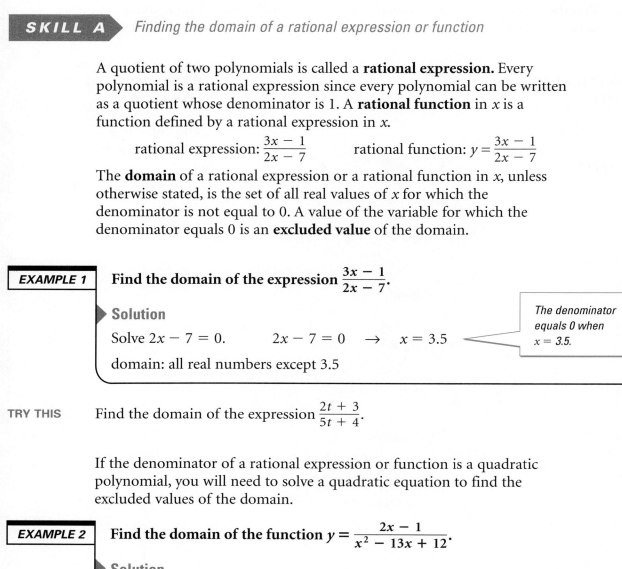

EXAMPLE 1 **Find the domain of the expression $\dfrac{3x-1}{2x-7}$.**

Solution

Solve $2x-7=0$. $2x-7=0 \rightarrow x=3.5$

The denominator equals 0 when $x=3.5$.

domain: all real numbers except 3.5

TRY THIS Find the domain of the expression $\dfrac{2t+3}{5t+4}$.

If the denominator of a rational expression or function is a quadratic polynomial, you will need to solve a quadratic equation to find the excluded values of the domain.

EXAMPLE 2 **Find the domain of the function $y=\dfrac{2x-1}{x^2-13x+12}$.**

Solution

Set the denominator equal to 0 and then solve for x.
$x^2-13x+12=0$
$(x-12)(x-1)=0$ ⟵ *Factor.*
$x-12=0$ or $x-1=0$ ⟵ *Apply the Zero-Product Property.*
$x=12$ or $x=1$ ⟵ *Solve for x.*

domain: all real numbers except 1 and 12

TRY THIS Find the domain of each function.

a. $y=\dfrac{3x^2-2x-1}{6x^2+x-1}$ **b.** $y=\dfrac{3x^2}{x^2+4}$

EXERCISES

Write the equation you would solve to find the excluded values of the domain of each rational expression or function. Do not solve.

1. $\dfrac{6}{y-2}$

2. $\dfrac{3}{x}-5$

3. $y=\dfrac{3n+7}{6n^2-11n-35}$

PRACTICE

Find the domain of each rational expression or function.

4. $\dfrac{1}{x-5}$

5. $\dfrac{4r-1}{r+3}$

6. $\dfrac{2a-1}{5a+1}$

7. $\dfrac{5x+7x^2}{x^2}$

8. $\dfrac{3x^2+2x}{x^2}$

9. $\dfrac{t^2-5t+6}{(3t+2)(2t-3)}$

10. $\dfrac{2x^2-5x+2}{(3x-1)(2x+5)}$

11. $\dfrac{4n-7}{n^2-5n+4}$

12. $\dfrac{x^2+2x-3}{x^2+4x-5}$

13. $y=\dfrac{x^2+3x-2}{(4x+5)(2x+5)}$

14. $y=\dfrac{x^2+5x-4}{(4x+3)(4x+3)}$

15. $y=\dfrac{x^2-x-3}{16x^2-1}$

16. $y=\dfrac{x^2-3}{x^2-4}$

17. $y=\dfrac{x^2+1}{9x^2+25}$

18. $y=\dfrac{x^2-1}{4x^2+x+9}$

19. $y=\dfrac{x^2+7x-1}{14x^2+17x+5}$

20. $y=\dfrac{5x^2+x-2}{10x^2-29x+21}$

21. $y=\dfrac{7x^2+x-1}{6x^2-35x+49}$

22. Show that the domain of $y=\dfrac{2m-1}{7m^2+10m+5}$ is the set of all real numbers.

23. Critical Thinking **a.** Use a number line to find the distance between the excluded values of $y=\dfrac{4x-9}{36x^2-100}$.

b. Let a and b represent real numbers with $a\neq 0$. Use a number line to find the distance between the excluded values of $y=\dfrac{1}{a^2x^2-b^2}$.

MIXED REVIEW

Factor. (7.3 Skill A)

24. $x^2+15x+50$

25. $c^2+16c+63$

26. $z^2+18x+77$

Factor. (7.3 Skill B)

27. $a^2+12a-13$

28. $n^2-10n-39$

29. $p^2-21p-100$

9.2 LESSON

Simplifying Rational Expressions

SKILL A *Using monomial factoring to simplify rational expressions*

Recall that when you simplify a fraction, you first factor the numerator and the denominator. Then you divide both the numerator and the denominator by all common factors.

$$\frac{60}{75} = \frac{3 \cdot 4 \cdot 5}{3 \cdot 5 \cdot 5} = \frac{{}^1\cancel{3} \cdot 4 \cdot \cancel{5}^1}{{}_1\cancel{3} \cdot {}_1\cancel{5} \cdot 5} = \frac{4}{5}$$

Simplify a rational expression in the same way. A common factor can be a number, a variable, or an expression.

EXAMPLE 1 Simplify $\dfrac{3}{3x - 15}$. **Identify any excluded values.**

▶ **Solution**

$$\frac{3}{3x - 15} = \frac{\cancel{3}^1}{{}_1\cancel{3}(x - 5)} \qquad \longleftarrow \text{Factor the denominator.}$$

$$= \frac{1}{x - 5} \qquad \longleftarrow \text{Divide both the numerator and the denominator by 3.}$$

The excluded value is 5 because the denominator, $x - 5$, equals 0 when $x = 5$.

TRY THIS Simplify $\dfrac{7}{21r - 49}$. Identify any excluded values.

EXAMPLE 2 Simplify $\dfrac{4w + 8}{4w - 8}$. **Identify any excluded values.**

▶ **Solution**

$$\frac{4w + 8}{4w - 8} = \frac{\cancel{4}^1(w + 2)}{{}_1\cancel{4}(w - 2)} \qquad \longleftarrow \text{Factor 4 from the numerator and the denominator.}$$

$$= \frac{w + 2}{w - 2} \quad w \neq 2 \qquad \longleftarrow \text{Divide both the numerator and denominator by 4.}$$

TRY THIS Simplify. Identify any excluded values. **a.** $\dfrac{6c + 2}{2c + 6}$ **b.** $\dfrac{8n - 8}{2n + 6}$

EXAMPLE 3 Simplify $\dfrac{3x - 9}{5x - 15}$. **Identify any excluded values.**

▶ **Solution**

$$\frac{3x - 9}{5x - 15} = \frac{3(x - 3)^1}{5(x - 3)} \qquad \longleftarrow \text{Factor the numerator and the denominator.}$$

$$= \frac{3}{5} \qquad \longleftarrow \text{Divide the numerator and the denominator by } (x - 3).$$

Although the expression simplifies to $\dfrac{3}{5}$, the factor $(x - 3)$ was originally in the denominator, so 3 is an excluded value.

TRY THIS Simplify. Identify any excluded values. **a.** $\dfrac{-9m + 108}{3m - 36}$ **b.** $\dfrac{8n^2 + 2n}{2n}$

EXERCISES

KEY SKILLS

Simplify each rational expression. Identify any excluded values.

1. $\dfrac{4(x + 4)}{4(x - 4)}$

2. $\dfrac{4(x + 5)}{5(x + 5)}$

3. $\dfrac{2 \cdot 5}{5(x - 1)}$

PRACTICE

Simplify. Identify any excluded values.

4. $\dfrac{21}{7y + 35}$

5. $\dfrac{-40}{10b + 50}$

6. $\dfrac{2d + 4}{d + 2}$

7. $\dfrac{3h - 15}{h - 5}$

8. $\dfrac{3z - 18}{3z + 12}$

9. $\dfrac{10a + 20}{5a + 15}$

10. $\dfrac{4b - 12}{2b + 14}$

11. $\dfrac{11c - 33}{11c + 22}$

12. $\dfrac{2v + 10}{5v + 25}$

13. $\dfrac{-7v + 21}{3v - 9}$

14. $\dfrac{-3z + 9}{10z - 30}$

15. $\dfrac{z + 8}{3z + 24}$

Find a such that each equation is true for all values of n.

16. $\dfrac{n - a}{2n - 6} = \dfrac{1}{2}$

17. $\dfrac{3n + 9}{5n - a} = \dfrac{3}{5}$

18. $\dfrac{8n + a}{3n + 15} = \dfrac{8}{3}$

Find the unknown numerator or denominator.

19. $\dfrac{\text{numerator}}{2x - 6} = \dfrac{1}{2}$

20. $\dfrac{2x - 6}{\text{denominator}} = \dfrac{2}{5}$

21. **Critical Thinking** Consider $\dfrac{ax + b}{cx + d}$, where a, b, c, and d are nonzero real numbers.
    a. What must be true of a, b, c, and d if the value of the expression is 1?
    b. What must be true of a, b, c, and d if the expression can be simplified to a constant?

MIXED REVIEW

Factor. (7.4 Skill A)

22. $10a^2 + 25a + 15$

23. $4a^2 + 8a + 4$

24. $24 + 10x - x^2$

25. $30 + x - x^2$

26. $64 + 16x + x^2$

27. $9 - 6x + x^2$

In Chapter 7, you learned how to factor a quadratic expression into a product of linear factors. This skill is often needed to simplify a rational expression.

EXAMPLE 1 **Simplify $\dfrac{a^2 + 5a + 6}{a^2 - 4}$. Identify any excluded values.**

▶ **Solution**

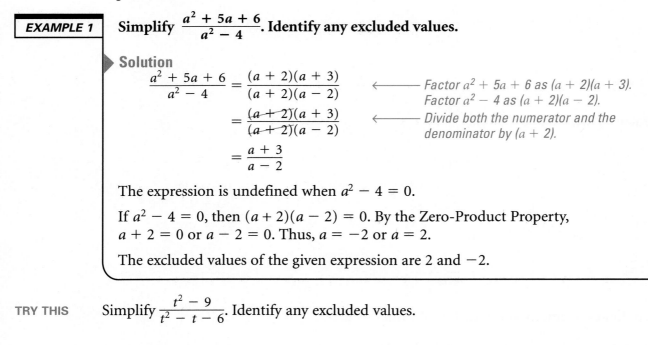

$$\frac{a^2 + 5a + 6}{a^2 - 4} = \frac{(a + 2)(a + 3)}{(a + 2)(a - 2)} \qquad \longleftarrow \text{Factor } a^2 + 5a + 6 \text{ as } (a + 2)(a + 3).$$
$$\text{Factor } a^2 - 4 \text{ as } (a + 2)(a - 2).$$
$$= \frac{\cancel{(a + 2)}(a + 3)}{\cancel{(a + 2)}(a - 2)} \qquad \longleftarrow \text{Divide both the numerator and the denominator by } (a + 2).$$
$$= \frac{a + 3}{a - 2}$$

The expression is undefined when $a^2 - 4 = 0$.

If $a^2 - 4 = 0$, then $(a + 2)(a - 2) = 0$. By the Zero-Product Property, $a + 2 = 0$ or $a - 2 = 0$. Thus, $a = -2$ or $a = 2$.

The excluded values of the given expression are 2 and -2.

TRY THIS Simplify $\dfrac{t^2 - 9}{t^2 - t - 6}$. Identify any excluded values.

In Lesson 7.5, you learned how to factor some types of cubic polynomials. This skill is often useful in simplifying rational expressions involving cubic numerators or denominators.

EXAMPLE 2 **Simplify $\dfrac{c^3 - 10c^2 + 25c}{c^3 - 6c^2 + 5c}$. Identify any excluded values.**

▶ **Solution**

$$\frac{c^3 - 10c^2 + 25c}{c^3 - 6c^2 + 5c} = \frac{c(c^2 - 10c + 25)}{c(c^2 - 6c + 5)} \qquad \longleftarrow \text{Factor } c \text{ from both the numerator and the denominator.}$$
$$= \frac{c(c - 5)(c - 5)}{c(c - 5)(c - 1)} \qquad \longleftarrow \text{Factor } c^2 - 10c + 25 \text{ as } (c - 5)(c - 5).$$
$$\text{Factor } c^2 - 6c - 5 \text{ as } (c - 5)(c - 1).$$
$$= \frac{\cancel{c}\cancel{(c - 5)}(c - 5)}{\cancel{c}\cancel{(c - 5)}(c - 1)} \qquad \longleftarrow \text{Divide by the common factors, } c \text{ and } (c - 5).$$
$$= \frac{c - 5}{c - 1}$$

The expression is undefined when $c^3 - 6c^2 + 5c = 0$.

If $c^3 - 6c^2 + 5c = 0$, then $c(c - 5)(c - 1) = 0$.

By the Zero-Product Property, $c = 0$, $c - 5 = 0$, or $c - 1 = 0$.
The excluded values of the given expression are 0, 1, and 5.

TRY THIS Simplify $\dfrac{p^3 + 2p^2 + p}{p^3 + 8p^2 + 7p}$. Identify any excluded values.

EXERCISES

> ## KEY SKILLS

Write the numerator and the denominator in factored form. Do not simplify.

1. $\dfrac{z^2 + 2z + 1}{z^2 - 1}$

2. $\dfrac{y^2 - 25}{y^2 - 6y + 5}$

3. $\dfrac{n^2 + 8n + 15}{n^2 + 6n + 5}$

> ## PRACTICE

Simplify. Identify any excluded values.

4. $\dfrac{z - 3}{z^2 - 9}$

5. $\dfrac{z + 4}{z^2 - 16}$

6. $\dfrac{4v^2 - 25}{2v + 5}$

7. $\dfrac{9u^2 - 49}{3u - 7}$

8. $\dfrac{a^2 + 14a + 49}{a^2 - 49}$

9. $\dfrac{c^2 - 100}{c^2 - 20c + 100}$

10. $\dfrac{c^2 + 2c - 3}{c^2 - 11c + 10}$

11. $\dfrac{m^2 + 3m - 18}{m^2 - 11m + 24}$

12. $\dfrac{p^2 + p - 56}{p^2 + 13p + 40}$

13. $\dfrac{s^3 - s}{s^2 - 1}$

14. $\dfrac{t^2 - 1}{t^3 - t}$

15. $\dfrac{k^2 - 1}{k^3 - k^2}$

16. $\dfrac{x^3 - 2x^2}{x^3 - 2x}$

17. $\dfrac{y^3 - 2y^2 + y}{y^3 - 2y^2 + 3y}$

18. $\dfrac{y^3 + 4y^2 - 12y}{y^3 - 4y^2 + 4y}$

19. $\dfrac{h^3 - 2h^2 - 35h}{h^3 + 6h^2 + 5h}$

20. $\dfrac{t^3 + 11t^2 + 28t}{t^3 + 9t^2 + 14t}$

21. $\dfrac{a^3 - 20a^2 + 64a}{a^3 - 3a^2 - 4a}$

22. **Critical Thinking** Let x be any real number in the domain of $\dfrac{x^4 + 2ax^3 + a^2x^2}{x^4 - a^2x^2}$ and let a be a fixed nonzero real number. Simplify $\dfrac{x^4 + 2ax^3 + a^2x^2}{x^4 - a^2x^2}$. Identify the excluded values of the expression.

> ## MIXED REVIEW

Simplify. Write the answers with positive exponents only. Assume that all variables are nonzero. (6.5 Skill B)

23. $\dfrac{4a^3b^4}{16ab^2}$

24. $\dfrac{25d^2f^2}{10d^3f^3}$

25. $\dfrac{24r^9s^2}{12r^9s^2}$

26. $\dfrac{7x^4y^2}{3.5x^3y^3}$

27. $\dfrac{(2xy^2)^3}{8x^2y^4}$

28. $\dfrac{9a^2b^4}{(3a^2b)^3}$

29. $\dfrac{(4m^2n^3)^3}{(4m^2n^3)^2}$

30. $\dfrac{(-3r^3t^3)^3}{(-3r^3t^3)^4}$

A rational expression may contain more than one variable. To simplify such a rational expression, look for common factors of the numerator and the denominator. Then divide the numerator and the denominator by their greatest common factor (GCF).

When simplifying expressions in the remainder of this chapter, assume that variables do not represent any excluded values. That is, assume the denominator does not equal 0.

EXAMPLE 1   Simplify $\dfrac{ab}{3ab^2 + a^2b}$.

▶ **Solution**

$$\dfrac{ab}{3ab^2 + a^2b} = \dfrac{ab}{ab(3b + a)} \quad \longleftarrow \text{ Factor } ab \text{ from each term in the denominator.}$$

$$= \dfrac{{}^{1}ab}{{}_{1}ab(3b + a)} \quad \longleftarrow \text{ Divide the numerator and the denominator by their GCF, } ab.$$

$$= \dfrac{1}{3b + a}$$

TRY THIS   Simplify $\dfrac{2a^2b^3}{2a^2b^3 + 4a^2b^2}$.

EXAMPLE 2   Simplify $\dfrac{8r^3s - 4r^2s^2}{8r^3s - 20r^2s}$.

▶ **Solution**

$$\dfrac{8r^3s - 4r^2s^2}{8r^3s - 20r^2s} = \dfrac{4r^2s(2r - s)}{4r^2s(2r - 5)} \quad \longleftarrow \text{ Factor } 4r^2s \text{ from each term in the numerator and the denominator.}$$

$$= \dfrac{2r - s}{2r - 5} \quad \longleftarrow \text{ Divide the numerator and the denominator by their GCF, } 4r^2s.$$

TRY THIS   Simplify $\dfrac{-5c^2d^3 - 5cd^4}{-5c^2d^3 + 5cd^4}$.

In Example 3, you will use the Opposite of a Difference (or Multiplicative Property of -1) to factor: $(a-b) = -(b - a)$.

EXAMPLE 3   Simplify $\dfrac{g^2 - y^2}{4y^2 - 4g^2}$.

▶ **Solution**

$$\dfrac{g^2 - y^2}{4y^2 - 4g^2} = \dfrac{g^2 - y^2}{4(y^2 - g^2)}$$

$$= \dfrac{g^2 - y^2}{-4(g^2 - y^2)} \quad \longleftarrow \text{ Recognize that } (y^2 - g^2) = -(g^2 - y^2) \text{ and rewrite.}$$

$$= -\dfrac{1}{4} \quad \longleftarrow \text{ Divide by the common factor, } (g^2 - y^2).$$

TRY THIS   Simplify $\dfrac{4u^2 - a^2}{5a^2 - 20u^2}$.

EXERCISES

Identify the common factors in each rational expression.

1. $\dfrac{a^2b(2a - b)}{a^2b(2a + b)}$

2. $\dfrac{(3r - z)(4r - z)}{(2r - z)(3r - z)}$

3. $\dfrac{3(4q^2 - 81)}{-5(4q^2 - 81)}$

PRACTICE

Simplify.

4. $\dfrac{bc^2 - bc}{b}$

5. $\dfrac{s}{s^2 + sc}$

6. $\dfrac{x^2y^2 - xy}{xy}$

7. $\dfrac{w^2z^2}{w^2z^2 + wz}$

8. $\dfrac{a - b}{b - a}$

9. $\dfrac{ab - b}{ba - b}$

10. $\dfrac{pq^2 - q}{q^2 - q}$

11. $\dfrac{nb - b}{b^2 - b}$

12. $\dfrac{mn^2 - mn}{n^2 - 1}$

13. $\dfrac{p^2q^2 - pq}{pq^2 - q}$

14. $\dfrac{q^3a^2 - q^3a}{aq^3 - aq^2}$

15. $\dfrac{r^2s^2 - r^3s^2}{s^3r^2 - s^2r^2}$

16. $\dfrac{r^2s^2 - 4}{4 - r^2s^2}$

17. $\dfrac{4a^2c^2 - 9a^2}{8a^2c^2 - 18a^2}$

18. $\dfrac{6c^2 - 7ac - 5a^2}{9c^2 - 25a^2}$

19. $\dfrac{9y^2 + 30yz + 25z^2}{9y^2 - 25z^2}$

20. $\dfrac{3m^2 - mz - 2z^2}{6m^2 - 5mz - 6z^2}$

21. $\dfrac{6a^2 - 7ac - 20c^2}{3a^2 + 13ac + 12c^2}$

22. **Critical Thinking** What must be true of x and y so that
$\dfrac{4x^2 - 81y^2}{4x^2 + 36xy + 81y^2} = -\dfrac{7}{11}$?

23. **Critical Thinking** Find the excluded values of $\dfrac{ab}{3ab^2 + a^2b}$ in terms of a and b.

MIXED REVIEW

Simplify. (previous courses)

24. $\dfrac{21}{25} \cdot \dfrac{10}{14}$

25. $\dfrac{18}{35} \cdot \dfrac{49}{6}$

26. $\dfrac{100}{121} \cdot \dfrac{33}{10}$

27. $\dfrac{45}{64} \cdot \dfrac{16}{90}$

28. $\left(\dfrac{4}{5}\right)^2$

29. $\left(\dfrac{2}{3}\right)^3$

30. $\dfrac{1}{5} \cdot \dfrac{5}{18} \cdot \dfrac{6}{7}$

31. $\dfrac{14}{48} \cdot \dfrac{4}{7} \cdot \dfrac{21}{16}$

9.3 LESSON

Multiplying and Dividing Rational Expressions

Multiplying rational expressions

Use the following rule to multiply two fractions or two rational expressions.

> ### Multiplying Fractions
> If a, b, c, and d represent real numbers with $b \neq 0$ and $d \neq 0$, then
> $$\frac{a}{b} \cdot \frac{c}{d} = \frac{ac}{bd}.$$

For example, $\dfrac{6}{21} \cdot \dfrac{15}{14} = \dfrac{6 \cdot 15}{21 \cdot 14}$. To simplify the product of two fractions such as $\dfrac{6 \cdot 15}{21 \cdot 14}$, first factor the numerator and the denominator so that they only contain prime numbers. Then divide both the numerator and denominator by their common factors.

$$\frac{6}{21} \cdot \frac{15}{14} = \frac{6 \cdot 15}{21 \cdot 14} = \frac{\overset{1}{2} \cdot \overset{1}{3} \cdot 3 \cdot 5}{3 \cdot 7 \cdot \underset{1}{2} \cdot 7} = \frac{15}{49}$$

EXAMPLE 1 **Simplify $\dfrac{3x^2}{4y^3} \cdot \dfrac{16y^4}{9x^3}$. Write the answer with positive exponents.**

▶ **Solution**

$$\frac{3x^2}{4y^3} \cdot \frac{16y^4}{9x^3} = \left(\frac{3}{4} \cdot \frac{16}{9}\right)\left(\frac{x^2}{x^3}\right)\left(\frac{y^4}{y^3}\right) \quad \longleftarrow \text{Separate like terms.}$$

$$= \frac{4}{3} \cdot \frac{1}{x} \cdot \frac{y}{1} \quad \longleftarrow \text{Apply the Quotient Property of Exponents.}$$

$$= \frac{4y}{3x} \quad \longleftarrow \text{Write a single fraction.}$$

TRY THIS Simplify $\dfrac{49z^7}{81a^5} \cdot \dfrac{9a^4}{14z^6}$. Write the answer with positive exponents.

EXAMPLE 2 **Simplify $\dfrac{2x - 10}{x^2 + 6x + 9} \cdot \dfrac{4x + 12}{x^2 - 25}$.**

▶ **Solution**

$$\frac{2x - 10}{x^2 + 6x + 9} \cdot \frac{4x + 12}{x^2 - 25} = \frac{2(x - 5)}{(x + 3)(x + 3)} \cdot \frac{4(x + 3)}{(x - 5)(x + 5)}$$

$$= \frac{2(x - 5)}{(x + 3)(x + 3)} \cdot \frac{4(x + 3)}{(x - 5)(x + 5)}$$

$$= \frac{8}{(x + 3)(x + 5)}, \text{ or } \frac{8}{x^2 + 8x + 15}$$

> Note that the excluded values of the answer include all the excluded values of the original expressions: -3, -5, and $+5$.

TRY THIS Simplify $\dfrac{-3y - 21}{y^2 - 10y + 25} \cdot \dfrac{4y - 20}{y^2 - 49}$.

EXERCISES

Simplify. Write the answers with positive exponents only.

1. $\left(\frac{5}{7} \cdot \frac{14}{15}\right)\left(\frac{a^5}{a^2}\right)\left(\frac{b}{b^3}\right)$

2. $\left(\frac{3}{8} \cdot \frac{24}{1}\right)\left(\frac{r}{r^2}\right)\left(\frac{s}{s^2}\right)$

3. $\left(\frac{1}{15} \cdot \frac{25}{1}\right)\left(\frac{n^3}{n^3}\right)\left(\frac{m^4}{m^2}\right)$

4. $\frac{3x(x-5)}{(x+1)(x-5)} \cdot \frac{4(x+1)}{6x(x-1)}$

5. $\frac{(x+1)(x+1)}{(x-1)(x+1)} \cdot \frac{4(x-1)}{6x(x+1)}$

6. $\frac{3}{(x+3)(x-3)} \cdot \frac{4(x+2)}{5x(x+2)}$

PRACTICE

Simplify. Write the answers with positive exponents only.

7. $\frac{4n^3}{5m^2} \cdot \frac{55m}{12n^2}$

8. $\frac{11p^3}{7q^2} \cdot \frac{21p^3}{33q^2}$

9. $\frac{13a^3}{7k^2} \cdot \frac{21k^3}{26a^4}$

10. $\frac{25z^4}{34x^2} \cdot \frac{17x^3}{50z^4}$

11. $\frac{y^4}{32t^2} \cdot \frac{16t^2}{5y^5}$

12. $\frac{6w^4}{25t^2} \cdot \frac{50t^4}{3w^4}$

13. $\frac{3(x-1)}{x^2-2x+1} \cdot \frac{2x-2}{x-3}$

14. $\frac{3d^2-27}{d^2-6d+5} \cdot \frac{d-5}{d^2-9}$

15. $\frac{t+1}{t^2+6t+5} \cdot \frac{t-5}{t^2-25}$

16. $\frac{(c-6)(c-5)}{c^2-36} \cdot \frac{(c+5)(c+6)}{c^2-25}$

17. $\frac{(x-1)(x-7)}{x^2-49} \cdot \frac{(x+7)(x+1)}{x^2-25}$

18. $\frac{n^2+3n+2}{n^2-49} \cdot \frac{n^2-6n-7}{n^2-3n-4}$

19. $\frac{m^2+8m-9}{m^2+9m+8} \cdot \frac{m^2+2m+1}{m^2+10m+9}$

20. $\frac{z^2+11z+10}{z^2+10z+9} \cdot \frac{z^2+11z+18}{z^2+12z+20}$

21. $\frac{s^2-3s-28}{s^2+s-2} \cdot \frac{s^2+2s-3}{s^2-5s-14}$

22. $\frac{y^2-9}{y^2-1} \cdot \frac{y^2+2y+1}{y^2+6y+9} \cdot \frac{y^2+4y-5}{y^2-2y-3}$

23. $\frac{v^2+14v+49}{v^2-12+36} \cdot \frac{v^2-13v+42}{v^2+7v} \cdot \frac{v^2-36}{v^2-49}$

MIXED REVIEW

Simplify. (previous courses)

24. $\frac{18}{11} \div \frac{54}{22}$

25. $\frac{1}{3} \div \frac{3}{14}$

26. $\frac{4}{7} \div \frac{8}{14}$

27. $\frac{13}{21} \div \frac{52}{7}$

Simplify. (2.6 Skill A)

28. $\frac{1}{7}(7x+14)$

29. $\frac{1}{-2}(4x-6)$

30. $\frac{-1}{8}(-4w+16)$

31. $-\frac{1}{3}(-21t-18)$

To divide one rational number or expression by another, multiply by the reciprocal of the divisor. Recall that the reciprocal of a fraction $\frac{x}{y}$ is $\frac{y}{x}$.

Dividing Fractions

If a, b, c, and d are real numbers with $b \neq 0$, $c \neq 0$, and $d \neq 0$, then

$$\frac{a}{b} \div \frac{c}{d} = \frac{a}{b} \times \frac{d}{c}.$$

For example, $\frac{12}{25} \div \frac{4}{15} = \frac{12}{25} \times \frac{15}{4}$.

$$\frac{12}{25} \div \frac{4}{15} = \frac{12}{25} \cdot \frac{15}{4} = \frac{\overset{3}{\cancel{12}}}{\underset{5}{\cancel{25}}} \cdot \frac{\overset{3}{\cancel{15}}}{\underset{1}{\cancel{4}}} = \frac{3}{5} \cdot \frac{3}{1} = \frac{9}{5}$$

EXAMPLE 1 Simplify $\dfrac{4x^3}{15y^3} \div \dfrac{12x^4}{35y^4}$. **Write the answer with positive exponents only.**

▸ **Solution**

$$\frac{4x^3}{15y^3} \div \frac{12x^4}{35y^4} = \frac{4x^3}{15y^3} \cdot \frac{35y^4}{12x^4} \qquad \longleftarrow \text{\textit{Multiply by } } \tfrac{35y^4}{12x^4}\text{\textit{, the reciprocal of }} \tfrac{12x^4}{35y^4}.$$

$$= \left(\frac{4 \cdot 35}{15 \cdot 12}\right)\left(\frac{x^3}{x^4}\right)\left(\frac{y^4}{y^3}\right) \qquad \longleftarrow \text{\textit{Separate like terms.}}$$

$$= \frac{7}{9} \cdot \frac{1}{x} \cdot \frac{y}{1} \qquad \longleftarrow \text{\textit{Apply the Quotient Property of Exponents.}}$$

$$= \frac{7y}{9x}$$

TRY THIS Simplify $\dfrac{5t}{8s^3} \div \dfrac{5t^3}{18s^2}$. Write the answer with positive exponents only.

EXAMPLE 2 Simplify $\dfrac{n^2 + 2n - 3}{n^2 - 16} \div \dfrac{n^2 - 9}{n^2 - 8n + 16}$.

▸ **Solution**

$$\frac{n^2 + 2n - 3}{n^2 - 16} \div \frac{n^2 - 9}{n^2 - 8n + 16} = \frac{n^2 + 2n - 3}{n^2 - 16} \cdot \frac{n^2 - 8n + 16}{n^2 - 9}$$

$$= \frac{(n + 3)(n - 1)}{(n - 4)(n + 4)} \cdot \frac{(n - 4)(n - 4)}{(n + 3)(n - 3)}$$

$$= \frac{(n - 1)(n - 4)}{(n + 4)(n - 3)}, \text{ or } \frac{n^2 - 5n + 4}{n^2 + n - 12}$$

TRY THIS Simplify $\dfrac{m^2 + 2m + 1}{m^2 - 36} \div \dfrac{m^2 - 1}{m^2 - 12m + 36}$.

EXERCISES

KEY SKILLS

Write the reciprocal of each rational expression.

1. $\dfrac{2x + 3}{3x + 1}$

2. $\dfrac{x^2 - 1}{x^2 - 25}$

3. $\dfrac{x^2 + 7x + 6}{x^2 - 5x + 6}$

4. $\dfrac{x^2 + 2x + 1}{x^2 - 9x - 10}$

PRACTICE

Simplify. Write the answers with positive exponents only.

5. $\dfrac{a}{3} \div \dfrac{a}{4}$

6. $\dfrac{b}{5} \div \dfrac{b}{6}$

7. $\dfrac{x^2}{2} \div \dfrac{x^3}{6}$

8. $\dfrac{y^3}{2} \div \dfrac{y^2}{3}$

9. $\dfrac{4a^3}{5b} \div \dfrac{6a^4}{25b^2}$

10. $\dfrac{12x^3}{5n} \div \dfrac{18x^4}{35n^2}$

11. $\dfrac{4y^3}{7p^6} \div \dfrac{18y^4}{49p^2}$

12. $\dfrac{22z^2}{4a^4} \div \dfrac{33z^6}{16a^5}$

13. $\dfrac{2(x + 3)}{(x + 4)} \div \dfrac{3(x + 3)}{5(x + 4)}$

14. $\dfrac{5(y - 1)}{(y + 1)} \div \dfrac{3(y - 1)}{7(y + 1)}$

15. $\dfrac{5a - 15}{3a + 3} \div \dfrac{a - 3}{2a + 2}$

16. $\dfrac{3n - 30}{n - 1} \div \dfrac{n - 10}{n - 1}$

17. $\dfrac{5k + 15}{5k + 25} \div \dfrac{k + 3}{2k^2 + 10k}$

18. $\dfrac{5p - 5}{5p - 15} \div \dfrac{p - 1}{3p^2 - 9p}$

19. $\dfrac{9d + 9}{2d - 20} \div \dfrac{d + 1}{5d^2 - 50d}$

20. $\dfrac{5b + 15}{3b + 33} \div \dfrac{b + 3}{3b^2 + 33b}$

21. $\dfrac{c - 6}{c^2 + 7c} \div \dfrac{3c - 18}{3c + 21}$

22. $\dfrac{w - 8}{7w^2 - 56w} \div \dfrac{w - 1}{3w - 24}$

23. $\dfrac{5g^2 - 15g}{g - 2} \div \dfrac{3g - 9}{2g - 4}$

24. $\dfrac{v^2 + v}{v - 8} \div \dfrac{v + 1}{2v - 16}$

25. $\dfrac{r^2 + 5r}{r - 1} \div \dfrac{r + 5}{r - 1}$

26. $\dfrac{x^2 + 2x - 3}{x^2 - 49} \div \dfrac{x^2 - 2x + 1}{x^2 + 6x - 7}$

27. $\dfrac{x^2 + 3x - 10}{x^2 - 4x - 21} \div \dfrac{x^2 - 3x + 2}{x^2 - 2x - 15}$

28. $\dfrac{y^2 + 10y + 25}{y^2 - 6y - 7} \div \dfrac{y^2 - 25}{y^2 - 4y - 5}$

29. $\dfrac{u^2 - 8u + 15}{u^2 - 6u + 5} \div \dfrac{u^2 + 2u - 15}{u^2 + 4u - 5}$

30. $\dfrac{h^2 - 6h + 9}{h^2 + 2h - 15} \div \dfrac{h^2 - 9}{h^2 + 4h - 5}$

31. $\dfrac{n^2 - 9}{n^2 + 2n + 1} \div \dfrac{n^2 - 9}{n^2 - 1}$

32. Critical Thinking Find a such that $\dfrac{x^2 - a}{x^2 + x - 6} \times \dfrac{x^2 + 4x + 3}{x^2 + 3x + 2} = 1$.

MIXED REVIEW

Simplify. (2.6 Skill A)

33. $\dfrac{3}{4}(8a + 3) + a$

34. $3 + \dfrac{1}{6}(24a - 3)$

35. $\dfrac{1}{2}(10t - 50) + \dfrac{1}{3}(3t)$

36. $\dfrac{3}{8}(r - 8) + \dfrac{4r}{8}$

37. $\dfrac{1}{3}(a + 2) + \dfrac{1}{2}(a + 2)$

38. $\dfrac{5}{7}(2r - 2) + \dfrac{1}{3}(2r - 2)$

In Skill B of this lesson, you learned to simplify a quotient of two rational expressions. You rewrote the division as multiplication by the reciprocal of the divisor. This technique still applies when you are simplifying rational expressions that involve both multiplication and division.

EXAMPLE 1 **Simplify** $\dfrac{2ab}{6a^2b} \cdot \dfrac{18ab^2}{12ab} \div \dfrac{a^3b}{2ab^3}.$

▶ **Solution**

$$\dfrac{2ab}{6a^2b} \cdot \dfrac{18ab^2}{12ab} \div \dfrac{a^3b}{2ab^3} = \dfrac{2ab}{6a^2b} \cdot \dfrac{18ab^2}{12ab} \cdot \dfrac{2ab^3}{a^3b}$$ ← *Change division to multiplication by the reciprocal.*

$$= \dfrac{2 \cdot 18 \cdot 2}{6 \cdot 12 \cdot 1} \cdot \dfrac{a \cdot a \cdot a}{a^2 \cdot a \cdot a^3} \cdot \dfrac{b \cdot b^2 \cdot b^3}{b \cdot b \cdot b}$$ ← *Separate like terms.*

$$= 1 \cdot \dfrac{a^3}{a^6} \cdot \dfrac{b^6}{b^3}$$

$$= 1 \cdot \dfrac{1}{a^3} \cdot \dfrac{b^3}{1}$$ ← *Apply the Quotient Property of Exponents.*

$$= \dfrac{b^3}{a^3}$$

TRY THIS Simplify $\dfrac{1}{6rs} \cdot \dfrac{6r^2s^2}{18rs^2} \div (r^4s^4).$

Recall that the reciprocal of 4 is $\dfrac{1}{4}$. Similarly, the reciprocal of $x - 1$ is $\dfrac{1}{x - 1}$. This fact is important when you divide by a polynomial, as in the example below.

EXAMPLE 2 **Simplify** $\dfrac{x^2 - 3x + 2}{x^2 - 9} \div \dfrac{x^2 - 4x + 4}{x^2 - 6x + 9} \div (x - 1).$

▶ **Solution**

$$\dfrac{x^2 - 3x + 2}{x^2 - 9} \div \dfrac{x^2 - 4x + 4}{x^2 - 6x + 9} \div (x - 1)$$

$$\dfrac{x^2 - 3x + 2}{x^2 - 9} \cdot \dfrac{x^2 - 6x + 9}{x^2 - 4x + 4} \cdot \dfrac{1}{x - 1}$$ ← *Change division to multiplication by the reciprocal.*

$$\dfrac{(x - 1)(x - 2)}{(x - 3)(x + 3)} \cdot \dfrac{(x - 3)(x - 3)}{(x - 2)(x - 2)} \cdot \dfrac{1}{x - 1}$$ ← *Factor and divide by common factors.*

$$\dfrac{x - 3}{(x + 3)(x - 2)}, \text{ or } \dfrac{x - 3}{x^2 + x - 6}$$

TRY THIS Simplify $\dfrac{x^2 + 4x + 4}{x^2 - 16} \div (x + 2) \div (x + 2).$

EXERCISES

Write each expression as a product.

1. $\dfrac{2m^2n}{5mn^2} \cdot \dfrac{15mn^2}{12m^2n^3} \div \dfrac{5m^3n}{2mn}$

2. $\dfrac{3x^2y}{6xy^2} \div \dfrac{15xy^2}{12x^2y^3} \div \dfrac{x^3y}{2xy}$

3. $\dfrac{x^2 - 5x + 6}{x^2 - 9} \div \dfrac{x^2 - 4x + 4}{x^2 - 6x + 9} \div (x - 2)$

4. $\dfrac{x^2 - 5x + 6}{x^2 - 9} \cdot \dfrac{x^2 - 6x + 5}{x^2 - 25} \div \dfrac{x - 2}{x - 5}$

PRACTICE

Simplify. Write the answers with positive exponents only.

5. $\dfrac{a}{4} \div \dfrac{a}{8} \cdot \dfrac{a^2}{2}$

6. $\dfrac{4n}{n} \div \dfrac{n}{4} \div \dfrac{n^2}{4}$

7. $\dfrac{5}{w} \cdot \dfrac{2}{w} \div \dfrac{15}{w}$

8. $\dfrac{c^2}{-3} \div \dfrac{c^3}{9} \div \dfrac{c}{12}$

9. $\dfrac{x^3y}{x^2y^2} \cdot \dfrac{5x^2y^2}{2x^2y^3} \div \dfrac{5x^3y^2}{2}$

10. $\dfrac{ab}{a^2b^2} \cdot \dfrac{5a^2b^2}{2a^3b^3} \div \dfrac{5ab}{10a^3b^3}$

11. $\dfrac{d^2c^2}{dc} \div \dfrac{7d^2c^2}{2dc} \div \dfrac{7dc}{10d^3c^3}$

12. $\dfrac{dc^3}{dc} \div \dfrac{d^2}{2c} \div \dfrac{dc}{10d^2c^3}$

13. $\dfrac{x^2y^3}{x^2y} \div \dfrac{xy^2}{2xy} \cdot \dfrac{xy}{12x^2y^3}$

14. $\dfrac{r^2s^3}{r^2s} \div \dfrac{r^2s^3}{r^2s} \cdot \dfrac{rs}{r^2s^3}$

15. $\dfrac{(ab)^2}{a^3b^2} \cdot \dfrac{a^2b^3}{(ab)^3} \cdot \dfrac{ab}{b}$

16. $\dfrac{(mn)^5}{m^2n^3} \cdot \dfrac{m^3n^4}{(mn)^4} \cdot \dfrac{(mn)^2}{mn}$

17. $\dfrac{(pq)^2}{p^2q^3} \div \dfrac{p^3q^4}{(pq)^3} \div (p^4q^4)$

18. $\dfrac{x^2 - 2x - 15}{x^2 - 9} \div \dfrac{x^2 - 25}{x^2 - 10x + 21} \div (x - 7)$

19. $\dfrac{x^2 - x - 2}{x^2 - 1} \div \dfrac{x^2 - 7x + 10}{x^2 - 2x + 1} \div (x - 1)$

20. $\dfrac{v^2 - 3v + 2}{v^2 - v - 2} \cdot \dfrac{v^2 - 2v + 1}{v^2 - 1} \div \dfrac{v + 1}{v - 1}$

21. $\dfrac{p^2 - 5p - 14}{p^2 - 3p - 10} \cdot \dfrac{p^2 - 6p + 5}{p^2 - 2p + 1} \div \dfrac{p - 7}{p - 1}$

22. $\dfrac{c^2 - 9}{c^2 + 3c - 10} \div \dfrac{c^2 + c - 6}{c^2 - 7c + 10} \cdot \dfrac{c - 2}{c - 3}$

23. $\dfrac{a^2 - 7a - 30}{a^2 - 7a + 10} \div \dfrac{a^2 + a - 6}{a^2 + 3a - 10} \cdot \dfrac{a - 2}{a - 10}$

24. $\dfrac{(a + b)^3}{(r + s)^3} \cdot \dfrac{(r + s)^2}{(a + b)^2} \div \dfrac{a + b}{r + s}$

25. $\dfrac{(m + n)^2}{(x + y)^2} \cdot \dfrac{(x + y)^3}{(m + n)^3} \div \dfrac{m + n}{x + y}$

MIXED REVIEW

Simplify. (previous courses)

26. $\dfrac{7}{11} + \dfrac{19}{22}$

27. $\dfrac{1}{35} + \dfrac{6}{7}$

28. $\dfrac{5}{18} + \dfrac{5}{24}$

29. $\dfrac{5}{14} + \dfrac{6}{49}$

30. $\dfrac{18}{25} + \dfrac{7}{15}$

31. $\dfrac{17}{24} + \dfrac{5}{6}$

32. $\dfrac{9}{14} + \dfrac{12}{21}$

33. $\dfrac{17}{20} + \dfrac{7}{10}$

9.4 Adding Rational Expressions

LESSON

SKILL A *Adding rational expressions with like denominators*

To add two fractions, first determine whether the denominators are like or unlike. If the two fractions have the same denominator, then you can use the rule below to add the fractions.

Adding Fractions with Like Denominators

Let a, b, and c be real numbers with $c \neq 0$. Then $\dfrac{a}{c} + \dfrac{b}{c} = \dfrac{a+b}{c}$.

EXAMPLE 1 **Simplify.** **a.** $\dfrac{2}{9x} + \dfrac{4}{9x}$ **b.** $\dfrac{z-5}{2} + \dfrac{z+3}{2}$

▶ **Solution**

Write the sum of the numerators over the like denominator. Then simplify further, if necessary.

a. $\dfrac{2}{9x} + \dfrac{4}{9x} = \dfrac{6}{9x}$

$= \dfrac{2}{3x}$

b. $\dfrac{z-5}{2} + \dfrac{z+3}{2} = \dfrac{(z-5)+(z+3)}{2}$

$= \dfrac{2z-2}{2}$

$= \dfrac{2(z-1)}{2}$

$= z - 1$

TRY THIS Simplify. **a.** $\dfrac{2}{15n} + \dfrac{-7}{15n}$ **b.** $\dfrac{n-18}{-3} + \dfrac{2n+3}{-3}$

The rule for addition applies whether the denominators are numbers or polynomials.

EXAMPLE 2 **Simplify** $\dfrac{x^2 - 3x}{x-5} + \dfrac{x-15}{x-5}$.

▶ **Solution**

$\dfrac{x^2-3x}{x-5} + \dfrac{x-15}{x-5} = \dfrac{(x^2-3x)+(x-15)}{x-5}$ ⟵——— Add the numerators.

$= \dfrac{x^2-2x-15}{x-5}$ ⟵——— Combine like terms.

$= \dfrac{(x-5)(x+3)}{x-5}$ ⟵——— Factor the numerator.

$= \dfrac{(x-5)(x+3)}{x-5}$ ⟵——— Divide both the numerator and the denominator by $(x-5)$.

$= x + 3$

TRY THIS Simplify $\dfrac{a^2}{a+9} + \dfrac{-81}{a+9}$.

356 Chapter 9 Rational Expressions, Equations, and Functions

EXERCISES

Write the numerator of each sum in simplified form. Do not simplify further.

1. $\dfrac{3n - 5}{3n^2} + \dfrac{2n - 1}{3n^2}$

2. $\dfrac{5z^2}{5(z + 1)} + \dfrac{-5z^2 + 3z}{5(z + 1)}$

3. $\dfrac{5a^2 - 3a}{-3(a - 4)} + \dfrac{a^2 + 3a}{-3(a - 4)}$

Simplify.

4. $\dfrac{1}{2a} + \dfrac{5}{2a}$

5. $\dfrac{1}{9n} + \dfrac{2}{9n}$

6. $\dfrac{-5}{14k} + \dfrac{-3}{14k}$

7. $\dfrac{-11}{27k} + \dfrac{2}{27k}$

8. $\dfrac{13}{22b^2} + \dfrac{-2}{22b^2}$

9. $\dfrac{72}{25c^2} + \dfrac{-22}{25c^2}$

10. $\dfrac{w - 5}{3} + \dfrac{5w - 4}{3}$

11. $\dfrac{3n - 12}{-5} + \dfrac{7n - 3}{-5}$

12. $\dfrac{z - 12}{3z} + \dfrac{5z + 12}{3z}$

13. $\dfrac{t + 2}{6t} + \dfrac{11t - 2}{6t}$

14. $\dfrac{v + 2}{2v^2} + \dfrac{v - 2}{2v^2}$

15. $\dfrac{4u - 5}{-3u^2} + \dfrac{8u + 5}{-3u^2}$

16. $\dfrac{5y + 5}{y + 3} + \dfrac{-3y + 1}{y + 3}$

17. $\dfrac{5x + 53}{x + 6} + \dfrac{4x + 1}{x + 6}$

18. $\dfrac{5m + 19}{2m + 5} + \dfrac{5m + 6}{2m + 5}$

19. $\dfrac{q - 9}{2q - 3} + \dfrac{5q}{2q - 3}$

20. $\dfrac{a}{2a - 3} + \dfrac{9a - 15}{2a - 3}$

21. $\dfrac{b + 3}{-2b + 7} + \dfrac{-3b + 4}{-2b + 7}$

22. $\dfrac{2x + 3}{x^2 - 25} + \dfrac{-x + 2}{x^2 - 25}$

23. $\dfrac{3y - 3}{4y^2 - 36} + \dfrac{-y - 3}{4y^2 - 36}$

24. $\dfrac{2y - 1}{y^2 + 4y + 4} + \dfrac{-y + 3}{y^2 + 4y + 4}$

25. $\dfrac{5y + 1}{y^2 - 6y + 9} + \dfrac{-4y - 4}{y^2 - 6y + 9}$

26. $\dfrac{x^2 + 5x}{x + 3} + \dfrac{5x + 21}{x + 3}$

27. $\dfrac{x^2 + x}{x - 4} + \dfrac{x - 24}{x - 4}$

28. $\dfrac{5x^2 - 5x - 2}{2x - 3} + \dfrac{x^2 - 2x - 1}{2x - 3}$

29. $\dfrac{7a^2 + 10a - 2}{3a - 1} + \dfrac{2a^2 - a - 2}{3a - 1}$

30. $\dfrac{10v^2 - 5v - 6}{4v - 3} + \dfrac{10v^2 - 2v}{4v - 3}$

31. Critical Thinking Find m and n such that $\dfrac{mr + ms}{2x - 1} + \dfrac{nr - ns}{2x - 1} = \dfrac{r - 2s}{2x - 1}$.

Find each difference. (6.4 Skill B)

32. $(x^2 - 3x + 1) - (4x + 1)$

33. $(a^2 - 3a) - (-a^2 - 4a + 1)$

34. $(n^2 - 5n) - (4n^2 - n)$

35. $(b^2 - 4b) - (4b^3 + 2b)$

36. $(x^4 - x^2) - (x^3 - x)$

37. $n^4 - (n^3 - n^2 - n + 1)$

To add two fractions or rational expressions with unlike denominators, write new equivalent fractions or rational expressions that have the same denominator, and then add.

The **least common multiple** (LCM) of two numbers is the smallest number that is a multiple of both numbers. The **least common denominator** (LCD) of two fractions is the least common multiple of their denominators.

$$\frac{1}{3} + \frac{4}{5} = \frac{1}{3}\left(\frac{5}{5}\right) + \frac{4}{5}\left(\frac{3}{3}\right) = \frac{5}{15} + \frac{12}{15} = \frac{17}{15}, \text{ or } 1\frac{2}{15}$$

In the example above, 15 is the least common multiple of 3 and 5, and thus 15 is the least common denominator of $\frac{1}{3}$ and $\frac{4}{5}$.

EXAMPLE 1 **Simplify.**   **a.** $\dfrac{3}{a} + \dfrac{4}{a^2}$ **b.** $\dfrac{x-1}{2} + \dfrac{3}{x}$

▶ **Solution**

a. $\dfrac{3}{a} + \dfrac{4}{a^2} = \dfrac{3}{a} \cdot \dfrac{a}{a} + \dfrac{4}{a^2}$

$= \dfrac{3a}{a^2} + \dfrac{4}{a^2}$

$= \dfrac{3a + 4}{a^2}$

b. $\dfrac{x-1}{2} + \dfrac{3}{x} = \dfrac{x-1}{2} \cdot \dfrac{x}{x} + \dfrac{3}{x} \cdot \dfrac{2}{2}$

$= \dfrac{x(x-1) + 3 \cdot 2}{2x}$

$= \dfrac{x^2 - x + 6}{2x}$

TRY THIS   Simplify.   **a.** $\dfrac{2}{b^2} + \dfrac{-5}{b^3}$ **b.** $\dfrac{2n-1}{3} + \dfrac{3}{2n}$

EXAMPLE 2 **Simplify** $\dfrac{5y+1}{3y} + \dfrac{2}{y^2} + \dfrac{1}{2y^2}.$

▶ **Solution**

$\dfrac{5y+1}{3y} + \dfrac{2}{y^2} + \dfrac{1}{2y^2} = \dfrac{5y+1}{3y} \cdot \dfrac{2y}{2y} + \dfrac{2}{y^2} \cdot \dfrac{6}{6} + \dfrac{1}{2y^2} \cdot \dfrac{3}{3}$ ⟵ *The LCD of 3y, y^2, and $2y^2$ is $6y^2$.*

$= \dfrac{2y(5y+1) + 2(6) + 1(3)}{6y^2}$

$= \dfrac{10y^2 + 2y + 12 + 3}{6y^2}$

$= \dfrac{10y^2 + 2y + 15}{6y^2}$

TRY THIS   Simplify $\dfrac{5p+1}{3p} + \dfrac{2}{p} + \dfrac{1}{2p^2}.$

EXERCISES

KEY SKILLS

Write the least common denominator for each group of rational expressions.

1. $\dfrac{2z+7}{4z}$ and $\dfrac{z+3}{5z^2}$

2. $\dfrac{3a+1}{3a}, \dfrac{2a}{4}$, and $\dfrac{a-1}{a^2}$

3. $\dfrac{3n-5}{3n^2}, \dfrac{2n-1}{2n}$, and $\dfrac{2n-1}{5n^2}$

PRACTICE

Simplify.

4. $\dfrac{3}{2a} + \dfrac{5}{4a}$

5. $\dfrac{5}{2z} + \dfrac{-7}{8z}$

6. $\dfrac{-1}{3z} + \dfrac{3}{7z}$

7. $\dfrac{-2}{3b} + \dfrac{-5}{4b}$

8. $\dfrac{4}{5d^2} + \dfrac{1}{6d^2}$

9. $\dfrac{7}{6w^2} + \dfrac{5}{7w^2}$

10. $\dfrac{2}{3} + \dfrac{3a-2}{a}$

11. $\dfrac{2}{5} + \dfrac{2t+1}{t}$

12. $\dfrac{4}{5z} + \dfrac{2z-3}{z}$

13. $\dfrac{1}{2n} + \dfrac{n+2}{3n}$

14. $\dfrac{s}{2} + \dfrac{5s+1}{5s}$

15. $\dfrac{b}{3} + \dfrac{b+1}{5b}$

16. $\dfrac{d+1}{2} + \dfrac{d+1}{d}$

17. $\dfrac{m+1}{3} + \dfrac{m-1}{5m}$

18. $\dfrac{w+3}{3w} + \dfrac{w-2}{w^2}$

19. $\dfrac{3v+2}{3v^2} + \dfrac{v-2}{v}$

20. $\dfrac{p+2}{2p} + \dfrac{p-1}{p} + \dfrac{1}{p^2}$

21. $\dfrac{r+2}{5r^2} + \dfrac{2r-3}{r} + \dfrac{1}{r}$

22. $\dfrac{b}{2} + \dfrac{b+1}{b} + \dfrac{b+2}{b^2}$

23. $\dfrac{2}{a^3} + \dfrac{2a+1}{a} + \dfrac{2a+3}{a^2}$

24. $3 + \dfrac{2a+1}{a} + \dfrac{2a+3}{a^2}$

25. $7 + \dfrac{4z-5}{z} + \dfrac{4z-3}{z^2}$

26. $\dfrac{1}{x} + \dfrac{4x-5}{x^2} + \dfrac{x+7}{x^3}$

27. $\dfrac{u-1}{u} + \dfrac{u+1}{u^2} + \dfrac{u+3}{u^3}$

28. $\dfrac{1}{a} + \dfrac{1}{a^2} + \dfrac{1}{a^3} + \dfrac{1}{a^4}$

29. $\dfrac{1}{(3s)^2} + \dfrac{2}{s^2} + \dfrac{3}{(3s)^3}$

30. $\dfrac{5}{(-2r)^2} + \dfrac{3}{r^2} + \dfrac{1}{(-2r)^3}$

31. Critical Thinking   Let n represent a natural number.
Find a formula for $\dfrac{1}{x} + \dfrac{1}{x^2} + \dfrac{1}{x^3} + \cdots + \dfrac{1}{x^n}$.

MIXED REVIEW

Solve each inequality. (4.5 Skill B)

32. $-4(r+5) - 6 < 0$

33. $6 - 4(s-1) \geq 3(s+1)$

34. $2(t-5) - 6(t-5) < 0$

35. $\dfrac{1}{3}x - 2x + 1 \leq \dfrac{1}{3}(3-x)$

36. $5 - 3(d+3) - 3 \geq \dfrac{2}{5}d$

37. $\dfrac{3}{7}(y-4) - \dfrac{4}{7}(y-4) > 2$

To add $\frac{2}{m}$ and $\frac{1}{m+3}$, you need to find a common denominator for the pair of rational expressions.

EXAMPLE 1

Simplify $\frac{2}{m} + \frac{1}{m+3}$.

▶ **Solution**

The least common denominator (LCD) of $\frac{2}{m}$ and $\frac{1}{m+3}$ is $m(m+3)$.

Multiply each expression by the equivalent of 1 that makes the two denominators the same.

$$\frac{2}{m} + \frac{1}{m+3} = \frac{2}{m} \cdot \frac{(m+3)}{(m+3)} + \frac{1}{(m+3)} \cdot \frac{m}{m}$$

$$= \frac{2(m+3)}{m(m+3)} + \frac{m}{m(m+3)} \qquad \longleftarrow \text{The denominators are now the same.}$$

$$= \frac{2(m+3)+m}{m(m+3)} \qquad \longleftarrow \text{Add the numerators.}$$

$$= \frac{3m+6}{m^2+3m} \qquad \longleftarrow \text{Simplify.}$$

TRY THIS Simplify $\frac{1}{b-1} + \frac{5}{b+1}$.

You may need to factor each denominator before looking for the LCD. In Example 2, factor $b^2 + b$ as $b(b+1)$ and $b^2 - b$ as $b(b-1)$.

EXAMPLE 2

Simplify $\frac{1}{b^2+b} + \frac{1}{b^2-b}$.

▶ **Solution**

$$\frac{1}{b^2+b} + \frac{1}{b^2-b}$$

$$\frac{1}{b(b+1)} + \frac{1}{b(b-1)} \qquad \longleftarrow \text{Factor each denominator.}$$

$$\frac{1}{b(b+1)} \cdot \frac{b-1}{b-1} + \frac{1}{b(b-1)} \cdot \frac{b+1}{b+1} \qquad \longleftarrow \frac{b-1}{b-1} = 1 \text{ and } \frac{b+1}{b+1} = 1$$

$$\frac{(b-1)+(b+1)}{b(b-1)(b+1)} \qquad \longleftarrow \text{Write one denominator: } b(b-1)(b+1)$$

$$\frac{2b}{b(b-1)(b+1)} \qquad \longleftarrow \text{Add the numerators: } (b-1)+(b+1)$$

$$\frac{2}{(b-1)(b+1)}, \text{ or } \frac{2}{b^2-1}$$

TRY THIS Simplify $\frac{2}{2c^2+c} + \frac{2}{2c^2-c}$.

EXERCISES

Write the denominators in factored form. Then find the LCD. Do not simplify further.

1. $\dfrac{1}{z^2 + 3z} + \dfrac{1}{z + 3}$

2. $\dfrac{6a}{a^2 - 25} + \dfrac{5a}{2a - 10}$

3. $\dfrac{6k + 1}{k^2 - 1} + \dfrac{5k + 3}{k^2 - k}$

Simplify.

4. $\dfrac{1}{n} + \dfrac{1}{n + 1}$

5. $\dfrac{2}{z} + \dfrac{1}{z - 1}$

6. $\dfrac{3}{2k + 1} + \dfrac{5}{k}$

7. $\dfrac{5}{3y - 5} + \dfrac{3}{y}$

8. $\dfrac{2}{x^2} + \dfrac{1}{x - 2}$

9. $\dfrac{5}{r - 5} + \dfrac{5}{r^2}$

10. $\dfrac{1}{b + 1} + \dfrac{1}{b - 1}$

11. $\dfrac{2}{d + 1} + \dfrac{3}{d + 2}$

12. $\dfrac{5}{d - 1} + \dfrac{2}{d - 3}$

13. $\dfrac{5}{q + 2} + \dfrac{2}{q + 5}$

14. $\dfrac{5s + 1}{s + 1} + \dfrac{2s - 3}{s + 2}$

15. $\dfrac{2t - 1}{t + 3} + \dfrac{2t + 1}{t + 2}$

16. $\dfrac{3v - 1}{v - 3} + \dfrac{3v + 1}{v + 3}$

17. $\dfrac{2u + 1}{u - 1} + \dfrac{2u + 1}{u + 1}$

18. $\dfrac{2a}{a^2 - 1} + \dfrac{2a}{a + 1}$

19. $\dfrac{3g}{g^2 - 4} + \dfrac{g}{g - 2}$

20. $\dfrac{c}{c^2 - 1} + \dfrac{c - 2}{c^2 - c}$

21. $\dfrac{z}{z^2 + 2z} + \dfrac{5z}{z^2 - 2z}$

22. $\dfrac{2}{h^2 + 4h + 4} + \dfrac{5}{h^2 + 2h}$

23. $\dfrac{3}{b^2 + 4b + 4} + \dfrac{2}{b^2 - 4}$

24. $\dfrac{1}{b^2 + 2b + 1} + \dfrac{1}{b^2 + b}$

25. $\dfrac{1}{a + 1} + \dfrac{1}{(a + 1)^2} + \dfrac{1}{(a + 1)^3}$

26. $\dfrac{1}{n - 1} + \dfrac{2}{(n - 1)^2} + \dfrac{3}{(n - 1)^3}$

27. $\dfrac{3}{m} + \dfrac{2}{m + 1} + \dfrac{1}{(m + 1)^2}$

28. Critical Thinking Find A such that $\dfrac{7x - 19}{x^2 - 5x + 4} = \dfrac{A}{x - 4} + \dfrac{4}{x - 1}$.

Solve each system of equations. (5.3 Skill A)

29. $\begin{cases} 2x + 7y = 0 \\ -2x + 6y = 0 \end{cases}$

30. $\begin{cases} 2a + 7b = 3 \\ 2a - 7b = 4 \end{cases}$

31. $\begin{cases} -3m + 2n = 6 \\ 3m + 2n = 4 \end{cases}$

32. $\begin{cases} 3u - 5v = 2 \\ 3u + 5v = 3 \end{cases}$

33. $\begin{cases} -3b - 4d = -1 \\ 3b + 5d = -1 \end{cases}$

34. $\begin{cases} 11w - 6y = 35 \\ 10w + 6y = -14 \end{cases}$

9.5
LESSON

Subtracting Rational Expressions

SKILL A *Subtracting rational expressions with like denominators*

When you subtract $\dfrac{b}{c}$ from $\dfrac{a}{c}$, you subtract b from a and write the difference over c.

$$\frac{a}{c} - \frac{b}{c} = \frac{a}{c} + \frac{-b}{c} = \frac{a + (-b)}{c} = \frac{a - b}{c}$$

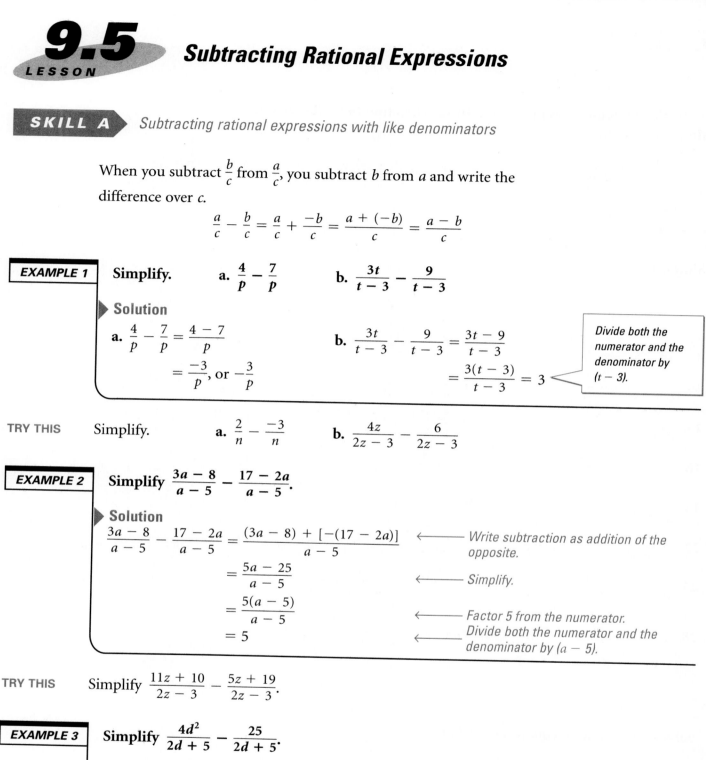

EXAMPLE 1   **Simplify.**   **a.** $\dfrac{4}{p} - \dfrac{7}{p}$ **b.** $\dfrac{3t}{t-3} - \dfrac{9}{t-3}$

▶ **Solution**

a. $\dfrac{4}{p} - \dfrac{7}{p} = \dfrac{4-7}{p}$

$= \dfrac{-3}{p}$, or $-\dfrac{3}{p}$

b. $\dfrac{3t}{t-3} - \dfrac{9}{t-3} = \dfrac{3t-9}{t-3}$

$= \dfrac{3(t-3)}{t-3} = 3$

> *Divide both the numerator and the denominator by (t − 3).*

TRY THIS   **Simplify.**   **a.** $\dfrac{2}{n} - \dfrac{-3}{n}$ **b.** $\dfrac{4z}{2z-3} - \dfrac{6}{2z-3}$

EXAMPLE 2   **Simplify** $\dfrac{3a-8}{a-5} - \dfrac{17-2a}{a-5}.$

▶ **Solution**

$\dfrac{3a-8}{a-5} - \dfrac{17-2a}{a-5} = \dfrac{(3a-8) + [-(17-2a)]}{a-5}$ ⟵ *Write subtraction as addition of the opposite.*

$= \dfrac{5a-25}{a-5}$ ⟵ *Simplify.*

$= \dfrac{5(a-5)}{a-5}$ ⟵ *Factor 5 from the numerator.*

$= 5$ ⟵ *Divide both the numerator and the denominator by (a − 5).*

TRY THIS   **Simplify** $\dfrac{11z+10}{2z-3} - \dfrac{5z+19}{2z-3}.$

EXAMPLE 3   **Simplify** $\dfrac{4d^2}{2d+5} - \dfrac{25}{2d+5}.$

▶ **Solution**

$\dfrac{4d^2}{2d+5} - \dfrac{25}{2d+5} = \dfrac{4d^2-25}{2d+5}$

$= \dfrac{(2d+5)(2d-5)}{2d+5}$ ⟵ *Factor $4d^2 - 25$ as (2d + 5)(2d − 5).*

$= 2d-5$

TRY THIS   **Simplify** $\dfrac{9t^2+49}{3t-7} - \dfrac{42t}{3t-7}.$

EXERCISES

KEY SKILLS

Write each subtraction as addition of the opposite. Do not simplify further.

1. $\dfrac{3t - 1}{t} - \dfrac{2t - 2}{t}$

2. $\dfrac{3y^2 - 1}{y + 5} - \dfrac{3y^2 + 2}{y + 5}$

3. $\dfrac{3v^2 - 3}{v - 7} - \dfrac{2v^2 + 3}{v - 7}$

PRACTICE

Simplify.

4. $\dfrac{5}{s} - \dfrac{7}{s}$

5. $\dfrac{5}{2v} - \dfrac{3}{2v}$

6. $\dfrac{2t}{t - 7} - \dfrac{14}{t - 7}$

7. $\dfrac{a}{7a + 7} - \dfrac{-1}{7a + 7}$

8. $\dfrac{4r + 1}{2r + 1} - \dfrac{-1}{2r + 1}$

9. $\dfrac{9w + 1}{3w - 1} - \dfrac{4}{3w - 1}$

10. $\dfrac{5b - 3}{b + 2} - \dfrac{b - 11}{b + 2}$

11. $\dfrac{6n - 18}{n - 2} - \dfrac{n - 8}{n - 2}$

12. $\dfrac{11x - 2}{3x - 1} - \dfrac{2x + 1}{3x - 1}$

13. $\dfrac{11r + 2}{5r + 2} - \dfrac{6r}{5r + 2}$

14. $\dfrac{6n^2 + 3n}{2n - 1} - \dfrac{6n^2 + 2n}{2n - 1}$

15. $\dfrac{2s^2}{3s - 2} - \dfrac{2s^2 - 4s + 1}{3s - 2}$

16. $\dfrac{10d^2 - 6}{d + 1} - \dfrac{2d^2 + 2}{d + 1}$

17. $\dfrac{14p^2 - 200}{2p + 7} - \dfrac{2p^2 + 94}{2p + 7}$

18. $\dfrac{5q^2 - 2q + 1}{q^2 + 6q + 9} - \dfrac{5q^2 - q + 4}{q^2 + 6q + 9}$

19. $\dfrac{4z^2 + 6z + 1}{4z^2 + 4z + 1} - \dfrac{4z^2 - 2z - 3}{4z^2 + 4z + 1}$

20. $\dfrac{6v^2 - v}{9v^2 - 25} - \dfrac{15}{9v^2 - 25}$

21. $\dfrac{5n^2 + n}{4n^2 - 9} - \dfrac{n^2 - 7n - 3}{4n^2 - 9}$

22. **Critical Thinking** **a.** Let a and b represent nonzero real numbers. Assume that $a \neq b$. Prove that $\dfrac{b}{a - b} + 1 = \dfrac{a}{a - b}$. Justify each step.

   b. Simplify $\dfrac{rx + s}{ax + b} - \dfrac{ux + v}{ax + b}$. Assume all of the variables represent positive real numbers.

   c. Find r and s such that $\dfrac{rx + s}{3x - 5} - \dfrac{2x + 7}{3x - 5} = 1$ for all values of x.

   d. Find a and b such that $\dfrac{ax - b}{2x + 1} - \dfrac{(b - 1)x + 2a}{2x + 1} = \dfrac{2x - 5}{2x + 1}$ for all values of x.

MIXED REVIEW

Simplify. (2.3 Skill A)

23. $\dfrac{5}{6} - \dfrac{1}{3}$

24. $\dfrac{9}{14} - \dfrac{1}{2}$

25. $\dfrac{5}{7} - \dfrac{3}{5}$

26. $\dfrac{5}{11} - \dfrac{13}{33}$

27. $\dfrac{1}{3} - \dfrac{11}{18}$

28. $\dfrac{2}{5} - \dfrac{6}{7}$

29. $\dfrac{3}{11} - \dfrac{15}{33}$

30. $\dfrac{2}{7} - \dfrac{5}{9}$

You subtract rational expressions with unlike denominators in the same way that you subtract fractions with unlike denominators.

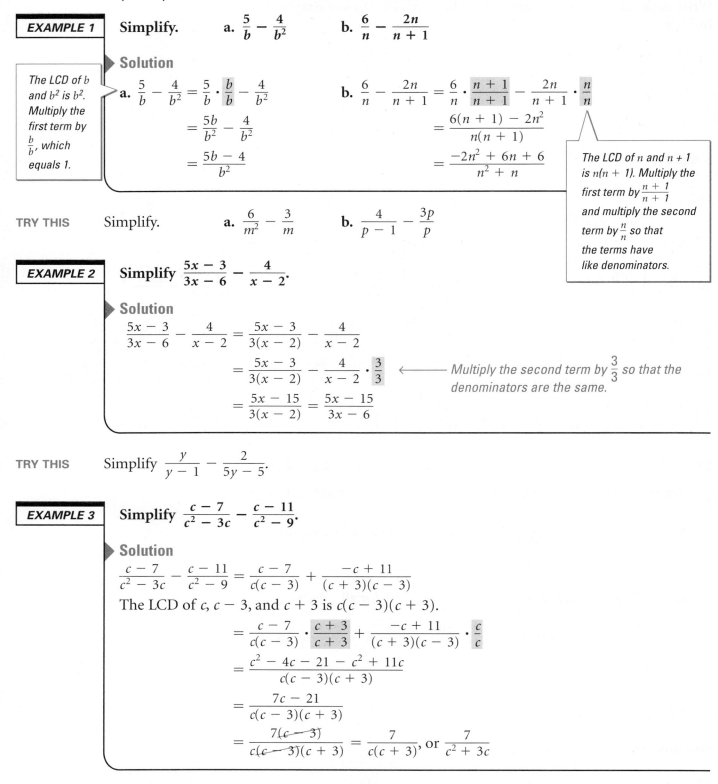

EXAMPLE 1 **Simplify.** **a.** $\dfrac{5}{b} - \dfrac{4}{b^2}$ **b.** $\dfrac{6}{n} - \dfrac{2n}{n+1}$

▶ **Solution**

The LCD of b and b^2 is b^2. Multiply the first term by $\dfrac{b}{b}$, which equals 1.

a. $\dfrac{5}{b} - \dfrac{4}{b^2} = \dfrac{5}{b} \cdot \dfrac{b}{b} - \dfrac{4}{b^2}$

$= \dfrac{5b}{b^2} - \dfrac{4}{b^2}$

$= \dfrac{5b - 4}{b^2}$

b. $\dfrac{6}{n} - \dfrac{2n}{n+1} = \dfrac{6}{n} \cdot \dfrac{n+1}{n+1} - \dfrac{2n}{n+1} \cdot \dfrac{n}{n}$

$= \dfrac{6(n+1) - 2n^2}{n(n+1)}$

$= \dfrac{-2n^2 + 6n + 6}{n^2 + n}$

The LCD of n and $n+1$ is $n(n+1)$. Multiply the first term by $\dfrac{n+1}{n+1}$ and multiply the second term by $\dfrac{n}{n}$ so that the terms have like denominators.

TRY THIS Simplify. **a.** $\dfrac{6}{m^2} - \dfrac{3}{m}$ **b.** $\dfrac{4}{p-1} - \dfrac{3p}{p}$

EXAMPLE 2 **Simplify** $\dfrac{5x-3}{3x-6} - \dfrac{4}{x-2}.$

▶ **Solution**

$\dfrac{5x-3}{3x-6} - \dfrac{4}{x-2} = \dfrac{5x-3}{3(x-2)} - \dfrac{4}{x-2}$

$= \dfrac{5x-3}{3(x-2)} - \dfrac{4}{x-2} \cdot \dfrac{3}{3}$ ← *Multiply the second term by $\dfrac{3}{3}$ so that the denominators are the same.*

$= \dfrac{5x-15}{3(x-2)} = \dfrac{5x-15}{3x-6}$

TRY THIS Simplify $\dfrac{y}{y-1} - \dfrac{2}{5y-5}.$

EXAMPLE 3 **Simplify** $\dfrac{c-7}{c^2-3c} - \dfrac{c-11}{c^2-9}.$

▶ **Solution**

$\dfrac{c-7}{c^2-3c} - \dfrac{c-11}{c^2-9} = \dfrac{c-7}{c(c-3)} + \dfrac{-c+11}{(c+3)(c-3)}$

The LCD of c, $c-3$, and $c+3$ is $c(c-3)(c+3)$.

$= \dfrac{c-7}{c(c-3)} \cdot \dfrac{c+3}{c+3} + \dfrac{-c+11}{(c+3)(c-3)} \cdot \dfrac{c}{c}$

$= \dfrac{c^2 - 4c - 21 - c^2 + 11c}{c(c-3)(c+3)}$

$= \dfrac{7c - 21}{c(c-3)(c+3)}$

$= \dfrac{7(c-3)}{c(c-3)(c+3)} = \dfrac{7}{c(c+3)}, \text{ or } \dfrac{7}{c^2+3c}$

TRY THIS Simplify $\dfrac{a}{a^2+8a+16} - \dfrac{a}{a^2+4a}.$

EXERCISES

Tell whether the fractions have like or unlike denominators. If the fractions have unlike denominators, find the LCD. Do not perform the subtraction.

1. $\dfrac{5}{3x + 6} - \dfrac{4}{3(x + 2)}$ **2.** $\dfrac{4}{3x - 6} - \dfrac{1}{x - 2}$ **3.** $\dfrac{10}{5x - 15} - \dfrac{1}{x + 3}$ **4.** $\dfrac{7}{5x - 15} - \dfrac{4}{3x - 9}$

Simplify.

5. $\dfrac{2}{3x} - \dfrac{1}{x}$ **6.** $\dfrac{7}{a} - \dfrac{4}{5a}$ **7.** $\dfrac{4}{3z} - \dfrac{2}{5z}$ **8.** $\dfrac{3}{2v} - \dfrac{6}{7v}$

9. $\dfrac{1}{x} - \dfrac{1}{x - 1}$ **10.** $\dfrac{1}{n} - \dfrac{1}{n + 2}$ **11.** $\dfrac{3z}{z - 2} - \dfrac{5}{2z}$ **12.** $\dfrac{7w}{w - 1} - \dfrac{5}{2w}$

13. $\dfrac{2x - 3}{5x - 15} - \dfrac{1}{x - 3}$ **14.** $\dfrac{m + 5}{m + 2} - \dfrac{3}{2m + 4}$ **15.** $\dfrac{z - 2}{z + 2} - \dfrac{7}{5z + 10}$

16. $\dfrac{3v - 2}{2v - 10} - \dfrac{v}{v - 5}$ **17.** $\dfrac{3u - 2}{2u - 6} - \dfrac{u + 2}{3u + 9}$ **18.** $\dfrac{3u + 4}{2u + 6} - \dfrac{u + 1}{4u + 12}$

19. $\dfrac{x - 1}{x + 1} - \dfrac{x + 1}{x}$ **20.** $\dfrac{c + 1}{c - 1} - \dfrac{c + 1}{c}$ **21.** $\dfrac{a - 1}{a + 1} - \dfrac{a + 1}{a - 1}$

22. $\dfrac{3s + 1}{s - 1} - \dfrac{s - 3}{s + 2}$ **23.** $\dfrac{2b}{b^2 - 1} - \dfrac{b}{b^2 - b}$ **24.** $\dfrac{2d}{d^2 + 2d} - \dfrac{d + 1}{d^2 - 4}$

25. $\dfrac{y - 1}{y^2 + 2y + 1} - \dfrac{y + 1}{y^2 - 1}$ **26.** $\dfrac{t - 2}{t^2 - 9} - \dfrac{t + 2}{t^2 + 6t + 9}$ **27.** $\dfrac{g - 1}{4 - g^2} - \dfrac{g + 1}{g^2 + 4g + 4}$

28. Let a and b represent nonzero real numbers. Simplify $\dfrac{a}{b} - \dfrac{b}{a}$.

29. Let a and b represent nonzero real numbers such that $a \neq \pm b$. Simplify $\dfrac{a}{a - b} - \dfrac{b}{a + b}$.

30. Critical Thinking Show that if n is any positive real number, then $\dfrac{1}{n} - \dfrac{1}{n + 1}$ is positive.

Simplify. Write answers with positive exponents only. Assume that all variables are nonzero. (9.3 Skill A)

31. $\dfrac{4a^2}{-2b^3} \cdot \dfrac{4b}{12a}$ **32.** $\dfrac{15m^2}{-21n^3} \cdot \dfrac{-7n^2}{15m^2}$ **33.** $\dfrac{24m^2}{-11n^3} \cdot \dfrac{11n^3}{24m^2}$ **34.** $\dfrac{3u^3}{21v^4} \cdot \dfrac{7v^3}{u^3}$

35. $\dfrac{3y^4}{-z} \cdot \dfrac{-z^3}{-7y^3}$ **36.** $\dfrac{3a^2}{b^3} \cdot \dfrac{5b}{-7a^4}$ **37.** $\dfrac{-5d^5}{c^5} \cdot \dfrac{5c^4}{-5d^4}$ **38.** $\dfrac{2.5s^4}{3.5r^5} \cdot \dfrac{7r^4}{5s^4}$

Two triangles are *similar* if the ratios of the lengths of the corresponding sides are equal. Triangles *ABC* and *XYZ* shown below are similar. The ratio of each pair of corresponding sides is 1.25.

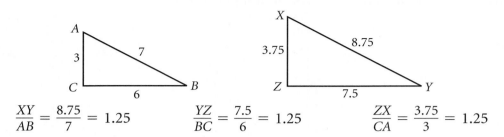

$$\frac{XY}{AB} = \frac{8.75}{7} = 1.25 \qquad \frac{YZ}{BC} = \frac{7.5}{6} = 1.25 \qquad \frac{ZX}{CA} = \frac{3.75}{3} = 1.25$$

You can use what you know about operations on rational expressions to solve problems involving other similar figures. For figures with four or more sides, the measures of corresponding angles must also be equal in order for the figures to be similar.

EXAMPLE

Refer to the diagram at right. All angles in the diagram are right angles. The lengths of the sides of the outer T are proportional to the corresponding sides of the inner T. The dimensions of the inner T are all $\frac{1}{2}$ of the corresponding dimensions in the outer T. Write an expression for the shaded area A.

▶ **Solution**

The area of the shaded region equals the area of the outer **T** minus the area of the inner **T**.

1. Find the area of the outer **T**.

Use a formula.

$$x\left(\frac{x}{4}\right) + \left(\frac{x}{2}\right)\left(\frac{x}{4}\right) = \frac{x^2}{4} + \frac{x^2}{8}$$

$$= \frac{2x^2}{8} + \frac{x^2}{8} = \frac{3x^2}{8} \text{ square units}$$

2. Find the area of the inner **T**.

$$\left(\frac{1}{2} \cdot x\right)\left(\frac{1}{2} \cdot \frac{x}{4}\right) + \left(\frac{1}{2} \cdot \frac{x}{2}\right)\left(\frac{1}{2} \cdot \frac{x}{4}\right)$$

$$= \frac{x^2}{16} + \frac{x^2}{32} = \frac{2x^2}{32} + \frac{x^2}{32} = \frac{3x^2}{32} \text{ square units}$$

3. Subtract the result of Step **2** from the result of Step **1** to find the shaded area A.

$$A = \frac{3x^2}{8} - \frac{3x^2}{32} = \frac{3x^2}{8} \cdot \frac{4}{4} - \frac{3x^2}{32} = \frac{12x^2 - 3x^2}{32} = \frac{9x^2}{32}$$

Therefore, the shaded area is $\frac{9x^2}{32}$ square units.

TRY THIS

Rework the Example given that each of the dimensions of the inner **T** are one-third those of the corresponding dimensions in the outer **T**.

EXERCISES

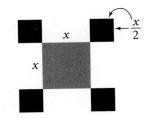

KEY SKILLS

Refer to the diagram at right. Assume the four smaller squares are all the same size.

1. Write expressions for the area of the large square and an expression for the area of each of the smaller squares.

2. Write an expression for the area of all five squares together.

3. How would the answer to Exercise 2 change if each small square has sides one third as long as the large square?

PRACTICE

Refer to the Example on the previous page.

4. Show that the grey-shaded portion of the outer **T** is 75% of the entire area of the outer **T**.

5. Let n represent a natural number. Suppose that the dimensions of the inner **T** are $\frac{1}{n}$ times as long as the corresponding dimensions of the outer **T**. Find a formula for the shaded area.

6. Refer to Exercise 5. What percent of the area of the outer **T** is the shaded area? Justify your answer.

MID-CHAPTER REVIEW

7. Find the excluded values of $y = \dfrac{3x + 2}{x^2 + 10x + 21}$. **(9.1 Skill A)**

Simplify. (9.2–9.4)

8. $\dfrac{-2x - 10}{4x + 8}$

9. $\dfrac{4u^2 - 9}{2u^2 - 9u + 9}$

10. $\dfrac{6m^3n^2 + 3m^2n^3}{6m^3n^2 - 3m^2n^3}$

11. $\dfrac{x + 1}{x^2 - 4} \cdot \dfrac{x - 2}{x^2 + 2x + 1}$

12. $\dfrac{a - 3}{a^2 - 4} \div \dfrac{a^2 - 6a + 9}{a + 2}$

13. $\dfrac{6u^3v^2}{uv^3} \div \dfrac{7u^2v^3}{2uv} \cdot \dfrac{14u^2v}{12u^3v^3}$

14. $\dfrac{n^2 - 6n}{n - 4} + \dfrac{8}{n - 4}$

15. $\dfrac{5}{3z^2} + \dfrac{1}{6z}$

16. $\dfrac{3}{t^2 + t} + \dfrac{t}{t^2 - 1}$

17. $\dfrac{k^2 + 4k}{k + 3} - \dfrac{-3}{k + 3}$

18. $\dfrac{z + 5}{2z - 14} - \dfrac{3z + 8}{3z - 21}$

19. The lengths of the sides of the outer and inner rectangles at right are proportional. The dimensions of the inner rectangle are two-thirds of the corresponding dimensions of the outer rectangle. Write an expression for the shaded area. **(9.5 Skill C)**

Simplifying and Using Complex Fractions

SKILL A *Simplifying complex fractions*

A *complex fraction* is a quotient of two rational expressions written as a fraction. The expressions at right are both complex fractions.

To simplify a complex fraction, you may need to simplify parts of the expression one at a time.

numerical complex fraction:
$$\dfrac{\dfrac{2}{3}}{\dfrac{4}{5}}$$

algebraic complex fraction:
$$\dfrac{\dfrac{2x+5}{4x-1}}{\dfrac{x+7}{3x+9}}$$

EXAMPLE 1 **Simplify.** **a.** $\dfrac{\dfrac{18}{25}}{\dfrac{2}{15}}$ **b.** $\dfrac{\dfrac{3p}{20}}{\dfrac{4p}{15}}$ **c.** $\dfrac{\dfrac{21z^2}{25}}{\dfrac{7z}{50}}$

▶ **Solution**

a.
$$\dfrac{\dfrac{18}{25}}{\dfrac{2}{15}} = \dfrac{18}{25} \div \dfrac{2}{15}$$
$$= \dfrac{18}{25} \cdot \dfrac{15}{2}$$
$$= \dfrac{27}{5}, \text{ or } 5\dfrac{2}{5}$$

b.
$$\dfrac{\dfrac{3p}{20}}{\dfrac{4p}{15}} = \dfrac{3p}{20} \div \dfrac{4p}{15}$$
$$= \dfrac{3p}{20} \cdot \dfrac{15}{4p}$$
$$= \dfrac{9}{16}$$

c.
$$\dfrac{\dfrac{21z^2}{25}}{\dfrac{7z}{50}} = \dfrac{21z^2}{25} \div \dfrac{7z}{50}$$
$$= \dfrac{21z^2}{25} \cdot \dfrac{50}{7z}$$
$$= 6z$$

TRY THIS **Simplify.** **a.** $\dfrac{\dfrac{10}{21}}{\dfrac{5}{28}}$ **b.** $\dfrac{\dfrac{11a}{30}}{\dfrac{33a}{50}}$ **c.** $\dfrac{\dfrac{35k^2}{16}}{\dfrac{7k}{48}}$

EXAMPLE 2 **Simplify** $\dfrac{1}{\dfrac{b}{a} + \dfrac{a}{b}}$.

▶ **Solution**

$$\dfrac{1}{\dfrac{b}{a} + \dfrac{a}{b}} = \dfrac{1}{\dfrac{b}{a} \cdot \dfrac{b}{b} + \dfrac{a}{b} \cdot \dfrac{a}{a}}$$

⟵ *The LCD of $\dfrac{b}{a}$ and $\dfrac{a}{b}$ is ab. Multiply $\dfrac{b}{a}$ by $\dfrac{b}{b}$ and multiply $\dfrac{a}{b}$ by $\dfrac{a}{a}$ so that their denominators are both ab.*

$$= \dfrac{1}{\dfrac{a^2 + b^2}{ab}}$$

$$= \dfrac{ab}{a^2 + b^2}$$

TRY THIS **Simplify** $\dfrac{ab}{\dfrac{b}{a} - \dfrac{a}{b}}$.

EXERCISES

Write each complex fraction as a product. Then write the numerators and the denominators in factored form. Do not simplify further.

1. $\dfrac{\frac{2}{15}}{\frac{12}{35}}$

2. $\dfrac{\frac{3x}{49}}{\frac{12x}{21}}$

3. $\dfrac{\frac{2(a-1)}{12}}{\frac{3a-3}{4}}$

4. $\dfrac{\frac{5t+25}{3t-3}}{\frac{4t+20}{5t-5}}$

Simplify.

5. $\dfrac{\frac{3}{5}}{\frac{9}{25}}$

6. $\dfrac{\frac{8}{21}}{\frac{16}{35}}$

7. $\dfrac{\frac{3n^2}{7}}{\frac{5n}{28}}$

8. $\dfrac{\frac{6r^2}{35}}{\frac{3r}{70}}$

9. $3 + \dfrac{1}{2 + \frac{3}{2}}$

10. $3 - \dfrac{1}{3 - \frac{1}{3}}$

11. $\dfrac{\frac{2b+6}{5b-15}}{\frac{3b+9}{25b-75}}$

12. $\dfrac{\frac{4r+4}{7r+14}}{\frac{7r+7}{2r+4}}$

13. $\dfrac{\frac{3a+21}{7a-14}}{\frac{7a+49}{7a-14}}$

14. $\dfrac{\frac{4z+6}{3z-36}}{\frac{4z+16}{3z-36}}$

15. $\dfrac{\frac{1}{a+b}}{\frac{1}{a-b}}$

16. $\dfrac{\frac{a}{a-b}}{\frac{b}{a+b}}$

17. **Critical Thinking**   Find an expression in simplified form for the sum of the reciprocals of two nonzero real numbers, r and s, divided by the difference of the reciprocals of r and s.

Solve each equation. (2.8 Skill B)

18. $5 - 3(r - 9) = 4(r + 1)$

19. $3(x - 3) + 5(x - 3) = 0$

20. $4 - (3y + 1) = 4y$

21. $3d + 5 - 4d = 4(d - 1)$

22. $4g - 5 = 4 - 7(g + 1)$

23. $4x + 5x + 6x = 12(x + 7)$

Solve each equation. (7.5 Skill C)

24. $9f^2 + 27f - 22 = 0$

25. $3w^2 + 38w + 99 = 0$

26. $27w^2 - 69w - 70 = 0$

27. $16h^2 - 8h + 1 = 0$

28. $121z^2 - 49 = 0$

29. $49a^2 + 140a + 100 = 0$

Light-bulb filaments are resistors in an electric circuit. The diagram at right shows two resistors in a parallel arrangement.

The total resistance R_T is given by the formula $R_T = \dfrac{1}{\dfrac{1}{A} + \dfrac{1}{B}}$.

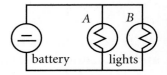

EXAMPLE 1

A parallel circuit has two resistances A and B. Show that the total effective resistance R_T is given by $R_T = \dfrac{AB}{A + B}$.

▶ **Solution**

$$R_T = \frac{1}{\dfrac{1}{A} + \dfrac{1}{B}} = \frac{1}{\dfrac{1}{A} \cdot \dfrac{B}{B} + \dfrac{1}{B} \cdot \dfrac{A}{A}} = \frac{1}{\dfrac{B + A}{AB}} = \frac{AB}{A + B}$$

TRY THIS

A parallel circuit has resistances A, B, and C. Show that the total resistance, $R_T = \dfrac{1}{\dfrac{1}{A} + \dfrac{1}{B} + \dfrac{1}{C}}$, can be written as $R_T = \dfrac{ABC}{BC + AC + AB}$.

EXAMPLE 2

Jack drove from his shop to a job site at an average speed of 55 miles per hour and returned along the same highway at an average speed of 45 miles per hour. Find his average speed for the entire trip.

▶ **Solution**

Use a formula.

1. Let d represent the length of the trip one way and let t_1 and t_2 represent his travel time to and from the site, respectively. Use the relationship $\text{rate} = \dfrac{\text{distance}}{\text{time}}$.

average speed over the entire trip: $\dfrac{\text{total distance (in miles)}}{\text{total time (in hours)}} \rightarrow \dfrac{d + d}{t_1 + t_2}$

Because $d = 55t_1 = 45t_2$, then $t_1 = \dfrac{d}{55}$ and $t_2 = \dfrac{d}{45}$.

2. Simplify $\dfrac{2d}{\dfrac{d}{55} + \dfrac{d}{45}}$.

$$\frac{2d}{\dfrac{d}{55} + \dfrac{d}{45}} = \frac{2d}{\dfrac{9d + 11d}{495}} = 2d \times \frac{495}{\underset{10}{20d}} = 49.5 \text{ miles/hour}$$

Jack's average speed over the entire trip is 49.5 miles per hour.

TRY THIS

Rework Example 2 using 44 miles per hour and 50 miles per hour.

EXERCISES

KEY SKILLS

In Exercises 1 and 2, use the formula distance (d) = rate (r) × time (t).

1. Which properties justify *If $d = rt$, then $t = \dfrac{d}{r}$*?

2. Which properties justify *If $t_1 = \dfrac{d}{55}$, $t_2 = \dfrac{d}{45}$, and $s = \dfrac{d + d}{t_1 + t_2}$, then $s = \dfrac{2d}{\dfrac{d}{55} + \dfrac{d}{45}}$*?

PRACTICE

3. A parallel circuit has resistors of 15 ohms and of 12 ohms. Simplify $\dfrac{1}{\dfrac{1}{12} + \dfrac{1}{15}}$, an expression for the total resistance.

Electrical resistance is measured in ohms (Ω). Find the total effective resistance to the nearest tenth for resistances A and B in a parallel circuit.

4. A: 12 Ω and B: 9 Ω

5. A: 10 Ω and B: 12 Ω

6. A: 8 Ω and B: 14 Ω

7. A: 10 Ω and B: 10 Ω

8. Show that if resistances A and B are equal, then the total resistance is one half of one of them.

Refer to Example 2. Find the average speed for the entire trip.

9. to the site: 40 miles per hour   from the site: 45 miles per hour

10. to the site: 40 miles per hour   from the site: 50 miles per hour

11. Show that if Jack drives to and from a job site at r miles per hour, then his average speed for the entire trip is r miles per hour.

12. **Critical Thinking** Write and simplify a formula for the average speed for a trip of d miles one way and d miles back if the driver travels at r_1 miles per hour to the destination and r_2 miles per hour from the destination.

MIXED REVIEW APPLICATIONS

An object propelled upward from an initial altitude h_0 in feet and given an initial velocity v_0 in feet per second will have altitude h in feet given by $h = -16t^2 + v_0t + h_0$.

A ball thrown directly upward from the edge of a cliff 84 feet above sea level is given an initial velocity of 96 feet per second. Find the amount of time it takes the ball to have the specified altitude above sea level. Give answers to the nearest tenth of a second. (8.7 Skill B)

13. 130 feet

14. 140 feet

15. 200 feet

16. 180 feet

9.7 Solving Rational Equations

LESSON

SKILL A *Solving proportions involving rational expressions*

A **rational equation** is an equation containing at least one rational expression. A solution to a rational equation is any number that satisfies the equation.

Recall that a proportion is the equality of two ratios. Any proportion containing a variable is a rational equation. You can use *cross products* to solve proportions.

Cross Product Property of Proportions

If a, b, c, and d are real numbers with $b \neq 0$
and $d \neq 0$, and $\frac{a}{b} = \frac{c}{d}$, then $ad = bc$.

Cross products:
ad and bc

The equation $\frac{3}{z} = \frac{4}{5}$ is a proportion. Using the property above, you can write $3 \cdot 5 = 4z$. Then divide both sides by 4 to solve for z.

EXAMPLE 1 Solve $\frac{3}{x-1} = \frac{8}{x}$.

▶ **Solution**

$$\frac{3}{x-1} = \frac{8}{x}$$
$$3x = 8(x-1) \quad \longleftarrow \text{Apply the Cross Product Property of Proportions.}$$
$$x = \frac{8}{5}, \text{ or } 1\frac{3}{5}$$

TRY THIS Solve. **a.** $\frac{y+2}{7} = \frac{y}{5}$ **b.** $\frac{-3}{n-2} = \frac{4}{n}$

EXAMPLE 2 Solve $\frac{a}{2} = \frac{2}{a+3}$.

▶ **Solution**

$$\frac{a}{2} = \frac{2}{a+3}$$
$$a(a+3) = 2(2) \quad \longleftarrow \text{Apply the Cross Product Property of Proportions.}$$
$$a^2 + 3a - 4 = 0$$
$$(a+4)(a-1) = 0$$
$$a = -4 \text{ or } a = 1 \quad \longleftarrow \text{Apply the Zero-Product Property and solve for } a.$$

The solutions are -4 and 1.

TRY THIS Solve. **a.** $\frac{n-6}{-2} = \frac{4}{n}$ **b.** $\frac{r-4}{-2} = \frac{2}{r}$

EXERCISES

Write the equation that results after you apply the Cross Product
Property of Proportions. Do not solve.

1. $\frac{5}{z} = \frac{2}{7}$

2. $\frac{3}{x-2} = \frac{2}{x+2}$

3. $\frac{5}{a-1} = \frac{2}{a}$

4. $\frac{3}{x-1} = \frac{2x}{5}$

Solve each proportion.

5. $\frac{5}{7} = \frac{w}{21}$

6. $\frac{s}{8} = \frac{8}{32}$

7. $\frac{5}{9} = \frac{t}{6}$

8. $\frac{z}{3} = \frac{13}{5}$

9. $\frac{6}{b} = \frac{11}{2}$

10. $\frac{6}{7} = \frac{4}{d}$

11. $\frac{5}{y} = \frac{12}{7}$

12. $\frac{1}{9} = \frac{13}{n}$

13. $\frac{1}{3n} = \frac{3}{n-1}$

14. $\frac{5}{3m} = \frac{2}{m-2}$

15. $\frac{1}{4} = \frac{3k+1}{k-2}$

16. $\frac{4p-1}{5} = \frac{3p+1}{2}$

17. $\frac{y-1}{y} = \frac{y}{4}$

18. $\frac{v-4}{v} = \frac{v}{-2}$

19. $\frac{x}{2} = \frac{3x-4}{x}$

20. $\frac{a}{10} = \frac{2}{a+1}$

Solve each equation for the specified variable.

21. $\frac{ax}{b} = \frac{c}{d}$ for x

22. $\frac{z}{t} = \frac{r}{z}$ for z

23. $\frac{s}{t} = \frac{rn}{an+b}$ for n

The *geometric mean* between positive numbers a and b is defined as the
positive solution to the equation $\frac{a}{x} = \frac{x}{b}$. Find each geometric mean.
Write the answers in simplest radical form.

24. 2 and 3

25. 10 and 20

26. 12 and 24

27. 7 and 70

28. Show that the geometric mean of a and b is $\sqrt{ab}$.

29. **Critical Thinking**

   a. Show that if $\frac{a}{b} = \frac{c}{d}$, then $a = \frac{bc}{d}$.

   b. Show that if $\frac{a}{b} = \frac{c}{d}$ is true, then $\frac{b}{a} = \frac{d}{c}$ is also true.

   c. Describe how the information in parts **a** and **b** can make solving
   proportions easier.

Find the range of the function represented by each equation. (7.7 Skill A)

30. $y = 2x^2 - 12x + 7$

31. $y = -5x^2 + 10x + 1$

32. $y = 3.5x^2 - 7x$

33. $y = -3x^2 + 3x - 11$

34. $y = 4x^2 + 9x - 13$

35. $y = -5x^2 + 15x - 9$

EXAMPLE 1

Solve.   **a.** $x + \dfrac{2}{x} = 3$ **b.** $\dfrac{a}{2} - \dfrac{1}{3a} = \dfrac{1}{6}$

▶ **Solution**

a.
$$x + \frac{2}{x} = 3$$
$$\frac{x^2}{x} + \frac{2}{x} = 3$$
$$\frac{x^2 + 2}{x} = 3$$
$$x^2 + 2 = 3x$$
$$x^2 - 3x + 2 = 0$$
$$(x - 2)(x - 1) = 0$$
$$x = 1 \quad \text{or} \quad x = 2$$

b.
$$\frac{a}{2} - \frac{1}{3a} = \frac{1}{6}$$
$$\frac{3a^2}{6a} - \frac{2}{6a} = \frac{1}{6}$$
$$\frac{3a^2 - 2}{6a} = \frac{1}{6}$$
$$3a^2 - 2 = a$$
$$3a^2 - a - 2 = 0$$
$$(3a + 2)(a - 1) = 0$$
$$a = -\frac{2}{3} \quad \text{or} \quad a = 1$$

TRY THIS   **Solve.**   **a.** $2d - \dfrac{6}{d} = 1$ **b.** $\dfrac{5s}{3} + \dfrac{6}{s} = 7$

An **extraneous solution** to a given equation is a number found as a solution to an equation that does not satisfy the given equation. An extraneous solution can be a number that causes the denominator to equal zero, which is not allowed.

EXAMPLE 2

Solve $\dfrac{2}{z^2 - 2z} - \dfrac{1}{z - 2} = 1$**. Identify extraneous solutions.**

▶ **Solution**

$$\frac{2}{z^2 - 2z} - \frac{1}{z - 2} = 1$$

$$\frac{2}{z(z - 2)} - \frac{z}{z(z - 2)} = 1 \qquad \longleftarrow \text{Write using a common denominator.}$$

$$\frac{2 - z}{z(z - 2)} = 1 \qquad \longleftarrow \text{Subtract.}$$

$$2 - z = z(z - 2)$$
$$2 - z = z^2 - 2z$$
$$z^2 - z - 2 = 0 \qquad \longleftarrow \text{Write a quadratic equation in standard form.}$$
$$(z - 2)(z + 1) = 0 \qquad \longleftarrow \text{Factor.}$$
$$z = 2 \quad \text{or} \quad z = -1 \qquad \longleftarrow \text{Apply the Zero-Product Property and solve for } z.$$

Substitute 2 and -1 for z in the original equation to see if the numbers are solutions. Only -1 satisfies the original equation. The solution is -1.

Check: $\dfrac{2}{(2)^2 - 2(2)} - \dfrac{1}{(2) - 2} \overset{?}{=} 1 \quad \longrightarrow \quad \dfrac{2}{0} - \dfrac{1}{0} \neq 1 ✗$

$\dfrac{2}{(-1)^2 - 2(-1)} - \dfrac{1}{(-1) - 2} \overset{?}{=} 1 \quad \longrightarrow \quad \dfrac{2}{3} - \left(-\dfrac{1}{3}\right) = 1 ✔$

TRY THIS   **Solve** $\dfrac{1}{a^2 - a} - \dfrac{1}{a - 1} = \dfrac{1}{2}$**. Identify extraneous solutions.**

EXERCISES

Substitute the given numbers into the given equation. Which are actual solutions and which, if any, are extraneous?

1. n: -1 and $\frac{5}{2}$

$n - \frac{3}{2n + 1} = 2$

2. s: 0 and 3

$\frac{3s}{s^2 - s} - \frac{1}{s - 1} = 1$

3. p: 1

$\frac{1}{p^2 - p} - \frac{1}{p - 1} = 0$

PRACTICE

Solve each equation. Identify extraneous solutions.

4. $x + \frac{6}{x} = -5$

5. $x - \frac{10}{x} = 3$

6. $x + \frac{1}{6x} = \frac{5}{6}$

7. $9a - \frac{2}{a} = 3$

8. $2x + \frac{1}{x - 1} = -1$

9. $2y + \frac{3}{y + 2} = 1$

10. $2r - \frac{1}{3r + 1} = 1$

11. $3t - \frac{1}{2t - 1} = 4$

12. $\frac{k}{k^2 + k} + \frac{1}{k + 1} = \frac{-1}{2k}$

13. $\frac{2v}{v^2 + 3v} - \frac{2}{v + 3} = \frac{2}{v}$

14. $\frac{3y}{y^2 - 2y} + \frac{5}{y - 2} = \frac{2}{y - 2}$

15. $\frac{x}{x - 3} + \frac{2x}{x + 3} = \frac{18}{x^2 - 9}$

Solve each equation for x. For what value(s) of a will there be a solution?

16. $\frac{1}{x} + \frac{1}{a} = 1$

17. $\frac{1}{x} + \frac{1}{a + 1} = 1$

18. $\frac{1}{x(a + 1)} - \frac{a}{x} = 1$

19. Suppose that you choose a number n, take its reciprocal, and add it to the reciprocal of $n + 1$. How many values of n will give the sum $\frac{3}{2}$?

20. Critical Thinking For what value(s) of a will $\frac{1}{n} + \frac{1}{n + 1} = \frac{a}{n}$ always have a solution for n, except $n \neq 0$ and $n \neq -1$?

MIXED REVIEW

Suppose that $a > 0$. Use laws of exponents and $\sqrt{a} = a^{\frac{1}{2}}$ to prove each statement. (8.1 Skill C)

21. $\sqrt{81a} = 9\sqrt{a}$

22. $\sqrt{100a} = 10\sqrt{a}$

23. $\sqrt{\frac{a}{25}} = \frac{\sqrt{a}}{5}$

24. $\sqrt{\frac{a}{49}} = \frac{\sqrt{a}}{7}$

25. $\sqrt{\frac{64a}{121}} = \frac{8\sqrt{a}}{11}$

26. $\sqrt{\frac{16a}{9}} = \frac{4\sqrt{a}}{3}$

If it takes 10 days to complete a task and the worker spends 1 day on it, then the worker completes $\frac{1}{10}$ of the work in one day. This unit rate is called the *work rate*. If the worker spends t days on the task and the work rate is constant, then the worker completes $\frac{t}{10}$ of the task. This can be expressed as a formula:

$$\text{portion of task completed} = \text{rate} \times \text{time}$$

EXAMPLE 1

After working 5 days at a constant rate, a student completed 60%, or $\frac{3}{5}$, of a project. How long would it take the student to complete the entire project?

▶ **Solution**

1. Let t represent the time it would take to complete the whole project.

Write an equation.

2. Solve $\frac{3}{5} = \frac{5}{t}$. $t = \frac{25}{3}$, or $8\frac{1}{3}$

The project would take $8\frac{1}{3}$ days to complete.

TRY THIS   Rework Example 1 if 75% of a project is completed in 6 days.

If two workers, worker A and worker B, work together, then you can use the equation below to solve combined-work problems.

$$\frac{\text{work time}}{\text{time needed}_A} + \frac{\text{work time}}{\text{time needed}_B} = \text{portion of task completed}$$

EXAMPLE 2

Louis takes twice as long as Alan to complete a school project. It takes them 15 hours to complete the project together. How long would it take each student to complete the project if he works alone?

▶ **Solution**

1. Let t and $2t$ represent time needed for Alan and Louis to complete the project working alone, respectively.

Write an equation.

2. Write an equation.

Alan: $\dfrac{\text{work time}}{\text{time needed}_A} \rightarrow \dfrac{15}{t}$ Louis: $\dfrac{\text{work time}}{\text{time needed}_B} \rightarrow \dfrac{15}{2t}$

$$\frac{15}{t} + \frac{15}{2t} \rightarrow 100\%, \text{ or } 1$$

3. Solve $\frac{15}{t} + \frac{15}{2t} = 1$: $t = \frac{45}{2}$, or $22\frac{1}{2}$.

It would take Alan $22\frac{1}{2}$ hours to complete the project. It would take Louis 45 hours.

TRY THIS   Rework Example 2 given that it takes Louis three times as long as Alan to complete the project.

EXERCISES

KEY SKILLS

For Exercises 1 and 2, find each work rate. Then write an equation for work, *w*, in terms of time spent working, *t*.

1. 12 days to complete the work

2. 15 days to complete the work

3. Amit needs 1.5 times the number of hours it takes Juan to complete a project. Together, it takes them 18 hours to complete it. Write an equation you can use to find how long it would take each student working alone to complete the project. Do not solve.

PRACTICE

For Exercises 4 and 5, calculate how long would it take the worker to complete the project. Assume that work rate is constant.

4. After 7 days, 35% of the work is completed.

5. After 4.5 days, 40% of the work is completed.

How long would it take each student to complete the project if he or she works alone? Assume that work rate is constant.

6. Sam takes 1.6 times as long as Nina to do the project working alone. Together, it takes them 15 days to complete the project.

7. Lee takes the same amount of time as Kim to do the project working alone. Together, it takes them 18 hours to complete the project.

Let *t* represent the amount of time it takes Students *A* and *B* working together to complete a task. Let *a* and *b* represent the respective amounts of time it takes them working alone.

8. Solve $\frac{t}{a} + \frac{t}{b} = 1$ for *t*.

9. Solve $\frac{t}{a} + \frac{t}{b} = 1$ for *a*.

MIXED REVIEW APPLICATIONS

Solve each problem. (2.8 Skill C)

10. How much of $5000 should an investor put into accounts paying 5% and 7.5% simple interest to earn $300 interest in one year?

11. A motorist drove 420 miles in 8 hours. During part of the trip she drove 50 miles per hour and during the rest of it she drove 54 miles per hour. How far did she drive at each speed?

12. A chemist wants to mix a 5% salt solution and a 10% salt solution to make 500 milliliters of an 8% salt solution. How much of the 5% and 10% solutions are needed?

Inverse Variation

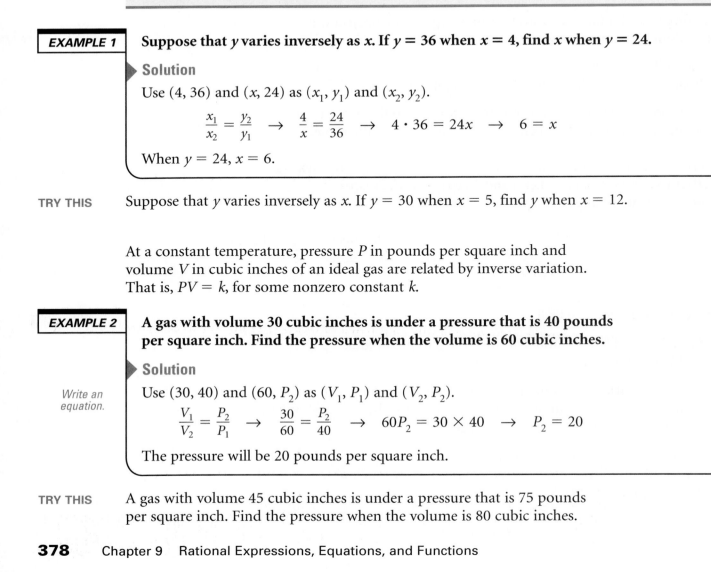

A relationship between y and x is an *inverse-variation* relationship if y varies with the reciprocal, or multiplicative inverse, of x. That is, as one variable gets larger, the other variable gets smaller.

Inverse Variation

If there is a fixed nonzero number k such that $xy = k$, then you can say that y *varies inversely as x*. The relationship between x and y is called an **inverse-variation** relationship and k is called the *constant of variation*.

The equation $xy = k$ can also be written as $y = \dfrac{k}{x}$.

Alternative Form of Inverse Variation

If (x_1, y_1) and (x_2, y_2) satisfy $xy = k$, then $x_1y_1 = x_2y_2$ and $\dfrac{x_1}{x_2} = \dfrac{y_2}{y_1}$.

EXAMPLE 1

Suppose that y varies inversely as x. If $y = 36$ when $x = 4$, find x when $y = 24$.

Solution

Use $(4, 36)$ and $(x, 24)$ as (x_1, y_1) and (x_2, y_2).

$$\dfrac{x_1}{x_2} = \dfrac{y_2}{y_1} \rightarrow \dfrac{4}{x} = \dfrac{24}{36} \rightarrow 4 \cdot 36 = 24x \rightarrow 6 = x$$

When $y = 24$, $x = 6$.

TRY THIS

Suppose that y varies inversely as x. If $y = 30$ when $x = 5$, find y when $x = 12$.

At a constant temperature, pressure P in pounds per square inch and volume V in cubic inches of an ideal gas are related by inverse variation. That is, $PV = k$, for some nonzero constant k.

EXAMPLE 2

A gas with volume 30 cubic inches is under a pressure that is 40 pounds per square inch. Find the pressure when the volume is 60 cubic inches.

Solution

Write an equation.

Use $(30, 40)$ and $(60, P_2)$ as (V_1, P_1) and (V_2, P_2).

$$\dfrac{V_1}{V_2} = \dfrac{P_2}{P_1} \rightarrow \dfrac{30}{60} = \dfrac{P_2}{40} \rightarrow 60P_2 = 30 \times 40 \rightarrow P_2 = 20$$

The pressure will be 20 pounds per square inch.

TRY THIS

A gas with volume 45 cubic inches is under a pressure that is 75 pounds per square inch. Find the pressure when the volume is 80 cubic inches.

EXERCISES

In Exercises 1 and 2, speed s varies inversely with elapsed time t.

1. Samantha and Derek plan to bicycle a distance of 8 miles. Copy and complete the time/speed table below.

t	$\frac{1}{4}$ hour	$\frac{1}{2}$ hour	1 hour	2 hours	3 hours	4 hours
s						

2. How would you find the bicycling speed they need if they want to travel 8 miles in $1\frac{1}{2}$ hours?

In Exercises 3–6, y varies inversely as x. Find the value of the indicated variable.

3. If $y = 12$ when $x = 4$, find y when $x = 12$.

4. If $y = 10$ when $x = 2$, find y when $x = 6$.

5. If $y = 36$ when $x = 5$, find x when $y = 12$.

6. If $y = 9$ when $x = 1$, find x when $y = 10$.

7. Show that if (x_1, y_1) and (x_2, y_2) satisfy $xy = k$, then $\frac{x_1}{x_2} = \frac{y_2}{y_1}$.

Solve each problem.

8. Given a fixed number as the area of a rectangle, length varies inversely as width. What happens to the length of the rectangle if area is constant and width is doubled?

9. The volume of a gas is 45 cubic inches when the pressure is 42 pounds per square inch. Given constant temperature, find the pressure when the volume is cut by 50%.

10. Two boxes balance on a seesaw when box weight and distance from the fulcrum satisfy an inverse-variation relationship. How far from the fulcrum should Box B be placed to balance Box A if Box A weighs 45 pounds, Box B weighs 60 pounds and Box A is 4.5 feet from the fulcrum?

In Exercises 11 and 12, y varies directly as x. (3.3 Skill B)

11. If $y = 18$ when $x = 6$, find y when $x = 2.5$.

12. If $y = 100$ when $x = 25$, find y when $x = 5$.

SKILL B Using a graph to represent an inverse-variation relationship

If y varies inversely as x and the constant of variation is k, then $y = \frac{k}{x}$. In particular, the function represented by $y = \frac{1}{x}$ is called the *reciprocal function.*

EXAMPLE 1 **Graph $xy = 4$.**

Solution

Write y in terms of x: $y = \frac{4}{x}$

Make tables of ordered pairs.

Make a table.

x	-8	-4	-2	-1	-0.5
y	-0.5	-1	-2	-4	-8

x	0.5	1	2	4	8
y	8	4	2	1	0.5

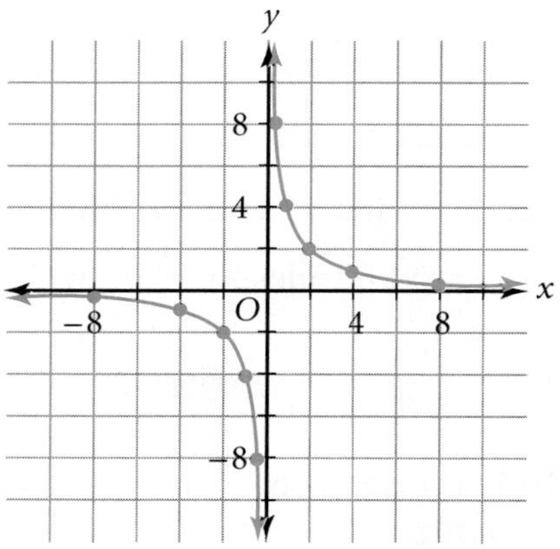

Sketch smooth curves. The graph gets closer and closer to each axis but does not touch or cross either axis.

TRY THIS Graph $xy = 16$.

EXAMPLE 2 **The graph at right represents an equation of the form $xy = k$.**

a. Does the graph represent a function?
b. What are the domain and range?
c. Find y when $x = 250$.
d. Is (10, 0.25) on the graph? Justify your response.

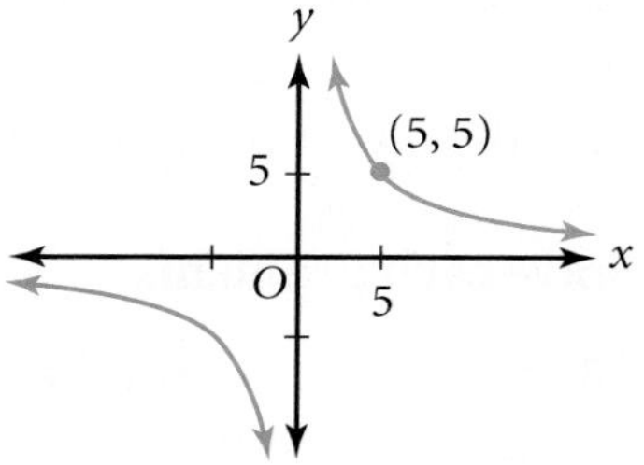

Solution

a. Apply the vertical-line test. No vertical line intersects the graph in more than one point. So, the graph represents a function.

b. domain: all real numbers except 0 range: all real numbers except 0

c. Use the point given on the graph to find k: $k = (5)(5) = 25$.
Substitute 250 for x to find y: $250y = 25 \rightarrow y = \frac{1}{10}$.

d. No, (10, 0.25) is not on the graph, because $10 \cdot 0.25 \neq 5 \cdot 5$.

TRY THIS The graph at right represents an equation of the form $xy = k$.

a. Does the graph represent a function?
b. What are the domain and range?
c. Find y when $x = 120$.
d. Is (80, 0.25) on the graph? Justify your response.

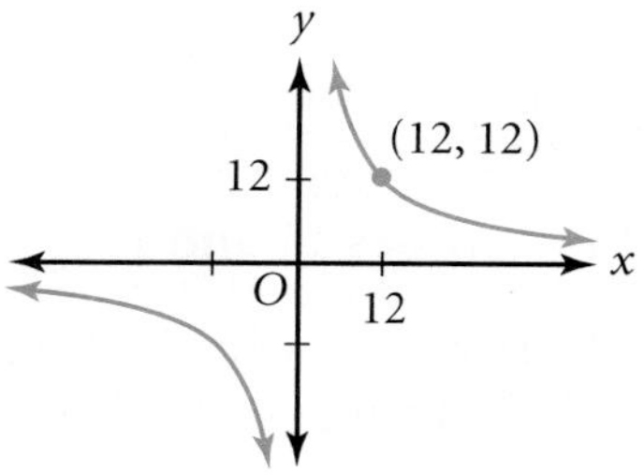

EXERCISES

KEY SKILLS

Refer to the graph at right. The graph represents an equation of the form $y = \dfrac{k}{x}$. Point $P(7, 7)$ is on the graph.

1. What is the value of k?

2. Explain why $Q(-7, -7)$ is also on the graph.

3. Explain why there is no point on the graph that corresponds to $x = 0$.

PRACTICE

Graph each equation.

4. $xy = 1$

5. $xy = 9$

6. $xy = 36$

Use each graph in Exercises 7–9 to answer the questions below.
a. Does the graph represent a function?
b. What is the domain? the range?
c. What is the value of y when $x = 150$?
d. Is the point $(100, 0.25)$ on the graph? Justify your response.

7.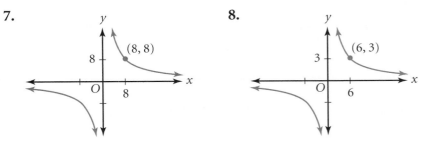

8.

9.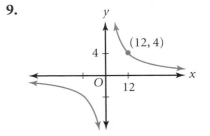

10. **Critical Thinking**   Let $y = \dfrac{1}{n}$, where n is a natural number. Is $\dfrac{1}{n + 1}$ greater than or less than $\dfrac{1}{n}$? Justify your response. What does your conclusion tell you about the graph of $y = \dfrac{1}{n}$ as n increases?

MIXED REVIEW

Give a counterexample to disprove each statement. (4.7 Skill C)

11. The product $(3n - 1)(2n + 3)$ is always less than -6.

12. The product $(-2n - 1)(2n - 3)$ is always greater than 5.

13. The product $(-3n)(n + 1)$ is always positive.

Recall from Lesson 6.2 that y varies directly as the square of x if there is a nonzero number k such that $y = kx^2$.

> ### Inverse Variation as the Square
>
> If there is a fixed nonzero number k such that $x^2y = k$, then you can say that y *varies inversely as the square of* x. The relationship between x and y is called an **inverse-square-variation** relationship and k is called the *constant of variation*.

If y varies *inversely* as the square of x, you can say that y varies *directly* as the square of the *reciprocal* of x. This can be expressed symbolically as shown below.

$$y = k\left(\frac{1}{x}\right)^2, y = k \cdot \frac{1}{x^2}, \text{ or } y = \frac{k}{x^2}$$

EXAMPLE 1

If y **varies inversely as the square of** x **and** $y = \frac{3}{100}$ **when** $x = 10$, **find** x **when** $y = \frac{1}{3}$.

▶ Solution

If $y = \frac{k}{x^2}$, $x = 10$, and $y = \frac{3}{100}$, then $k = 3$. Solve $\frac{1}{3} = \frac{3}{x^2}$.

$$\frac{1}{3} = \frac{3}{x^2} \rightarrow x^2 = 9 \rightarrow x = \pm 3$$

If $y = \frac{1}{3}$, then $x = \pm 3$.

TRY THIS If y varies inversely as the square of x and $y = \frac{4}{100}$ when $x = 10$, find y when $x = 16$.

EXAMPLE 2

The illumination, I, of a point varies inversely as the square of d, the distance from the light source. How does the illumination of a point compare with that of a point twice the distance from the source?

▶ Solution

The variables I and d are related by $I = \frac{k}{d^2}$, where $k \neq 0$.

Point A: d units from source   Point B: $2d$ units from source

$$I_A = \frac{k}{d^2} \qquad\qquad I_B = \frac{k}{(2d)^2}$$

Simplify the ratio $\frac{I_B}{I_A}$. $\frac{I_B}{I_A} = \frac{\frac{k}{(2d)^2}}{\frac{k}{d^2}} = \frac{k}{(2d)^2} \cdot \frac{d^2}{k} = \frac{1}{4}$

If distance is doubled, illumination is one quarter of what it was.

TRY THIS Rework Example 2 for the case where distance is tripled.

EXERCISES

In Exercises 1 and 2, $y = \frac{1}{x^2}$.

1. Copy and complete the table below.

x	1	2	3	4	5
y	?	?	?	?	?

2. What can you say from the table in Exercise 1 about the values of y as x increases?

PRACTICE

In Exercises 3–6, y varies inversely as the square of x.

3. If $y = 0.05$ when $x = 10$, find y when $x = 15$.

4. If $y = 3$ when $x = 4$, find y when $x = 12$.

5. If $y = \frac{1}{7}$ when $x = 5$, find x when $y = \frac{1}{28}$.

6. If $y = 2$ when $x = 8$, find x when $y = \frac{2}{9}$.

In Exercises 7–9, refer to Example 2 on the previous page.

7. Fill in the steps that show that $\frac{I_B}{I_A} = \frac{1}{4}$.

8. Rework Example 2 for the case where distance is quadrupled.

9. Suppose that distance d is multiplied by r, where $r > 0$. How does illumination at distance rd compare with the illumination at distance d?

10. Points P and Q are situated so that the illumination at P is one-sixteenth that of the illumination at Q. How are the distances of P and Q from the light source related?

MIXED REVIEW APPLICATIONS

A car has a stopping distance, d, that varies directly as the square of the car's speed, s. Round to the nearest hundredth. (Lesson 6.2 Skill B)

If $d = 120$ ft when $s = 30$ mph, find d when:

11. $s = 15$ mph

12. $s = 40$ mph

13. $s = 50$ mph

14. $s = 60$ mph

 **Proportions and Deductive Reasoning**

LESSON

EXAMPLE 1

Let $\dfrac{x - a}{x} = b$. Use deductive reasoning to write a formula for x in terms

of a and b. State any restrictions on the values of a, b, and x.

▶ **Solution**

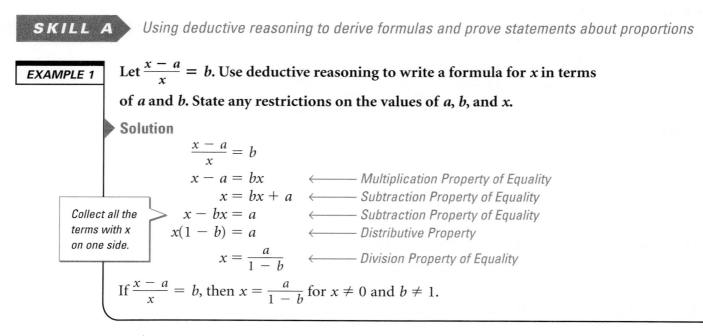

$$\frac{x - a}{x} = b$$

$$x - a = bx \qquad \longleftarrow \text{ Multiplication Property of Equality}$$

$$x = bx + a \qquad \longleftarrow \text{ Subtraction Property of Equality}$$

$$x - bx = a \qquad \longleftarrow \text{ Subtraction Property of Equality}$$

$$x(1 - b) = a \qquad \longleftarrow \text{ Distributive Property}$$

$$x = \frac{a}{1 - b} \qquad \longleftarrow \text{ Division Property of Equality}$$

Collect all the terms with x on one side.

If $\dfrac{x - a}{x} = b$, then $x = \dfrac{a}{1 - b}$ for $x \neq 0$ and $b \neq 1$.

TRY THIS

Let $\dfrac{x + a}{x} = b$. Use deductive reasoning to write a formula for x in terms

of a and b. State restrictions on variables.

Recall from Lesson 9.7 the Cross Product Property of Proportions:

If a, b, c, and d are real numbers, $b \neq 0$, $d \neq 0$, and $\dfrac{a}{b} = \dfrac{c}{d}$, then $ad = bc$.

You can prove the *converse* statement of this property:

If a, b, c, and d are real numbers, $b \neq 0$, $d \neq 0$, and $ad = bc$, then $\dfrac{a}{b} = \dfrac{c}{d}$.

EXAMPLE 2

Prove: If a, b, c, and d are real numbers, and $b \neq 0$, $d \neq 0$, and $ad = bc$,

then $\dfrac{a}{b} = \dfrac{c}{d}$.

▶ **Solution**

$$ad = bc$$

$$\frac{ad}{bd} = \frac{bc}{bd} \qquad \longleftarrow \text{ Division Property of Equality}$$

$$\frac{a}{b} \cdot \frac{d}{d} = \frac{b}{b} \cdot \frac{c}{d} \qquad \longleftarrow \text{ Definition of multiplication of fractions}$$

$$\frac{a}{b} \cdot 1 = 1 \cdot \frac{c}{d} \qquad \longleftarrow \text{ Multiplicative inverse}$$

$$\frac{a}{b} = \frac{c}{d} \qquad \longleftarrow \text{ Multiplicative identity}$$

TRY THIS

Prove: If a, b, c, and d are nonzero real numbers and $\dfrac{a}{b} = \dfrac{c}{d}$, then $\dfrac{a}{c} = \dfrac{b}{d}$.

EXERCISES

State any restrictions on the values of a, b, c, and x. Do not prove the statement.

1. If $\frac{x-a}{b} = c$, then $x = a + bc$.

2. If $\frac{x+a}{x-b} = c$, then $x = \frac{a+bc}{c-1}$.

Use deductive reasoning to solve for x in terms of the other variables. State any restrictions on the values of a, b, c, d, and x.

3. $\frac{ax}{b} = \frac{c}{d}$

4. $\frac{a}{bx} = \frac{c}{d}$

5. $\frac{a}{bx+c} = 1$

6. $\frac{a}{bx-c} = 1$

7. $\frac{x+a}{x+b} = c$

8. $\frac{a}{bx+c} = d$

9. $\frac{ax-b}{ax+b} = c$

10. $\frac{a}{bx} = \frac{cx}{d}$

11. $\frac{1}{ax+b} = ax + b$

Give the justification for each step. Assume a, b, c, and d represent nonzero real numbers.

12. If $\frac{a}{b} = \frac{c}{d}$, then $bd \cdot \frac{a}{b} = bd \cdot \frac{c}{d}$.

13. $\frac{ad}{bd} = \frac{a}{b} \cdot \frac{d}{d}$

Prove each statement below. Assume a, b, c, and d represent nonzero real numbers.

14. If $\frac{1}{b} = \frac{c}{d}$, then $\frac{d}{b} = c$.

15. If $\frac{1}{b} = \frac{c}{d}$, then $\frac{b}{d} = \frac{1}{c}$.

16. If $\frac{a}{b} = \frac{c}{d}$, then $\frac{a+b}{b} = \frac{c+d}{d}$.

17. If $\frac{a}{b} = \frac{c}{d}$, then $\frac{a-b}{b} = \frac{c-d}{d}$.

18. Prove: If a, b, c, and d are real numbers, $b \neq 0$, $d \neq 0$, and $\frac{a}{b} = \frac{c}{d}$, then $\frac{a^2}{b^2} = \frac{c^2}{d^2}$.

19. **Critical Thinking** Is the statement below always, sometimes or never true? If the statement is sometimes true, when is it true?

If a, b, c, and d are real numbers, and $b \neq 0$, $d \neq 0$, and $\frac{a}{b} = \frac{c}{d}$, then $\frac{a}{d} = \frac{c}{b}$.

Find the domain of each rational expression or function. (9.1 Skill A)

20. $\frac{3x^2-5}{4x^2-x}$

21. $\frac{7a^2+1}{-4a^2+1}$

22. $\frac{v+7}{4v^2+v-5}$

23. $y = \frac{-7x+3}{2x+11}$

24. $y = \frac{1}{4x^2+28x+49}$

25. $y = \frac{2x^2-12}{4x^2-16x}$

Identify any excluded values of the domain.

1. $\dfrac{2.5x^2 - x + 1}{4x + 3}$

2. $\dfrac{a^2 - 1}{4a^2 - 9}$

3. $y = \dfrac{5a + 2}{5a + 2}$

Simplify. Identify any excluded values.

4. $\dfrac{11}{22n - 55}$

5. $\dfrac{-3d + 6}{9d - 18}$

6. $\dfrac{15v + 60}{15v + 30}$

7. $\dfrac{t^2 - 25}{t^2 - 2t - 15}$

8. $\dfrac{a^2 - 8a + 7}{a^2 - 2a + 1}$

9. $\dfrac{m^2 + 8m - 9}{m^2 + 7m - 18}$

Simplify.

10. $\dfrac{p^2 t^2 + 4pt}{8pt + pt^2}$

11. $\dfrac{r^2 s^2 - 9}{2rs - 6}$

12. $\dfrac{3a^2 b - 12a}{a^2 b^2 - 16}$

Write the answer with positive exponents only.

13. $\dfrac{30h^3}{17g^2} \cdot \dfrac{34g}{15h^2}$

14. $\dfrac{-3x + 9}{x^2 + 5x + 6} \cdot \dfrac{x^2 + 6x + 9}{9 - x^2}$

15. $\dfrac{16a^2}{45z^2} \div \dfrac{4a^4}{9z^2}$

16. $\dfrac{2x - 6}{x^2 - 5x + 6} \div \dfrac{5x - 15}{x^2 - 6x + 9}$

17. $\dfrac{3a - 6}{a^2 - 25} \div \dfrac{a^2 - 1}{a^2 - 4a - 5} \cdot \dfrac{a^2 - 4a + 3}{a^2 - 8a + 15}$

18. $\dfrac{5g + 2}{3g^2} + \dfrac{4g + 7}{3g^2}$

19. $\dfrac{15z^2}{3z + 1} + \dfrac{11z + 2}{3z + 1}$

20. $\dfrac{2h + 1}{3h} + \dfrac{5}{6h^2}$

21. $\dfrac{1}{2} + \dfrac{1}{3z} + \dfrac{4}{5z^2}$

22. $\dfrac{3c + 1}{3c^2 - c} + \dfrac{2}{c - 1}$

23. $\dfrac{1}{x^2 + 2x + 1} + \dfrac{1}{x^2 - 1}$

24. $\dfrac{4y^3 + 1}{3y^3} + \dfrac{y^3 + 4}{3y^3}$

25. $\dfrac{4a^2 + 2a}{2a + 1} - \dfrac{-1}{2a + 1}$

26. $\dfrac{u}{7u + 14} - \dfrac{u}{u + 2}$

27. $\dfrac{n + 3}{n^2 + 6n + 9} - \dfrac{n - 3}{n^2 - 9}$

28. The lengths of the sides of the outer and inner rectangles below are in proportion. The dimensions of the inner rectangle are all one quarter of the corresponding dimensions in the outer rectangle. Write an expression for the shaded area.

Simplify.

29. $\dfrac{\dfrac{24z^2}{49n^3}}{\dfrac{81z^4}{14n^4}}$

30. $\dfrac{\dfrac{4a^2 - 8ab + 4b^2}{a^2 - 2ab + b^2}}{\dfrac{2a - 2b}{a + b}}$

31. In a parallel electrical circuit, the total effective resistance of two resistances is given in ohms by $\dfrac{1}{\dfrac{1}{9} + \dfrac{1}{12}}$. Write this as a single fraction.

Solve.

32. $\dfrac{-4}{3z} = \dfrac{7}{5z - 2}$

33. $\dfrac{s}{s - 1} = \dfrac{-3}{s - 5}$

34. $c - \dfrac{4}{c} = 3$

35. $\dfrac{x}{x - 3} + \dfrac{2x}{x + 3} = \dfrac{18}{x^2 - 9}$

36. Ken needs twice as long as Maria to complete a project. Together, it takes them 24 hours to complete it. How long would it take each student working alone?

37. Suppose that y varies inversely as x. If $y = 15$ when $x = 3$, find x when $y = 10$.

38. The graph of an equation of the form $xy = k$ $(k \neq 0)$ contains the point $(4, 12)$. Does the graph represent a function? What are its domain and range? Find y when $x = 50$. Is $(5, 50)$ on the graph? Justify your response.

39. Suppose that y varies inversely as the square of x. If $y = 16$ when $x = 1$, find x when $y = 25$.

40. Let $\dfrac{ax + b}{x} = 1$. Use deductive reasoning to write a formula for x in terms of a and b. State the restrictions on variables.

CHAPTER

10

Radical Expressions, Equations, and Functions

What You Already Know

By this stage in the course, you have learned concepts and skills that involve: square roots, functions of various kinds, integer exponents, solving linear and quadratic equations, and solving linear inequalities.

Now it will be time to bring all this experience together to learn more about square-root expressions and start to learn about rational exponents.

What You Will Learn

In Chapter 10, you will use what you learned about solving linear inequalities in one variable to find the domain and range for a square-root expression. You will also use what you learned about solving linear and quadratic equations to help you solve equations involving square roots.

The set of real numbers of the form $a + b\sqrt{2}$, where a and b are rational numbers, is a special set of real numbers. In Lesson 10.3, you will see how to work with such numbers and discover that this set is closed under both addition and under multiplication. In Lesson 10.4, your study is extended to numbers of the form $a + b\sqrt{p}$, where a and b are rational numbers.

The last two lessons are somewhat independent of one another. You will see how to extend the concept of exponent so that rational numbers are allowed as exponents. Lastly, you will use your critical-thinking skills to consider the truth of statements about square roots.

VOCABULARY

compound interest	index
conjugate	Property of Equality of Squares
cube-root radical	radicand
Density Property of Real Numbers	rationalizing the denominator
Distance Formula	square-root equation
field	square-root expression
fourth-root radical	square-root function

The diagram below shows how mathematical skills and mathematical reasoning are interrelated with the skills and concepts in Chapter 10. Notice that this chapter involves radical expressions and equations with an emphasis on square-root expressions and equations.

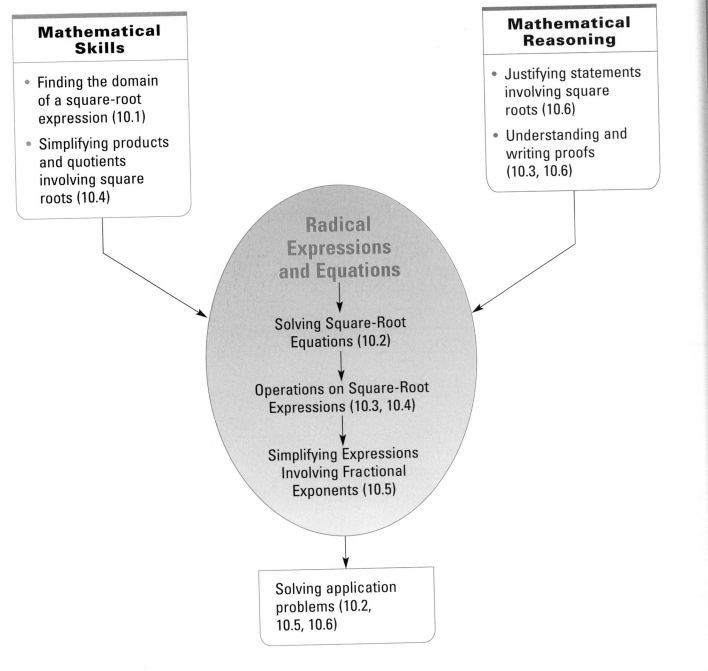

Mathematical Skills

- Finding the domain of a square-root expression (10.1)
- Simplifying products and quotients involving square roots (10.4)

Mathematical Reasoning

- Justifying statements involving square roots (10.6)
- Understanding and writing proofs (10.3, 10.6)

Radical Expressions and Equations

Solving Square-Root Equations (10.2)

Operations on Square-Root Expressions (10.3, 10.4)

Simplifying Expressions Involving Fractional Exponents (10.5)

Solving application problems (10.2, 10.5, 10.6)

10.1 Square-Root Expressions and Functions

SKILL A *Finding the domain of a square-root expression or function*

A **square-root expression** is an expression that contains a square-root symbol. The expressions $\sqrt{20}$ and $\sqrt{2x-15}$ are examples of square-root expressions. The expression under the square-root symbol, or *radical*, is called the **radicand.** For example, in $\sqrt{2x-15}$, the radicand is $2x-15$.

A **square-root function** is a function defined by a square-root expression. For example, $y=\sqrt{x}$ and $y=3\sqrt{x}-7$ are square-root functions.

The square root of a negative number is not defined in the set of real numbers. Thus, the domain of $\sqrt{x}$ is the set of all nonnegative real numbers.

The Domain and Range of $y=\sqrt{x}$

The domain of $y=\sqrt{x}$ is the set of all nonnegative real numbers.

The range of $y=\sqrt{x}$ is the set of all nonnegative real numbers.

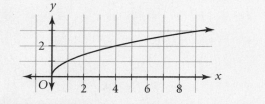

EXAMPLE

a. Find the domain of $\sqrt{-2x+15}$.

b. Find the domain and range of $y=\sqrt{3(x-5)}+1$.

▶ **Solution**

a. Because the square root of a negative number is not defined in the set of real numbers, the radicand must be greater than or equal to 0. Therefore, solve the inequality $-2x+15 \geq 0$.

$$-2x+15-\mathbf{15} \geq 0-\mathbf{15} \quad \longleftarrow \text{ Subtraction Property of Inequality}$$
$$-2x \geq -15$$

> *Reverse the inequality when you divide by a negative number.*

$$x \leq \frac{15}{2}, \text{ or } 7.5 \quad \longleftarrow \text{ Division Property of Inequality}$$

The domain of $\sqrt{-2x+15}$ is the set of all real numbers 7.5 or less.

b. The radicand cannot be negative, so solve $3(x-5)+1 \geq 0$.

$$3(x-5)+1 \geq 0$$
$$3x-15+1 \geq 0 \quad \longleftarrow \text{ Distributive Property}$$
$$3x \geq 14$$
$$x \geq \frac{14}{3}$$

The domain of $y=\sqrt{3(x-5)}+1$ is $x \geq \frac{14}{3}$. Because y can have any nonnegative value, the range is the set of all nonnegative real numbers.

TRY THIS

a. Find the domain of $\sqrt{3x+5}$.

b. Find the domain and range of $y=\sqrt{2(x-1)}+3$.

EXERCISES

Write the inequality that you would solve to find the domain of each expression or function. Do not solve.

1. $\sqrt{3m}$

2. $\sqrt{2(3t + 1)}$

3. $y = \sqrt{3(x - 4)} - 12$

PRACTICE

Find the domain of each square-root expression.

4. $\sqrt{4x}$

5. $\sqrt{15n}$

6. $\sqrt{-3r}$

7. $\sqrt{-7a}$

8. $\sqrt{2x - 9}$

9. $\sqrt{2a + 9}$

10. $\sqrt{5(d - 3)}$

11. $\sqrt{4(m + 5)}$

12. $\sqrt{-3(z + 2)} + 6$

Find the domain and range of each square-root function.

13. $y = \sqrt{11t}$

14. $y = \sqrt{-x}$

15. $y = \sqrt{3(g + 3)}$

16. $y = \sqrt{-2(k - 3)} + 5$

17. $y = \sqrt{7(s - 1)} - 2$

18. $y = \sqrt{-(3t - 2)} + 4$

19. $y = \sqrt{2(x - 3) + 2(x + 1)}$

20. $y = \sqrt{3(t + 1) - (t - 1)}$

21. $y = \sqrt{-(r + 3) - 2(r - 3)}$

22. Let a and b represent positive real numbers. Find the domain and range of $y = \sqrt{ax + b}$.

23. Find a such that the domain of $y = \sqrt{ax - 3}$ is the set of all real numbers greater than or equal to 5.

24. Find the domain and range of $y = \sqrt{2x - 9} + 3$. Justify your response.

25. **Critical Thinking**   Show that all the numbers in the domain of $y = \sqrt{2x - 7}$ are also in the domain of $y = \sqrt{2x - 5}$. Then show that the domain of $y = \sqrt{2x - 5}$ contains numbers that are not in the domain of $y = \sqrt{2x - 7}$.

MIXED REVIEW

Solve. Write the solutions in simplest radical form. (8.2 Skill A)

26. $7x^2 = 28$

27. $-6n^2 = -150$

28. $4p^2 + 3 = 28$

29. $3x^2 + 5 = 14$

30. $-2a^2 + 11 = -61$

31. $7v^2 - 11 = 38$

32. $11 = 11d^2$

33. $24 = 3u^2 - 3$

34. $-7 = -9t^2 + 2$

Solving Square-Root Equations

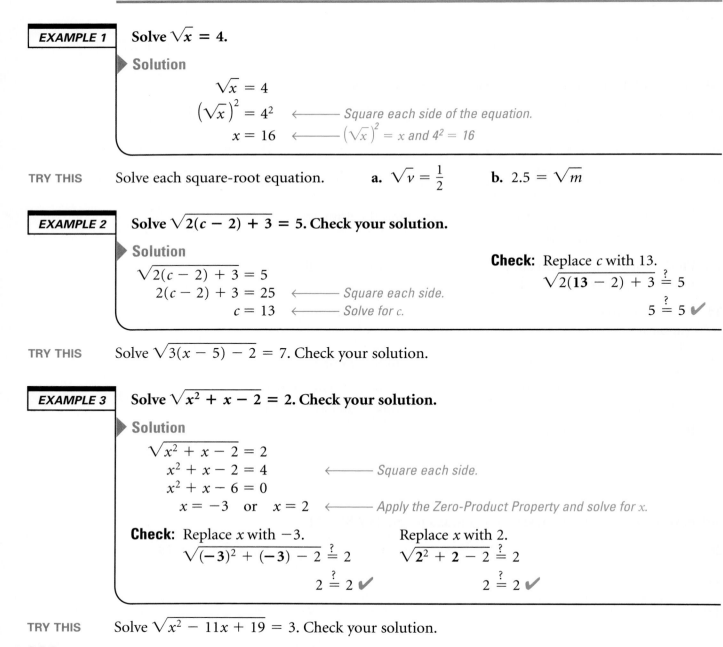

SKILL A *Solving equations in which one side is a square-root expression*

A **square-root equation** is an equation that contains a square-root expression. For example, $\sqrt{x} = 3$ and $\sqrt{x^2 + 2x} = 1$ are square-root equations. To solve such equations, apply the following property.

> **Property of Equality of Squares**
> If $r = s$, then $r^2 = s^2$. In words, if two expressions r and s are equal, then their squares, r^2 and s^2, are also equal.

EXAMPLE 1 Solve $\sqrt{x} = 4$.

▶ **Solution**
$$\sqrt{x} = 4$$
$$\left(\sqrt{x}\right)^2 = 4^2 \quad \longleftarrow \text{ Square each side of the equation.}$$
$$x = 16 \quad \longleftarrow \left(\sqrt{x}\right)^2 = x \text{ and } 4^2 = 16$$

TRY THIS Solve each square-root equation. **a.** $\sqrt{v} = \frac{1}{2}$ **b.** $2.5 = \sqrt{m}$

EXAMPLE 2 Solve $\sqrt{2(c - 2)} + 3 = 5$. **Check your solution.**

▶ **Solution**
$$\sqrt{2(c - 2)} + 3 = 5$$
$$2(c - 2) + 3 = 25 \quad \longleftarrow \text{ Square each side.}$$
$$c = 13 \quad \longleftarrow \text{ Solve for } c.$$

Check: Replace c with 13.
$$\sqrt{2(13 - 2)} + 3 \stackrel{?}{=} 5$$
$$5 \stackrel{?}{=} 5 \checkmark$$

TRY THIS Solve $\sqrt{3(x - 5)} - 2 = 7$. Check your solution.

EXAMPLE 3 Solve $\sqrt{x^2 + x - 2} = 2$. **Check your solution.**

▶ **Solution**
$$\sqrt{x^2 + x - 2} = 2$$
$$x^2 + x - 2 = 4 \quad \longleftarrow \text{ Square each side.}$$
$$x^2 + x - 6 = 0$$
$$x = -3 \quad \text{or} \quad x = 2 \quad \longleftarrow \text{ Apply the Zero-Product Property and solve for } x.$$

Check: Replace x with -3.
$$\sqrt{(-3)^2 + (-3) - 2} \stackrel{?}{=} 2$$
$$2 \stackrel{?}{=} 2 \checkmark$$

Replace x with 2.
$$\sqrt{2^2 + 2 - 2} \stackrel{?}{=} 2$$
$$2 \stackrel{?}{=} 2 \checkmark$$

TRY THIS Solve $\sqrt{x^2 - 11x + 19} = 3$. Check your solution.

EXERCISES

KEY SKILLS

Substitute the given values of x into the equation. Are the values of x solutions to the given square-root equation?

1. $x = 4$ or $x = -1$
$\sqrt{x^2 - 3x} = 2$

2. $x = -3$ or $x = 12$
$\sqrt{x^2 - 9x} = 6$

3. $x = 4$ or $x = 1$
$\sqrt{x^2 - 5x} = -2$

PRACTICE

Solve each square-root equation. Check your solution.

4. $\sqrt{t} = 6$

5. $\sqrt{c} = 9$

6. $10 = \sqrt{z}$

7. $15 = \sqrt{d}$

8. $10 = 2\sqrt{d}$

9. $-6 = -3\sqrt{d}$

10. $\sqrt{3x - 1} = 2$

11. $\sqrt{5n - 7} = 2$

12. $\sqrt{5n} = 5$

13. $\sqrt{-3p} = 2$

14. $\sqrt{-3(k + 1)} - 5 = 2$

15. $\sqrt{2(z + 1)} - z = 5$

16. $3\sqrt{5k + 1} = 27$

17. $-4\sqrt{4n - 3} = -12$

18. $1 = \sqrt{-3(s - 3) + (s - 1)}$

19. $5 = \sqrt{4(b - 1) - 2(b - 2)}$

20. $\sqrt{x^2 + 3} = 2$

21. $\sqrt{x^2 - 9} = 4$

22. $\sqrt{(v - 1)^2 + 84} = 10$

23. $5 = \sqrt{(u + 3)^2 - 11}$

24. $\sqrt{g^2 - 2g + 1} = 2$

25. $\sqrt{z^2 - 3z - 3} = 5$

26. $6 = \sqrt{n^2 + n + 6}$

27. $4 = \sqrt{q^2 - 4q + 19}$

28. Let a represent a real number. Solve $\sqrt{x^2 - a^2} = 1$ for x.

29. Let b represent a real number. Solve $\sqrt{x^2 + b^2} = 1$ for x.

30. Critical Thinking Let n represent a real number. For what value(s) of n will $\sqrt{x^2 - (x + n)^2} = 5$ not have a solution? Justify your response.

MIXED REVIEW

Multiply. (6.6 Skill B)

31. $(x + 1)(2x - 5)$

32. $(3y - 2)(y + 7)$

33. $(2n - 4)(2n + 2)$

34. $(1 + bx)(1 + ax)$

35. $(1 + bx)(1 - ax)$

36. $(1 - bx)(1 + ax)$

37. $(a + bx)(a + bx)$

38. $(a - bx)(a + bx)$

39. $(a + bx)(a - bx)$

You can use square-root equations to solve distance problems.

EXAMPLE 1

From an altitude of h kilometers, the approximate distance d in kilometers you can see to the horizon is given by $d = 113.14\sqrt{h}$. To see 25 kilometers to the horizon, how high up will you need to be?

▶ **Solution**

$d = 113.14\sqrt{h} \quad \rightarrow \quad 25 = 113.14\sqrt{h}$ ←———— *Replace d with 25.*

$$\frac{25}{113.14} = \sqrt{h}$$

$$h = \left(\frac{25}{113.14}\right)^2 \approx 0.05 \quad ←——— \text{\textit{Square each side of the equation. Use a calculator to find the approximate value of h.}}$$

You will need to be about 0.05 kilometers, or 50 meters, above Earth.

TRY THIS   Rework Example 1 if the desired distance is to be 150 kilometers.

If you apply the Pythagorean Theorem to find the distance between two points $P(x_1, y_1)$ and $Q(x_2, y_2)$ in the coordinate plane, you can derive what is called the *Distance Formula.*

The Distance Formula

If $P(x_1, y_1)$ and $Q(x_2, y_2)$ are two points in the coordinate plane, then the distance between them is given by the formula below.

$$PQ = \sqrt{(x_2 - x_1)^2 + (y_2 - y_1)^2}$$

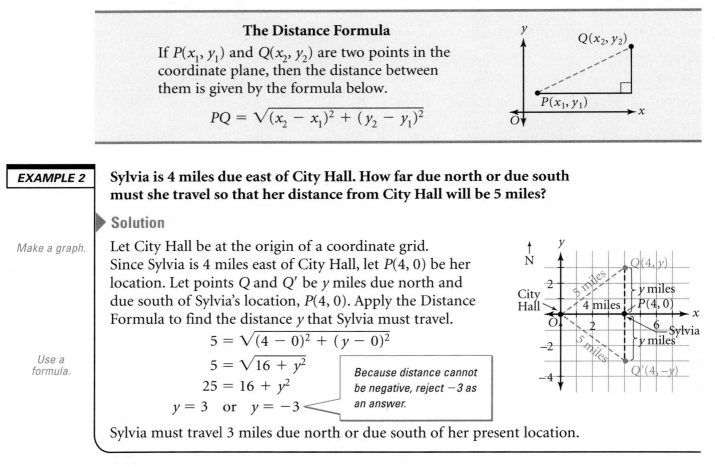

EXAMPLE 2

Sylvia is 4 miles due east of City Hall. How far due north or due south must she travel so that her distance from City Hall will be 5 miles?

▶ **Solution**

Make a graph.

Let City Hall be at the origin of a coordinate grid. Since Sylvia is 4 miles east of City Hall, let $P(4, 0)$ be her location. Let points Q and Q' be y miles due north and due south of Sylvia's location, $P(4, 0)$. Apply the Distance Formula to find the distance y that Sylvia must travel.

$$5 = \sqrt{(4 - 0)^2 + (y - 0)^2}$$

$$5 = \sqrt{16 + y^2}$$

Use a formula.

$$25 = 16 + y^2$$

$$y = 3 \quad \text{or} \quad y = -3 \longleftarrow \boxed{\text{\textit{Because distance cannot be negative, reject -3 as an answer.}}}$$

Sylvia must travel 3 miles due north or due south of her present location.

TRY THIS   Rework Example 2 if the desired distance is to be 10 miles.

EXERCISES

Use the distance formula to write an equation for the distance between points *P* and *Q*. Do not solve.

1. $P(0, 4), Q(1, 7)$

2. $P(2, 0), Q(3, 1)$

3. $P(-1, -1), Q(3, 5)$

4. $P(6, 1), Q(4, 4)$

PRACTICE

Refer to Example 1 on the previous page. Find each altitude needed to see the given distance to the horizon. Round each altitude to the nearest hundredth of a kilometer.

5. 30 kilometers

6. 24 kilometers

7. 36 kilometers

8. 154 kilometers

Find the coordinates of *Q* and *Q*′ that meet the specified conditions.

9. 12 units from $O(0, 0)$ and directly above or below $P(4, 0)$

10. 15 units from $O(0, 0)$ and directly left or right of $P(0, 4)$

11. 8 units from $C(2, 2)$ and directly above or below $Z(7, 2)$, as shown in the graph

12. 8 units from $K(4, 1)$ and directly left or right of $L(4, 6)$, as shown in the graph

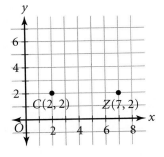

13. Solve the equation in Example 1 for altitude *h*.

MIXED REVIEW APPLICATIONS

If each student can complete a class project in the given amount of time, how long will it take the two students to complete the project if they work together? Round answers to the nearest whole unit. (9.7 Skill C)

14. John: 20 hours; Viola: 24 hours

15. Carol: 7 days; Carmen: 14 days

16. Ken: 34 hours; Reema: 48 hours

17. Jackie: 10 days; Bill: 12 days

A square-root equation may have a square-root expression on both sides of the equal sign. To solve such an equation, apply the Property of Equality of Squares; that is, square each side of the equation. Then solve the equation that results.

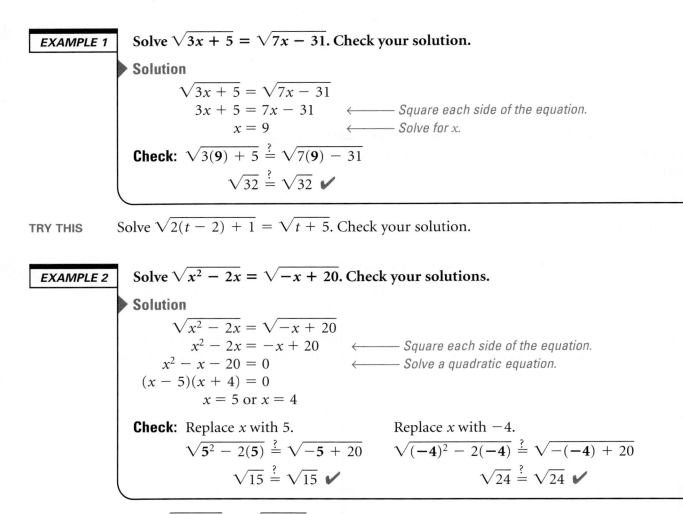

EXAMPLE 1 Solve $\sqrt{3x + 5} = \sqrt{7x - 31}$. Check your solution.

▶ **Solution**

$$\sqrt{3x + 5} = \sqrt{7x - 31}$$
$$3x + 5 = 7x - 31 \quad\longleftarrow\text{ Square each side of the equation.}$$
$$x = 9 \quad\longleftarrow\text{ Solve for } x.$$

Check: $\sqrt{3(9) + 5} \overset{?}{=} \sqrt{7(9) - 31}$
$$\sqrt{32} \overset{?}{=} \sqrt{32} \ ✔$$

TRY THIS Solve $\sqrt{2(t - 2) + 1} = \sqrt{t + 5}$. Check your solution.

EXAMPLE 2 Solve $\sqrt{x^2 - 2x} = \sqrt{-x + 20}$. Check your solutions.

▶ **Solution**

$$\sqrt{x^2 - 2x} = \sqrt{-x + 20}$$
$$x^2 - 2x = -x + 20 \quad\longleftarrow\text{ Square each side of the equation.}$$
$$x^2 - x - 20 = 0 \quad\longleftarrow\text{ Solve a quadratic equation.}$$
$$(x - 5)(x + 4) = 0$$
$$x = 5 \text{ or } x = 4$$

Check: Replace x with 5.   Replace x with -4.
$$\sqrt{5^2 - 2(5)} \overset{?}{=} \sqrt{-5 + 20} \qquad \sqrt{(-4)^2 - 2(-4)} \overset{?}{=} \sqrt{-(-4) + 20}$$
$$\sqrt{15} \overset{?}{=} \sqrt{15} \ ✔ \qquad\qquad\qquad \sqrt{24} \overset{?}{=} \sqrt{24} \ ✔$$

TRY THIS Solve $\sqrt{x^2 + 2x} = \sqrt{7x - 4}$. Check your solution(s).

- A square-root equation may have no solution in the set of real numbers. The equation $\sqrt{x} + 1 = 0$, for example, has no real solution. (However, recall from Lesson 8.2 that $\sqrt{x} + 1 = 0$ *can* be solved using imaginary numbers.)

- A square-root equation may have an *extraneous solution*, a solution that does not satisfy the given equation. The equation $\sqrt{x} = x - 2$ has the solutions 1 and 4 when you solve it by squaring each side of the equation. However, when you substitute both values back into the original equation, you will find that 1 is *not* a solution to the given equation.

$$\sqrt{4} \overset{?}{=} 4 - 2 \qquad\qquad \sqrt{1} \overset{?}{=} 1 - 2$$
$$2 \overset{?}{=} 2 \ ✔ \qquad\qquad\quad 1 \overset{?}{=} -1 \ ✗$$

EXERCISES

Write the equation that results from squaring each side of the given equation. Identify the equation as linear or quadratic. Do not solve.

1. $\sqrt{3(x-2)+5} = \sqrt{3x-1}$　　**2.** $\sqrt{(s-1)^2+1} = \sqrt{2s^2}$　　　**3.** $\sqrt{r^2-2r} = \sqrt{3r}$

PRACTICE

Solve each square-root equation. Check your solution.

4. $\sqrt{3d} = \sqrt{-2d}$　　　　　　　　　　　**5.** $\sqrt{3a} = \sqrt{a-5}$

6. $\sqrt{x} = \sqrt{2x-3}$　　　　　　　　　　　**7.** $\sqrt{b+2} = \sqrt{2-b}$

8. $\sqrt{3n+2} = \sqrt{4n}$　　　　　　　　　　**9.** $\sqrt{4(k-1)} = \sqrt{5k}$

10. $\sqrt{8} = \sqrt{5(z+2)}$　　　　　　　　　**11.** $\sqrt{4(g+1)-5} = \sqrt{6-(g-2)}$

12. $\sqrt{3-2(v-2)} = \sqrt{6v-2}$　　　　　**13.** $\sqrt{n^2} = \sqrt{4n-4}$

14. $\sqrt{k^2} = \sqrt{6k-9}$　　　　　　　　　　**15.** $\sqrt{z^2} = \sqrt{-8z-16}$

16. $\sqrt{p^2+25} = \sqrt{-10p}$　　　　　　　**17.** $\sqrt{a^2+12} = \sqrt{8a}$

18. $\sqrt{c^2-15} = \sqrt{-2c}$　　　　　　　　**19.** $\sqrt{x^2-10x+10} = \sqrt{x-14}$

Show that each square-root equation has no solution.

20. $\sqrt{x-3} = \sqrt{x-4}$　　　　　　　　　**21.** $\sqrt{3t+12} = \sqrt{3(4+t)} - 1$

In Exercises 22–24, let *a*, *b*, *c*, and *d* represent real numbers. Solve each equation for *x*.

22. $\sqrt{ax} = \sqrt{b+c}$　　　　　**23.** $\sqrt{ax+b} = \sqrt{cx+d}$　　　　**24.** $\sqrt{ax^2+b} = \sqrt{bx^2+a}$

25. **Critical Thinking**　Triangle *OAB* has vertices $O(0, 0)$, $A(6, r)$, and $B(r, 6)$ for some nonzero real number *r*. Show that the triangle always has two sides that are equal in length.

26. Solve $\sqrt{n+1} = 1 + \sqrt{n-12}$. Check your solution(s).

MIXED REVIEW

Write each expression in simplest radical form.　(8.1 Skill B)

27. $\sqrt{147}$　　　　　**28.** $\sqrt{320}$　　　　　**29.** $\sqrt{891}$　　　　　**30.** $\sqrt{625}$

31. $\sqrt{\dfrac{3}{100}}$　　　　　**32.** $\sqrt{\dfrac{7}{225}}$　　　　　**33.** $\sqrt{\dfrac{98}{121}}$　　　　　**34.** $\sqrt{\dfrac{75}{169}}$

10.3 LESSON
Numbers of the form $a + b\sqrt{2}$

Adding and subtracting numbers of the form $a + b\sqrt{2}$

If a and b are rational numbers, then all numbers of the form $a + b\sqrt{2}$ are members of a subset of the real numbers. This subset of the real numbers follows all of the basic properties of real numbers for addition and multiplication that you learned in Lesson 1.2 Skill A: Closure, Associative, Identity, and Inverse.

To add two numbers of the form $a + b\sqrt{2}$, group like terms.

EXAMPLE 1 **Simplify** $(3 + 5\sqrt{2}) + (-5 + 7\sqrt{2})$.

▶ **Solution**
$$(3 + 5\sqrt{2}) + (-5 + 7\sqrt{2}) = (3 + (-5)) + (5\sqrt{2} + 7\sqrt{2})$$
$$= -2 + (5 + 7)\sqrt{2}$$
$$= -2 + 12\sqrt{2}$$

TRY THIS Simplify. **a.** $(-2 + 5\sqrt{2}) + (2 + (-3)\sqrt{2})$ **b.** $(4 + 5\sqrt{2}) + (3 + (-7)\sqrt{2})$

Every number of the form $a + b\sqrt{2}$, has an opposite, or additive inverse, $(-a) + (-b)\sqrt{2}$. To subtract one number of the form $a + b\sqrt{2}$ from another number of this form, add the opposite.

EXAMPLE 2 **Simplify** $(3 + 5\sqrt{2}) - (-4 + 6\sqrt{2})$.

▶ **Solution**
$$(3 + 5\sqrt{2}) - (-4 + 6\sqrt{2}) = (3 + 5\sqrt{2}) + (4 - 6\sqrt{2})$$ ⟵ Write subtraction as
$$= (3 + 4) + (5\sqrt{2} - 6\sqrt{2})$$ addition of the opposite.
$$= 7 + (5 - 6)\sqrt{2}$$
$$= 7 - \sqrt{2}$$

TRY THIS Simplify. **a.** $(3 + 5\sqrt{2}) - (7 - 3\sqrt{2})$ **b.** $(7 + 5\sqrt{2}) - (2 + 5\sqrt{2})$

The sum of two numbers of the form $a + b\sqrt{2}$ is another number of the form $a + b\sqrt{2}$. Thus, the set of numbers of the form $a + b\sqrt{2}$ is *closed* under addition. That is, the set follows the Closure Property of Addition. For example, the sum $(3 + 5\sqrt{2}) + (-5 + 7\sqrt{2})$ equals $-2 + 12\sqrt{2}$, which is another number of the form $a + b\sqrt{2}$.

EXERCISES

Simplify each number.

1. $\frac{3}{4} + 0\sqrt{2}$ 2. $0 + 5\sqrt{2}$ 3. $0 + 1\sqrt{2}$ 4. $0 - 1\sqrt{2}$

PRACTICE

Simplify. Write answers in the form $a + b\sqrt{2}$.

5. $3\sqrt{2} + 2\sqrt{2}$ 6. $3\sqrt{2} - 2\sqrt{2}$

7. $7\sqrt{2} + (-2)\sqrt{2}$ 8. $4\sqrt{2} - (-3)\sqrt{2}$

9. $-11\sqrt{2} + (-9)\sqrt{2}$ 10. $-\sqrt{2} - (-13)\sqrt{2}$

11. $\left(7 + 3\sqrt{2}\right) + \left(5 + \sqrt{2}\right)$ 12. $\left(4 - 4\sqrt{2}\right) + \left(-7 + 2\sqrt{2}\right)$

13. $\left(5 + 4\sqrt{2}\right) - \left(-3 + 4\sqrt{2}\right)$ 14. $\left(5 + 4\sqrt{2}\right) - \left(5 - 4\sqrt{2}\right)$

15. $\left(5 + 4\sqrt{2}\right) + \left(-5 - 4\sqrt{2}\right)$ 16. $\left(7 + 4\sqrt{2}\right) + \left(-7 - 3\sqrt{2}\right)$

17. $\left(-7 - 5\sqrt{2}\right) - \left(7 - 3\sqrt{2}\right)$ 18. $\left(-11 - 5\sqrt{2}\right) + \left(5 - 11\sqrt{2}\right)$

19. $\left(3 - 3\sqrt{2}\right) + \left(5 - 5\sqrt{2}\right) + \left(2 - 2\sqrt{2}\right)$ 20. $\left(3 - 5\sqrt{2}\right) + \left(6 - 3\sqrt{2}\right) - \left(7 - 3\sqrt{2}\right)$

Two expressions $a + b\sqrt{2}$ and $c + d\sqrt{2}$ are *equal* if $a = c$ and $b = d$. Find r and s such that each equation is true.

21. $\left(3 + 7\sqrt{2}\right) + \left(r + s\sqrt{2}\right) = -1 + 5\sqrt{2}$

22. $\left(6 - 5\sqrt{2}\right) - \left(r + s\sqrt{2}\right) = 7 + 2\sqrt{2}$

23. $\left[(r + s) - (r - s)\sqrt{2}\right] + (r + s)\sqrt{2} = 6 + 4\sqrt{2}$

24. $(r^2 + 4) + (s^2 + 9)\sqrt{2} = -4r + 6s\sqrt{2}$

Let a and b represent rational numbers. Prove each statement using properties of addition, multiplication, and equality.

25. $\left(a + b\sqrt{2}\right) + \left[(-a) + (-b)\sqrt{2}\right] = 0$ 26. $\left(a + b\sqrt{2}\right) + \left(a - b\sqrt{2}\right)$ is a rational number.

MIXED REVIEW

Multiply. (6.6 Skill B)

27. $(2 + 3a)(4 + 2a)$ 28. $(-2 + 3a)(-5 + 2a)$ 29. $(-2 - 2a)(-3 + 2a)$

30. $(5 + 3b)(5 + 3b)$ 31. $(1 + b)(-2 + b)$ 32. $(3 - 4b)(-1 + 3b)$

When you multiply or divide numbers of the form $a + b\sqrt{2}$, the result is another number of the form $a + b\sqrt{2}$.

EXAMPLE 1

Simplify $\left(3 + 4\sqrt{2}\right)\left(-5 + 2\sqrt{2}\right)$.

▶ **Solution**

$$\left(3 + 4\sqrt{2}\right)\left(-5 + 2\sqrt{2}\right) = 3\left(-5 + 2\sqrt{2}\right) + 4\sqrt{2}\left(-5 + 2\sqrt{2}\right)$$

> Apply the Distributive Property.

$$= -15 + 6\sqrt{2} - 20\sqrt{2} + 8\sqrt{2} \cdot \sqrt{2}$$

$$= -15 + 6\sqrt{2} - 20\sqrt{2} + 8(2) \qquad \longleftarrow \sqrt{2} \cdot \sqrt{2} = 2$$

$$= (-15 + 16) + (6 - 20)\sqrt{2}$$

$$= 1 - 14\sqrt{2}$$

TRY THIS Simplify. **a.** $6\sqrt{2}\left(5 - 3\sqrt{2}\right)$ **b.** $\left(7 + 5\sqrt{2}\right)\left(-1 + (-5)\sqrt{2}\right)$

To simplify a quotient that has a square root in the denominator, multiply the numerator and the denominator by a number that will eliminate the square root sign from the denominator. To simplify $\left(a + b\sqrt{2}\right) \div \left(c + d\sqrt{2}\right)$, multiply the numerator and the denominator by $c - d\sqrt{2}$, which is the **conjugate** of $c + d\sqrt{2}$. When you multiply a number by its conjugate, the result has no radical signs. For example, the conjugate of $3 + \sqrt{2}$ is $3 - \sqrt{2}$ and $\left(3 + \sqrt{2}\right)\left(3 - \sqrt{2}\right) = 7$.

EXAMPLE 2

Simplify.   **a.** $\left(2 + 9\sqrt{2}\right) \div \left(5\sqrt{2}\right)$ **b.** $\left[5 + (-4)\sqrt{2}\right] \div \left(3 + \sqrt{2}\right)$

▶ **Solution**

a. $\dfrac{2 + 9\sqrt{2}}{5\sqrt{2}} = \dfrac{2 + 9\sqrt{2}}{5\sqrt{2}}\left(\dfrac{\sqrt{2}}{\sqrt{2}}\right)$

$$= \dfrac{\left(2 + 9\sqrt{2}\right)\left(\sqrt{2}\right)}{5 \cdot 2}$$

$$= \dfrac{2\sqrt{2} + 9 \cdot 2}{10}$$

$$= \dfrac{18 + 2\sqrt{2}}{10}$$

$$= \dfrac{9 + \sqrt{2}}{5}$$

b. $\dfrac{5 + (-4)\sqrt{2}}{3 + \sqrt{2}} = \dfrac{5 + (-4)\sqrt{2}}{3 + \sqrt{2}}\left(\dfrac{3 - \sqrt{2}}{3 - \sqrt{2}}\right)$

$$= \dfrac{\left(5 - 4\sqrt{2}\right)\left(3 - \sqrt{2}\right)}{\left(3 + \sqrt{2}\right)\left(3 - \sqrt{2}\right)}$$

$$= \dfrac{\left(5 - 4\sqrt{2}\right)\left(3 - \sqrt{2}\right)}{9 - 3\sqrt{2} + 3\sqrt{2} - 2}$$

$$= \dfrac{15 - 5\sqrt{2} - 12\sqrt{2} + 4 \cdot 2}{7}$$

$$= \dfrac{23 - 17\sqrt{2}}{7}$$

TRY THIS Simplify. **a.** $\dfrac{5}{3\sqrt{2}}$ **b.** $\dfrac{5 + 7\sqrt{2}}{3\sqrt{2}}$ **c.** $\dfrac{2 + 5\sqrt{2}}{1 + 3\sqrt{2}}$

- The product of two numbers of the form $a + b\sqrt{2}$ is another number of the form $a + b\sqrt{2}$. Thus, the set of numbers of the form $a + b\sqrt{2}$ is *closed* under multiplication.

- Every nonzero number of the form $a + b\sqrt{2}$, where a and b are rational numbers, has a reciprocal, or multiplicative inverse, that is also in the set of numbers of the form $a + b\sqrt{2}$.

EXERCISES

Complete the simplification of each product.

1. $\left(2 - 4\sqrt{2}\right)\left(3 + 5\sqrt{2}\right) = 2 \cdot 3 + 2 \cdot 5\sqrt{2} - 4\sqrt{2} \cdot (3) - 4\sqrt{2} \cdot 5\sqrt{2}$

2. $\left(2 + 5\sqrt{2}\right)\left(3 - 2\sqrt{2}\right) = 2 \cdot 3 - 2 \cdot 2\sqrt{2} + 5\sqrt{2} \cdot (3) - 5\sqrt{2} \cdot 2\sqrt{2}$

Simplify. Write the answers in the form $a + b\sqrt{2}$.

3. $3\left(2 + 5\sqrt{2}\right)$

4. $4\left(2 - 5\sqrt{2}\right)$

5. $-2\left(1 + 7\sqrt{2}\right)$

6. $-3\left(1 - 5\sqrt{2}\right)$

7. $2\sqrt{2}\left(1 + \sqrt{2}\right)$

8. $2\sqrt{2}\left(3 - 2\sqrt{2}\right)$

9. $-5\sqrt{2}\left(1 - 3\sqrt{2}\right)$

10. $-4\sqrt{2}\left(-3 - 5\sqrt{2}\right)$

11. $\dfrac{1}{\sqrt{2}}$

12. $\dfrac{3}{5\sqrt{2}}$

13. $\dfrac{-4}{5\sqrt{2}}$

14. $\dfrac{4}{-3\sqrt{2}}$

15. $\left(3 + 5\sqrt{2}\right)\left(3 - \sqrt{2}\right)$

16. $\left(3 - \sqrt{2}\right)\left(7 - \sqrt{2}\right)$

17. $\left(-2 + 4\sqrt{2}\right)\left(7 - 3\sqrt{2}\right)$

18. $\left(-2 - 3\sqrt{2}\right)\left(5 - 6\sqrt{2}\right)$

19. $\left(2 - 5\sqrt{2}\right)^2$

20. $\left(3 + 7\sqrt{2}\right)^2$

21. $\dfrac{3 + \sqrt{2}}{6 + \sqrt{2}}$

22. $\dfrac{3 + 5\sqrt{2}}{6 - \sqrt{2}}$

23. $\dfrac{-3 + 2\sqrt{2}}{3 - \sqrt{2}}$

24. $\dfrac{7 - 2\sqrt{2}}{3 + 2\sqrt{2}}$

25. $\dfrac{5 - 2\sqrt{2}}{5 - 7\sqrt{2}}$

26. $\dfrac{3 - 7\sqrt{2}}{1 - 7\sqrt{2}}$

Find $\left(r + s\sqrt{2}\right)^3$ by using $\left(r + s\sqrt{2}\right)^2\left(r + s\sqrt{2}\right)$. Write the answers in the form $a + b\sqrt{2}$.

27. $\left(1 + \sqrt{2}\right)^3$

28. $\left(3 - \sqrt{2}\right)^3$

29. $\left(2 + 3\sqrt{2}\right)^3$

30. $\left(5 - 3\sqrt{2}\right)^3$

Let a and b represent rational numbers. Prove each statement.

31. $\left(a + b\sqrt{2}\right)\left(a - b\sqrt{2}\right) = a^2 - 2b^2$

32. $\dfrac{1}{a + b\sqrt{2}} = \dfrac{a}{a^2 - 2b^2} - \dfrac{b}{a^2 - 2b^2}\left(\sqrt{2}\right)$ $a \neq 0$ and $b \neq 0$

Simplify. (previous courses)

33. $4\frac{2}{3} + 4\frac{4}{7}$

34. $4\frac{2}{3} + 11\frac{1}{5}$

35. $7\frac{2}{5} + 7\frac{2}{5}$

36. $3\frac{3}{11} + 5\frac{1}{2}$

37. $7\frac{5}{9} - 5\frac{2}{3}$

38. $10\frac{1}{10} - 6\frac{9}{10}$

39. $11\frac{4}{9} - 11\frac{3}{11}$

40. $1\frac{1}{3} - \frac{4}{13}$

What follows is a summary of what you learned so far in this lesson.

Addition, Multiplication, and Numbers of the Form $a + b\sqrt{2}$

- Addition and multiplication of these numbers are closed.
- Every number of this form has an additive inverse and every nonzero number of this form has a multiplicative inverse.
- The Commutative, Associative, and Identity properties of addition and multiplication are true for these numbers.

Any set of numbers that satisfies these properties and the Distributive Property is called a **field.** The set of real numbers, the set of numbers of the form $a + b\sqrt{2}$, and the set of rational numbers are all *fields.*

EXAMPLE 1

Prove that addition of numbers of the form $a + b\sqrt{2}$, where a and b are rational numbers, is closed.

▶ **Solution**

To prove that addition of numbers of the form $a + b\sqrt{2}$ is closed, we must show that the sum of any two numbers of that form is another number of that form. Let r, s, t, and u be rational numbers.

$$\left(r + s\sqrt{2}\right) + \left(t + u\sqrt{2}\right)$$
$$= \left(r + t\right) + \left(s\sqrt{2} + u\sqrt{2}\right) \quad \longleftarrow \text{ Commutative and Associative Properties of Addition}$$
$$= (r + t) + (s + u)\sqrt{2} \quad \longleftarrow \text{ Distributive Property}$$

Since addition and multiplication of rational numbers are closed, then $r + t$ and $s + u$ must also be rational numbers. Therefore, the sum has the form $a + b\sqrt{2}$, where $a = r + t$ and $b = s + u$.

TRY THIS

Prove that $\left(-7 + 3\sqrt{2}\right) + \left(1 + (-5)\sqrt{2}\right)$ has the form $a + b\sqrt{2}$.

You can solve simple equations involving numbers of the form $a + b\sqrt{2}$.

EXAMPLE 2

Solve $\left(2 + 3\sqrt{2}\right)x + \left(5 - 3\sqrt{2}\right) = 4 - 2\sqrt{2}$.

▶ **Solution**

$$\left(2 + 3\sqrt{2}\right)x + \left(5 - 3\sqrt{2}\right) = 4 - 2\sqrt{2}$$
$$\left(2 + 3\sqrt{2}\right)x = -1 + \sqrt{2} \quad \longleftarrow \text{ Subtract } 5 - 3\sqrt{2} \text{ from each side.}$$
$$x = \frac{-1 + \sqrt{2}}{2 + 3\sqrt{2}} \quad \longleftarrow \text{ Divide each side by } 2 + 3\sqrt{2}.$$
$$x = \frac{-1 + \sqrt{2}}{2 + 3\sqrt{2}}\left(\frac{2 - 3\sqrt{2}}{2 - 3\sqrt{2}}\right) \quad \longleftarrow \text{ Multiply both the numerator and}$$
$$\qquad\qquad\qquad\qquad\qquad\qquad\quad \text{the denominator by } 2 - 3\sqrt{2},$$
$$x = \frac{4}{7} + \frac{-5}{14}\sqrt{2} \qquad\qquad\qquad \text{the conjugate of } 2 + 3\sqrt{2}.$$

TRY THIS

Solve $\left(1 + 2\sqrt{2}\right)x + \left(3 - 2\sqrt{2}\right) = 4 - 2\sqrt{2}$.

EXERCISES

Write the additive inverse of each number.

1. $-2.5 + 5\sqrt{2}$

2. $-4 - 7\sqrt{2}$

3. $4 - 1.4\sqrt{2}$

Show that each sum or product has the form $a + b\sqrt{2}$.

4. $\left(2 + \sqrt{2}\right) + \left(3 - 7\sqrt{2}\right)$

5. $\left(2 - \sqrt{2}\right) + \left(5 + 4\sqrt{2}\right)$

6. $\left(4 - 3\sqrt{2}\right)\left(3 - \sqrt{2}\right)$

Solve each equation. Write the solutions in the form $a + b\sqrt{2}$.

7. $n + \left(2 - 4\sqrt{2}\right) = 1 - 3\sqrt{2}$

8. $m - \left(2 - 3\sqrt{2}\right) = 6 - 5\sqrt{2}$

9. $\left(1 + 2\sqrt{2}\right)k = 4 + 3\sqrt{2}$

10. $\dfrac{d}{1 + 2\sqrt{2}} = 2 - \sqrt{2}$

11. $2n + 3 = 7 - 2\sqrt{2}$

12. $2t + \left(5 + \sqrt{2}\right) = 4 - 3\sqrt{2}$

13. Find n such that $nx = 3 + 5\sqrt{2}$ has $2 + \sqrt{2}$ as the solution for x.

Find the domain of each expression or function. (10.1 Skill A)

14. $\sqrt{3 - 4(a + 1)}$

15. $y = \sqrt{4(x + 5) - 2(x + 6)}$

Solve each equation. (10.2 Skill A)

16. $\sqrt{6(t - 7)} + 5 = 3$

17. $\sqrt{12v - 4v^2} = 3$

18. Let O represent the origin of a coordinate system. Find the coordinates of the points Q and Q' directly above and below $P(-3, 0)$ and such that $OQ = OQ' = 5$. (10.2 Skill B)

Solve each equation. (10.2 Skill C)

19. $\sqrt{3(r - 1)} - 4 = \sqrt{10 - 2(r + 5)}$

20. $\sqrt{6m^2 + 21m} = \sqrt{10m + 35}$

Simplify. (10.3 Skills A and B)

21. $\left(4 - 6\sqrt{2}\right) - \left(11 + 11\sqrt{2}\right)$

22. $\left(-1 - \sqrt{2}\right)\left(10 + 10\sqrt{2}\right)$

23. $\dfrac{5 - \sqrt{2}}{3\sqrt{2}}$

Solve each equation. (10.3 Skill C)

24. $\left(3 + 2\sqrt{2}\right)q + \left(3 + \sqrt{2}\right) = 3 + 4\sqrt{2}$

25. $\left(3 + 5\sqrt{2}\right)v - \left(3 - \sqrt{2}\right) = -3 + 2\sqrt{2}$

10.4

Operations on Square-Root Expressions

SKILL A ▶ *Using properties of square roots to simplify products and quotients*

Recall the Product Property of Square Roots from Lesson 8.1. It can be restated in the form given below.

If a and b are nonnegative real numbers, then $\sqrt{a} \cdot \sqrt{b} = \sqrt{ab}$.

EXAMPLE 1 **Simplify $\sqrt{3} \cdot \sqrt{12}$. Write the answer in simplest radical form.**

▶ **Solution**

$$\sqrt{3} \cdot \sqrt{12} = \sqrt{3 \cdot 12} = \sqrt{36} = \sqrt{6^2} = 6$$

Notice that an answer in "simplest radical form" does not have to contain a radical.

TRY THIS Simplify $\sqrt{54} \cdot \sqrt{6}$. Write the answer in simplest radical form.

Recall the Quotient Property of Square Roots from Lesson 8.1. It can be restated in the form given below.

If a and b are positive real numbers, then $\dfrac{\sqrt{a}}{\sqrt{b}} = \sqrt{\dfrac{a}{b}}$.

EXAMPLE 2 **Simplify. Write the answers in simplest radical form.**   **a.** $\dfrac{\sqrt{405}}{\sqrt{5}}$ **b.** $\dfrac{\sqrt{300}}{\sqrt{15}}$

▶ **Solution**

a. $\dfrac{\sqrt{405}}{\sqrt{5}} = \sqrt{\dfrac{405}{5}} = \sqrt{81} = 9$ **b.** $\dfrac{\sqrt{300}}{\sqrt{15}} = \sqrt{\dfrac{300}{15}} = \sqrt{20} = \sqrt{4 \cdot 5} = 2\sqrt{5}$

TRY THIS Simplify. Write the answers in simplest radical form.   **a.** $\dfrac{\sqrt{75}}{\sqrt{3}}$ **b.** $\dfrac{\sqrt{252}}{\sqrt{21}}$

To simplify the expressions in Example 3, multiply each expression by a clever choice for 1 to make the denominator an integer. This process of changing a fraction with a radical in the denominator to an *equivalent* fraction with a rational denominator is called **rationalizing the denominator.**

EXAMPLE 3 **Simplify by rationalizing the denominator.**   **a.** $\dfrac{\sqrt{25}}{\sqrt{3}}$ **b.** $\dfrac{\sqrt{27}}{\sqrt{8}}$

▶ **Solution**

a. $\dfrac{\sqrt{25}}{\sqrt{3}} = \dfrac{5}{\sqrt{3}} = \dfrac{5}{\sqrt{3}}\left(\dfrac{\sqrt{3}}{\sqrt{3}}\right) = \dfrac{5\sqrt{3}}{3}$ **b.** $\dfrac{\sqrt{27}}{\sqrt{8}} = \dfrac{3\sqrt{3}}{2\sqrt{2}}\left(\dfrac{\sqrt{2}}{\sqrt{2}}\right) = \dfrac{3\sqrt{6}}{4}$

TRY THIS Simplify by rationalizing the denominator.   **a.** $\dfrac{\sqrt{36}}{\sqrt{7}}$ **b.** $\dfrac{\sqrt{75}}{\sqrt{24}}$

EXERCISES

Write each expression in simplest radical form.

1. $\sqrt{5^2} \cdot \sqrt{7}$

2. $\sqrt{5^2} \cdot \sqrt{5}$

3. $\sqrt{3^2} \cdot \sqrt{5^2}$

4. $\sqrt{2^2} \cdot \sqrt{3^2} \cdot \sqrt{11}$

PRACTICE

Simplify each expression.

5. $\sqrt{49} \cdot \sqrt{2}$

6. $\sqrt{144} \cdot \sqrt{3}$

7. $\sqrt{121} \cdot \sqrt{3}$

8. $\sqrt{64} \cdot \sqrt{5}$

9. $\sqrt{6} \cdot \sqrt{24}$

10. $\sqrt{10} \cdot \sqrt{40}$

11. $\sqrt{7} \cdot \sqrt{14}$

12. $\sqrt{5} \cdot \sqrt{10}$

13. $\dfrac{\sqrt{200}}{\sqrt{2}}$

14. $\dfrac{\sqrt{27}}{\sqrt{3}}$

15. $\dfrac{\sqrt{343}}{\sqrt{7}}$

16. $\dfrac{\sqrt{605}}{\sqrt{5}}$

17. $\dfrac{\sqrt{60}}{\sqrt{20}}$

18. $\dfrac{\sqrt{112}}{\sqrt{14}}$

19. $\dfrac{\sqrt{105}}{\sqrt{35}}$

20. $\dfrac{\sqrt{96}}{\sqrt{12}}$

21. $\dfrac{\sqrt{5}}{\sqrt{3}}$

22. $\dfrac{\sqrt{2}}{\sqrt{7}}$

23. $\dfrac{\sqrt{5}}{\sqrt{8}}$

24. $\dfrac{\sqrt{50}}{\sqrt{27}}$

25. $\dfrac{\sqrt{3}}{\sqrt{200}}$

26. $\dfrac{\sqrt{2000}}{\sqrt{72}}$

27. $\dfrac{\sqrt{343}}{\sqrt{75}}$

28. $\dfrac{\sqrt{512}}{\sqrt{12}}$

Critical Thinking   Write each expression as a single square-root expression in simplest form. State any restrictions on the variables. (*Hint:* Remember that an expression under a square-root sign must be nonnegative.)

29. $\dfrac{\sqrt{x^2 + 5x + 6}}{\sqrt{x^2 + 6x + 9}}$

30. $\dfrac{\sqrt{a^2 + 10a + 25}}{\sqrt{a^2 - 25}}$

31. $\dfrac{\sqrt{2z^2 + 19z - 33}}{\sqrt{z^2 + 22z + 121}}$

32. **Critical Thinking**   If $4 \le x \le 16$, between what two numbers will the value of $\dfrac{\sqrt{x^3}}{\sqrt{2}}$ be found?

MIXED REVIEW

Evaluate each expression. (8.1 Skill C)

33. $1^{\frac{1}{2}}$

34. $169^{\frac{1}{2}}$

35. $225^{\frac{1}{2}}$

36. $256^{\frac{1}{2}}$

37. $100^{\frac{1}{2}}$

38. $\left(\dfrac{9}{16}\right)^{\frac{1}{2}}$

39. $\left(\dfrac{81}{25}\right)^{\frac{1}{2}}$

40. $\left(\dfrac{1}{49}\right)^{\frac{1}{2}}$

If p is a prime number, then $\sqrt{p}$ is an irrational number. For example, $\sqrt{2}$, $\sqrt{3}$, and $\sqrt{5}$ are irrational numbers. Using properties of addition and muliplication and the Distributive Property, you can simplify expressions involving $\sqrt{p}$.

EXAMPLE 1   **Simplify.**   **a.** $-11\sqrt{7} + 8\sqrt{7}$ **b.** $\left(10 + 7\sqrt{3}\right) + \left(5 + 4\sqrt{3}\right).$

▶ **Solution**

Combine like terms by using the Distributive Property.

a. $-11\sqrt{7} + 8\sqrt{7} = (-11 + 8)\sqrt{7} = -3\sqrt{7}$

b. $\left(10 + 7\sqrt{3}\right) + \left(5 + 4\sqrt{3}\right) = (10 + 5) + (7 + 4)\sqrt{3} = 15 + 11\sqrt{3}$

TRY THIS   Simplify.   **a.** $-2\sqrt{5} + \left((-3)\sqrt{5}\right)$ **b.** $\left(-1 + 3\sqrt{5}\right) + \left(-4 - 3\sqrt{5}\right)$

In Examples 2 and 3, you need to write the expressions in simplest radical form before you can combine like terms.

EXAMPLE 2   **Simplify $\sqrt{75} - \sqrt{108}$.**

▶ **Solution**

$$\begin{aligned}
\sqrt{75} - \sqrt{108} &= \sqrt{25 \cdot 3} - \sqrt{36 \cdot 3} \\
&= 5\sqrt{3} - 6\sqrt{3} &&\longleftarrow \text{\textit{Write in simplest radical form.}} \\
&= (5 - 6)\sqrt{3} &&\longleftarrow \text{\textit{Factor } } \sqrt{3} \text{ \textit{ from each term.}} \\
&= -\sqrt{3}
\end{aligned}$$

TRY THIS   Simplify.   **a.** $\sqrt{3} + 3\sqrt{75}$ **b.** $-2\sqrt{54} + 3\sqrt{24}$

Recall from Lesson 2.3 that $-(a + b) = -a - b$.

EXAMPLE 3   **Simplify $\left(2 + 3\sqrt{12}\right) - \left(-5 + \sqrt{27}\right)$.**

▶ **Solution**

$$\begin{aligned}
\left(2 + 3\sqrt{12}\right) - \left(-5 + \sqrt{27}\right) &= \left(2 + 3\sqrt{12}\right) + \left(5 - \sqrt{27}\right) &&\longleftarrow -\left(-5 + \sqrt{27}\right) = 5 - \sqrt{27} \\
&= \left(2 + 6\sqrt{3}\right) + \left(5 - 3\sqrt{3}\right) &&\longleftarrow \text{\textit{Write } } 3\sqrt{12} \text{ \textit{ and } } \sqrt{27} \text{ \textit{ in}} \\
&= (2 + 5) + \left(6\sqrt{3} - 3\sqrt{3}\right) &&\quad\;\; \text{\textit{simplest radical form.}} \\
&= 7 + (6 - 3)\sqrt{3} \\
&= 7 + 3\sqrt{3}
\end{aligned}$$

TRY THIS   Simplify.   **a.** $\left(10 + 7\sqrt{75}\right) - \left(5 + \sqrt{3}\right)$ **b.** $\left(-3 + \sqrt{75}\right) - \left(4 - 3\sqrt{27}\right)$

EXERCISES

Simplify using mental math.

1. $\left(6 + \sqrt{11}\right) - 5\sqrt{11}$

2. $\left(7 - 5\sqrt{2}\right) + 5\sqrt{2}$

3. $\left(7 + \sqrt{3}\right) + \left(-7 + 5\sqrt{3}\right)$

4. $\left(6 + \sqrt{5}\right) + \left(6 - 5\sqrt{5}\right)$

5. $\left(-3 + \sqrt{5}\right) + \left(7 + 4\sqrt{5}\right)$

6. $\left(-7 + \sqrt{7}\right) - \left(-7 - \sqrt{7}\right)$

PRACTICE

Simplify. Write the answers in simplest radical form. (Recall that simplest radical form may consist of numbers with no radical sign.)

7. $\left(-1 + 2\sqrt{5}\right) + \left(-6 - 2\sqrt{5}\right)$

8. $\left(-7 - 2\sqrt{2}\right) + \left(1 - 2\sqrt{2}\right)$

9. $\left(-17 - 7\sqrt{7}\right) + \left(2 + 7\sqrt{7}\right)$

10. $\left(-4 + 2\sqrt{11}\right) + \left(-8 - 2\sqrt{11}\right)$

11. $\left(7 + 4\sqrt{5}\right) - \left(1 - \sqrt{5}\right)$

12. $\left(-7 + \sqrt{7}\right) - \left(3 + 4\sqrt{7}\right)$

13. $\left(3 - 4\sqrt{3}\right) - \left(-3 - 4\sqrt{3}\right)$

14. $\left(2.5 - 7\sqrt{13}\right) - \left(4.5 - 7\sqrt{13}\right)$

15. $\sqrt{27} + \sqrt{27}$

16. $\sqrt{24} + \sqrt{54}$

17. $\sqrt{125} - \sqrt{5}$

18. $\sqrt{98} - \sqrt{8}$

19. $\left(4 - 4\sqrt{27}\right) + \left(3 - \sqrt{147}\right)$

20. $\left(4 + 8\sqrt{125}\right) + \left(-7 + 3\sqrt{5}\right)$

21. $\left(1 - 7\sqrt{32}\right) - \left(7 + 7\sqrt{128}\right)$

22. $\left(3 - 2\sqrt{32}\right) - \left(-3 + 2\sqrt{128}\right)$

23. $3 - 2\sqrt{32} + 2\sqrt{128} - 3\sqrt{8}$

24. $\left(-2 - 7\sqrt{5}\right) + 2\sqrt{125} - 3\sqrt{625}$

Let a and b represent rational numbers with $b > 0$. Simplify each expression. Write the answers in simplest radical form.

25. $\left(a + 8\sqrt{2b^2}\right) + \left(3a + 3\sqrt{8b^2}\right)$

26. $\left(a^2 + 8\sqrt{3b^2}\right) - \left(a^2 + 3\sqrt{27b^2}\right)$

27. **Critical Thinking** Suppose that $x > 0$. Simplify the expression.
$\sqrt{x} + \sqrt{x^3} + \sqrt{x^5} + \sqrt{x^7} + \sqrt{x^9} + \sqrt{x^{11}} + \sqrt{x^{13}} + \sqrt{x^{15}} + \sqrt{x^{17}}$.

MIXED REVIEW

Simplify. (10.3 Skill B)

28. $\left(3 - 2\sqrt{2}\right)\left(-3 + 2\sqrt{2}\right)$

29. $\left(1 - \sqrt{2}\right)\left(7 + 7\sqrt{2}\right)$

30. $\left(5 - 5\sqrt{2}\right)\left(5 - 5\sqrt{2}\right)$

31. $\left(-7 - \sqrt{2}\right)\left(1 + 7\sqrt{2}\right)$

32. $\left(3.5 - \sqrt{2}\right)\left(2 + \sqrt{2}\right)$

33. $\left(3 - 2\sqrt{2}\right)\left(2.5 + 3\sqrt{2}\right)$

34. $\dfrac{2}{3 - 5\sqrt{2}}$

35. $\dfrac{1 + \sqrt{2}}{1 - \sqrt{2}}$

36. $\dfrac{1 + 3\sqrt{2}}{\sqrt{2}}$

To multiply one radical expression by another, apply the Distributive Property, simplify the resulting products, and then combine like terms.

EXAMPLE 1 | **Simplify $\sqrt{3}(5 + 2\sqrt{3})$.**

> **Solution**

$$\sqrt{3}(5 + 2\sqrt{3}) = \sqrt{3} \cdot 5 + \sqrt{3} \cdot 2\sqrt{3} \quad \longleftarrow \text{Distribute } \sqrt{3} \text{ to each term.}$$
$$= 5\sqrt{3} + 2 \cdot 3 \quad \longleftarrow \sqrt{3} \cdot \sqrt{3} = 3$$
$$= 6 + 5\sqrt{3}$$

TRY THIS   Simplify each expression.   **a.** $\sqrt{7}(10\sqrt{7} + 2\sqrt{98})$ **b.** $\sqrt{2}(10\sqrt{2} - 5)$

EXAMPLE 2 | **Simplify $(2 - 5\sqrt{2})(3\sqrt{2} + 1)$.**

> **Solution**

$$(2 - 5\sqrt{2})(3\sqrt{2} + 1) = 2(3\sqrt{2} + 1) - 5\sqrt{2}(3\sqrt{2} + 1) \quad \longleftarrow \text{Apply the Distributive Property.}$$
$$= 6\sqrt{2} + 2 - 15 \cdot 2 - 5\sqrt{2}$$
$$= \sqrt{2} - 28, \text{ or } -28 + \sqrt{2}$$

TRY THIS   Simplify each expression.   **a.** $(6\sqrt{3} - 5)(3 + \sqrt{3})$ **b.** $(5 + 4\sqrt{3})(5 - 4\sqrt{3})$

Recall from Skill A that in order to simplify a quotient that has a radical in the denominator, you need to rationalize the denominator. In Example 3, the denominator of the quotient is $5 - 3\sqrt{2}$. To eliminate the square root from the denominator, multiply the quotient by $\dfrac{5 + 3\sqrt{2}}{5 + 3\sqrt{2}}$, which equals 1.

EXAMPLE 3 | **Simplify $\dfrac{4 + \sqrt{2}}{5 - 3\sqrt{2}}$.**

> **Solution**

$$\frac{4 + \sqrt{2}}{5 - 3\sqrt{2}} = \frac{4 + \sqrt{2}}{5 - 3\sqrt{2}}\left(\frac{5 + 3\sqrt{2}}{5 + 3\sqrt{2}}\right) \quad \longleftarrow \frac{5 + 3\sqrt{2}}{5 + 3\sqrt{2}} = 1$$
$$= \frac{20 + 12\sqrt{2} + 5\sqrt{2} + 3 \cdot 2}{25 + 15\sqrt{2} - 15\sqrt{2} - 9 \cdot 2}$$
$$= \frac{26 + 17\sqrt{2}}{7}, \text{ or } \frac{26}{7} + \frac{17}{7}\sqrt{2}$$

TRY THIS   Simplify each expression.   **a.** $\dfrac{5 + \sqrt{5}}{1 + 3\sqrt{5}}$ **b.** $\dfrac{1 + \sqrt{7}}{5 - 2\sqrt{7}}$

It is important to note that when you rationalize a denominator, you are not changing the value of the original expression. Because you are multiplying the numerator and the denominator by the same quantity, you are really just multiplying by 1: $\dfrac{a}{a} = 1$.

EXERCISES

KEY SKILLS

Apply the Distributive Property to each expression. Do not simplify further.

1. $\sqrt{5}(5 - 3\sqrt{5})$

2. $(1 + \sqrt{3})(1 - 3\sqrt{3})$

3. $(2 - \sqrt{7})(7 - 3\sqrt{7})$

What expression would you multiply the given quotient by in order to eliminate the radical from the denominator?

4. $\dfrac{1}{3 - 4\sqrt{5}}$

5. $\dfrac{-3}{2 + 5\sqrt{3}}$

6. $\dfrac{-3 + 4\sqrt{3}}{1 - 3\sqrt{3}}$

PRACTICE

Simplify. Write the answers in simplest radical form.

7. $\sqrt{2}(3 + 5\sqrt{2})$

8. $3\sqrt{11}(6 - \sqrt{11})$

9. $(1 - 3\sqrt{2})(1 - \sqrt{2})$

10. $(1 + 2\sqrt{13})(3 - 2\sqrt{13})$

11. $(4 + 3\sqrt{5})(4 - 3\sqrt{5})$

12. $(3 - 2\sqrt{7})(3 + 2\sqrt{7})$

13. $(3 + 3\sqrt{3})(5 - 3\sqrt{3})$

14. $(-5 + \sqrt{7})(-5 - 3\sqrt{7})$

15. $\sqrt{5}(-2 + \sqrt{5})(-5 + \sqrt{5})$

16. $\dfrac{1}{1 + \sqrt{2}}$

17. $\dfrac{3}{1 - \sqrt{11}}$

18. $\dfrac{\sqrt{3}}{2 + \sqrt{3}}$

19. $\dfrac{\sqrt{5}}{3 - \sqrt{5}}$

20. $\dfrac{1 - \sqrt{5}}{2 - \sqrt{5}}$

21. $\dfrac{4 + \sqrt{7}}{3 + \sqrt{7}}$

22. **Critical Thinking** Let a and b represent rational numbers with $ab \neq 0$. Simplify the expression $\dfrac{a + b\sqrt{p}}{a - b\sqrt{p}}$.

23. **Critical Thinking** **a.** Is $\sqrt{a^2 + b^2} \leq a + b$ always true? Justify your response.
 b. What does your answer indicate about squaring both sides as a method of solving an inequality?

MIXED REVIEW

Multiply. (6.5 Skill A)

24. $a^2 \cdot a^3$

25. $n^0 \cdot n^5$

26. $m^1 \cdot m^2 \cdot m^3$

27. $z^3 \cdot z^1 \cdot z^3$

28. $(2h^3)(3h^2)$

29. $(-5p^3)(-5p^4)$

30. $(4r^2)(r^3)(11r)$

31. $\left(\dfrac{3}{4}b^2\right)(b^3)\left(\dfrac{14}{33}b^4\right)$

10.5 Powers, Roots, and Rational Exponents

LESSON

SKILL A *Simplifying expressions that involve exponents of the form $\frac{1}{n}$*

In Lesson 8.1, you learned that the equations below are all related.

$$5 \times 5 = 25 \qquad 5^2 = 25 \qquad 5 = \sqrt{25} \qquad 25^{\frac{1}{2}} = 5$$

Using the equations above, you can form patterns as follows.

$$5 \times 5 \times 5 = 125 \qquad 5^3 = 125 \qquad 5 = \sqrt[3]{125} \qquad 125^{\frac{1}{3}} = 5$$

$$5 \times 5 \times 5 \times 5 = 625 \qquad 5^4 = 625 \qquad 5 = \sqrt[4]{625} \qquad 625^{\frac{1}{4}} = 5$$

The symbols $\sqrt[3]{}$ and $\sqrt[4]{}$ are called *cube-root* and *fourth-root radicals*, respectively. The numbers 3 and 4 are each called the **index** of the radical. Notice the relationship in the equations above between the radical expression and the exponential expression with a rational-number exponent.

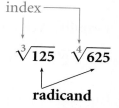

index

$$\sqrt[3]{125} \qquad \sqrt[4]{625}$$

radicand

$$\sqrt{25} = 25^{\frac{1}{2}} \qquad \sqrt[3]{125} = 125^{\frac{1}{3}} \qquad \sqrt[4]{625} = 625^{\frac{1}{4}}$$

Definition of the Exponent $\frac{1}{n}$

If $a > 0$ and n is a positive integer, $a^{\frac{1}{n}} = \sqrt[n]{a}$.

EXAMPLE 1 Write each radical expression as an exponential expression and write each exponential expression as a radical expression.

    a. $\sqrt[3]{10}$ **b.** $\sqrt[4]{24}$ **c.** $78^{\frac{1}{2}}$ **d.** $75^{\frac{1}{4}}$

▶ **Solution**

    a. $\sqrt[3]{10} = 10^{\frac{1}{3}}$ **b.** $\sqrt[4]{24} = 24^{\frac{1}{4}}$ **c.** $78^{\frac{1}{2}} = \sqrt{78}$ **d.** $75^{\frac{1}{4}} = \sqrt[4]{75}$

TRY THIS Write each radical expression as an exponential expression and write each exponential expression as a radical expression.

    a. $\sqrt[4]{3}$ **b.** $\sqrt[3]{30}$ **c.** $190^{\frac{1}{4}}$ **d.** $75^{\frac{1}{3}}$

EXAMPLE 2 Evaluate.   **a.** $\sqrt[3]{8}$ **b.** $\sqrt[4]{81}$ **c.** $16^{\frac{1}{4}}$ **d.** $64^{\frac{1}{3}}$

▶ **Solution**

    a. $\sqrt[3]{8} = \sqrt[3]{2^3} = 2^{3 \cdot \frac{1}{3}} = 2^1 = 2$ **b.** $\sqrt[4]{81} = \sqrt[4]{3^4} = 3^{4 \cdot \frac{1}{4}} = 3^1 = 3$ $(a^m)^n = a^{mn}$

    c. $16^{\frac{1}{4}} = (2^4)^{\frac{1}{4}} = 2^{4 \cdot \frac{1}{4}} = 2$ **d.** $64^{\frac{1}{3}} = (4^3)^{\frac{1}{3}} = 4^{3 \cdot \frac{1}{3}} = 4$

TRY THIS Evaluate.   **a.** $\sqrt[3]{27}$ **b.** $\sqrt[4]{256}$ **c.** $16^{\frac{1}{4}}$ **d.** $125^{\frac{1}{3}}$

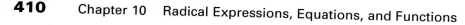

EXERCISES

Write each equation using a radical and using an exponential expression with a rational-number exponent.

1. $7^2 = 49$
2. $2^6 = 64$
3. $10^3 = 1000$
4. $3^5 = 243$

Write each radical expression as an exponential expression and write each radical expression as an exponential expression.

5. $\sqrt[3]{13}$
6. $\sqrt{5}$
7. $100^{\frac{1}{3}}$
8. $101^{\frac{1}{2}}$

9. $\sqrt[4]{43}$
10. $\sqrt[3]{5}$
11. $90^{\frac{1}{3}}$
12. $60^{\frac{1}{2}}$

13. $\sqrt[5]{55}$
14. $\sqrt[6]{6}$
15. $33^{\frac{1}{3}}$
16. $222^{\frac{1}{2}}$

Evaluate each expression.

17. $(2^3)^{\frac{1}{3}}$
18. $(5^4)^{\frac{1}{4}}$
19. $(7^4)^{\frac{1}{4}}$
20. $(7^6)^{\frac{1}{6}}$

21. $\sqrt[3]{1}$
22. $\sqrt[4]{1}$
23. $\sqrt[3]{512}$
24. $\sqrt[4]{10,000}$

Simplify. Assume that the variables represent positive numbers.

25. $(a^2)^{\frac{1}{2}}$
26. $(u^4)^{\frac{1}{4}}$
27. $(n^3)^{\frac{1}{3}}$
28. $(y^{10})^{\frac{1}{10}}$

29. $(a^3)^{\frac{1}{3}}(b^4)^{\frac{1}{4}}$
30. $(y^5)^{\frac{1}{5}}(z^6)^{\frac{1}{6}}$
31. $\sqrt[5]{r^5 s^5}$
32. $\sqrt[3]{a^3 b^3 c^3}$

33. Critical Thinking How are real numbers r and s related if $(rs)^{\frac{1}{4}} = 1$?

34. Critical Thinking How are real numbers m and n related if $\sqrt[3]{3m + n} = 1$?

Write each expression in simplest radical form. (8.1 Skill B)

35. $\sqrt{275}$
36. $\sqrt{117}$
37. $\sqrt{484}$
38. $\sqrt{363}$

39. $\sqrt{\frac{1}{100}}$
40. $\sqrt{\frac{121}{100}}$
41. $\sqrt{\frac{24}{49}}$
42. $\sqrt{\frac{125}{64}}$

Suppose that m and n are integers and $n \neq 0$. You know that a^m represents the mth power of a. From Skill A of this lesson, you know that $a^{\frac{1}{n}}$ represents $\sqrt[n]{a}$, the nth root of a. You can now define the rational number, $\frac{m}{n}$, as an exponent as follows.

> **Definition of the Exponent $\frac{m}{n}$**
>
> If a is a positive number, m and n are integers, and $n \neq 0$, then
> $$a^{\frac{m}{n}} = \sqrt[n]{a^m}, \text{ or } \left(\sqrt[n]{a}\right)^m.$$

If the bases of exponential expressions represent positive numbers and the exponents are rational numbers, then all of the properties of exponents you learned in Chapter 6 continue to be true.

EXAMPLE 1

Write each expression as a radical expression in simplest form.

a. $27^{\frac{2}{3}}$ b. $2^{\frac{4}{3}}$

▶ **Solution**

a. **Method 1:** Use the definition of the exponent $\frac{m}{n}$.
$$27^{\frac{2}{3}} = \left(\sqrt[3]{27}\right)^2 = \left(\sqrt[3]{3^3}\right)^2 = 3^2 = 9$$

Method 2: Use the Power-of-a-Power Property of Exponents.
$$27^{\frac{2}{3}} = (3^3)^{\frac{2}{3}} = (3)^{3 \times \frac{2}{3}} = 3^2 = 9$$

b. **Method 1:** Use the definition of the exponent $\frac{m}{n}$.
$$2^{\frac{4}{3}} = \sqrt[3]{2^4} = \sqrt[3]{2^3 \cdot 2^1} = \sqrt[3]{2^3} \cdot \sqrt[3]{2^1} = 2 \cdot \sqrt[3]{2} = 2\sqrt[3]{2}$$

Method 2: Write $\frac{4}{3}$ as $1 + \frac{1}{3}$. Then use the Product Property of Exponents.
$$2^{\frac{4}{3}} = 2^{1 + \frac{1}{3}} = 2^1 \cdot 2^{\frac{1}{3}} = 2 \cdot \sqrt[3]{2} = 2\sqrt[3]{2}$$

TRY THIS   Simplify each expression.   a. $32^{\frac{4}{5}}$ b. $5^{\frac{3}{2}}$

Just as you can simplify exponential expressions whose bases are numbers, you can simplify exponential expressions whose bases are variables.

EXAMPLE 2

Write $(x^2)\sqrt[3]{x^4}$ as an expression with a rational exponent.

▶ **Solution**
$$(x^2)\sqrt[3]{x^4} = x^2 \cdot x^{\frac{4}{3}} = x^{2 + \frac{4}{3}} \quad \longleftarrow \quad a^m \cdot a^n = a^{m+n}$$
$$= x^{\frac{10}{3}}$$

TRY THIS   Write $(a^3)\sqrt{a^5}$ as an expression with a rational exponent.

EXERCISES

KEY SKILLS

Write each radical expression as an exponential expression with a rational exponent.

1. $\sqrt[3]{5^2}$ **2.** $\sqrt[4]{7^3}$ **3.** $\left(\sqrt[4]{3}\right)^3$ **4.** $\left(\sqrt[3]{10}\right)^2$

Write each exponential expression as a radical expression.

5. $11^{\frac{2}{3}}$ **6.** $5^{\frac{3}{2}}$ **7.** $3^{2\frac{1}{2}}$ **8.** $7^{3\frac{2}{3}}$

PRACTICE

Write each expression as a radical expression in simplest form.

9. $125^{\frac{4}{3}}$ **10.** $8^{\frac{5}{3}}$ **11.** $81^{\frac{7}{4}}$ **12.** $32^{\frac{8}{5}}$

13. $3^{\frac{3}{2}}$ **14.** $7^{\frac{4}{3}}$ **15.** $2^{\frac{5}{3}}$ **16.** $5^{\frac{9}{4}}$

Write each expression as an expression with a rational exponent. Assume all of the variables represent positive real numbers.

17. $x\sqrt{x}$ **18.** $a^3\sqrt{a}$ **19.** $(t^2)\sqrt[3]{t}$ **20.** $(v^2)\sqrt[3]{v^2}$

21. $n^{\frac{4}{3}} \cdot n^{\frac{1}{3}}$ **22.** $k^{\frac{1}{2}} \cdot k^{\frac{1}{3}}$ **23.** $b^{\frac{1}{2}} \cdot \sqrt[3]{b^2}$ **24.** $c^{\frac{3}{2}} \cdot \sqrt{c^3}$

25. Let $t > 0$. Write the expression at right as an exponential expression with a single rational exponent.
$$\frac{(t^5)\sqrt[4]{t^3}}{(t^2)\sqrt[3]{t^2}}$$

26. Let $a > 0$ and let n be a nonzero integer. Prove that $a^{\frac{n+1}{n}} = a\sqrt[n]{a}$.

27. Let $a > 0$ and let n be a nonzero integer. Prove that $a^{\frac{n-1}{n}} = a\left(\sqrt[n]{\frac{1}{a}}\right)$.

MIXED REVIEW

Evaluate each formula for the given values of the variables. (1.2 Skill C)

28. $A = P(1 + rt)$; $P = 1000$, $r = 0.05$, and $t = 4$

29. $A = P(1 + rt)$; $P = 2000$, $r = 0.05$, and $t = 3$

30. $A = P(1 + r)^t$; $P = 2000$, $r = 0.05$, and $t = 2$

31. $A = P(1 + r)^t$; $P = 4000$, $r = 0.04$, and $t = 4$

32. $A = P(1 + r)^t$; $P = 5000$, $r = 0.04$, and $t = 2$

33. $A = P(1 + r)^t$; $P = 8500$, $r = 0.06$, and $t = 3$

Suppose that you invest *P* dollars in a bank account that pays interest compounded annually. Then after *t* years, the amount *A* in the account (in dollars) is given by the *compound-interest formula* below, where *r* represents the annual interest rate in decimal form.

$$A = P(1 + r)^t$$

If *t* is a positive integer, then the money remains in the account for a whole number of years. How can you find the amount in the account if *t* is a period of time such as 1 year and 6 months? The example below illustrates how to answer that question.

EXAMPLE

Suppose that an investor puts $5000 into an account that pays 4.5% interest compounded annually. Assuming that no money is withdrawn and no additional money is put into the account, how much money will be in the account after 1 year and 6 months?

▶ **Solution**

1. Substitute the numbers into the compound-interest formula. Since time is 1 year and 6 months, use $1\frac{1}{2}$, or 1.5, for *t*. Rewrite 4.5% as .045.

Use a formula.

$$A = 5000(1 + .045)^{1.5} \longleftarrow \text{ Let } P = 5000, r = .045, \text{ and } t = 1.5.$$

2. Use a calulator and enter a key sequence like the one below.

exponent key
↓

5000 $\boxed{(}$ 1 $\boxed{+}$ .045 $\boxed{)}$ $\boxed{\wedge}$ 1.5 $\boxed{\text{ENTER}}$

3. The calculator display will show 5341.268868.

The account will have $5341.27 in it after 1 year and 6 months.

TRY THIS Rework the Example given an initial deposit of $6000, an annual interest rate of 5.5% and a time of 2 years and 3 months.

Interest may be compounded every 6 months, every 3 months, or even monthly. In these situations, you can use the compound-interest formula given below to find the amount *A* in the account.

$$A = P\left(1 + \frac{r}{n}\right)^{nt}$$

n: number of compound periods in one year
t: number of years
r: annual rate of interest in decimal form

Refer to the Example. If the interest had been compounded monthly rather than annually, then the amount in the account could be found as shown below.

$$A = 5000\left(1 + \frac{.045}{12}\right)^{12 \cdot 1.5} = \$5348.48$$

EXERCISES

Suppose money is placed into an account that pays 6% as the annual interest rate and the money is in the account for 3 years. Suppose also that compound interest is applied every 6 months.

1. How many times in one year will interest be paid?

2. What percent will be applied each six months?

3. How many times will the interest be applied over 3 years?

PRACTICE

Suppose that $1500 is deposited into an account that pays compound interest at 5% annually. To the nearest cent, how much money will be in the account after each amount of time?

4. 1 year and 9 months

5. 2 years and 6 months

6. 3 years and 3 months

7. 3 years and 8 months

Mr. Tyge invests $2400 into an account that pays 6% compound interest annually. He intends to keep the money there for 2.5 years.

8. a. How much will be in the account if interest is paid annually?

   b. How much will be in the account if interest is paid semi-annually?

   c. By how much will the amounts in parts a and b differ?

9. Suppose that interest is compounded annually. Will the amount Mr. Tyge have after 5 years be double the amount he will have after 2.5 years?

10. Suppose that interest is compounded annually and that Mr. Tyge puts the money into an account that pays 12% instead of 6%. Will the amount in the first account be double that of the second account after 2.5 years? Justify your response.

MIXED REVIEW APPLICATIONS

A ball thrown directly upward from the edge of a cliff 130 feet above sea level is given an initial velocity of 128 feet per second. To the nearest hundredth of a second, find the amount of time it takes the ball to have the specified altitude above sea level. (8.7 Skill B)

11. 180 feet

12. 210 feet

13. 300 feet

14. 240 feet

15. 130 feet

16. 90 feet

10.6 Assessing and Justifying Statements About Square Roots

SKILL A *Exploring and proving statements that involve square roots*

Using problem-solving strategies and properties of square roots, you can analyze statements and answer questions about them.

EXAMPLE 1 **Prove or disprove: If x is any real number, then $\sqrt{2x} - \sqrt{x}$ is positive.**

▶ Solution

The statement is false. The expression $\sqrt{2x} - \sqrt{x}$ is not defined in the set of real numbers unless $x \geq 0$. Thus, the statement cannot be true for all real numbers x. Furthermore, if $x = 0$, then $\sqrt{2(0)} - \sqrt{0} = 0$, which is not positive.

TRY THIS Prove or disprove: If x is any real number, then $\sqrt{x} - \sqrt{x-1}$ is positive.

EXAMPLE 2 **Find and correct the error in the reasoning below.**
$$\sqrt{3^2 + 4^2} = \sqrt{3^2} + \sqrt{4^2} = 3 + 4 = 7 \quad ✗$$

▶ Solution

Writing $\sqrt{3^2} + \sqrt{4^2}$ in place of $\sqrt{3^2 + 4^2}$ is incorrect. It assumes that $\sqrt{a + b} = \sqrt{a} + \sqrt{b}$ for all real numbers a and b. However, $\sqrt{a + b} = \sqrt{a} + \sqrt{b}$ is not always true. The correct solution is shown below.

$$\sqrt{3^2 + 4^2} = \sqrt{9 + 16} = \sqrt{25} = 5 \quad ✔$$

TRY THIS Find and correct the error in the reasoning below.
$$\sqrt{5^2 - 4^2} = \sqrt{5^2} - \sqrt{4^2} = 5 - 4 = 1 \quad ✗$$

EXAMPLE 3 **Is the statement below sometimes, always, or never true?**
If a is any positive number, then $\sqrt{2a} = 2\sqrt{a}$.

▶ Solution

If $\sqrt{2a} = 2\sqrt{a}$, then, by squaring each side of the equation, $\left(\sqrt{2a}\right)^2 = \left(2\sqrt{a}\right)^2$. Therefore, $2a = 4a$. The only solution to this equation is 0. Since a represents a *positive* number, 0 is excluded from consideration. Thus, the statement is never true.

TRY THIS Is the statement below sometimes, always, or never true?
If x is any positive number, then $x > \sqrt{x}$.

EXERCISES

Let x represent any real number and let $\sqrt{x+5} = \sqrt{x} + \sqrt{5}$.

1. Find three values of x for which the equation is undefined.

2. Find three values of x for which the equation is defined but is false.

3. Find one value of x for which the equation is true.

PRACTICE

Prove or disprove.

4. If x is any real number, then $\sqrt{x+3} > \sqrt{x} + \sqrt{3}$.

5. Let p and q represent squares of positive integers, that is, $p = a^2$ and $q = b^2$ for some positive integers a and b. Then $\sqrt{pq}$ is a positive integer.

Find and correct any error.

6. $\dfrac{12\sqrt{3} - 18\sqrt{5}}{6} = \dfrac{12\sqrt{3}}{6} - 18\sqrt{5} = 2\sqrt{3} - 18\sqrt{5}$ ✗

7. $\dfrac{\sqrt{10(3+7)}}{5} = \sqrt{2(3+7)} = \sqrt{2(10)} = \sqrt{20}$ ✗

In Exercises 8 and 9, is the given statement sometimes, always, or never true? Justify your response.

8. If r represents any positive real number, then $r^2 \geq \sqrt{r}$.

9. If t represents a positive real number, then $\sqrt{\dfrac{t}{2}} = \dfrac{\sqrt{t}}{2}$.

10. For which positive integers n will $\sqrt{n}$ be an even integer?

MIXED REVIEW

In Exercises 11–16, y varies inversely as x. (9.8 Skill A)

11. If $y = 12.5$ when $x = 8$, find x when $y = 25$.   12. If $y = 49$ when $x = 7$, find x when $y = 7$.

13. If $y = 0.25$ when $x = 24$, find x when $y = 1$.   14. If $y = \dfrac{1}{3}$ when $x = 15$, find y when $x = 25$.

15. If $y = 0.2$ when $x = 2$, find y when $x = 2.5$.   16. If $y = 14$ when $x = 0.4$, find y when $x = 0.7$.

The relationship between a vehicle's speed s (in miles per hour) and the distance d (in feet) it takes the vehicle to stop once the brakes are applied can be modeled by the following square-root equation.

$$s = \sqrt{30fd}$$

In the equation above, the variable f represents the *coefficient of friction*. Its value depends on road surface, such as concrete or asphalt, and road conditions, such as wet or dry.

f	wet	dry
concrete	0.4	0.8
asphalt	0.5	1.0

EXAMPLE 1

Find the speed of a vehicle that was traveling on wet concrete and required 200 feet to stop after the brakes were applied. Approximate your answer to the nearest hundredth.

▶ **Solution**

Evaluate $s = \sqrt{30fd}$ with $f = 0.4$ and $d = 200$.

$$s = \sqrt{30(0.4)(200)}$$

Use a formula.

$$= \sqrt{30 \cdot \frac{4}{10} \cdot 200}$$

$$= \sqrt{2400} \qquad \longleftarrow \text{Simplify.}$$

$$s \approx 48.99 \qquad \longleftarrow \text{Use a calculator.}$$

The vehicle was traveling about 48.99 miles per hour.

TRY THIS

Rework Example 1 given that the vehicle was traveling on dry concrete.

In Example 2, you are asked to compare the speed of a vehicle given one set of road conditions with the speed of the same vehicle under different road conditions.

EXAMPLE 2

Two vehicles take the same distance to stop after the brakes are applied. One is traveling on wet concrete. The other is traveling on dry concrete. Compare the speeds of the two vehicles before braking.

▶ **Solution**

Use a ratio to compare $\sqrt{30(0.8)d}$ with $\sqrt{30(0.4)d}$.

$$\frac{\text{speed on dry concrete}}{\text{speed on wet concrete}} \quad \rightarrow \quad \frac{\sqrt{30(0.8)d}}{\sqrt{30(0.4)d}} = \sqrt{\frac{30(0.8)d}{30(0.4)d}} = \sqrt{2}$$

For the vehicle to stop in the same distance on dry concrete, the vehicle can travel at $\sqrt{2}$, or about 1.4, times the speed of the vehicle on wet concrete.

TRY THIS

Rework Example 2 given dry and wet asphalt.

EXERCISES

The speed that a vehicle travels before braking on dry asphalt is given by the formula $s = \sqrt{30d}$. The graph of this relationship is shown at right.

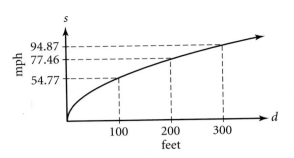

1. In general terms, what change in speed would increase the stopping distance from 100 feet to 200 feet to 300 feet?

2. What change in speed would double the stopping distance from 100 feet to 200 feet?

PRACTICE

To the nearest hundredth of a mile per hour, approximate the speed a vehicle was traveling, given each stopping distance and road condition.

3. 50 feet, wet asphalt

4. 50 feet, dry concrete

5. 200 feet, wet concrete

6. 260 feet, dry asphalt

Compare the speeds of two vehicles. Assume each pair of vehicles travels under the same set of conditions.

7. The stopping distance of a vehicle is 1.2 times that of the other vehicle.

8. The stopping distance of a vehicle is 2.4 times that of the other vehicle.

9. **Critical Thinking**   Let r represent a positive number. One vehicle takes d feet to stop. A second vehicle takes rd feet to stop under the same road conditions. Which motorist has the greater speed? Justify your response.

MIXED REVIEW APPLICATIONS

At a constant temperature, pressure P (in pounds per square inch) and volume V (in cubic inches) of an ideal gas are related by the inverse-variation equation $PV = k$, for some nonzero constant k. (9.8 Skill C)

10. A gas with volume 45 cubic inches is under a pressure that is 60 pounds per square inch. Find the volume when the pressure is 90 pounds per square inch.

11. A gas with volume 40 cubic inches is under a pressure that is 45 pounds per square inch. Find the volume when the pressure is 75 pounds per square inch.

The graph of the square-root function $y = \sqrt{x}$ is shown below.

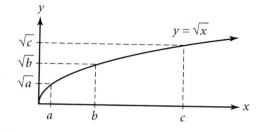

Notice that if $a < b < c$, then $\sqrt{a} < \sqrt{b} < \sqrt{c}$. For example, since $1 < 2 < 4$, you can say that $1 < \sqrt{2} < 2$. This means that a decimal approximation of $\sqrt{2}$ is a rational number between 1 and 2. Using a numerical method based on the reasoning above, you can approximate any square root to any degree of accuracy.

EXAMPLE

Approximate $\sqrt{47}$ to the nearest hundredth without using a calculator.

▶ **Solution**

Step 1: Find the two perfect squares that lie on either side of 47 on the number line: 36 and 49. Therefore, the square root of 47 will lie between the square roots of 36 and 49, which are 6 and 7.

$$36 < 47 < 49$$
$$\sqrt{36} < \sqrt{47} < \sqrt{49}$$
$$6 < \sqrt{47} < 7$$

Step 2: 47 is closer to 49 than to 36, and so 7 is a closer approximation of $\sqrt{47}$. Divide 47 by 7. Stop when the quotient has one more decimal place than the divisor.

$$\begin{array}{r} 6.7 \\ 7\overline{)47.0} \end{array}$$

Step 3: Find the average of 7 and 6.7. The average should have the same number of decimal places as the quotient.

$$\frac{7 + 6.7}{2} = \frac{13.7}{2} \approx 6.8$$

Repeat Steps **2** and **3** using the average 6.8 as the divisor.

Step 4: Divide 47 by 6.8.

$$\begin{array}{r} 6.91 \\ 6.8\overline{)47.00} \end{array}$$

Step 5: Find the average of 6.8 and 6.91.

Thus, $\sqrt{47} \approx 6.85$.

$$\frac{6.8 + 6.91}{2} = \frac{13.71}{2} \approx 6.85$$

TRY THIS   Approximate $\sqrt{75}$ to the nearest hundredth.

If you continue to divide and average as in Steps 4 and 5, you will get closer and closer approximations of the irrational number $\sqrt{47}$.

The process of finding new numbers within smaller and smaller intervals illustrates the **Density Property of Real Numbers**, which states that between any two real numbers there is another real number.

EXERCISES

Answer each question and justify your response.

1. Is $\sqrt{5}$ between 2.2 and 2.3 or between 2.3 and 2.4?

2. Suppose that you square 3.3 and find 10.89 while you are trying to approximate $\sqrt{11}$. Would your next estimate be greater than 3.3 or less than 3.3?

3. Given that $4^2 = 16$, $4.5^2 = 20.25$, and $5^2 = 25$, will you find $\sqrt{20}$ between 4 and 4.5 or between 4.5 and 5?

PRACTICE

Use the method in the Example to approximate each square root to the nearest hundredth. Then use a calculator to check your approximation.

4. $\sqrt{5}$ 　　　　 5. $\sqrt{11}$ 　　　　 6. $\sqrt{50}$ 　　　　 7. $\sqrt{45}$

Given the extended inequality below, locate each square root between two consecutive numbers in the inequality.

$$11 \; < \; 12 \; < \; 13 \; < \; 14 \; < \; 15 \; < \; 16 \; < \; 17 \; < \; 18$$

8. $\sqrt{260}$ 　　　　 9. $\sqrt{205}$ 　　　　 10. $\sqrt{123}$ 　　　　 11. $\sqrt{230}$

12. Order $\sqrt{73}$, $\sqrt{37}$, $\sqrt{30.7}$, $\sqrt{3.07}$, $\sqrt{7.3}$, and $\sqrt{3.7}$ from least to greatest.

13. Without doing any calculations, order $\dfrac{1}{\sqrt{73}}$, $\dfrac{1}{\sqrt{37}}$, $\dfrac{1}{\sqrt{30.7}}$, $\dfrac{1}{\sqrt{3.07}}$, $\dfrac{1}{\sqrt{7.3}}$, and $\dfrac{1}{\sqrt{3.7}}$ from least to greatest.

14. **Critical Thinking**　Use guess-and-check to determine which of the statements below is true for all positive real numbers x.

$$\sqrt{x+2} \leq \sqrt{x} + \sqrt{2} \qquad \sqrt{x+2} = \sqrt{x} + \sqrt{2} \qquad \sqrt{x+2} \geq \sqrt{x} + \sqrt{2}$$

MIXED REVIEW

Solve each equation. Write the solutions in simplest radical form. (Answers may contain imaginary terms.) (8.2 Skill C)

15. $5x^2 + 21 = -54$ 　　　　　　　　　 16. $3n^2 + 14 = -13$

17. $-2n^2 - 8 = 0$ 　　　　　　　　　　 18. $11k^2 + 176 = 0$

19. $5a^2 + 1 = -9$ 　　　　　　　　　　 20. $2d^2 + 31 = -9$

21. $3y^2 + 50 = 11$ 　　　　　　　　　　 22. $-7v^2 + 114 = 16$

Find the domain of each square-root expression.

1. $\sqrt{-7(3-d)} + 2$

2. $\sqrt{2(n+3) - (3-n)}$

Find the domain and range of each square-root function.

3. $y = \sqrt{-3(4x - 3x + 1)} - 2$

4. $y = \sqrt{\dfrac{6(x+7) - 12}{3}}$

Solve each square-root equation. Check your solution.

5. $\sqrt{4 - 5(x+1)} - 3 = 1$

6. $\sqrt{5(p+3) - (-1)(2-p)} = 5$

7. $\sqrt{y^2 + 5y} = 6$

8. $\sqrt{\dfrac{a^2 + 8a}{3}} = 4$

9. Let $O(0, 0)$ represent the origin of the coordinate plane. Find the coordinates of Q and Q' that are directly above and below $P(5, 0)$, respectively, and that are 10 units from the origin.

Solve each square-root equation. Check your solution.

10. $\sqrt{4(g-5)} = \sqrt{3g+4}$

11. $\sqrt{\dfrac{4x+1}{3}} = \sqrt{\dfrac{3x-2}{5}}$

12. $\sqrt{12z^2 - 66z} = \sqrt{14z - 77}$

13. $\sqrt{15a^2} = \sqrt{34a - 15}$

Simplify. Write the answers in the form $a + b\sqrt{2}$, where a and b are rational numbers.

14. $\left(-2 - 5\sqrt{2}\right) + \left(-2 - 11\sqrt{2}\right)$

15. $\left(-10 + 7\sqrt{2}\right) - \left(8 - 10\sqrt{2}\right)$

16. $\left(-1 + 7\sqrt{2}\right)\left(1 - \sqrt{2}\right)$

17. $\left(4 - 7\sqrt{2}\right)\left(4 - 2\sqrt{2}\right)$

18. $\dfrac{10 - 2\sqrt{2}}{3\sqrt{2}}$

19. $\dfrac{-1 - 5\sqrt{2}}{1 - \sqrt{2}}$

Solve. Write the answers in the form $a + b\sqrt{2}$, where a and b are rational numbers.

20. $\left(1 - \sqrt{2}\right)x + 5 = 3\sqrt{2}$

21. $\left(\dfrac{1}{2} + \sqrt{2}\right)n + \left(2 - \sqrt{2}\right) = -\sqrt{2}$

Simplify. Write the answers in simplest radical form.

22. $\sqrt{50}\sqrt{2}$

23. $\dfrac{\sqrt{192}}{\sqrt{3}}$

24. $\dfrac{\sqrt{500}}{\sqrt{50}}$

25. $\dfrac{\sqrt{3}}{\sqrt{18}}$

26. $7\sqrt{13} + \left(-5 + 13\sqrt{13}\right)$

27. $\left(5 + 7\sqrt{7}\right) + \left(-5 + 7\sqrt{7}\right)$

28. $-10\sqrt{17} - \left(-5 - \sqrt{17}\right)$

29. $\left(3 - 3\sqrt{13}\right) - \left(13 + 13\sqrt{13}\right)$

30. $-2\sqrt{5}\left(-1 + 11\sqrt{5}\right)$

31. $(3 - \sqrt{5})(-2 + 10\sqrt{5})$

32. $\dfrac{-3 + 9\sqrt{7}}{\sqrt{7}}$

33. $\dfrac{-5 - 8\sqrt{11}}{5 + \sqrt{11}}$

Write each radical expression as an exponential expression and write each exponential expression as a radical expression.

34. $\sqrt[3]{7}$

35. $\sqrt[4]{29}$

36. $11^{\frac{1}{3}}$

37. $15^{\frac{1}{4}}$

Simplify each expression.

38. $\sqrt[5]{1}$

39. $\sqrt[3]{64}$

40. $27^{\frac{1}{3}}$

41. $625^{\frac{1}{4}}$

Write each expression in simplest radical form.

42. $4^{\frac{3}{2}}$

43. $1000^{\frac{1}{3}}$

44. $32^{\frac{2}{5}}$

45. $27^{\frac{2}{3}}$

Write each expression as an expression with a rational exponent. Assume all variables represent positive real numbers.

46. $a\sqrt[3]{a}$

47. $(b^2)\sqrt[3]{b}$

48. $(c^3)\sqrt[3]{c^2}$

49. $(z^5)\sqrt[5]{z}$

50. If \$3500 is invested in an account paying 5% compound interest annually, how much will be in the account after 2 years and 3 months?

51. Is the statement below sometimes, always, or never true?
   If a is any positive real number, then $\sqrt{6.25a^2} = 2.5a$.

52. Let $s = \sqrt{30(0.8)d}$, where s represents speed of a vehicle (in miles per hour) when the brakes are applied and d represents the stopping distance of the vehicle (in feet). To the nearest hundredth of a mile per hour, approximate the speed of a vehicle that takes 240 feet to stop.

53. Using inequalities, approximate $\sqrt{7}$ to the nearest hundredth.

Glossary

A

absolute value   The absolute value of any real number x, written $|x|$, is the distance from x to 0 on a number line. If x is greater than or equal to 0, then $|x| = x$. If x is less than 0, then $|x| = -x$. (44)

acceleration   The rate of change of velocity with respect to time. (334)

Addition Property of Equality   If a, b, and c are real numbers and $a = b$, then $a + c = b + c$. (20, 62)

Addition Property of Inequality   If a, b, and c are real numbers and $a < b$, then $a + c < b + c$. This statement is also true if $<$ is replaced by $>$, $\leq$ or $\geq$. (148)

additive identity   The number 0 is the additive identity because 0 added to any number equals that number. (8)

additive inverses   Two numbers are additive inverses if their sum is 0. For example, $3 + (-3) = 0$, so 3 and -3 are additive inverses, or opposites.

algebra   The branch of mathematics in which operations of arithmetic are generalized by the use of letters to represent unknown or varying quantities. (4)

algebraic expression   An expression that contains numbers, variables, and operation symbols. (12)

Associative Property of Addition   If a, b, and c are real numbers, then $(a + b) + c = a + (b + c)$. (8)

Associative Property of Multiplication   If a, b, and c are real numbers, then $(a \times b) \times c = a \times (b \times c)$. (8)

axis of symmetry (of a parabola)   The vertical line which divides a parabola into two parts that are mirror images of each other. (286)

B

boundary line   The graph of the related linear equation for a linear inequality. The line is the edge of the solution region of the inequality and is either solid or dashed. (206)

base (of an exponential expression)   The number that is raised to an exponent. In an expression of the form x^a, x is the base. (220)

binomial   A polynomial with exactly two terms. (230)

C

closure   A set of numbers is said to be closed under an operation if for every pair of numbers in the set, the result of the operation is also a member of the set. For example, the natural numbers are closed under addition and multiplication but not under subtraction and division; the difference of two natural numbers can be a negative number and the quotient can be a fraction. (8)

coefficient   The number used as a factor in a monomial, such as 6 in $6a$. If only a variable is written, the coefficient is understood to be 1. (66)

collinear   Points are collinear if they lie on the same line. (108)

common binomial factor   A binomial that appears as a factor in more than one term of a polynomial. (268)

Commutative Property of Addition   For all real numbers a and b, $a + b = b + a$. (8)

Commutative Property of Multiplication   For all real numbers a and b, $a \times b = b \times a$. (8)

completing the square   A technique for modifying a quadratic polynomial to obtain a perfect-square trinomial; it provides a general method for solving quadratic equations. (314)

complex fraction   A fraction whose numerator or denominator (or both) contains a fraction. (368)

composite number   A number that has factors other than 1 and itself. (260)

compound inequality   A pair of inequalities joined by *or* (disjunction) or *and* (conjunction). (146)

compound interest   Interest paid on earned interest. The formula for calculating compound interest is $A = P(1 + r)^t$, where A is the total amount of principal and interest, P is the original amount of principal, r is the rate of interest for the compounding period, and t is the number of compounding periods. (414)

conclusion   The consequence that follows the conditions in a conditional statement. If the statement is in *If-then* form, the conclusion is the part that follows *then*. (84)

conditional statement   A statement in which a specific conclusion follows a given hypothesis or hypotheses. The statement may be written in *If-then* form. The part following *If* is the hypothesis and part following *then* is the conclusion. (84)

congruent polygons   Polygons that can be made to coincide. They have the same size and shape and their corresponding sides and angles are equal. (238)

conjugates (radical)   A pair of expressions of the forms $a + b\sqrt{c}$ and $a - b\sqrt{c}$. The product of two conjugates contains no radical sign. (400)

conjunction   A pair of inequalities joined by *and*. A conjunction can be written in the form $a < x < b$. (170)

consistent system   A system of equations or inequalities that has at least one solution. (202)

constant of variation   A fixed, nonzero number k, in a direct variation $\left(\frac{y}{x} = k\right)$ or in a inverse variation $\left(y = \frac{k}{x}\right)$. (102)

constant   A term in an algebraic expression that does not contain any variables. (230)

constant of variation   The constant, k, in an equation of direct variation $\left(\frac{y}{x} = k\right)$ or of indirect variation $(xy = k)$. (378)

continuous graph   An unbroken graph. (116)

converse   A statement obtained by interchanging the parts of a conditional statement. Given *If p, then q*, the converse is *If q, then p*. The converse of a true statement is not necessarily true. (384)

coordinate axes   The vertical and horizontal lines in a coordinate plane that intersect at the origin, (0, 0). Both axes represent number lines. The horizontal axis usually represents the values of the independent variable and the vertical axis usually represents values of the the dependent variable. (92)

coordinate plane   A plane that is divided into four regions by a horizontal and a vertical number line.

counterexample   An example used to show that a statement is not always true. (16)

Cross-Product Property of Proportions   If a, b, c, and d are real numbers with $b \neq 0$ and $d \neq 0$, and $\frac{a}{b} = \frac{c}{d}$, then $ad = bc$. (372)

cross products   In a proportion $\frac{a}{b} = \frac{c}{d}$, the cross products are bc (the product of the means) and ad (the product of the extremes). (372)

cube-root radical   The symbol $\sqrt[3]{}$, which indicates to take the cube root of the expression under it. The cube root of a is the number that, when used as a factor 3 times, gives a as the product. (410)

cubic equation   An equation that can be written in the form $ax^3 + bx^2 + cx + d = 0$, where $a \neq 0$. (284)

cubic polynomial   A polynomial whose degree is three. (230)

D

data   (plural) Factual information, such as measurements or statistics. (116)

deductive reasoning The process of reasoning logically from a hypothesis to a conclusion, using only certain clearly stated assumptions (axioms), defined terms, and statements (theorems) that have been previously proven. These elements are put in a sequence that follows the rules of formal logic. (84, 134, 326)

degree (of a polynomial) The greatest exponent in all of the terms of a simplified polynomial. (230)

Density Property of Real Numbers The density property state that between any two real numbers, there is another real number. (420)

dependent system A system of equations that has infinitely many solutions. (204)

dependent variable In a function of two variables, the variable of the range. (100)

descending order (of a polynomial) A polynomial in one variable is said to be in descending order when the exponents of the terms decrease from left to right. (230)

difference of two squares A binomial of the form $a^2 - b^2$. It can be factored as $(a - b)(a + b)$. (250, 266)

direct variation If y varies directly as x, then $y = kx$, where k is a fixed nonzero number called the *constant of variation*. (102)

discriminant The expression under the radical in the quadratic formula, $b^2 - 4ac$. (320)

discrete graph A graph consisting of points that are not connected. (116)

disjunction A pair of inequalities joined by *or*. (172)

distance formula If $P(x_1, y_1)$ and $Q(x_2, y_2)$ are two points in the coordinate plane, then the distance between them is given by the formula $PQ = \sqrt{(x_2 - x_1)^2 + (y_2 - y_1)^2}$. (394)

Distributive Property For all real numbers a, b, and c, $a(b + c) = ab + ac$ and $(b + c)a = ba + ca$. (8)

Division Property of Equality If a, b, and c are real numbers and $a = b$ and $c \neq 0$, then $\frac{a}{c} = \frac{b}{c}$. (30, 64)

Division Property of Inequality Let a, b, and c be real numbers. If c is positive and $a < b$, then $\frac{a}{c} < \frac{b}{c}$. If c is negative and $a < b$, then $\frac{a}{c} > \frac{b}{c}$. Similar statements can be written using the other inequality symbols. (154)

domain The set of all numbers which can be assigned to the independent variable in a function with two variables. The domain of a rational expression is restricted to values that do not cause the denominator to equal 0. (96, 342)

double root If the polynomial member of an equation has two identical factors, then the equation has a double root. For example, in $y = (x - r)^2$, r is a double root. This double root will be represented in the graph of the equation as touching, but not crossing, the x-axis. (288)

E

elimination method A method of solving a system of equations by multiplying and combining the equations in the system in order to eliminate a variable. (196)

equation A mathematical statement which uses the equal symbol ($=$) to show that two expressions are equivalent. (14)

equivalent equations Equations that have the same solution, such as $5x = 20$ and $2x = 8$. (70)

evaluate (an algebraic expression) Substitute a given value for the variable(s) in the expression and simplify. (232)

excluded value A value that is excluded from the domain of a function because no range value exists for it. The term is commonly used in reference to values in rational functions that cause the denominator to equal 0. (342)

exponent The number that tells how many times the *base* is used as a factor. In an expression of the form x^a, a is the exponent. (10, 220)

exponential expression An algebraic expression in which the exponent is a variable and the base is a fixed number. (220)

extraneous solution A solution, found in the process of solving an equation, that is not a solution to the original equation. (374, 396)

F

factor (noun) A divisor of a number or of an expression. (verb) To write a number or an expression as the product of two or more numbers or expressions. (260)

factored completely The description of a polynomial which has been written as a product of factors that cannot be factored any further. It is the equivalent of prime factorization of numbers. (278)

field Any set that has the Associative, Closure, Commutative, Distributive, Identity and Inverse Properties for any two operations. (402)

FOIL (method) A mnemonic for the procedure of multiplying two binomials. Add the products of the First terms, Outer terms, Inner terms, and Last terms. (248)

formula An equation that expresses a relationship between two or more quantities. (12)

fourth-root radical A radical with the number 4 as an *index*, $\sqrt[4]{}$. (410)

function A relation in which each member of the first set, the domain, is assigned exactly one member of the second set, the range. Functions can be expressed in the form of algebraic equations which give rules for assigning values in the range to values in the domain. For example, $y = 5x$ is a function that pairs each domain value, x, with a range value 5 times greater. (96)

G

greatest common factor (GCF) The largest number that is a divisor of two or more given numbers. Also, the monomial with all the integer

and variable factors common to every term of a polynomial. (262)

horizontal-addition format (in the addition and subtraction of polynomials) The placement of expressions to be added side by side; the calculations are performed by grouping and combining like terms. (234)

horizontal line A line parallel to the x-axis. The equation of the horizontal line that contains the point $(0, s)$ is $y = s$. (112)

hypotenuse In a right triangle, the side opposite the right angle. (308)

hypothesis The condition(s) given in a conditional statement. If the statement is in *If-then* form, the hypothesis is the part that follows *If*. (84)

I

Identity An equation that is true regardless of the value of the variable(s). (14)

Identity Property of Addition For all real numbers a, $a + 0 = a$ and $0 + a = a$; 0 is the identity element for addition. (8)

Identity Property of Multiplication For all real numbers a, $a \times 1 = a$ and $1 \times a = a$; 1 is the identity element for multiplication. (8)

imaginary unit The number i, which represents the quantity $\sqrt{-1}$. (310)

inconsistent system A system of equations that has no solution. (202)

independent system A consistent system of equations that has exactly one solution. (204)

independent variable In a function, the variable of the domain. (100)

index The number used with the radical sign to indicate the root. For example, in the expression $\sqrt[3]{5}$, the index 3 indicates that the expression represents the cube root of 5. (410)

indirect reasoning   The process by which a statement is proved to be true by showing that the negation of the statement is false. (86)

inductive reasoning   Reasoning that proceeds from individual or specific cases to general rules or theories. The scientific method is based upon inductive reasoning, and therefore scientific conclusions can never actually be proven. They can only appear to be supported more or less strongly by collected evidence. On the other hand, mathematical conclusions are based upon deductive reasoning and can be proved to be true or false. (134)

inequality   A statement that the value of one expression is not equal to the value of another. Strict inequality is shown by the symbols $>$ (is greater than) or $<$ (is less than). The symbols $\leq$ (is less than or equal to) and $\geq$ (is greater than or equal to) allow for the possibility of equality. (142)

infinite number of digits   An unlimited number of digits. (40, 420)

integers   The set of whole numbers and their opposites: $\ldots -4, -3, -2, -1, 0, 1, 2, 3, 4, \ldots$ (40)

inverse operations   Operations that "undo" one another. For example, addition and subtraction are inverse operations, as are multiplication and division. (32)

Inverse Property of Addition   For every real number a, there is an opposite real number, $-a$, such that $a + (-a) = -a + a = 0$. (8)

Inverse Property of Multiplication   For every real number a, there is an opposite real number $\frac{1}{a}$ such that $a \times \frac{1}{a} = \frac{1}{a} \times a = 1$. (8)

inverse-square variation   The relationship between x and y in the equation $x^2 y = k$. y is said to vary inversely as the square of x. (382)

inverse variation   The relationship between x and y in $xy = k$, where k is a constant. The variation is inverse because when x increases, y decreases. An alternative form of inverse variation can be written as $x_1 y_1 = x_2 y_2$, when (x_1, y_1) and (x_2, y_2) both satisfy $xy = k$. (378)

irrational numbers   The set of all nonrepeating and non-terminating decimals. Irrational numbers cannot be written in the form $\frac{p}{q}$ where p and q are integers and $q \neq 0$. (42)

L

leading coefficient   The numerical part of the leading term of a polynomial. (230)

leading term   In a polynomial, the term with the greatest exponent. (230)

least common denominator (LCD)   The smallest number that is a multiple of the denominators of two or more fractions. (358)

least common multiple (LCM)   The smallest number that is a multiple of two or more given numbers. (76, 358)

leg   In a right triangle, one of the sides adjacent to the right angle. (308)

like terms   Monomials whose variable parts are the same, including the powers to which they are raised. (66)

linear equation in two variables (standard form)   An equation that can be written in the form $Ax + By = C$, where A, B, and C are real numbers, and A and B are not both 0. The graph of a linear equation is a straight line in the coordinate plane. (112)

linear inequality in standard form   An inequality of the form $Ax + By < C$, where A, B, and C are real numbers, and A and B are not both 0. (The $<$ may be replaced by $>$, $\geq$, or $\leq$.) (212)

linear function   Any function that can be defined by a linear equation, and usually written in the form $y = mx + b$. (132)

linear inequality   An inequality formed when the equal sign ($=$) of a linear equation is replaced by an inequality symbol: $>$, $<$, $\geq$, or $\leq$. (206)

linear pattern   A pattern that can be modeled by a linear function. (132)

linear (polynomial) A polynomial whose degree is one. (230)

logic The study of the structure of statements and the formal laws of reasoning. (4)

mathematical expression A quantity made up of variables, numbers, operation symbols, and inclusion symbols. (4)

mathematical proof A logical argument, or sequence of statements, used to show that a statement is true. (18)

maximum The greatest value (of a function, or of a situation). (158)

minimum The least value (of a function, or of a situation). (158)

monomial An algebraic expression that is either a constant, a variable, or a product of a constant and one or more variables; one term in a polynomial. (66, 230)

Multiplication Property of Equality If a, b, and c are real numbers and $a = b$, then $ac = bc$. (28, 64)

Multiplication Property of Inequality Let a, b, and c be real numbers. If c is positive and $a < b$, then $ac < bc$. If c is negative and $a < b$, then $ac > bc$. Similar statements can be written using the other inequality symbols. (154)

Multiplication Property of 0 If $a = 0$ or $b = 0$, then $ab = 0$. (280)

multiplicative identity The number 1, because $a \times 1 = 1 \times a = a$. (8)

multiplicative inverse A reciprocal. If $a \neq 0$, then $\frac{1}{a}$ is the multiplicative inverse, or reciprocal, of a. (58)

Multiplicative Property of −1 For all real numbers a, $-1(a) = -a$. (52)

natural numbers The numbers that are used in counting: 1, 2, 3, … (40)

negation Turning a statement into its opposite. The negation of "it is raining" is "it is not raining." In symbolic logic, the negation of statement p is written $\sim p$. (86)

negative exponent If a and n are real numbers and $a \neq 0$, then $a^{-n} = \frac{1}{a^n}$. (222)

numerical expression An expression that contains numbers and operation symbols, but no variables. (10)

open sentence An equation that contains at least one variable. (14)

opposite of a difference For all real numbers a and b, $-(a - b) = b - a$. (52)

opposite of a sum For all real numbers a and b, $-(a + b) = -a - b$. (52)

opposites Two numbers that lie on opposite sides of zero and are the same distance from zero on a number line, such as 3 and −3. Also called *additive inverses*. (8, 44)

order of operations When simplifying an expression, the order of operations is: 1) simplify expressions within grouping symbols, 2) simplify expressions involving exponents, 3) multiply and divide from left to right, and 4) add and subtract from left to right. (10)

ordered pair A pair of numbers, usually written in the form (x, y), which correspond to a point in the coordinate plane. (92)

origin On the real number line, the origin is the number 0. In the coordinate plane, the origin is the point where the axes intersect, $(0, 0)$. (44, 92)

P

parabola   The U-shaped graph of a quadratic function. (286)

parallel lines   Lines in the same plane that never intersect. Two lines are parallel if they have the same slope. (126)

parallelogram   A quadrilateral in which both pairs of opposite sides are parallel. (131)

parameter   The variable in parametric equations upon which the x- and y-values are dependent. Frequently, the parameter represents time. (136)

parametric equations   Equations by which the coordinates of a graph may be described as two functions of a single variable. That variable is called the parameter, and the x- and y-values are both dependent upon it. (136)

pentagon   A 5-sided closed figure. (214)

perfect square   A rational number whose square roots are also rational numbers. (300)

perfect-square trinomial   The result of squaring a binomial; a trinomial of the form $a^2 + 2ab + b^2$ or $a^2 - 2ab + b^2$, which can be factored as $(a + b)^2$ or $(a - b)^2$, respectively. (250, 266)

perpendicular lines   Lines that intersect at right angles. Two lines in a coordinate plane are perpendicular if the product of their slopes is -1. (128)

point-slope form   The point-slope form of a linear equation is $y - y_1 = m(x - x_1)$ where the coordinates x_1 and y_1 are taken from a given point (x_1, y_1), and m is the slope. (118)

polynomial   The sum of one or more monomials. (230)

power function   A function of the form $y = kx^n$, where $k \neq 0$ and n is a positive integer. (226)

Power-of-a-Fraction Property   If a, b, and n are real numbers and $b \neq 0$, then $\left(\dfrac{a}{b}\right)^n = \dfrac{a^n}{b^n}$. (222)

Power-of-a-Power Property   If a, m, and n are real numbers and $a \neq 0$, then $(a^m)^n = a^{mn}$. (220)

Power-of-a-Product Property   If a, b, and m are real numbers, then $(ab)^m = a^m b^m$. (220)

prime factorization   A product that contains only prime numbers or powers of prime numbers as factors. (260)

prime number   Any natural number greater than 1 whose only factors are 1 and itself. (260)

principal square root   The positive square root of a positive number. (300)

Product Property of Square Roots   If a and b are positive real numbers, then $\sqrt{ab} = \sqrt{a}\sqrt{b}$. (302)

profit   In business, the amount remaining after expenses are subtracted from income. (238)

Property of Equality of Squares   If two expressions are equal, then their squares are also equal: if $r = s$, then $r^2 = s^2$. (392)

proportion   A mathematical statement that two ratios are equal: $\dfrac{a}{b} = \dfrac{c}{d}$. (104)

pure imaginary number   A number of the form bi, where $b \neq 0$. (310)

Pythagorean Theorem   In any right triangle, the square of the length of the hypotenuse, c, equals the sum of the squares of the lengths, a and b, of the legs: $a^2 + b^2 = c^2$. (308)

quadrants   The four sections of the coordinate plane formed by the coordinate axes. (92)

quadratic equation   An equation that can be written in the form $ax^2 + bx + c = 0$, where $a \neq 0$. (280)

quadratic formula   A formula that can be used to solve quadratic equations: If $ax^2 + bx + c = 0$, then $x = \dfrac{-b \pm \sqrt{b^2 - 4ac}}{2a}$. (318)

quadratic function   Any function defined by a quadratic equation. (286)

quadratic pattern   A pattern that can be modeled by a quadratic equation. (332)

quadratic polynomial   A polynomial whose degree is two. (230)

quadrilateral   A four-sided closed figure. Squares, rectangles, trapezoids, and kites are all examples of quadrilaterals. (130)

Quotient Property of Exponents   If a, m, and n are real numbers and $a \neq 0$, then $\dfrac{a^m}{a^n} = a^{m-n}$. (222)

Quotient Property of Square Roots   If a and b represent positive real numbers, then $\sqrt{\dfrac{a}{b}} = \dfrac{\sqrt{a}}{\sqrt{b}}$. (302)

R

radical sign   The sign that shows a number is a radical. For example, $\sqrt{a}$ is used to denote the *principal square root* of a. (300, 390)

radicand   The expression under a radical sign. For example, in $\sqrt{x-1}$, $x - 1$ is the radicand. (390)

range   The second numbers in a set of ordered pairs. In a function, there is only one number in the range assigned to a number in the domain. (96)

rate   A ratio that compares quantities of different kinds of units, such as miles per hour or minutes per day. (102)

rate of change   In a linear function, the ratio of the change in the y-values to the change in the x-values. This relationship is represented by the slope of the graph of the function. (110)

ratio   A quotient that compares two quantities. A ratio of 5 boys to 2 girls can be written as $\dfrac{5}{2}$. (102)

rational equation   An equation that contains at least one rational expression. (372)

rational expression   An algebraic expression that is or can be expressed as the quotient of two polynomials. (342)

rational function   A function defined by a rational equation. (342)

rationalizing the denominator   The process of changing a fraction with an irrational denominator to an equivalent fraction with a rational denominator. (404)

rational numbers   The set of numbers that can be written in the form $\dfrac{p}{q}$, where p and q are integers and $q \neq 0$. (40)

real numbers   The set of numbers that consists of all the rational and irrational numbers. It does not include the sets imaginary numbers. (42)

reciprocal   For any nonzero number a, the reciprocal is $\dfrac{1}{a}$. A number multiplied by its reciprocal equals 1. Also called the *multiplicative inverse*. (8)

reciprocal function   The function in which the dependent variable is the reciprocal of the independent variable, $y = \dfrac{1}{x}$. This function is a form of the inverse-variation function in which $k = 1$. (380)

Reflexive Property of Equality   For any real number a, $a = a$ (a number is equal to itself). (18)

relation   A pairing of the elements in two sets. A relation can be represented by a set of ordered pairs, (x, y). (96)

relatively prime numbers   Two or more numbers that have no common factors other than 1. (261)

repeating decimal   A number in decimal form that has a string of one or more digits that repeat infinitely, such as 0.5555… or 4.123412341234… These repeating decimals can be indicated by a bar over the repeated numbers, as in $0.\overline{5}$ or $4.\overline{1234}$. (40)

replacement set   The set of all numbers that satisfy the conditions as possible replacements for the variable. (14)

right triangle   A triangle that has a 90° angle. (130)

root (of an equation)   A solution to an equation. (280, 318)

S

scientific notation   A number written in scientific notation is a product of the form $a \times 10^n$, where $1 \leq a < 10$ and n is an integer. (224)

second differences   In a table of ordered pairs, the numerical difference between one entry and the next of one of the variables is called the first difference. The difference between one first difference and the next is called the second difference. Second differences of the dependent variable can be used to determine whether the pattern is quadratic or not. (328)

sequence   An ordered set of real numbers. Each number in a sequence is called a term. (132)

similar geometric figures   Geometric shapes in which the corresponding sides have the same ratio and corresponding angles have the same measurement. If corresponding sides have the same ratio in a triangle, the corresponding angles will be equal. (366)

simplest radical form   The form of a square-root radical expression when there are no perfect squares or fractions under the radical sign and no radical expressions in the denominator of a fraction. (302)

simplify   To carry out all indicated operations in an expression. (10)

slope   A measurement of the steepness of a line in the coordinate plane. Given two points on a line, (x_1, y_1) and (x_2, y_2), the slope, m, of the line is given by $m = \dfrac{\text{rise}}{\text{run}} = \dfrac{y_2 - y_1}{x_2 - x_1}$.

slope-intercept form   A linear equation in the form $y = mx + b$, where m is the slope of the line and b is the y-intercept. (114)

solution (of an absolute-value equation)   If $|x| = a$ and $a \geq 0$, then $x = a$ or $x = -a$. (176)

solution (of an absolute-value inequality)   If $|x| < a$ and $a > 0$, then $x < a$ and $x > -a$ (which can be written $-a < x < a$). If $|x| > a$ and $a > 0$, then $x > a$ or $x < -a$. (178)

solution (of an open sentence)   Any value of the variable(s) that makes the equation true. (14)

solution (of an equation in two variables)   An ordered pair of numbers (x, y) that makes the equation true. (94, 112)

solution (of a system of linear equations)   Any ordered pair that satisfies each of the equations in the system. Graphically, the solution is the point of intersection of the graphs. (188)

solution region   The set of all possible solutions of a linear inequality. In a graph, the solution region is indicated by shading the region. (206)

splitting the middle term   The process of factoring an expression of the form $ax^2 + bx + c$ when $a \neq 1$. In this process, the middle term is rewritten as the sum of two terms and then factoring by grouping is applied. (274)

square root   The number c is a square root of a if $c^2 = a$. The principle, or positive, square root of a can be represented as $\sqrt{a}$ or as $a^{\frac{1}{2}}$. (300)

square-root equation   An equation in one variable that contains a square-root expression, such as $\sqrt{5x + 2} = 12$. (392)

square-root expression   An expression that contains a radical, $\sqrt{}$. (343190)

square-root function   A function defined by a square-root expression, such as $y = \sqrt{3x}$ or $y = 4 - 2\sqrt{x - 5}$. (390)

standard form of a cubic equation   The form $ax^3 + bx^2 + cx + d = 0$, where $a \neq 0$. (284)

standard form of a quadratic equation   The form $ax^2 + bx + c = 0$, where $a \neq 0$. (280)

subset   A set that is contained in another set. A set that has fewer elements than the set that contains it is called a proper subset. (42)

substitution method   A procedure for solving a system of equations in which variables are replaced with known values or algebraic expressions. (190)

Substitution Principle   If $a = b$, then a may replace b in any statement containing a and the resulting statement will be true. (12)

Subtraction Property of Equality   If a, b, and c are real numbers and $a = b$, then $a - c = b - c$. (22, 62)

Subtraction Property of Inequality   If a, b, and c are real numbers and $a < b$, then $a - c < b - c$. This statement is also true if $<$ is replaced by $>$, $\le$ or $\ge$. (150)

Symmetric Property of Equality   For all numbers a and b, if $a = b$, then $b = a$. (18)

system of linear equations (or inequalities)   A set of linear equations (or inequalities) with the same variables. (188, 210)

T

term (of a polynomial)   Any monomial in a polynomial. (230)

term (of a sequence)   Each number in a sequence of numbers. (132)

terminating decimal   A decimal number that has a finite number of nonzero digits to the right of the decimal point, such as 0.25 or 0.333. (40)

tolerance   The acceptable range that a measurement may differ from a fixed standard. (180)

Transitive Property of Deductive Reasoning   If p, q, and r represent statements and if $p \Rightarrow q$ and $q \Rightarrow r$, then $p \Rightarrow r$. (84)

Transitive Property of Equality   For all numbers a, b, and c, if $a = b$ and $b = c$, then $a = c$. (18)

Transitive Property of Inequality   For all numbers a, b, and c, if $a < b$ and $b < c$, then $a < c$. This is also true if $<$ is replaced by $>$, $\le$, or $\ge$. (146)

trapezoid   A quadrilateral with exactly one pair of parallel sides. (130)

trinomial   A polynomial with exactly three terms. (230)

V

variable   A letter used to represent numbers in an algebraic expression. (12)

vertex (of a parabola)   The point of intersection of a parabola and its axis of symmetry; this point will represent either the minimum or maximum value of the function. (286)

vertical-addition format (in the addition or subtraction of polynomials)   The placement of the like terms above (or below) one another before calculation. (234)

vertical line   A line parallel to the y-axis. The equation of the vertical line that contains the point $(r, 0)$ is $x = r$. (112)

vertical-line test   A method of determining whether a graph represents a function: If no vertical line in the coordinate plane intersects the graph in more than one point, then the graph represents a function. (96)

W

whole numbers   The set of natural numbers and 0. (40)

work rate   The proportional amount of a task that is completed in one unit of time. For example, if a task takes a person 5 hours to complete, then the work rate is $\frac{1}{5}$ of the task per hour. (376)

X

x-axis   The horizontal axis in a coordinate plane, usually representing the values of the independent variable. (92)

x-coordinate The first number in an ordered pair. It represents the horizontal distance from the origin of a point in the coordinate plane. (92)

x-intercept The x-coordinate of the point where a graph crosses the x-axis. (112)

Y

y-axis The vertical axis in a coordinate plane, usually representing the values of the dependent variable. (92)

y-coordinate The second number in an ordered pair. It represents the vertical distance from the origin of a point in the coordinate plane. (92)

y-intercept The y-coordinate of the point where a graph crosses the y-axis. (112)

Z

zero exponent For any nonzero number a, $a^0 = 1$. (222)

Zero-Product Property If a and b are real numbers and $ab = 0$, then $a = 0$ or $b = 0$. (280)

Selected Answers

CHAPTER 1

Lesson 1.1 Skill A (pages 4–5)

Try This Example 1

a. $12 + 3$ b. $12 - 3$ c. 12×3 d. $12 \div 3$

Try This Example 2

$(4 + 3) - (10 \times 2)$

Exercises

Answers for Exercises 1–5 may vary. Sample answers are given.

1. 6.5 times 4.5   3. 6.5 divided by 4.5

5. the difference when the sum of 3.5 and 1 is taken from 12

7. $9 - 9$ 9. $3 + 3$ 11. $12 - 4$ 13. $18 \div 6$

15. $20 - (6 + 2)$ 17. $(3 + 3) - (0 + 7)$

19. $(5 \times 4) + (3 \times 6)$ 21. $3 \times 4 + \dfrac{20}{4}$

23. Sample answer: $\dfrac{1}{2}$ times y, $\dfrac{1}{2}$ multiplied by y, the product of $\dfrac{1}{2}$ and y, $\dfrac{1}{2} \times y$, $\dfrac{1}{2} \cdot y$, $\dfrac{1}{2}y$, $\left(\dfrac{1}{2}\right)(y)$

25. Sample answer: the sum of a b's, b times a, a times b, $a \times b$, $a \cdot b$, $(a)(b)$, ab, $b \times a$, $b \cdot a$, $(b)(a)$, ba

27. 101   29. 101.2   31. 101.02   33. 98.9   35. 99.09

37. 100   39. 110.11   41. 101.0101   43. 90.91

45. 99.11   47. 320   49. 3.2

Lesson 1.1 Skill B (pages 6–7)

Try This Example 1

$(2 \times 22.95) + (2 \times 44.50)$

Try This Example 2

$2(0.02 + 0.1)$, or $(2 \times 0.02) + (2 \times 0.1)$

Try This Example 3

$\dfrac{80 + 97 + 75}{3} = \dfrac{80}{3} + \dfrac{97}{3} + \dfrac{75}{3}$

Exercises

1. 12×5 3. $1200 - 35$

5. $(7 \times 14.95) + (3 \times 4.50)$

7. $(7 \times 20.14) + (3 \times 14.25)$

9. $2(12.5 + 12.5)$

11. $(5 + 3) + 6 = (3 + 6) + 5$

13. 7   15. 35   17. 252   19. 49

Lesson 1.2 Skill A (pages 8–9)

Try This Example 1

Associative Property of Multiplication

Try This Example 2

2

Exercises

1. Additive Identity

3. Commutative Property of Addition

5. Multiplicative Inverse

7. Commutative Property of Multiplication

9. Additive Identity

11. Multiplicative Inverse

13. 1; Multiplicative Inverse

15. $\dfrac{1}{3}$; Multiplicative Identity

17. $\dfrac{11}{2}$, or $5\dfrac{1}{2}$; Additive Identity

19. 15; Multiplicative Identity

21. 2; Multiplicative Inverse

23. 7; Multiplicative Identity, Multiplicative Inverse

25. 9; Additive Identity

27. Sample answer: $12(1 + 0.15)$

29. 9   31. 25   33. 10,000   35. 0   37. $\dfrac{9}{16}$ 39. $\dfrac{49}{16}$

41. 0.04   43. 0.000004

Lesson 1.2 Skill B (pages 10–11)

Try This Example 1

$6\dfrac{1}{8}$, or 6.125

Try This Example 2

64

Exercises

1. $4 + 3(7)$ **3.** $4^2 + 5 \times 4$ **5.** $\dfrac{9^2}{3 \times 2}$ **7.** 29   **9.** 28

11. 5   **13.** 2   **15.** $\dfrac{13}{2}$, or $6\dfrac{1}{2}$ **17.** $\dfrac{5}{7}$ **19.** 4   **21.** 4900

23. 108   **25.** 127   **27.** $\dfrac{17}{81}$ **29.** 1   **31.** $\dfrac{4}{3}$, or $1\dfrac{1}{3}$

33. $\dfrac{5}{4}$, or $1\dfrac{1}{4}$ **35.** $\dfrac{7}{10}$ **37.** $\dfrac{5}{6}$ **39.** $\dfrac{13}{12}$, or $1\dfrac{1}{12}$

41. $\dfrac{22}{15}$, or $1\dfrac{7}{15}$ **43.** $\dfrac{23}{30}$

Lesson 1.2 Skill C (pages 12–13)

Try This Example 1

30

Try This Example 2

$\dfrac{4}{7}$

Exercises

1. $3(5)^2 - 5(5) + 4$ **3.** $\dfrac{3(3) + 3(2)}{4(2)}$ **5.** 45   **7.** 180

9. 44   **11.** $\dfrac{10}{3}$, or $3\dfrac{1}{3}$ **13.** $\dfrac{11}{4}$, or $2\dfrac{3}{4}$ **15.** 5

17. $\dfrac{7}{4}$, or $1\dfrac{3}{4}$ **19.** $\dfrac{7}{6}$, or $1\dfrac{1}{6}$ **21.** $1260   **23.** 0.25

25. 0.05   **27.** 0.75   **29.** 0.375   **31.** $\dfrac{3}{10}$ **33.** $\dfrac{33}{10}$, or $3\dfrac{3}{10}$

35. $0.1\overline{6}$ **37.** $0.\overline{6}$ **39.** $\dfrac{33}{50}$

Lesson 1.3 Skill A (pages 14–15)

Try This Example

a. 6   **b.** none

Exercises

1. no   **3.** yes   **5.** yes   **7.** 3   **9.** none   **11.** 4

13. 3   **15.** 3   **17.** 1   **19.** none   **21.** 1   **23.** yes

25. No; for example, 0 is not a solution.

27. yes

29. No; for example, 0 is not a solution.

31. a. no   **b.** no   **c.** yes

33. $\dfrac{1}{15}$ **35.** $\dfrac{1}{3}$ **37.** $\dfrac{8}{15}$ **39.** 1   **41.** 1   **43.** $\dfrac{5}{3}$, or $1\dfrac{2}{3}$

45. $\dfrac{4}{3}$, or $1\dfrac{1}{3}$ **47.** $\dfrac{5}{6}$ **49.** $\dfrac{16}{49}$ **51.** $\dfrac{9}{49}$

Lesson 1.3 Skill B (pages 16–17)

Try This Example 1

Answers may vary. Sample answer: The numbers 1 and 3 are odd, but their sum, 4, is even.

Try This Example 2

Answers may vary. Sample answer:

$12 \div 4 = 3$ False

$4 \div 1 = 4$ False

$4 \div \dfrac{1}{2} = 8$ True

$\dfrac{1}{2} \div \dfrac{1}{4} = 2$ True

These examples show that the statement is sometimes true. It appears that the statement may be true only when a or b (or both) is less than 1.

Exercises

1. yes   **3.** no

5. Answers may vary. Sample answer: Consider 3 and 4. The product is even, not odd.

7. 0, 4, 8, 12, 16, . . .

9. all even whole numbers

11. sometimes true; $1 + 3 + 5 = 9$ is odd, but $2 + 3 + 5 = 10$ is even.

13. sometimes true; the digits of 12 add to a multiple of 3, and 12 is divisible by 6; but the digits of 15 add to a multiple of 3, and 15 is not divisible by 6.

15. 0.6; 60%   **17.** 0.375; 37.5%   **19.** 2.25; 225%

21. 0.16; $\dfrac{4}{25}$ **23.** 1.25; $\dfrac{5}{4}$ **25.** 0.165; $\dfrac{33}{200}$

Lesson 1.3 Skill C (pages 18–19)

Try This Example 1

$\dfrac{3}{4} = \dfrac{3}{4} \times \dfrac{25}{25}$ Identity Property of Multiplication

$\dfrac{3}{4} \times \dfrac{25}{25} = \dfrac{75}{100}$ Multiplication of fractions

$\frac{75}{100} = 75\%$ Definition of percent

$\frac{3}{4} = 75\%$ Transitive Property of Equality

Try This Example 2

$4a = a + a + a + a$ Definition of multiplication by 4
$= (a + a) + (a + a)$ Associative Property of Addition
$= 2a + 2a$ Definition of multiplication by 2

Exercises

1. Associative Property of Addition, definition of multiplication by 2, definition of multiplication by 4

Answers for Exercises 3–11 may vary. Sample answers are given.

3. $\frac{3}{5} = \frac{3}{5} \cdot \frac{20}{20}$ Identity Property of Multiplication

$\frac{3}{5} \cdot \frac{20}{20} = \frac{60}{100}$ Multiplication of fractions

$\frac{60}{100} = 60\%$ Definition of percent

$\frac{3}{5} = 60\%$ Transitive Property of Equality

5. $\frac{4}{10} = \frac{4}{10} \cdot \frac{10}{10}$ Identity Property of Multiplication

$\frac{4}{10} \cdot \frac{10}{10} = \frac{40}{100}$ Multiplication of fractions

$\frac{40}{100} = 40\%$ Definition of percent

$\frac{4}{10} = 40\%$ Transitive Property of Equality

7. $3(x + 3) + 4 = 3x + 9 + 4$ Distributive Property
$3x + 9 + 4 = 3x + 13$ Addition
$3(x + 3) + 4 = 3x + 13$
     Transitive Property of Equality

9. $5a = a + a + a + a + a$ Multiplication by 5
$a + a + a + a + a = (a + a + a) + (a + a)$
     Associative Property of Addition
$(a + a + a) + (a + a) = 3a + 2a$
     Multiplication by 3 and Multiplication by 2
$5a = 3a + 2a$
     Transitive Property of Equality

11. $3 + 4(y + 2) = 3 + 4y + 8$ Distributive Property
$3 + 4y + 8 = 4y + 3 + 8$
     Commutative Property of Addition
$4y + 3 + 8 = 4y + 11$ Addition
$3 + 4(y + 2) = 4y + 11$
     Transitive Property of Equality

13. even, since $2n + 2n = 2(n + n)$ by the Distributive Property

15. even, since $2n + 1 + 2n + 1 = 2(2n + 1)$ by the Distributive Property

17. $3 + 5$ **19.** $1 + 2 \times 10 = 21$ **21.** $\frac{25}{3}$, or $8\frac{1}{3}$

23. 82   **25.** 3   **27.** none

Lesson 1.4 Skill A (pages 20–21)

Try This Example 1

$d = 9$

Try This Example 2

$t = 114.5$

Exercises

1. 2.3   **3.** 3   **5.** $x = 6$ **7.** $q = 9$ **9.** $x = 7.5$

11. $d = 21.6$ **13.** $x = 11$ **15.** $k = 2.5$ **17.** $x = 10$

19. $z = \frac{32}{3}$, or $10\frac{2}{3}$ **21.** $x = 10\frac{1}{2}$ **23.** $x = 2.75$

25. $x = 18$ **27.** $x = 13$ **29.** $x = \frac{33}{4}$, or $8\frac{1}{4}$

31. $c = \frac{41}{5}$, or $8\frac{1}{5}$ **33.** $a = 15.6$ **35.** $c = 0$ **37.** 120

39. 30   **41.** 21.6   **43.** 66.67%   **45.** 133.33%

Lesson 1.5 Skill A (pages 22–23)

Try This Example 1

$a = 24.5$

Try This Example 2

$c = 4.5$

Try This Example 3

$c = 2.05$

Exercises

1. 6.4   **3.** 3.1   **5.** $t = 10$ **7.** $a = 0$ **9.** $z = 0.05$

11. $t = \frac{33}{7}$, or $4\frac{5}{7}$ **13.** $w = \frac{1}{3}$ **15.** $t = \frac{4}{5}$ **17.** $t = 3.2$

19. $z = 1$ **21.** $a = \frac{6}{5}$, or $1\frac{1}{5}$ **23.** $z = 11$ **25.** $\frac{1}{2}$

27. $\frac{3}{2}$, or $1\frac{1}{2}$ **29.** $\frac{5}{18}$ **31.** $\frac{23}{24}$ **33.** $\frac{5}{36}$ **35.** $\frac{7}{40}$

37. $\frac{24}{5}$, or $4\frac{4}{5}$ **39.** 14   **41.** $x = 21.5$ **43.** $w = 200$

Lesson 1.6 Skill A (pages 24–25)

Try This Example 1

a. $x = 18$ b. $x = 0$

Try This Example 2

$m = 3.3$

Exercises

1. $x + 2 = 8$ 3. $x = 6$ 5. $c = 8.6$ 7. $y = 5.6$

9. $w = 4$ 11. $a = 24.2$ 13. $d = 33.5$

15. $m = \frac{4}{3}$, or $1\frac{1}{3}$ 17. $z = 10$ 19. $h = 5$ 21. $n = 16\frac{4}{5}$

23. $z = \frac{25}{4}$, or $6\frac{1}{4}$ 25. $a = 0$ 27. $\frac{1}{3}$ 29. $\frac{3}{4}$ 31. $\frac{7}{8}$

33. $\frac{1}{2}, \frac{3}{5}, \frac{2}{3}, \frac{5}{7}$ 35. $\frac{1}{21}$, 0.05, 0.14, $\frac{8}{50}, \frac{11}{25}$, 0.45

Lesson 1.6 Skill B (pages 26–27)

Try This Example 1

16 bagels

Try This Example 2

4.6 tons

Exercises

1. A key word is *left*; let *s* represent the number of students originally in the room; $s - 7 = 14$

3. Let *t* represent the original temperature; $t - 6 = 72$; 78°F

5. Let *m* represent the number of marbles originally in the bag; $m - 12 = 18$; 30 marbles

7. Let *d* represent the amount originally in the bank; $d - 320.39 = 1245.35$; $1565.74

9. Let *s* represent the amount of sand; $s + 1.3 = 10$; 8.7 pounds

11. $(1 \times 19.95) + (2 \times 7.95)$

Lesson 1.7 Skill A (pages 28–29)

Try This Example 1

$t = 30.6$

Try This Example 2

$v = 12$

Exercises

1. 3   3. 2   5. $\frac{9}{4}$ 7. $t = 4$ 9. $a = 30$ 11. $v = 10.4$

13. $x = 49.0$ 15. $w = \frac{18}{4}$, or $4\frac{1}{2}$ 17. $z = \frac{9}{7}$, or $1\frac{2}{7}$

19. $r = \frac{7}{2}$, or $3\frac{1}{2}$ 21. $s = \frac{1}{2}$ 23. $c = 5$ 25. $z = 8$

27. $c = 6$ 29. $n = 12$ 31. $x = \frac{12}{35}$ 33. $t = 9$

35. $x = 18$ 37. $x = \frac{bc}{a}$ 39. $g = 29$ 41. $x = 1.1$

43. $a = 0.1$ 45. $x = \frac{46}{5}$, or $9\frac{1}{5}$

Lesson 1.8 Skill A (pages 30–31)

Try This Example 1

$c = 6$

Try This Example 2

2.5 feet

Exercises

1. 3   3. 5   5. $x = 8$ 7. $b = 9$ 9. $s = 3.5$

11. $d = \frac{13}{3}$, or $4\frac{1}{3}$ 13. $r = \frac{9}{8}$, or $1\frac{1}{8}$ 15. $p = \frac{19}{20}$

17. $x = \frac{6}{7}$ 19. $c = \frac{7}{2}$, or $3\frac{1}{2}$ 21. $x = 3$ 23. $k = 6$

25. $n = 10$ 27. $x = 4$ 29. $\frac{3}{2}$, or $1\frac{1}{2}$ feet   31. 2.1 feet

33. No; let *x* represent the length of each board. The solution to $4x = 17$ is $\frac{17}{4}$, or $4\frac{1}{4}$, which is not a whole number.

35. 13.5   37. $\frac{1}{4}$ 39. 4.2

Lesson 1.9 Skill A (pages 32–33)

Try This Example 1

$z = 9$

Try This Example 2

$a = 6$

Try This Example 3

$x = 15$

Exercises

1. Three times x equals 22.

3. Three times c divided by 4 equals 4.

5. Divide each side by 3.5; $x = 2$

7. Divide each side by 4.3; $k = 2$

9. Multiply each side by $\frac{3}{4}$ or divide each side by $\frac{4}{3}$; $m = 7.5$

11. Multiply each side by 3.5; $n = 3.5$

13. Multiply each side by 2.5; $p = 6.25$

15. Multiply each side by the sum $2.5 + 1.5$, or 4; $n = 32$

17. Divide each side by $\frac{3}{2}$; $n = \frac{4}{9}$

19. Divide each side by the sum $1\frac{1}{2} + 6\frac{1}{2}$, or 8; $n = 4$

21. $x = 4$ 23. $x = 16$ 25. $x = \frac{16}{3}$, or $5\frac{1}{3}$

27. $x = 6$ 29. $n = 21$ 31. $c = 5.4$ 33. $x = 3.9$

35. $t = 0.75$

Lesson 1.9 Skill B (pages 34–35)

Try This Example 1

$9

Try This Example 2

$213

Exercises

1. divide; $\frac{d}{5} = 24$

3. Let x represent the amount of money each worker will receive; $6x = 1120$, $x = 190$. Each worker will receive $190.

5. Let w represent the number of weeks needed; $25w = 290$; $w = 11.6$; 12 weeks

7. Let n represent the number of buses needed; $\frac{380}{n} = 32$; $n = 11\frac{7}{8}$; 12 buses will be needed.

9. 22 exercises

11. 75.9°F

CHAPTER 2

Lesson 2.1 Skill A (pages 40–41)

Try This Example 1

$\frac{5}{8}$

Try This Example 2

$0.2\overline{7}$

Exercises

1. $\frac{35}{100}$ 3. $\frac{13,755}{1000}$ 5. $\frac{1}{5}$ 7. $\frac{7}{5}$, or $1\frac{2}{5}$

9. $\frac{1}{4}$ 11. $\frac{5}{4}$, or $1\frac{1}{4}$ 13. $\frac{31}{200}$ 15. $\frac{111}{1000}$

17. 0.5; terminating   19. 4.6; terminating

21. $0.\overline{1}$; repeating   23. $4.\overline{6}$; repeating

25. $0.1\overline{6}$; repeating   27. $10.\overline{27}$; repeating

29. a. $\left(4\frac{1}{2}\right) \div \left(2\frac{3}{4}\right) = \frac{9}{2} \div \frac{11}{4} = \frac{9}{2} \times \frac{4}{11} = \frac{18}{11}$, a quotient of two integers with nonzero denominator

   b. $\left(4\frac{1}{2}\right) \div \left(2\frac{3}{4}\right) = 4.5 \div 2.75 = 1.\overline{63}$, a repeating decimal

31. $\frac{1}{200}, \frac{1}{20}, 0.2, 0.22, \frac{1}{2}$ 33. 12   35. 0.02   37. 100

39. 15%   41. 6%   43. 600%   45. $t = 15$

47. $y = 123.2$ 49. $y = 20$ 51. $x = 36$

Lesson 2.1 Skill B (pages 42–43)

Try This Example 1

natural number, whole number, integer, rational number, real number

Try This Example 2

False; sample counter-example: -2 is an integer but not a natural number.

Exercises

1. $-3, -12, -4$

3. $1.05, 1.\overline{41}$

5. rational, real

7. rational, real

9. integer, rational, real

11. rational, real

13. irrational, real

15. natural number, whole number, integer, rational, real

17. False; sample counterexample: −7 is an integer but not a counting number.

19. False; sample counterexample: 1 divided by 3 is $\frac{1}{3}$, which is not a counting number.

21. A ratio $\frac{p}{q}$ represents a rational number only when p and q are both integers. A ratio may be irrational if p, q, or both are not integers.

23. 1   25. $\frac{17}{6}$, or $2\frac{5}{6}$ 27. $\frac{5}{3}$, or $1\frac{2}{3}$ 29. 11

Lesson 2.1 Skill C (pages 44–45)

Try This Example 1

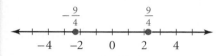

Try This Example 2

a. 3   b. 3   c. 3   d. −3

Exercises

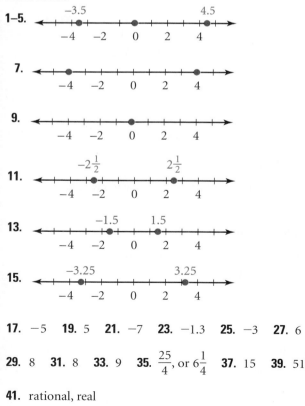

1–5.

7.

9.

11.

13.

15.

17. −5   19. 5   21. −7   23. −1.3   25. −3   27. 6

29. 8   31. 8   33. 9   35. $\frac{25}{4}$, or $6\frac{1}{4}$ 37. 15   39. 51

41. rational, real

440   Selected Answers

Lesson 2.2 Skill A (pages 46–47)

Try This Example 1

−1

Try This Example 2

−5

Exercises

1. −5   3. 0   5. −1   7. 1   9. 0   11. −4   13. −8

15. −6   17. 3   19. −10   21. 1   23. 2   25. −10

27. 1; Identity Property of Multiplication

29. 2; Inverse Property of Multiplication

31. $\frac{9}{10} = \frac{9}{10} \cdot \frac{10}{10} = \frac{90}{100}$

$90\% = \frac{90}{100}$; Therefore, by the Transitive Property of Equality, $\frac{9}{10} = 90\%$.

Lesson 2.2 Skill B (pages 48–49)

Try This Example 1
−4

Try This Example 2
−8

Try This Example 3

4

Exercises

1. subtract   3. add   5. 11   7. 8   9. −12   11. −123

13. −37   15. 0   17. −137   19. 1.5   21. 3.92   23. $-\frac{2}{3}$

25. $-\frac{1}{2}$ 27. −18   29. −25   31. −10   33. −130

35. 2.73   37. 0.15   39. −7   41. yes   43. yes

45. No; for example, 2 is not a solution.

47. irrational, real

49. integer, rational, real

51. rational, real

53. rational, real

Lesson 2.3 Skill A (pages 50–51)

Try This Example 1

−21

Try This Example 2

247.9

Try This Example 3

−11

Exercises

1. $12 + (−15)$ **3.** $1.2 + 0$ **5.** $14 + (−3)$

7. equal to 0   **9.** greater than 0   **11.** less than 0

13. equal to 0   **15.** greater than 0   **17.** 5   **19.** −15

21. −6   **23.** −25   **25.** 25   **27.** −5   **29.** −1   **31.** −3.95

33. −1   **35.** 1   **37.** 4   **39.** −1   **41.** −18

43. $x = 8.5$ **45.** $t = 1.6$ **47.** $t = 24$

Lesson 2.3 Skill B (pages 52–53)

Try This Example 1

27

Try This Example 2

4

Try This Example 3

60

Exercises

1. $−2 − 1$

3. $2 − (−3)$, or $2 + 3$

5. $4 + (−7)$, or $4 − 7$

7. 2   **9.** −25   **11.** −16   **13.** −18   **15.** 8   **17.** −31

19. $−2x − 3$ **21.** $5 − 2d$ **23.** $−x − 7$ **25.** −21

27. 0   **29.** 19   **31.** −2   **33.** −4   **35.** 2.5   **37.** −32

39. −15   **41.** 1   **43.** −3.1

Lesson 2.3 Skill C (pages 54–55)

Try This Example 1

$−3°F$

Try This Example 2

decrease of $26.27

Exercises

1. $6.75 − 1.50 + 0.42 + 1.00$

3. $450 + 275 − 45 − 120 − 230$

5. $2°F$ **7.** decrease of 6 points

9. decrease of $4.66   **11.** $225

Lesson 2.4 Skill A (pages 56–57)

Try This Example 1

a. 480   **b.** −6

Try This Example 2

$\frac{100}{9}$, or $11\frac{1}{9}$

Exercises

1. negative $(−)$ **3.** positive $(+)$ **5.** −12   **7.** 25

9. −44   **11.** 108   **13.** −5   **15.** $−\frac{4}{3}$, or $−1\frac{1}{3}$

17. −10   **19.** $\frac{64}{25}$, or $2\frac{14}{25}$ **21.** −24   **23.** −180

25. 240   **27.** negative   **29.** negative   **31.** negative

33. −1; strategies will vary. Sample strategy: Use the Commutative and Associative Properties of Multiplication to pair up multiplicative inverses.
$$\left(\frac{2}{3} \cdot \frac{3}{2}\right) \cdot \left(−\frac{5}{4} \cdot −\frac{4}{5}\right) \cdot \left(−\frac{5}{6} \cdot \frac{6}{5}\right) = 1 \cdot 1 \cdot (−1) = −1$$

35. −15.5   **37.** −11   **39.** −14.9   **41.** 0   **43.** −36

45. 80.1   **47.** −20.9   **49.** 0.2

Lesson 2.4 Skill B (pages 58–59)

Try This Example 1

a. −30   **b.** 30

Try This Example 2

−2

Exercises

1. negative $(−)$ **3.** positive $(+)$ **5.** −6   **7.** −9

9. −3   **11.** −9   **13.** −5   **15.** −1.25   **17.** −3

19. $−\frac{4}{3}$, or $−1\frac{1}{3}$ **21.** −1   **23.** $\frac{13}{4}$ **25.** 5   **27.** $\frac{3}{5}$

29. negative   **31.** negative   **33.** positive   **35.** 4

37. 16   **39.** 36   **41.** 64   **43.** 9   **45.** 19   **47.** $g = 10.05$

49. $r = 86.2$ **51.** $w = 35$ **53.** $a = 121$

Lesson 2.4 Skill C (pages 60–61)

Try This Example 1

26

Try This Example 2

1

Exercises

1. 25   **3.** -25 **5.** 38   **7.** 2   **9.** 6   **11.** $\frac{1}{2}$ **13.** $-\frac{2}{3}$

15. -6 **17.** 144   **19.** $\frac{9}{16}$ **21.** -3 **23.** $\frac{1}{10}$

25. $a = -1$ or $a = 9$

27. a. 0   **b.** 0

29. a. -2.3345 **b.** 2.3345

31. -8 **33.** -2.3 **35.** 15   **37.** -5.1

Lesson 2.5 Skill A (pages 62–63)

Try This Example 1

$a = -6$

Try This Example 2

$v = 1.5$, or $1\frac{1}{2}$

Exercises

1. a. Subtraction Property of Equality   **b.** 5

3. a. Subtraction Property of Equality   **b.** 4

5. $x = 16$ **7.** $y = 30$ **9.** $a = 1$ **11.** $b = 0$

13. $p = 0$ **15.** $x = -2.8$ **17.** $v = 4$ **19.** $x = 5.3$

21. $d = \frac{9}{2}$, or $4\frac{1}{2}$ **23.** $x = 1$ **25.** $y = -6.1$ **27.** 0.96

29. $0.41\overline{6}$ **31.** $\frac{1}{25}$ **33.** $\frac{1503}{400}$, or $3\frac{303}{400}$

35. natural number, whole number, integer, rational, real

37. irrational, real

39. 6   **41.** 6   **43.** -6 **45.** 6

Lesson 2.5 Skill B (pages 64–65)

Try This Example 1

$t = -4$

Try This Example 2

$a = -6$

Exercises

1. $6x = 12$ **3.** $3a = 18$ **5.** $\frac{13}{26}c = -1$, or $\frac{1}{2}c = -1$

7. $x = 7$ **9.** $x = -35$ **11.** $y = -21$ **13.** $x = -2$

15. $z = -4$ **17.** $z = 8$ **19.** $t = 20$ **21.** $t = 0.29$

23. $r = 9$ **25.** positive   **27.** negative   **29.** positive

31. $-\frac{a}{b} = \frac{-a}{b} \cdot \frac{-1}{-1} = \frac{a}{-b}$

33. Sample counterexample: $\frac{2}{1} = 2$, but $\frac{1}{2} \neq 2$.

Lesson 2.6 Skill A (pages 66–67)

Try This Example 1   Try This Example 2

a. $7x$ **b.** $-7r$ **c.** m $-x - 3$

Try This Example 3

$5z - 2$

Exercises

1. yes   **3.** yes   **5.** no   **7.** x **9.** $-10z$ **11.** $-2x$

13. $-3a$ **15.** t **17.** $5a - 5$ **19.** $7y - 12$

21. $-8d + 24$ **23.** $-6w$ **25.** $10 - 8x$ **27.** $20 - 2x$

29. $(2a - b)x + a$

31. $(a + c)x + (3c - a)$

33. $(a + c + d)x - (2a - bc - d)$

35. -25 **37.** 9   **39.** 0   **41.** x

Lesson 2.6 Skill B (pages 68–69)

Try This Example 1

a. $-6x - 6$ **b.** $-4z - 28$

Try This Example 2

$-10t + 7$

Try This Example 3

a. $-2z - 3$ **b.** $-2t + 1$

Exercises

1. $-6g + 6$ **3.** $6s - 12$ **5.** $-2x - 10$ **7.** $7t - 14$

9. $-12q - 18$ **11.** $2a - 6$ **13.** $10 - 5b$ **15.** $-4h + 15$

17. $-3y + 3$ **19.** $22m - 59$ **21.** $-9s + 45$

23. $5g - 2$ **25.** $-a - 2$ **27.** $-3z + 8$ **29.** $\frac{13}{2}z - 12$

31. $\frac{7}{2}a - 10$ **33.** $c + 21$ **35.** $-\frac{7}{3}u - 8$

37. $\frac{57}{5}k - 19$ **39.** 13 **41.** $x = -4.5$ **43.** $a = \frac{3}{2}$, or $1\frac{1}{2}$

45. $x = -40$ **47.** $x = 4$

Lesson 2.7 Skill A (pages 70–71)

Try This Example 1

$x = -\frac{5}{4}$

Try This Example 2

$x = \frac{1}{2}$

Exercises

1. Subtraction Property of Equality

3. Addition Property of Equality

5. $x = 2$ **7.** $x = 3$ **9.** $v = -2$ **11.** $f = -4$

13. $p = 6$ **15.** $b = \frac{5}{6}$ **17.** $t = -1$ **19.** $d = -\frac{8}{3}$, or $-2\frac{2}{3}$

21. $x = 9$ **23.** $x = -2$ **25.** $j = 18.75$

27. $q = -\frac{28}{3}$, or $-9\frac{1}{3}$ **29.** $n = \frac{54}{7}$, or $7\frac{5}{7}$ **31.** $d = 0$

33. $x = \frac{c}{a} - b$ **35.** $6n$ **37.** g **39.** $-20k$

41. $-5d + 5$ **43.** $-3k + 4$ **45.** $8u + 12$

Lesson 2.7 Skill B (pages 72–73)

Try This Example 1

$C = \frac{5}{9}(F - 32)$

Try This Example 2

$r = \frac{A}{P} - 1$

Exercises

1. Division Property of Equality

3. Division Property of Equality

5. $a = b - x$ **7.** $r = \frac{d}{t}$ **9.** $x = \frac{4 - a}{2}$

11. $x = \frac{y - b}{m}$ **13.** $y = \frac{12 - 2x}{3}$ **15.** $t = \frac{A - P}{Pr}$

17. $g = \frac{-f}{a}$ **19.** $f = -ag$

21. a. $2(x + y) = 12$
b. $x = 6 - y$
c. $y = 6 - x$

23. decrease of $\$76.03$

25. gain of 15 yards

Lesson 2.7 Skill C (pages 74–75)

Try This Example

18 feet by 27 feet

Exercises

1. Let ℓ represent Lin's age; $\ell + 2\ell = 27$, or $3\ell = 27$

3. 30 ounces **5.** 12 miles **7.** 30° and 60°

9. 26 nickels and 14 dimes **11.** 17°F

Lesson 2.8 Skill A (pages 76–77)

Try This Example 1

$c = \frac{18}{7}$, or $2\frac{4}{7}$

Try This Example 2

$r = 18$

Exercises

1. $-3x = 2x + 2$ **3.** $15t - 120 = 16t$ **5.** $x = -2$

7. $t = 0$ **9.** $t = 3$ **11.** $y = 11$ **13.** $x = 6$

15. $x = 2$ **17.** $y = 0.9$ **19.** $z = 1.48$ **21.** $a = -2$

23. $z = -\frac{3}{4}$ **25.** $x = \frac{1}{2}$ **27.** $x = \frac{14}{19}$

29. $x = \frac{-ac}{bc - ab}$, or $\frac{ac}{ab - bc}$ **31.** 1

33. 11 **35.** 9 **37.** none

Lesson 2.8 Skill B (pages 78–79)

Try This Example 1

$a = -5$

Try This Example 2

$x = 5$

Try This Example 3

$t = -\dfrac{6}{11}$

Exercises

1. $2x - 1$ **3.** $7z - 9$ **5.** $6t - 5$ **7.** $h = -4$

9. $z = -\dfrac{1}{2}$ **11.** $q = 3$ **13.** $w = -2$

15. $a = \dfrac{10}{7}$, or $1\dfrac{3}{7}$ **17.** $d = 7$ **19.** $w = -8$

21. $y = \dfrac{5}{2}$, or $2\dfrac{1}{2}$ **23.** $n = \dfrac{5}{9}$ **25.** $b = -\dfrac{3}{7}$ **27.** $r = -5$

29. For all real numbers s, combining like terms gives $2(s + 1) + 4(s + 1) + 6(s + 1) = 12(s + 1)$; thus $12(s + 1) = 12(s + 1)$. This is the Reflexive Property of Equality. So, the equation is always true, regardless of the value of s.

31. 400 **33.** 500 **35.** 640

Lesson 2.8 Skill C (pages 80–81)

Try This Example

Invest $3000 at 6% and $7000 at 8%.

Exercises

1. Let x represent the amount invested at 5%; $0.05x + 0.06(1500 - x) = 85$

3. $700 at 5% and $1300 at 6%

5. 4 hours at 50 miles per hour and 3 hours at 55 miles per hour

7. 120 child tickets and 80 adult tickets

9. $2x - 10 = 11$ Distributive Property
$2x = 21$ Addition Property of Equality
$x = \dfrac{21}{2}$ Division Property of Equality

11. $3x = -2$ Distributive Property
$x = -\dfrac{2}{3}$ Division Property of Equality

Lesson 2.9 Skill A (pages 82–83)

Try This Example 1

$4 - 10 \neq 10 - 4; -6$

Try This Example 2

$-2(x - 3) = -2x + 6; -2x + 6 = 18; -2x = 12; x = -6$

Exercises

1. -5 satisfies $3x = -15$ and $3x + 4 = -11$, but not $3(x + 4) = -11$

3. $-(3 + 4) = -3 - 4$, not $-3 + 4$; $-(3 + 4) + 5 = -3 - 4 + 5 = -2$

5. $\dfrac{3}{7} + \dfrac{2}{7} = \dfrac{3 + 2}{7}$, not $\dfrac{3 + 2}{7 + 7}$; $\dfrac{3}{7} + \dfrac{2}{7} = \dfrac{3 + 2}{7} = \dfrac{5}{7}$

7. $3x + 5 - 4x = 3x - 4x + 5; 3x - 4x + 5 = 19;$ $-x + 5 = 19; -x = 14; x = -14$

9. Associative Property of Addition; Commutative Property of Addition; Associative Property of Addition; Inverse Property of Addition; Identity Property of Addition

11. $x = \dfrac{y - 5}{-2}$, or $-\dfrac{1}{2}y + \dfrac{5}{2}$

13. $x = \dfrac{-5y + 12}{-3}$, or $\dfrac{5}{3}y - 4$

15. $x = -\dfrac{y}{3} + 3$ **17.** $y = \dfrac{3x + 24}{7}$, or $\dfrac{3}{7}x + \dfrac{24}{7}$

19. $y = \dfrac{-8x + 20}{7}$, or $-\dfrac{8}{7}x + \dfrac{20}{7}$

Lesson 2.9 Skill B (pages 84–85)

Try This Example

$\dfrac{x}{3} - 5 = 11 \Rightarrow \dfrac{x}{3} = 16$ Addition Property of Equality

$\dfrac{x}{3} = 16 \Rightarrow x = 48$ Multiplication Property of Equality

$\dfrac{x}{3} - 5 = 11 \Rightarrow x = 48$ Transitive Property of Deductive Reasoning

Exercises

1. p: $-2d + 7 = 11$; q: $-2d = 4$; r: $d = -2$

3. Distributive Property; Addition Property of Equality; Division Property of Equality; Transitive Property of Deductive Reasoning

5. Subtraction Property of Equality; Subtraction Property of Equality; Division Property of Equality; Symmetric Property of Equality; Transitive Property of Deductive Reasoning

7. $3(x - 5) = 5x \Rightarrow 3x - 15 = 5x$ (Distributive Property)
$3x - 15 = 5x \Rightarrow -15 = 2x$ (Subtraction Property of Equality)

$-15 = 2x \Rightarrow -\dfrac{7}{2} = x$

(Division Property of Equality)

$-\dfrac{7}{2} = x \Rightarrow x = -\dfrac{7}{2}$

(Symmetric Property of Equality)

$3(x - 5) = 5x \Rightarrow x = -\dfrac{15}{2}$

(Transitive Property of Deductive Reasoning)

9. $6(x - 3) + 4(x + 2) = 6 \Rightarrow$
$6x - 18 + 4x + 8 = 6$ (Distributive Property)
$6x - 18 + 4x + 8 = 6 \Rightarrow 10x - 10 = 6$
(Combining like terms)
$10x - 10 = 6 \Rightarrow 10x = 16$
(Addition Property of Equality)
$10x = 16 \Rightarrow x = 1.6$
(Division Property of Equality)
$6(x - 3) + 4(x + 2) = 6 \Rightarrow x = 1.6$
(Transitive Property of Deductive Reasoning)

11. $-6.1, -2, 0, 0.5, 9.2$

13. $3.6, 3.65, 3.7, 3.75, 3.754$

15. $<$ **17.** $<$

Lesson 2.9 Skill C (pages 87–87)

Try This Example 1

sometimes true; true when $a = 0$ or $a = 5$ but false otherwise

Try This Example 2

always true; $2(b - 1) - 7(b + 1) = -5b - 9$ and $-3b - 9 - 2b = -5b - 9$

Try This Example 3

$3(t - 1) = 3t$
$3t - 3 = 3t$
$3 = 0$ ✗
Therefore, $3(t - 1) \neq 3t$.

Exercises

1. true when $n = 0$ or $n = 3$ but false otherwise

3. $3(x + 5) + 2 = 3x + 20$
$3x + 17 = 3x + 20$
$17 = 20$ ✗
Therefore, $3(x + 5) + 2 \neq 3x + 20$.

5. sometimes true; true when $m = 0$ or $m = 4$ but false otherwise

7. sometimes true; true when $k = 0$ but false otherwise

9. sometimes true; true when y is any real number and $x = 1$

11. always true; $z(xy + x + y) = z(x[y + 1] + y) = zxy + zx + zy$

13. $3x + 6 = 3(x + 4) + 2$
$3x + 6 = 3x + 14$
$6 = 14$ ✗
Therefore, $3x + 6 \neq 3(x + 4) + 2$.

15. The first part of the statement is always true because no triangle can have all sides equal in length and no sides equal in length at the same time. The second part of the statement is always true since, if all sides are equal in length, then at least two sides are equal in length. The full statement is always true.

17. $x = -\dfrac{1}{2}$ **19.** $x = \dfrac{1}{2}$ **21.** $r = 3$ **23.** $m = -\dfrac{1}{2}$

CHAPTER 3

Lesson 3.1 Skill A (pages 92–93)

Try This Example

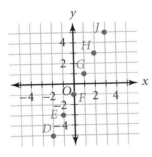

Points D and E are in Quadrant III; points G, H, and J are in Quadrant I; point F is on the y-axis and so is not in any quadrant.

Exercises

1. P is in a quadrant.   **3.** R is in a quadrant.

5. $F(5, 0)$ **7.** $X(-5, -3)$ **9.** $Z(5, 5)$

11. $H(1, 2)$ **13.** $L(-2, 5)$

15, 17, 19, 21.

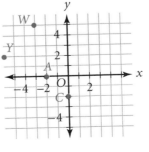

23. $a < 0$

25. One point is directly above (below) the other.

27. Both *r* and *s* must be positive. *X* is in Quadrant I.

29. 3   **31.** −3

Lesson 3.1 Skill B (pages 94–95)

Try This Example 1

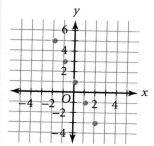

Try This Example 2

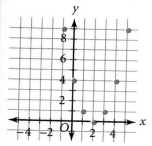

Exercises

1.

x	−3	−2	−1	0	1	2	3
y	14	11	8	5	2	−1	−4

3.

x	0	1	2	3	4	5	6
y	5	7	13	23	37	55	77

5.   **7.**

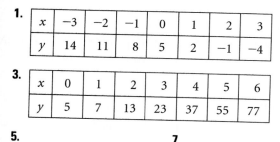

9.

x	−3	−2	−1	0	1	2	3
y	0	1	2	3	4	5	6

11.

x	0	1	2	3	4	5
y	0	2	4	6	8	10

13.

x	−3	−2	−1	0	1	2	3
y	1	0	−1	−2	−3	−4	−5

15.

x	−2	−1	0	1	2
y	7	1	−1	1	7

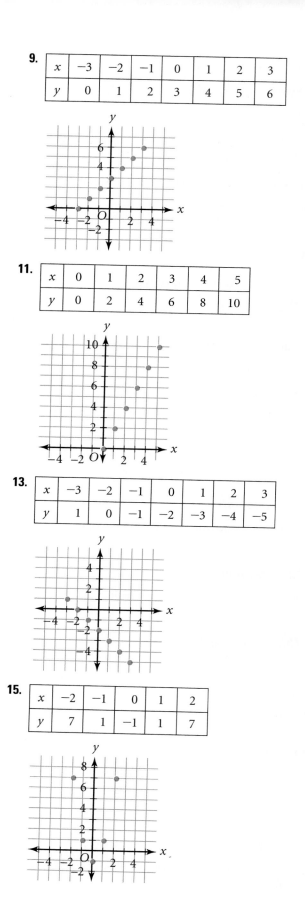

17.

x	0	1	2	3	4	5	6
y	−2	−2	−2	−2	−2	−2	−2

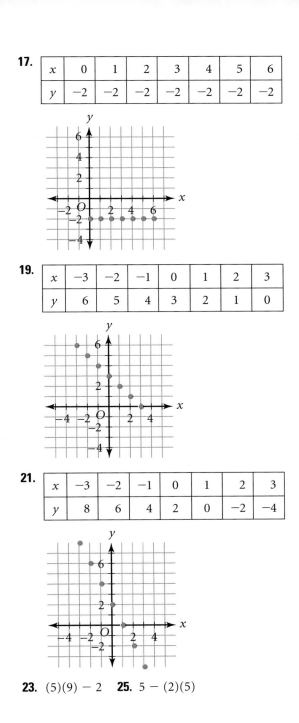

19.

x	−3	−2	−1	0	1	2	3
y	6	5	4	3	2	1	0

21.

x	−3	−2	−1	0	1	2	3
y	8	6	4	2	0	−2	−4

23. $(5)(9) - 2$ **25.** $5 - (2)(5)$

Lesson 3.2 Skill A (pages 96–97)

Try This Example 1

a. Yes; each member of the domain is assigned exactly one member of the range.
b. No; 2 is assigned both 4 and 10.

Try This Example 2

No; for example, the vertical line through $(3.5, 0)$ intersects the graph in two points.

Exercises

1. domain: 3, −4, 9
   range: 0, 1, 2

3. function

5. Yes; each member of the domain is assigned exactly one member of the range.

7. Yes; each member of the domain is assigned exactly one member of the range.

9. No; 16 is assigned both 4 and −4.

11. No; for example, the y-axis intersects the graph in two points.

13. Yes; there is no vertical line that intersects the graph in more than one point.

15. no value of y

17, 19.

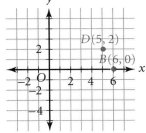

21.

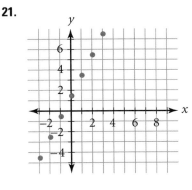

Lesson 3.2 Skill B (pages 98–99)

Try This Example 1

domain: −4, −3, −2, −1, 0, 1, 2, 3
range: 1

Try This Example 2

a. domain: all real numbers
   range: all real numbers
b. domain: all real numbers
   range: all nonnegative real numbers

Exercises

1. true   3. false

5. domain: $-3, -2, -1, 0, 1, 2, 3$
range: $-4, -3, -2, -1, 0, 1, 2, 3$

7. domain: all real numbers
range: all real numbers greater than or equal to -2

9. domain: all real numbers between -3 and 4, inclusive
range: all real numbers between -2 and 3, inclusive

11. domain: all real numbers
range: all real numbers

13. domain: all real numbers
range: all real numbers

15. domain: all real numbers
range: all nonpositive real numbers

17. $\frac{2}{5}$ **19.** $-\frac{1}{5}$ **21.** $-\frac{1}{7}$

Lesson 3.2 Skill C (pages 100–101)

Try This Example 1

independent variable: n
dependent variable: w

n	0	1	2	3	4	5
v	0	3	6	9	12	15

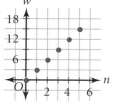

Try This Example 2

domain: all whole numbers
range: all nonnegative multiples of 5

Exercises

1. Multiply 5, 6, and 7 by 25.

3.

n	0	1	2	3	4	5	6
v	0	10	20	30	40	50	60

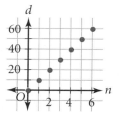

5.

n	1	2	3	4	5	6
c	24	48	72	96	120	144

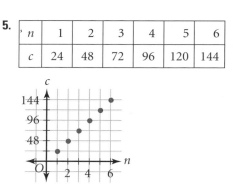

7. domain: all whole numbers
range: all nonnegative multiples of 6

9. $666.67 at 6% and $1333.33 at 12%

Lesson 3.3 Skill A (pages 102–103)

Try This Example

a. $t = 1.06c$
b. 1.06
c. The domain is any positive money amount in dollars and cents. The range is any positive money amount in dollars and cents.
d. $15.25

Exercises

1. dollars

3. **a.** $c = 1.15m$
  b. 1.15
  c. The domain is any positive money amount in dollars and cents. The range is any positive money amount in dollars and cents.
  d. $22.43

5. **a.** Let v represent the amount of water in the tank after t minutes; $v = 30t$
  b. 30
  c. domain: all nonnegative numbers
  range: all nonnegative numbers
  d. 165 gallons

7. **a.** 125 miles
  b. 6 inches

9. **a.** $498
  b. about 19.3 hours

11. 150%   **13.** $300

Lesson 3.3 Skill B (pages 104–105)

Try This Example 1

$39

Try This Example 2

562.5 gallons

Exercises

1. $n = 2$ **3.** $n = 4.8$ **5.** $y = 242$ **7.** $y = 28.8$

9. $y = 29.04$ **11.** 2.16 gallons; 3 gallons

13. 80 students   **15.** 155 stamps

Lesson 3.4 Skill A (pages 106–107)

Try This Example

a. $\frac{1}{7}$ **b.** -1

Exercises

1. falls from left to right   **3.** rise from left to right

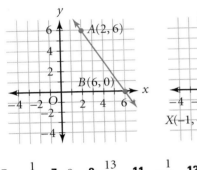

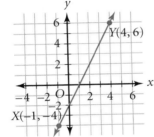

5. $-\frac{1}{5}$ **7.** 0   **9.** $\frac{13}{4}$ **11.** $-\frac{1}{2}$ **13.** 0   **15.** 0

17. 0   **19.** undefined   **21.** -10

23. a. If $-2(x + 5) = 16$, then $x + 5 = -8$ by the Division Property of Equality. If $x + 5 = -8$, then $x = -13$ by the Addition Property of Equality.
b. Therefore, by the Transitive Property of Equality, if $-2(x + 5) = 16$, then $x = -13$.

25. a. If $-4 = 3x + 5x$, then $-4 = 8x$ by combining like terms. If $-4 = 8x$, then $-\frac{1}{2} = x$ by the Division Property of Equality.

If $-\frac{1}{2} = x$, then $x = -\frac{1}{2}$ by the Symmetric Property of Equality.
b. Therefore, by the Transitive Property of Equality, if $-4 = 3x + 5x$, then $x = -\frac{1}{2}$.

27. a. If $-2(x - 1) + 3 = 7$, then $-2(x - 1) = 4$ by the Subtraction Property of Equality.
If $-2(x - 1) = 4$, then $x - 1 = -2$ by the Division Property of Equality.
If $x - 1 = -2$, then $x = -1$ by the Addition Property of Equality.
b. Therefore, by the Transitive Property of Equality, if $-2(x - 1) + 3 = 7$, then $x = -1$.

Lesson 3.4 Skill B (pages 108–109)

Try This Example 1

a.

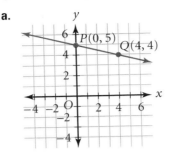

b. Method 1: $Q(4, 4)$
Method 2: Let $Q(x, y)$ be the coordinates of a second point on the line. Then $\frac{y - 5}{x - 0} = -\frac{1}{4}$. Choose a number besides 0 for x; for example, let $x = 4$.
Solve for y:
$$\frac{y - 5}{4 - 0} = -\frac{1}{4}$$
$$y - 5 = -1$$
$$y = 4$$
So, a second point on the line is $Q(4, 4)$.

Try This Example 2

slope of $\overleftrightarrow{AB}$: $\frac{3 - (-1)}{-1 - (-3)} = \frac{4}{2} = 2;$

slope of $\overleftrightarrow{BC}$: $\frac{7 - 3}{3 - (-1)} = \frac{4}{4} = 1;$

The points are not collinear.

Exercises

1. collinear

3. a.

b. Answers may vary. Sample answer: $Q(1, 2)$

5. a.

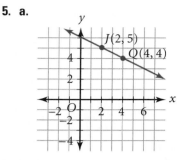

b. Answers may vary. Sample answer: $Q(4, 4)$

7. collinear

slope of $\overleftrightarrow{AB}$: $\frac{2-0}{2-0} = \frac{2}{2} = 1$

slope of $\overleftrightarrow{BC}$: $\frac{4-2}{4-2} = \frac{2}{2} = 1$

9. not collinear

slope of $\overleftrightarrow{PQ}$: $\frac{2-5}{2-6} = \frac{3}{4}$;

slope of $\overleftrightarrow{QR}$: $\frac{-2-2}{0-2} = \frac{4}{2} = 2$

11. parallel

13. If PQ has a slope of $\frac{m}{n}$, then it is the same as line l and Q is on l.

slope of $\overleftrightarrow{PQ}$: $\frac{(b+m)-b}{(a+n)-a} = \frac{b-b+m}{a-a+n} = \frac{m}{n}$

Since $\overleftrightarrow{PQ}$ contains $P(a, b)$ and has slope $\frac{m}{n}$, then it is line l.
Thus, $Q(a+n, b+m)$ is on line l.

15. Yes; every member of the domain is assigned exactly one member of the range.

17. Let s represent length and P represent perimeter. Then $P = 4s$; domain: all nonnegative real numbers; range: all nonnegative real numbers

19. $y = 77$

Lesson 3.4 Skill C (pages 110–111)

Try This Example

500 feet per minute (downward)

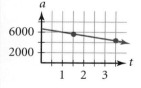

Exercises

1. positive   **3.** negative   **5.** 100 miles per hour

7. 2°F per minute (downward)

9. 125 gallons per minute

11. 420 feet per minute (downward)

13. 2 miles/4 minutes
8 miles/16 minutes
rate of change: $\frac{8-2}{16-4} = \frac{1}{2}$
8 miles/16 minutes
10 miles/24 minutes
rate of change: $\frac{10-8}{24-16} = \frac{1}{4}$
The speed is not constant over the interval from 4 minutes to 24 minutes.

15. $64.80   **17.** $210

Lesson 3.5 Skill A (pages 112–113)

Try This Example

x-intercept: -2
y-intercept: 6

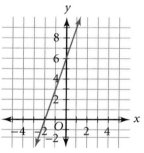

Exercises

1. x-intercept: 4
y-intercept: $-\frac{12}{5}$

3. x-intercept: none
y-intercept: $12\frac{3}{4}$

5. neither   **7.** vertical

9. x-intercept: 3
y-intercept: 3

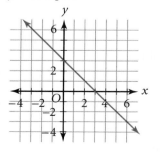

11. *x*-intercept: 4
  y-intercept: 8

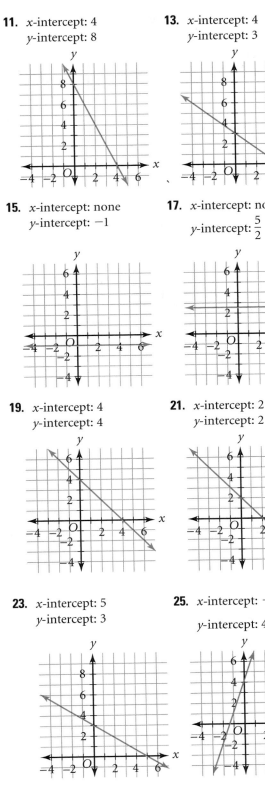

13. *x*-intercept: 4
  y-intercept: 3

15. *x*-intercept: none
  y-intercept: −1

17. *x*-intercept: none
  y-intercept: $\frac{5}{2}$

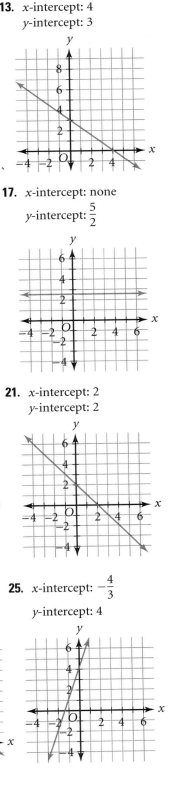

19. *x*-intercept: 4
  y-intercept: 4

21. *x*-intercept: 2
  y-intercept: 2

23. *x*-intercept: 5
  y-intercept: 3

25. *x*-intercept: $-\frac{4}{3}$
  y-intercept: 4

27. *x*-intercept: $-\frac{5}{3}$
  y-intercept: 5

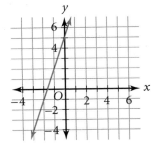

29. neither

31. a. Answers may vary. Sample answer: $y = x - 3$
  b. $x = 3$

33. $\frac{7}{2}$ **35.** $\frac{2}{5}$ **37.** $\frac{3}{2}$ **39.** 0

Lesson 3.5 Skill B (pages 114–115)

Try This Example 1

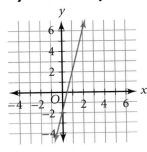

Try This Example 2

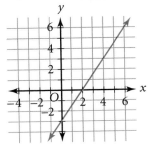

Try This Example 3

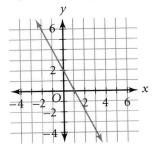

Exercises

1. $(3, 0)$ **3.** $(3, -4)$

5. 3 units up and 2 units to the right, or 3 units down and 2 units to the left

7. 1 unit up and 3 units to the right, or 1 unit down and 3 units to the left

9. 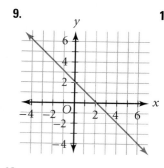   **11.**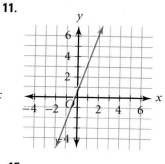

13. 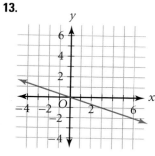   **15.**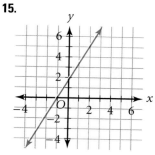

17. 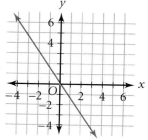   **19.**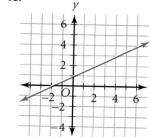

21. $y = \frac{1}{2}x + 4$ **23.** $y = \frac{2}{3}x - 2$

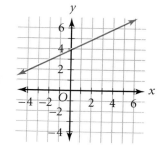   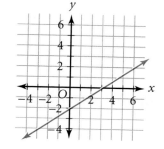

25. $y = -2x$ **27.** $y = 1.5x - 2$

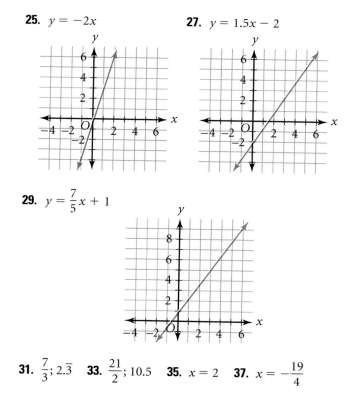

29. $y = \frac{7}{5}x + 1$

31. $\frac{7}{3}$; $2.\overline{3}$ **33.** $\frac{21}{2}$; 10.5 **35.** $x = 2$ **37.** $x = -\frac{19}{4}$

Lesson 3.5 Skill C (pages 116–117)

Try This Example 1

$V = -120t + 600$

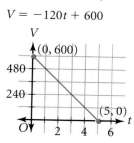

Try This Example 2

$5n + 10d = 200$, or $n = 40 - 2d$

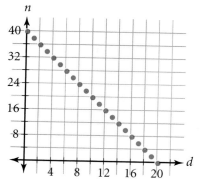

Exercises

1. $10d + 5n$

3. $1200 + 130t$, where t represents elapsed time in minutes

5. Let a represent altitude after t minutes; $a = 1200 - 300t$.

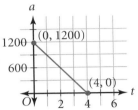

The graph is continuous.

7. Let v represent volume remaining and let t represent elapsed time; $v = 200 - 40t$.

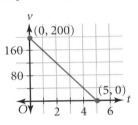

The graph is continuous.

9. Let p represent the original price and let s represent the sale price; $s = 0.75p$.

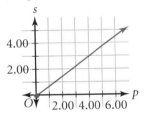

The graph is continuous.

11. Let v represent volume after t minutes; $v = 10t$; 55 cubic feet

Lesson 3.6 Skill A (pages 118–119)

Try This Example 1

$y + 3 = \frac{3}{5}(x - 2)$; $y = \frac{3}{5}x - \frac{21}{5}$

Try This Example 2

$y = 4$

Exercises

1.

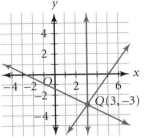

Answers may vary. Sample answer: Through a given point, you can draw several lines. No unique line is determined.

3. $y - 2 = x + 4$; $y = x + 6$

5. $y - 3 = 2(x - 5)$; $y = 2x - 7$

7. $y + 3 = -0.5(x - 0)$; $y = -0.5x - 3$

9. $y + 2 = \frac{3}{7}(x + 4)$; $y = \frac{3}{7}x - \frac{2}{7}$

11. $y + 9 = 2.5(x - 5)$; $y = 2.5x - 21.5$

13. $y = -1$ **15.** $y = 2$

17. $y - 10 = \frac{11}{12}(x - 4)$; $y = \frac{11}{12}x + \frac{19}{3}$

19. If the line contains $P(a, b)$ and has slope 0, then $y - b = 0(x - a)$. Thus, $y - b = 0$ and the equation has the form $y = b$.

21. -1 **23.** 10

25.

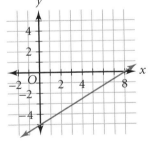

27. vertical; $(r, 0)$; undefined

Lesson 3.6 Skill B (pages 120–121)

Try This Example 1

302 miles

Try This Example 2

3720 bottles

Exercises

1. point: (4, 1900); slope: 350

3. Let t represent hours spent baking and c represent the number of cookies on hand; $c - 144 = 24\,(t - 0)$; 288

5. Let a represent the amount and let t represent the number of T-shirts; $a - 50 = 6(t - 5)$; $140

7. domain t: all positive real numbers; range d: all positive real numbers; $d = 50t + 520$; 820 miles

9. domain: all positive real numbers; range: all positive real numbers

11. domain: all whole numbers; range: all nonnegative multiples of 0.33

Lesson 3.7 Skill A (pages 122–123)

Try This Example 1

$y = -\dfrac{1}{3}x - \dfrac{2}{3}$

Try This Example 2

a. $y = 3$ **b.** $x = -1.4$

Exercises

1. vertical   **3.** neither   **5.** $y = -\dfrac{5}{4}x$ **7.** $y = -x - 5.5$

9. $y = \dfrac{7}{2}x + 7$ **11.** $y = 2.4$ **13.** $y = -2x + 5$

15. $y = \dfrac{9}{2}x - \dfrac{9}{2}$ **17.** above   **19.** above

21. (0, 2), (−4, 0)   **23.** (0, 4), (6, 0)

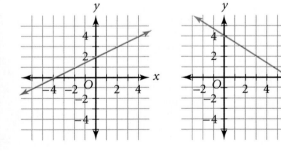

25.

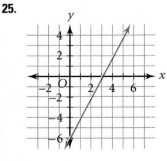

27. $y = -\dfrac{3}{2}x + \dfrac{5}{2}$

Lesson 3.7 Skill B (pages 124–125)

Try This Example 1

$60

Try This Example 2

a. Let d represent distance from home after t hours of driving; $d = 60x + 80$; the slope gives the constant driving speed.

b. 680 miles

Exercises

1. (4, 3.60) and (7, 6.30)

3. Let h represent the number of hours and let c represent the cost; $c = 25h + 20$; $145; the slope represents the cost per hour, $25; the y-intercept represents the fixed fee, $20.

5. Let t represent time and let h represent height; $h = -2t + 12$; 2 cm; the slope represents how fast the candle is burning, 2 cm/hr; the y-intercept represents the candle's original height, 12 cm.

7. $153; the slope gives the cost per day, $19; the y-intercept gives the fixed cost, $20; the cost for 7.5 days might be the same as the cost for 8 days if the rental company charges $19 for any part of a day.

9. 5 hours at 55 miles per hour and 2 hours at 50 miles per hour

Lesson 3.8 Skill A (pages 126–127)

Try This Example 1

no

Try This Example 2

$y = \dfrac{5}{4}x + \dfrac{19}{4}$

Exercises

1. not parallel   **3.** parallel $\left(2.5 = \dfrac{5}{2}\right)$

5. $y = -\dfrac{1}{2}x + \dfrac{3}{2}$ and $y = \dfrac{1}{2}x - \dfrac{5}{2}$; not parallel

7. not parallel; the slope of the first line is 0 and the slope of the second line is undefined.

9. $y = -\dfrac{7}{2}x$ **11.** $y = \dfrac{3}{8}x + \dfrac{31}{8}$ **13.** $y = -\dfrac{7}{8}x + \dfrac{145}{8}$

15. $y = \frac{5}{3}x + 6$ **17.** $y = -2x + 8$ **19.** -10

21. 0, 2.24, 2.26, 2.36, 2.62, 2.63

23. no   **25.** yes

Lesson 3.8 Skill B (pages 128–129)

Try This Example 1

Yes; $y = x - 5$ and $y = -x + 3$; the product of the slopes is -1. Therefore, the lines are perpendicular.

Try This Example 2

$y = \frac{2}{7}x - \frac{41}{7}$

Exercises

1. not perpendicular

3. perpendicular

5. Yes; one line is horizontal and the other is vertical.

7. Yes; the product of the slopes is -1.

9. No; the product of $\frac{7}{2}$ and $-\frac{4}{7}$ is not -1.

11. $x = 0$ **13.** $y = -\frac{1}{3}x + \frac{1}{3}$ **15.** $y = \frac{5}{6}x + \frac{19}{6}$

17. $a = -b$ **19.** $y = -\frac{11}{3}x + \frac{61}{3}; \left(\frac{61}{11}, 0\right)$

21.

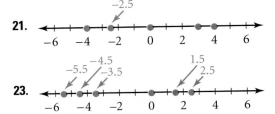

23.

25.

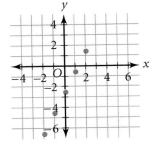

Lesson 3.8 Skill C (pages 130–131)

Try This Example 1

slopes: $\overleftrightarrow{AB}: -\frac{2}{5}, \overleftrightarrow{BC}: \frac{5}{3},$ and $\overleftrightarrow{AC}: \frac{3}{8}$; The triangle is not a right triangle because no pair of slopes has a product of -1.

Try This Example 2

slopes: $AB: \frac{3}{2}$, $BC: 0$, $CD: -3$, $AD: 0$

The figure is a trapezoid because exactly two of the sides are parallel.

Exercises

1. slope of $CF: \frac{2}{3}$; slope of $AC: -\frac{3}{2}$; slope of $LX: -\frac{3}{2}$

3. perpendicular

5. Yes; slope of $\overleftrightarrow{GH}: \frac{7}{10}$; slope of $\overleftrightarrow{GH}: -\frac{10}{7}$; So these lines are perpendicular.

7. No; no two slopes have a product of -1; slope of $\overleftrightarrow{CL}: -\frac{1}{5}$; slope of $\overleftrightarrow{LE}: -\frac{1}{2}$; slope of $\overleftrightarrow{CE}: -\frac{1}{8}$

9. Yes; exactly two slopes are equal. slope of $\overleftrightarrow{AB} = 0$; slope of $\overleftrightarrow{BC} = -\frac{4}{3}$; slope of $\overleftrightarrow{CD} = 0$; slope of $\overleftrightarrow{AD} = 2$

11. Yes; exactly two slopes are equal. slope of $\overleftrightarrow{AB} = \frac{1}{3}$; slope of $\overleftrightarrow{BC} = -\frac{2}{3}$; slope of $\overleftrightarrow{CD} = \frac{1}{3}$; slope of $\overleftrightarrow{AD} = \frac{4}{11}$

13. Figure $ABCD$ is a parallelogram because slope of $AB = 1$, slope of $BC = -1$, slope of $CD = 1$, and slope of $DA = -1$. AB is parallel to CD and BC is parallel to AD.

15. Figure $PQRS$ is a parallelogram and a rectangle.

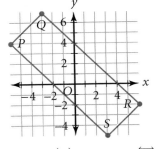

slope of $\overleftrightarrow{PQ}: 1$; slope of $\overleftrightarrow{QR}: -1$; slope of $\overleftrightarrow{RS}: 1$; slope of $\overleftrightarrow{SR}: -1$

17. Let t represent time in minutes and T represent temperature in degrees Celsius; $T = 5t + 25$

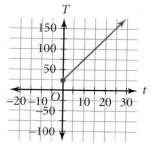

Lesson 3.9 Skill A (pages 132–133)

Try This Example 1

For each increase of 1 in the term number, the terms decrease by 4; -21

Try This Example 2

$y = 5x + 1$; 401

Exercises

1. The data shows a linear pattern; successive differences in x are 1 and successive differences in y are 8.

3.
$$y + 11 = 6(x - 1)$$
$(1, -11)$: $-11 + 11 = 6(1 - 1)$; $0 = 0$ ✔
$(2, -5)$: $-5 + 11 = 6(2 - 1)$; $6 = 6$ ✔
$(3, 1)$: $1 + 11 = 6(3 - 1)$; $12 = 12$ ✔
$(4, 7)$: $7 + 11 = 6(4 - 1)$; $18 = 18$ ✔
$(5, 13)$: $13 + 11 = 6(5 - 1)$; $24 = 24$ ✔

5. $-41, -51, -61$ **7.** $17.8, 22.8$

9. $y = -10x + 29$; -971 **11.** $y = 2.5x - 2.2$; 132.8

13. $y = \dfrac{11}{3}x - \dfrac{79}{6}$ **15.** $x = 3.5$ **17.** $y = 10x - 242$

Lesson 3.9 Skill B (pages 134–135)

Try This Example 1

$v = 3n$; 270

Try This Example 2

$t = -7n + 19$; -226

Exercises

1. $-7, -1, 5, 11, 17$ **3.** $t = 5n - 2$; 498

5. The last star is removed at the ninth step.

7. $t = -3n - 9.4$; -129.4 **9.** $\dfrac{43}{4}$; Subtract $\dfrac{5}{4}$.

11. $\dfrac{7}{8}$ **13.** 1   **15.** $y = -11x$

Lesson 3.9 Skill C (pages 136–137)

Try This Example

$x = 3n - 8$; $y = 2n - 5$; $(292, 195)$

Exercises

1. $x = 3n - 3$; $y = -4n + 4$

3. $x = 3n - 8$; $y = -2n + 8$; $(172, -112)$

5. $x = 2n - 6$; $y = 3$; $(114, 3)$

7. $(8, 22), (13, 27)$, and $(18, 32)$

CHAPTER 4

Lesson 4.1 Skill A (pages 142–143)

Try This Example 1

$-7 < -2$ and $-2 > -7$

Try This Example 2

$c \leq 12.95$

Exercises

1. $-3, -2, -1, 0, 1$

3. $-3, -2$

5. $0 < 3$ and $3 > 0$

7. $-3 < 3$ and $3 > -3$

9. $4 > 3$ and $3 < 4$

11. $-4 < 3$ and $3 > -4$

13. $-2 < 3.5$ and $3.5 > -2$

15. $-\dfrac{1}{2} < 4$ and $4 > -\dfrac{1}{2}$

17. $s > 55$ **19.** $h \geq 42$ **21.** $w > 125$ **23.** false

25. false   **27.** $x = 2$ **29.** $z = -8$ **31.** $x = 8$

Lesson 4.1 Skill B (pages 144–145)

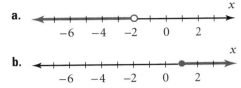

Try This Example 1

a.

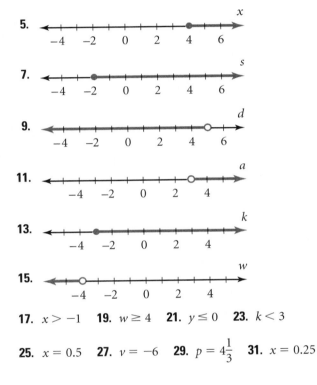

b.

Try This Example 2

a. $x \leq 5$ **b.** $x > -3$

Exercises

1. d **3.** a

5.

7.

9.

11.

13.

15.

17. $x > -1$ **19.** $w \geq 4$ **21.** $y \leq 0$ **23.** $k < 3$

25. $x = 0.5$ **27.** $v = -6$ **29.** $p = 4\frac{1}{3}$ **31.** $x = 0.25$

Lesson 4.1 Skill C (pages 146–147)

Try This Example 1

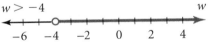

Try This Example 2

Exercises

1. c **3.** d

5.

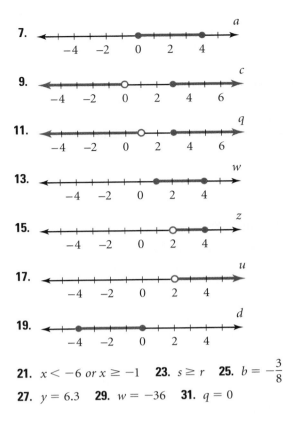

7.

9.

11.

13.

15.

17.

19.

21. $x < -6 \text{ or } x \geq -1$ **23.** $s \geq r$ **25.** $b = -\frac{3}{8}$

27. $y = 6.3$ **29.** $w = -36$ **31.** $q = 0$

Lesson 4.2 Skill A (pages 148–149)

Try This Example 1

$x > -1.5$

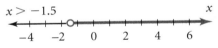

Try This Example 2

$w > -4$

Exercises

1. $x \geq -4$

3. $d \leq -2$

5. $x < 11$

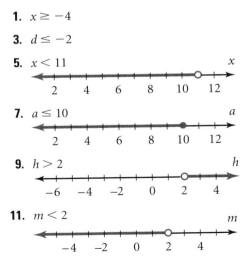

7. $a \leq 10$

9. $h > 2$

11. $m < 2$

13. $b < -2$

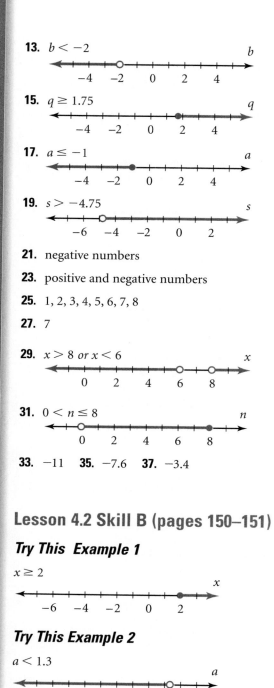

15. $q \geq 1.75$

17. $a \leq -1$

19. $s > -4.75$

21. negative numbers

23. positive and negative numbers

25. 1, 2, 3, 4, 5, 6, 7, 8

27. 7

29. $x > 8$ or $x < 6$

31. $0 < n \leq 8$

33. -11　　**35.** -7.6　　**37.** -3.4

Lesson 4.2 Skill B (pages 150–151)

Try This Example 1

$x \geq 2$

Try This Example 2

$a < 1.3$

Exercises

1. $n + 3.2 > -14$

3. $-2 + d < 4.7$

5. $n \geq 0$

7. $m > 6$

9. $k > 0$

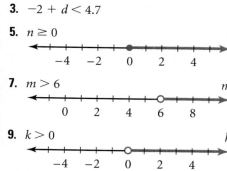

11. $c \geq 4.9$

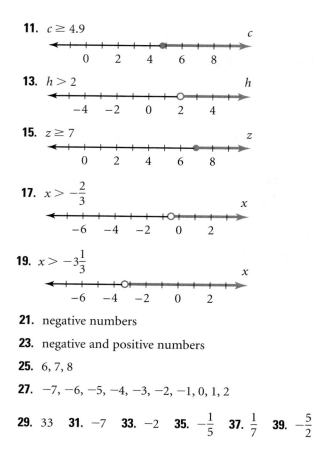

13. $h > 2$

15. $z \geq 7$

17. $x > -\dfrac{2}{3}$

19. $x > -3\dfrac{1}{3}$

21. negative numbers

23. negative and positive numbers

25. 6, 7, 8

27. $-7, -6, -5, -4, -3, -2, -1, 0, 1, 2$

29. 33　**31.** -7　**33.** -2　**35.** $-\dfrac{1}{5}$　**37.** $\dfrac{1}{7}$　**39.** $-\dfrac{5}{2}$

Lesson 4.2 Skill C (pages 152–153)

Try This Example 1

no more than 62 minutes

Try This Example 2

at least 40.2 tons and at most 82.2 tons

Exercises

1. 20 miles or more

3. at least 135 points

5. at least 11 miles

7. between 8 and 12 one-pound bags

9. width = 15 feet, length = 75 feet

Lesson 4.3 Skill A (pages 154–155)

Try This Example 1

a. $x > -2$

b. $t > 2$

Try This Example 2

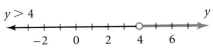

$y > 4$

Exercises

1. 4; no **3.** $-\dfrac{3}{2}$; yes

5. $r \le -2$

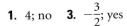

7. $x \le 6$

9. $p > -15$

11. $u > -\dfrac{5}{2}$

13. $n \le 0$

15. $x > 9$

17. $p > -15$

19. $w > 9$

21. $k < 2$

23. $s \le \dfrac{10}{9}$

25. $s \le -\dfrac{2}{3}$

27. If $\dfrac{7}{3} > \dfrac{x}{4}$, then $\dfrac{7}{3} \cdot 4 > \dfrac{x}{4} \cdot 4$ by the Multiplication Property of Inequality. Therefore, $\dfrac{28}{3} > x$ and $x < \dfrac{28}{3}$.
If $\dfrac{3}{7}x < 4$, then $\dfrac{7}{3} \cdot \dfrac{3}{7}x < \dfrac{7}{3} \cdot 4$ by the Multiplication Property of Inequality. Therefore, $x < \dfrac{28}{3}$.

29. -1 **31.** $-\dfrac{1}{4}$ **33.** $-\dfrac{1}{14}$ **35.** $\dfrac{9}{16}$ **37.** $-\dfrac{3}{16}$

Lesson 4.3 Skill B (pages 156–157)

Try This Example

a. $c \ge -2.5$

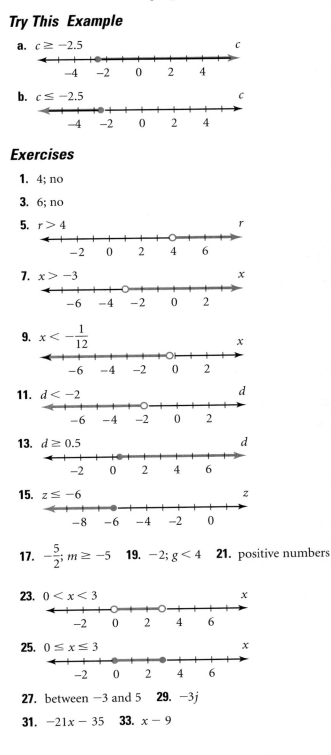

b. $c \le -2.5$

Exercises

1. 4; no

3. 6; no

5. $r > 4$

7. $x > -3$

9. $x < -\dfrac{1}{12}$

11. $d < -2$

13. $d \ge 0.5$

15. $z \le -6$

17. $-\dfrac{5}{2}$; $m \ge -5$ **19.** -2; $g < 4$ **21.** positive numbers

23. $0 < x < 3$

25. $0 \le x \le 3$

27. between -3 and 5 **29.** $-3j$

31. $-21x - 35$ **33.** $x - 9$

Lesson 4.3 Skill C (pages 158–159)

Try This Example 1

no more than 8 hours

Try This Example 2

between 30 and 66 flowers, inclusive

Exercises

1. minimum: 15 ounces
   maximum: 17 ounces

3. between 10 hours and 12 hours

5. between 0 and 160

7. a. between 0 and 11
   b. 1, 2, 3, 4, 5, 6, 7, 8, 9, 10

9. domain: whole numbers; range: nonnegative multiples of 2.3

11. domain: whole numbers; range: all nonnegative multiples of 24

Lesson 4.4 Skill A (pages 160–161)

Try This Example 1

Use the Subtraction Property of Inequality; $c > -7.5$

Try This Example 2

Use the Multiplication Property of Inequality; $z < \frac{1}{4}$

Try This Example 3

a. $j < 0$ b. $x \geq 4$

Exercises

1. Add 6 to each side.

3. Divide each side by 5.

5. a. Add $\frac{3}{4}$ to each side.
   b. no
   c. $t \geq 1$

7. a. Multiply each side by -3.
   b. yes
   c. $w < -8.4$

9. a. Multiply each side by -1.
   b. yes
   c. $y > 0$

11. a. Subtract 1.5 from each side.
    b. no
    c. $j \leq -3.5$

13. a. Add 6 to each side.
    b. no
    c. $z \leq -6.5$

15. a. Multiply each side by -5.
    b. yes
    c. $v < 60$

17. a. Multiply each side by $-\frac{11}{3}$.
    b. yes
    c. $d < 11$

19. a. Add h to each side.
    b. no
    c. $h < 13$

21. a. Add z to each side.
    b. no
    c. $z < 5$

23. $6 < x < 7$

25. a. $x > 4$
    b.

27. $a = 28$ 29. $x = -12$ 31. $x = 12$ 33. $n = 9$

Lesson 4.4 Skill B (pages 162–163)

Try This Example 1

less than 50 miles per hour

Try This Example 2

less than $75

Exercises

1. c   3. b   5. greater than 20

7. less than $250   9. less than $1380

11.

13. $d \leq -3.2$ 15. $t \geq -0.1$

17. $x < -\frac{17}{9}$ 19. $x \geq 14$

Lesson 4.5 Skill A (pages 164–165)

Try This Example 1

a. $t \geq 5$ b. $v \geq \frac{5}{4}$

Try This Example 2

a. $w \geq -2$ b. $z < -\frac{1}{2}$

Try This Example 3

$z < -4$

Exercises

1. $3n < -3$; divide each side by 3.

3. $2x \leq 18$; divide each side by 2.

5. $r < 1$ 7. $a < 6$ 9. $y > 2$ 11. $g < 1$ 13. $s < 2$

15. $u \leq 3$ 17. $w > 4$ 19. $q < 4$ 21. $c \leq 3$

23. $x > 2$

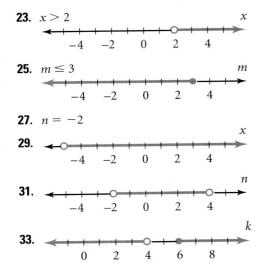

25. $m \leq 3$

27. $n = -2$

29.

31.

33.

Lesson 4.5 Skill B (pages 166–167)

Try This Example 1

$p < 2$

Try This Example 2

a. no solution   **b.** all real numbers

Exercises

1. $x > -4$ 3. $3x + 6 \leq 6$ 5. $c < -\dfrac{7}{2}$ 7. $q \leq -\dfrac{14}{5}$

9. $d < 8$ 11. $n \leq -2$ 13. $x > -\dfrac{1}{3}$ 15. $t < -\dfrac{1}{3}$

17. no solution   19. all real numbers

21. all real numbers   23. $y > 1$

25. If $3(4x + 4) - 11 = 12x + 2$, then $12x + 12 - 11 = 12x + 2$. This implies that $12x - 1 = 12x + 2$ and that $-1 = 2$. Since this is impossible, the given equation is never true.

27. If $3(t + t) - t = 5(t + 1)$, then $5t = 5t + 5$ and $0 = 5$. Since this is false, the given equation is never true.

Lesson 4.5 Skill C (pages 168–169)

Try This Example 1

at least 333 milliliters of solution A and no more than 167 milliliters of solution B

Try This Example 2

at least $857.14

Exercises

1. $0.20a + 0.30(800 - a) \leq 0.25(800)$

3. at least 333.33 milliliters of solution B and no more than 166.67 milliliters of solution A

5. $0.20a + 0.40(800 - a) \leq 0.15(800)$
   $0.20a - 0.40a + 320 \leq 120$
   $-0.20a \leq 200$
   $a \geq 1000$

   This means that at least 1000 milliliters of solution A are needed, which is impossible because only 800 milliliters of solution C are to be mixed.

7. solution A: 15% sugar
   solution B: 30% sugar
   solution C: 600 milliliters at no more than 20% sugar

9.

	Principal	+ Interest =	Amount
now	1200	$.06(1200) = 72$	1272
after deposit	$1200 + d$	$.06(1200 + d)$	≥ 2800

11. any integer less than or equal to 7

Lesson 4.6 Skill A (pages 170–171)

Try This Example 1

$-5 < b \leq 4$

Try This Example 2

$\dfrac{1}{2} < d < \dfrac{11}{2}$

Exercises

1. $-7 < x \leq 8$

3. $-2 < 2(d + 1) \leq 3$

5. $-2 \leq x < 5$

7. $-3 \leq g < -1$

9. $-4 \leq n \leq -3$

11. $-1 \leq b < 2$

13. $-\frac{7}{2} \le d \le -\frac{3}{2}$

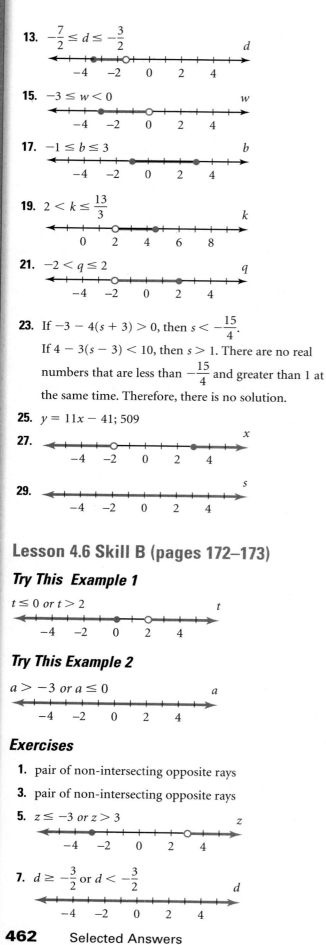

15. $-3 \le w < 0$

17. $-1 \le b \le 3$

19. $2 < k \le \frac{13}{3}$

21. $-2 < q \le 2$

23. If $-3 - 4(s + 3) > 0$, then $s < -\frac{15}{4}$. If $4 - 3(s - 3) < 10$, then $s > 1$. There are no real numbers that are less than $-\frac{15}{4}$ and greater than 1 at the same time. Therefore, there is no solution.

25. $y = 11x - 41$; 509

27.

29.

Lesson 4.6 Skill B (pages 172–173)

Try This Example 1

$t \le 0$ or $t > 2$

Try This Example 2

$a > -3$ or $a \le 0$

Exercises

1. pair of non-intersecting opposite rays

3. pair of non-intersecting opposite rays

5. $z \le -3$ or $z > 3$

7. $d \ge -\frac{3}{2}$ or $d < -\frac{3}{2}$

9. $f \ge 1$ or $f < -\frac{9}{2}$

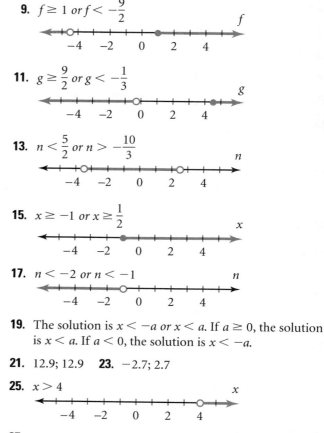

11. $g \ge \frac{9}{2}$ or $g < -\frac{1}{3}$

13. $n < \frac{5}{2}$ or $n > -\frac{10}{3}$

15. $x \ge -1$ or $x \ge \frac{1}{2}$

17. $n < -2$ or $n < -1$

19. The solution is $x < -a$ or $x < a$. If $a \ge 0$, the solution is $x < a$. If $a < 0$, the solution is $x < -a$.

21. 12.9; 12.9   23. -2.7; 2.7

25. $x > 4$

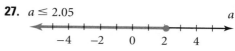

27. $a \le 2.05$

Lesson 4.6 Skill C (pages 174–175)

Try This Example 1

between 30 feet and 180 feet, inclusive

Try This Example 2

at least $1\frac{3}{7}$ pounds if the ratio is $2\frac{1}{2}$ to 1; at least $1\frac{1}{4}$ pounds if the ratio is 3 to 1

Exercises

1. Let a represent the number of apples, and let g represent the number of oranges. Then $3g < a$.

3. The smaller number can be any integer from 25 to 99, inclusive. The larger number can be any integer from 26 to 100, inclusive.

5. between $4\frac{4}{5}$ hours and $6\frac{2}{5}$ hours, inclusive

7. at least $3\frac{3}{5}$ pounds if the ratio is $1\frac{1}{2}$ to 1; at least 3 pounds if the ratio is 2 to 1.

9. 4.8 feet

Lesson 4.7 Skill A (pages 176–177)

Try This Example 1

  a. $x = -7.5$ or $x = -0.5$ **b.** $t = -5$ or $t = 5$

Try This Example 2

$x = -4$ or $x = \dfrac{2}{3}$

Try This Example 3

$z = 0$ or $z = -2$

Exercises

  1. $w = -3.6$ or $w = 3.6$ **3.** $z = 0$ **5.** $|x| = 8$

  7. $x = -2$ or $x = 8$ **9.** $d = -3$ or $d = 3$

11. $x = 0$ or $x = -4$ **13.** $g = -5$ or $g = 5$

15. $m = -2$ or $m = \dfrac{16}{3}$ **17.** $c = -\dfrac{19}{2}$ or $c = \dfrac{15}{2}$

19. $z = \dfrac{3}{2}$ or $z = \dfrac{13}{2}$ **21.** $k = 0$ or $k = 6$

23. $z = -\dfrac{2}{3}$ or $z = \dfrac{4}{3}$ **25.** $x = 10$ or $x = \dfrac{10}{3}$

27. $n = -5$ **29.** $n < -11$

31. The distance between x and 5 is 3.

33. $y \geq -5.1$ **35.** $p < -6$ **37.** $y \geq \dfrac{13}{3}$ **39.** $y < -1$

Lesson 4.7 Skill B (pages 178–179)

Try This Example 1

  a. $-5 < d < 1$

  b. $d < -5$ or $d > 1$

Try This Example 2

$-7 < n < 1$

Try This Example 3

$k \leq -\dfrac{7}{3}$ or $k \geq 1$

Exercises

  1. $x \leq 5$ and $x \geq -5$, or $-5 \leq x \leq 5$

  3. $3y < 9$ and $3y > -9$, or $-9 < 3y < 9$

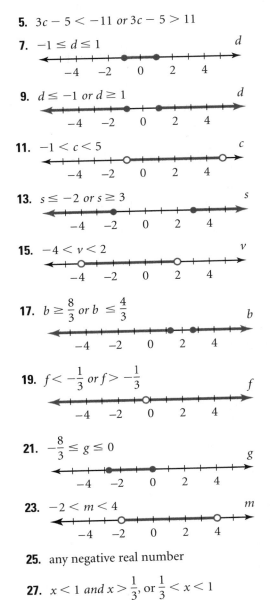

  5. $3c - 5 < -11$ or $3c - 5 > 11$

  7. $-1 \leq d \leq 1$

  9. $d \leq -1$ or $d \geq 1$

11. $-1 < c < 5$

13. $s \leq -2$ or $s \geq 3$

15. $-4 < v < 2$

17. $b \geq \dfrac{8}{3}$ or $b \leq \dfrac{4}{3}$

19. $f < -\dfrac{1}{3}$ or $f > -\dfrac{1}{3}$

21. $-\dfrac{8}{3} \leq g \leq 0$

23. $-2 < m < 4$

25. any negative real number

27. $x < 1$ and $x > \dfrac{1}{3}$, or $\dfrac{1}{3} < x < 1$

29. The distance between x and 0 is more than 3.

31. If $|ax + b| \leq c$, $-c \leq ax + b \leq c$. Thus, $-c - b \leq ax \leq c - b$, or $-b - c \leq ax \leq c - b$. If $a > 0$, then $\dfrac{-b - c}{a} \leq x \leq \dfrac{c - b}{a}$.

33. If $-(x - 4) = 3x - 1$, then $-x + 4 = 3x - 1$. Thus, $x = \dfrac{5}{4}$. The equation is sometimes true, when $x = \dfrac{5}{4}$.

Lesson 4.7 Skill C (pages 180–181)

Try This Example 1

acceptable: $12.48 \leq w \leq 12.52$
unacceptable: $w < 12.48$ or $w > 12.52$

Try This Example 2

between 4800 watches and 7200 watches, inclusive

Exercises

1. $|w - 5| \leq 0.2$

3. Let h represent height.
   acceptable: $16 \leq h \leq 20$
   unacceptable: $h < 16 \ or \ h > 20$

5. Let t represent thickness.
   acceptable: $1.1975 \leq t \leq 1.2025$
   unacceptable: $t < 1.1975 \ or \ t > 1.2025$

7. between 54 inches and 60 inches, inclusive

9. $1000 at 5% and $4000 at 6%

Lesson 4.8 Skill A (pages 182–183)

Try This Example 1

If $3(x - 4) > 3(x - 5)$, then $-12 > -15$. The inequality is always true.

Try This Example 2

If $r > 2r$, then $r < 0$. The statement is sometimes true.

Exercises

1. The inequality is always true.

3. The inequality is sometimes true.

5. If $3x < 3(x + 1)$, then $3x < 3x + 3$.
   This implies that $0 < 1$. The given statement is always true.

7. If $\frac{1}{2}r < r$, then $r < 2r$. Thus, $0 < r$. The given statement is sometimes true.

9. If $4(x - 3x) > 9 - 8x$, then $-8x > 9 - 8x$. This implies that $0 > 9$. This is false. The given statement is never true.

11. If $3z - 5 + 4z \geq z + 6z - 5$, then $-5 \geq -5$. The given statement is always true.

13. If $9 - 2(z + 2) < -2z$, then $5 - 2z < -2z$. This means $5 < 0$. This is false. Therefore, the given statement is never true.

15. If $500x > 250x^2$, then $2x > x^2$. The given statement is sometimes true.

17. If $-(x + 1) \leq x - 1$, then $-x - 1 \leq x - 1$. This implies that $x \geq 0$. The given statement is sometimes true.

19. If $x(-x) \leq 0$, then $-x^2 \leq 0$. This implies that $x^2 \geq 0$.
    If $x < 0$, then $-x^2 = x \cdot x > 0$ because the product of two negative numbers is positive. Thus, $x^2 > 0$ is true when $x < 0$.
    If $x = 0$, then $x^2 = 0^2 = 0$. Thus, $x^2 \geq 0$ is true when $x = 0$.
    If $x > 0$, then $x^2 = x \cdot x > 0$ because the product of two positive numbers is positive. Thus, $x^2 \geq 0$ is true when $x > 0$.
    Therefore, the given statement is always true.

21. If $x < 0$, then $x^2 > 0$ and $2x < 0$. Thus, $x^2 \geq 2x$ is true when $x < 0$.
    If $x = 0$, then $x^2 = 0$ and $2x = 0$. Thus, $x^2 \geq 2x$ is true when $x = 0$.
    If $x > 0$, then $x^2 \geq 2x$ implies that $x \geq 2$. Thus, $x^2 \geq 2x$ is sometimes true when $x > 0$.
    Therefore, the given statement is sometimes true.

23. If $x < 0$, then $\frac{x}{3} \leq \frac{3}{x}$ implies that $x^2 \geq 9$, or $-3 < x < 0$.
    Thus, $\frac{x}{3} \leq \frac{3}{x}$ is sometimes true when $x < 0$.
    If $x = 0$, then $\frac{3}{x}$ is undefined. Thus, $\frac{x}{3} \leq \frac{3}{x}$ is false when $x = 0$.
    If $x > 0$, then $\frac{x}{3} \leq \frac{3}{x}$ implies that $x^2 \leq 9$, or $0 < x < 3$.
    Thus, $\frac{x}{3} \leq \frac{3}{x}$ is sometimes true when $x > 0$.
    Therefore, the given statement is sometimes true.

25.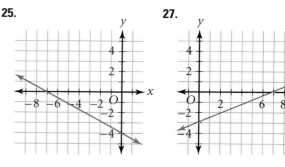

27.

29.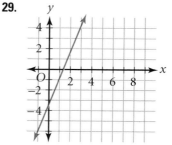

CHAPTER 5

Lesson 5.1 Skill A (pages 188–189)

Try This Example 1

$(5, -2)$

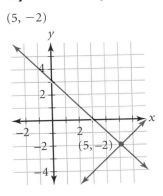

Try This Example 2

no solution

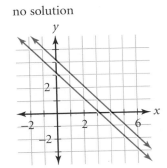

Exercises

1. no solution   **3.** one solution

5. none

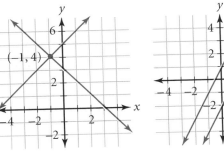

7. $(-3, -2)$

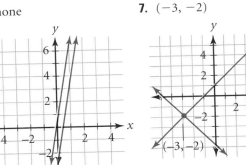

9. $(-1, 4)$

11. none

13. $(3, 4)$

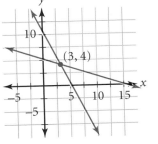

15. $\left(0, -\dfrac{1}{2}\right)$

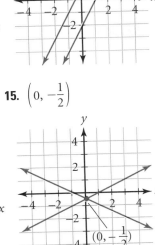

17. none

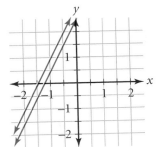

19. $y = \dfrac{4}{3}x - 2$: slope $\dfrac{4}{3}$ and y-intercept -2; $y = 2$: slope 0 and y-intercept 2. The lines intersect because they have different slopes and different y-intercepts.

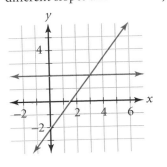

21. $y = \dfrac{1}{5}x + 1$: slope $\dfrac{1}{5}$ and y-intercept 1; $y = -2x + 4$: slope -2 and y-intercept 4. The lines intersect because they have different slopes and different y-intercepts.

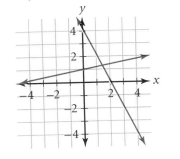

23. $-\dfrac{4}{5}x - 2$ **25.** $-9y + 4$ **27.** $5x + 2.5$

Lesson 5.2 Skill A (pages 190–191)

Try This Example

a. $(7, -2)$ **b.** $(-4, -5)$

Exercises

1. $-2x + 3(-3x + 1) = 1$

3. $-3x + \left(-\dfrac{1}{5}x + 3\right) = -5$

5. $(1, -4)$ **7.** $(-9, 13)$ **9.** $(7, -1)$ **11.** $(10, 3)$

13. $\left(1, \frac{2}{7}\right)$ **15.** $\left(-\frac{15}{11}, -\frac{12}{11}\right)$ **17.** $\left(\frac{b+a}{2}, \frac{b-a}{2}\right)$

19. $d \geq 0$ **21.** $g > \frac{14}{5}$ **23.** $a \leq -7$

25. $b < -22$ **27.** $c \leq -\frac{18}{7}$

Lesson 5.2 Skill B (pages 192–193)

Try This Example

a. $(8, -5)$ **b.** $\left(\frac{39}{17}, \frac{2}{17}\right)$

Exercises

1. $\begin{cases} -2x + 3y = 5 \\ x = -5y + 2 \end{cases}$ **3.** $\begin{cases} -2r - 3s = 2 \\ 7r + 3 = s \end{cases}$

5. $(1, -3)$ **7.** $\left(\frac{16}{5}, \frac{7}{5}\right)$ **9.** $(-3, 2)$ **11.** $\left(\frac{3}{5}, \frac{6}{5}\right)$

13. $(-42, 36)$ **15.** $\left(-6, \frac{9}{2}\right)$ **17.** $\left(-\frac{3}{5}, \frac{19}{10}\right)$

19. $-2k + 5$ **21.** $-4z - 6$ **23.** $-2n + 2$

25. $4w + 1$ **27.** $-12b - 25$

Lesson 5.2 Skill C (pages 194–195)

Try This Example 1

72 drops of vitamins and 5400 drops of water

Try This Example 2

105 feet by 120 feet

Exercises

1. $\begin{cases} w = 75v \\ w + v = 152 \cdot 12 \end{cases}$

3. about 76 drops of vitamins and 5320 drops of water

5. 40 adult tickets and 35 child tickets

7. 240 milliliters and 360 milliliters

9. at least 62.5 milliliters of the 12% solution and no more than 187.5 milliliters of the 20% solution

Lesson 5.3 Skill A (pages 196–197)

Try This Example 1

a. $(2, 3)$ **b.** $\left(\frac{5}{3}, 3\right)$

Try This Example 2

a. $(-5, -2)$ **b.** $(-2, 4)$

Exercises

1. Multiply the second equation by 3.

3. Multiply the first equation by -1 or multiply the second equation by -1.

5. $(-1, 1)$ **7.** $\left(1, \frac{1}{2}\right)$ **9.** $\left(4, \frac{1}{3}\right)$ **11.** $(0, 0)$

13. $(4, 6)$ **15.** $\left(\frac{12}{11}, -\frac{15}{11}\right)$ **17.** $\left(\frac{26}{17}, -\frac{11}{17b}\right)$

19. Multiply the first equation by -2.

$\begin{cases} 3x + 2y = -c \\ 6x + 4y = 1 - 2c \end{cases} \rightarrow \begin{cases} -6x - 4y = 2c \\ 6x + 4y = 1 - 2c \end{cases}$

Thus, $0 = 1$. Because this is false for all values of c, the given system has no solution.

21. 0

23.

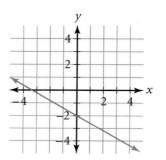

Lesson 5.3 Skill B (pages 198–199)

Try This Example 1

$\left(\frac{4}{3}, -\frac{1}{2}\right)$

Try This Example 2

$\left(\frac{35}{8}, -\frac{13}{8}\right)$

Exercises

1. Multiply the first equation by 2 and the second equation by 3.

3. Multiply the first equation by 2 and the second equation by 7.

5. $(0, -2)$ **7.** $(0, 0)$ **9.** $(8, -3)$ **11.** $(5, 7)$

13. $(4, -3)$ **15.** $(1, 3)$ **17.** $\left(\frac{40}{13}, -\frac{15}{13}\right)$

19. $\left(\frac{m-n}{3m-2n}, \frac{1}{3m-2n}\right)$; $m \neq \frac{2}{3}n$

21. parallel because both lines have slope 4 and different y-intercepts

23. not perpendicular because the product of the slopes is not -1

Lesson 5.3 Skill C (pages 200–201)

Try This *Example 1*

240 milliliters each of solution A and solution B

Try This *Example 2*

$a = 766\frac{2}{3}$, $b = -266\frac{2}{3}$; because b is negative and the amount of Solution A needed is more than the total amount, the solution cannot be made.

Exercises

1.

A	B	C
a	b	350
$0.18a$	$0.26b$	$(0.20)350$

3. 180 milliliters of solution A and 270 milliliters of solution B

5. 5 hours at 45 miles per hour and 3 hours at 55 miles per hour

7. $(4, -1)$ **9.** $(-5, -3)$ **11.** $(1, 1)$

Lesson 5.4 Skill A (pages 202–203)

Try This *Example 1*

consistent

Try This Example 2

$$\begin{cases} -3x - y = -2 \\ 6x + 2y = 10 \end{cases} \quad \begin{cases} y = -3x + 2 \\ y = -3x + 5 \end{cases}$$

The slopes, -3, are equal but the y-intercepts, 2 and 5, are different. So the lines must be parallel. The system is inconsistent.

Exercises

1. $\begin{cases} y = -x + 4 \\ y = -\frac{1}{2}x + 3 \end{cases}$ **3.** $\begin{cases} y = -\frac{7}{5}x + \frac{3}{5} \\ y = 4x - 12 \end{cases}$

5. consistent   **7.** inconsistent   **9.** consistent

11. no unique solution   **13.** $\left(-6, -\frac{5}{2}\right)$

15. $\left(\frac{1}{3}, \frac{8}{3}\right)$ **17.** $d = \frac{21}{2}$

19. 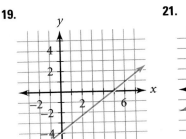   **21.**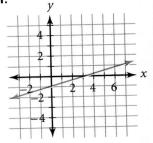

23.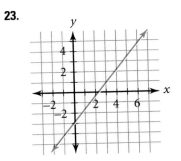

Lesson 5.4 Skill B (pages 204–205)

Try This *Example 1*

consistent and independent

Try This *Example 2*

inconsistent

Answers for Exercises 1 and 2 may vary. Sample answers are given.

Exercises

1.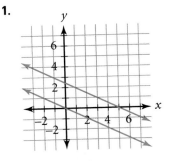

3. consistent and independent

5. consistent and independent

7. consistent and dependent

9. consistent and independent

11. inconsistent

13. consistent and dependent

15. consistent and dependent

17. inconsistent

19. $b = \dfrac{10}{3}$ and $e = \dfrac{50}{3}$

21. $c = 2.5$ and $f \neq -1$

23. $(9, 2)$ **25.** $(1, -1)$

Lesson 5.5 Skill A (pages 206–207)

Try This Example

a. 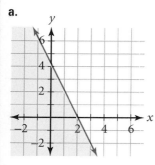   **b.**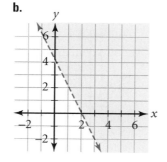

Exercises

1. $y \geq -\dfrac{5}{2}x + 5$ **3.** above

5. 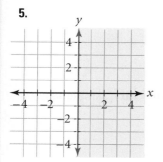   **7.**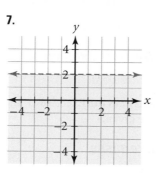

9. 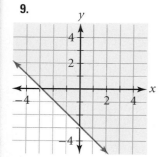   **11.**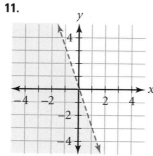

13. 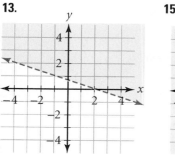   **15.**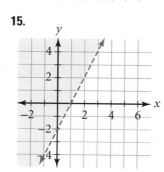

17. 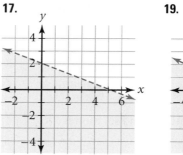   **19.**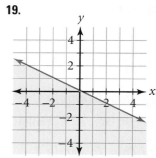

21.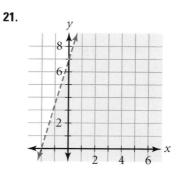

23. $y \geq \dfrac{4}{3}x + \dfrac{1}{3}$ **25.** $b = 0$ and $m < 5$

27. $1 < x < 3$

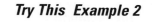

29. $y < -2$ or $y > 4$

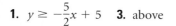

Lesson 5.5 Skill B (pages 208–209)

Try This Example 1   ### Try This Example 2

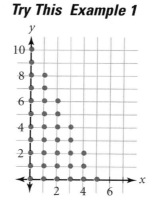   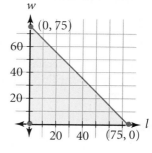

Exercises

1.

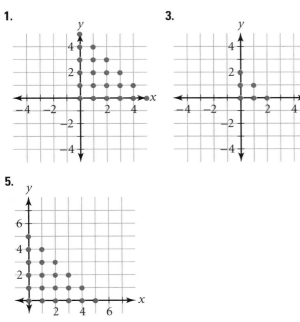

3.

5.

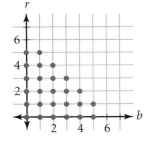

7. Let *r* represent the number of red chips and *b* represent the number of blue chips.

9. Let *l* represent length and *w* represent width. The ordered pairs $(w, 2w)$, where $0 < w < 35$, give rectangles in which the length is twice the width.

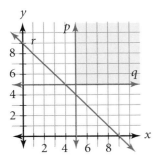

11. The restriction $x \geq 5$ means that all solutions must be to the right of the line $x = 5$. This is line *p*. The restriction $y \geq 5$ means that all solutions must be above the line $y = 5$. This is line *q*. However, the restriction that the sum be less than 9 means that all solutions must be below the line $x + y = 9$. This is line *r*. There are no such points satisfying all these restrictions.

13. solution A: 210 milliliters; solution B: 90 milliliters

Lesson 5.6 Skill A (pages 210–211)

Try This Example

a.

b.

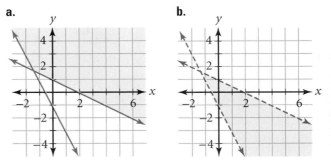

Exercises

1.

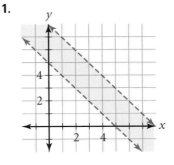

3. the second quadrant

5.

7.

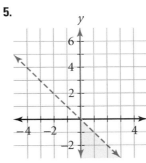

9.

11.

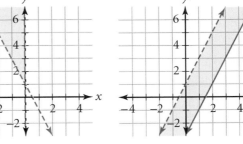

13.

15.

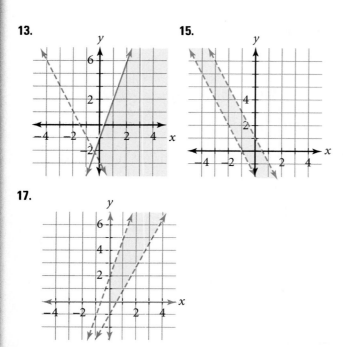

17.

19. $m_1 = m_2$ and $b_2 > b_1$ **21.** inconsistent

Lesson 5.6 Skill B (pages 212–213)

Try This Example 1

Try This Example 2

a. The solutions form a line. **b.** The solution areas do not intersect.

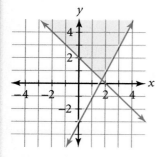

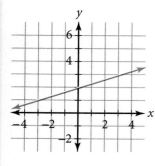

Exercises

1. $\begin{cases} y \ge \frac{2}{5}x \\ y \ge \frac{3}{5}x - \frac{2}{5} \end{cases}$ **3.** $\begin{cases} y \ge \frac{3}{4}x + \frac{1}{4} \\ y < \frac{2}{3}x + \frac{2}{3} \end{cases}$

5. **7.**

9. **11.**

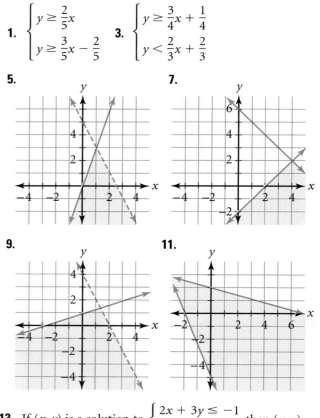

13. If (x, y) is a solution to $\begin{cases} 2x + 3y \le -1 \\ 2x + 3y \ge -1 \end{cases}$, then (x, y) must satisfy $2x + 3y = -1$. Thus, the solution to the system is all points on a line.

15. $a \le b$ **17.** 8 **19.** 9 **21.** 16 **23.** 1

25. 25 **27.** 8

Lesson 5.6 Skill C (pages 214–215)

Try This Example 1

$\begin{cases} x \ge 3 \\ x \le 7 \\ y \ge 0 \\ y \le \frac{1}{3}x + 3 \\ y \ge -\frac{3}{2}x + 5 \end{cases}$

Try This Example 2

$\begin{cases} c + s \le 40 \\ c \le s \\ c \ge 0 \\ s \ge 0 \end{cases}$

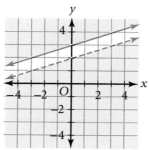

Exercises

1. $x = 1, x = 8,$ and $y = 0$

3. $\begin{cases} x \geq 0 \\ x \leq 5 \\ y \geq 0 \\ y \leq 6 \end{cases}$ **5.** $\begin{cases} y \leq \frac{1}{2}x + 2 \\ y \geq -\frac{3}{4}x + 3 \\ x \leq 4 \end{cases}$

7. $\begin{cases} c + s \leq 25 \\ s \geq 2c \\ c \geq 0 \\ s \geq 0 \end{cases}$

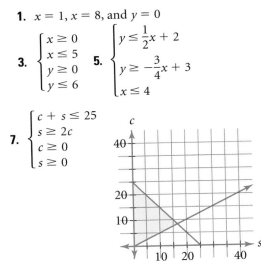

9. between 20 feet and 150 feet, inclusive

CHAPTER 6

Lesson 6.1 Skill A (pages 220–221)

Try This Example

a. 10^7 **b.** 3^8 **c.** 6^4

Exercises

1. $2 \cdot 2 \cdot 2$ **3.** $1 \cdot 1 \cdot 1 \cdot 1 \cdot 1$

5. Power-of-a-Product Property **7.** Product Property

9. 5^9 **11.** 3^6 **13.** 10^{35} **15.** 3^4 **17.** 27^6

19. 6^9 **21.** 3^4 **23.** 4^5 **25.** 2^8 **27.** 14^2 **29.** 6^3

31. $(-6)^4$ **33.** 3^8 **35.** $(-2)^6$

37. $(2x)^3 = 2x \cdot 2x \cdot 2x = 2^3 x^3 = 8x^3$
$2x^3 = x^3 + x^3$

39. n is an odd integer or $n = 0$.

41. $(11, 7)$ **43.** $(2, -1)$ **45.** $\left(\frac{3}{7}, \frac{15}{14}\right)$ **47.** $\left(\frac{27}{7}, -\frac{8}{7}\right)$

Lesson 6.1 Skill B (pages 222–223)

Try This Example 1

a. $\frac{2^3}{3^3}$, or $\frac{8}{27}$ **b.** 3^4, or 81 **c.** 5^4

Try This Example 2

a. 1 **b.** $\frac{1}{100}$ **c.** $\frac{1}{4^3}$

Exercises

1. Quotient Property

3. Definition of zero exponent

5. Power-of-a-Power Property

7. Product Property

9. $\frac{1}{9}$ **11.** $\frac{3^3}{5^3}$ **13.** 3^3 **15.** 2^2, or 4 **17.** $\frac{1}{2^3}$, or $\frac{1}{8}$

19. $\frac{1}{3^3}$ **21.** 3^2, or 9 **23.** 1 **25.** 4 **27.** $\frac{1}{7^2}$, or 49

29. 1 **31.** 2^9 **33.** $\frac{1}{(-3)^6}$ **35.** $\frac{1}{100}$

37. $3^4 \cdot 2^7$ **39.** $\frac{3^3}{2}$, or $\frac{27}{2}$

41. Each is the multiplicative inverse of the other.

43. 1.18 **45.** 0.078 **47.** 0.0257 **49.** 0.0183

Lesson 6.1 Skill C (pages 224–225)

Try This Example 1

a. 2.65×10^1 **b.** 1.4×10^{-3}

Try This Example 2

a. 7.38×10^5 **b.** 1.25×10^3

Exercises

1. yes **3.** no; 3.45×10^2 **5.** 1.24×10^1

7. 1.2×10^{-2} **9.** 1.005×10^4 **11.** 1.65×10^{-2}

13. 3.0×10^4 **15.** 1.95×10^5 **17.** 3.24×10^1

19. 2.0×10^2 **21.** 2.5×10^{-2} **23.** 5.0×10^{-2}

25. $1 \leq a < 2$ **27.** $y = 28.8$ **29.** $y = 1.3$

Lesson 6.2 Skill A (pages 226–227)

Try This Example 1

$y = 25$

Try This Example 2

$y = 216$

Exercises

1.

x	0	1	2	3	4	5
y	0	3	6	9	12	15

Yes, the differences are constant, 3.

3.

x	0	1	2	3	4	5
y	0	3	24	81	192	375

No, the differences are not constant.

5. $k = 2.5$; $y = 2.5x^2$; $y = 160$

7. $k = 5$; $y = 5x^3$; $y = 1080$

9. $k = 6$; $y = 6x^3$; $y = 162$

11. $\dfrac{y_2}{y_1} = \left(\dfrac{x_2}{x_1}\right)^2$ **13.** $4s + 4$ **15.** $7u - 21$ **17.** $1.4z - 8$

Lesson 6.2 Skill B (pages 228–229)

Try This Example

a. about 154.7 feet

b. No; $\dfrac{d_{0.5s}}{d_s} = \dfrac{\frac{130}{55^2}(0.5s)^2}{\frac{130}{55^2}s^2} = \dfrac{(0.5s)^2}{s^2} = \dfrac{1}{4}$

The stopping distance is reduced to $\dfrac{1}{4}$ of the original stopping distance when the speed is reduced by $\dfrac{1}{2}$.

Exercises

1. Answers are rounded to the nearest tenth of a foot.

s	0	5	10	15	20	25	30
d	0	0.9	3.6	8.2	14.5	22.7	32.7

3. 109.4 feet   **5.** 163.4 feet

7. $\dfrac{d_{rs}}{d_s} = \dfrac{k(rs)^2}{ks^2} = \dfrac{kr^2s^2}{ks^2} = \dfrac{r^2}{1} = r^2$

9. 144 feet   **11.** 1600 feet   **13.** 14,400 feet

15. fewer than 8 games

Lesson 6.3 Skill A (pages 230–231)

Try This Example 1

$-7y^3 + 13y^2 + 15y + 5$

Try This Example 2

cubic trinomial

Exercises

1. degree: 3
number of terms: 3
leading term: r^3
leading coefficient: 1

3. degree: 4
number of terms: 4
leading term: $3k^4$
leading coefficient: 3

5. $g^2 + 5g - 7$

7. $-3m^2 - m + 7$ **9.** $5c^4 + 8c^2 - 4$ **11.** $5y^3 + 2y$

13. cubic trinomial   **15.** quartic polynomial

17. quadratic binomial   **19.** cubic binomial

21. quadratic monomial   **23.** linear binomial

25. a is any real number other than -4 and $b = 6$.

27. 78   **29.** 57   **31.** $\dfrac{13}{36}$

Lesson 6.3 Skill B (pages 232–233)

Try This Example 1

47.1 cm^3; 188.4 cm^3; 423.9 cm^3

Try This Example 2

904.32 ft^2; 1073.88 ft^2; 1256 ft^2

Exercises

1. 16   **3.** 40   **5.** 2307.9 cm^3 **7.** 3402.975 cm^3

9. 69.08 ft^2 **11.** 244.92 ft^2 **13.** 2.14 in^3; 8.04 in^2

15. 22.44 in^3; 38.47 in^2 **17.** 463.01 in^3; 289.38 in^2

19. $\dfrac{1}{16}$ **21.** $\dfrac{1}{5}$ **23.** quadratic binomial

Lesson 6.4 Skill A (pages 234–235)

Try This Example 1

$$\begin{array}{r} 2.6a^3 - 1.5a^2 - 5a \\ 0.4a^3 - 5a + 3 \\ \hline 3.0a^3 - 1.5a^2 - 10a + 3 \end{array}$$

Try This Example 2

$(2.6 + 0.4)a^3 + (-1.5)a^2 + (-5-5)a + 3 = 3a^3 - 1.5a^2 - 10a + 3$

Exercises

1. $-x^2 - 5x - 4$ **3.** $y^3 - y^2 - 5y + 2$

5. $4b^2 - 3b - 1$ **7.** $z^3 + z$ **9.** $3k^4 + k^3 - k^2 - k$

11. $b - 1$ **13.** $6d^5 + d^2$ **15.** $-5x^3 - x^2 + 8$

17. when $a = -k$ and $b = -m$ **19.** 6.3×10^1

21. 3.6×10^6 **23.** 4.5×10^3 **25.** 5.0×10^{-2}

Lesson 6.4 Skill B (pages 236–237)

Try This Example 1

$3a^3 + 3.5a^2 + 0.5a$

Try This Example 2

$3n^3 - 3n^2 + 4n + 1$

Exercises

1. $-(a + b) = -a - b$

3. $-(a - b + c) = -a + b - c$

5. $(3z^3 + z^2 - z + 5) + (-4z^3 - z^2 + z - 5)$

7. $-6x^3 + 5x^2 + 3x - 1$

9. $7x^5 - 8x^4 + 3x - 5$

11. $-m^3 - 6m - 5$ **13.** $-4b^3$

15. $-2a^3 + 2a^2$ **17.** $3d^4 - d^3 - d^2 + 10d + 3$

19. $6x^2 - 3xy + 8y^2$ **21.** $2x^2 + 2$

23. $y = \frac{1}{2}x - \frac{3}{2}$ **25.** $y = 7$ **27.** $y = \frac{7}{4}x - \frac{7}{2}$

Lesson 6.4 Skill C (pages 238–239)

Try This Example 1

$32x$

Try This Example 2

a. $0.02p^2 + 1.39p - 610$ **b.** \$7424

Exercises

1. a. $7.2a$ **b.** $14.4a$

3. $14.4a + 48a = 62.4a$ **5.** $25x$ **7.** $22b$

9. profit of \$900   **11.** loss of \$162.50   **13.** 50mm

Lesson 6.5 Skill A (pages 240–241)

Try This Example 1

a. $3h^2$ **b.** $8n^3$ **c.** $4p^6$

Try This Example 2

a. $\frac{1}{16}t^{10}$ **b.** $3a^3$

Try This Example 3

a. $-a^5b^2$ **b.** $9m^3n^5$

Exercises

1. $10x^5$ **3.** c^6d^6 **5.** a^3 **7.** m^{10} **9.** $30z^5$

11. $5b^7$ **13.** $64a^3b^3$ **15.** $-8y^9z^{15}$ **17.** $6s^6$

19. $64t^{14}$ **21.** $5u^7$ **23.** $150a^5b^8$ **25.** $4m^6n^6$

27. $9d^8r^8$ **29.** $24m^{10}n^{14}$ **31.** $-15e^{11}r^9t^7$

33. 0 **35.** none   **37.** -10

Lesson 6.5 Skill B (pages 242–243)

Try This Example 1

a. b **b.** $\frac{1}{d}$ **c.** $-\frac{8}{5y}$

Try This Example 2

$5u^2v^4$

Exercises

1. a^4 **3.** $\frac{1}{m}$ **5.** d^5 **7.** $\frac{1}{w^2}$ **9.** $-4a$ **11.** $\frac{5}{2g}$

13. $8s^4$ **15.** $\frac{1}{a^2}$ **17.** $\frac{2}{q}$ **19.** $\frac{y^2z^2}{5}$ **21.** $8mn^7$

23. $\frac{1}{a^2b^2}$ **25.** $\frac{a}{c}$ **27.** qk **29.** $\frac{12y}{5}$ **31.** $\frac{b^2c^2}{6}$

33. $\frac{16p^4a^2}{9}$ **35.** f^8b^4 **37.** $y + 5 = -2.5(x + 3)$

39. $y = \frac{3}{7}x$ **41.** $y - 5 = 0$

Lesson 6.5 Skill C (pages 244–245)

Try This Example 1

$2400x^3$

Try This Example 2

$\frac{2d}{3}$

Exercises

1. length $= 2a$, width $= a$, height $= 2a$

3. $540x^3$ **5.** k^2 **7.** $\frac{5\ell}{17}$ **9.** about 148.8 feet

Lesson 6.6 Skill A (pages 246–247)

Try This Example 1

$2n^3 - 8n^2$

Try This Example 2

$10x^5 - 2x^4 - 10x^3$

Try This Example 3

$3r^4s^2 - r^5s^2$

Exercises

1. $3z^2(2z) + 3z^2(4)$ **3.** $uv(2uv^2) + uv(3u^2v)$

5. $4y^4 - 2y^7$ **7.** $12x^2 + 15x$ **9.** $-6a^3 + 2a^2$

11. $10c^5 + 5c^4$ **13.** $3q^5 - q^4 + 3q^3$

15. $2w^5 - 5w^4 - w^3$ **17.** $8k^8 + 16k^6 - 8k^5$

19. $r^2s^3 - r^2s^2$ **21.** $-6x^4y^3 + 10x^4y^4$

23. $14u^5v^2 + 8u^4v^4 - 8u^2v^5$ **25.** $12m^4 + 4m^3$

27. $a = 2$ **29.** $(2x + 3y)(a + b)$

31.

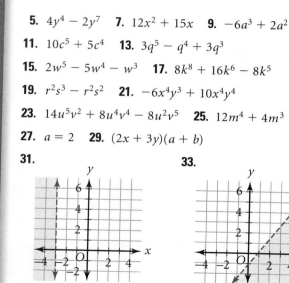

33.

35.

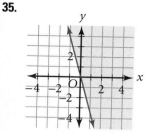

Lesson 6.6 Skill B (pages 248–249)

Try This Example 1
$-3z^2 + 13z + 10$

Try This Example 2
$3n^2 + 11n + 6$

Try This Example 3
$-15n^3 - 7n^2 + 11n + 6$

Exercises

1. $x^2 + 13x + 30$ **3.** $w^3 - 2w^2 - 15w$

5. $x^2 + 5x - 6$ **7.** $b^2 - b - 6$

9. $2z^2 - 4z - 6$ **11.** $8m^2 + 14m + 6$

13. $-v^2 - 8v - 16$ **15.** $-6p^2 + 21p - 15$

17. $x^3 + 2x^2 + 2x + 1$ **19.** $2r^3 - 3r^2 + 3r - 1$

21. $c^3 - 4c^2 + 9$ **23.** $15n^3 + 2n^2 + 43n - 28$

25. $-25h^3 + 25$ **27.** $2z^3 - 3z^2 - 11z + 6$

29. $3a^3b - 3ab^3$ **31.** $-4c^5d + 8c^4d^3 + 2c^3d^2 - 4c^2d^4$

33. $r = \dfrac{9}{2}$ **35.** 1 **37.** $-n^2 + 8n - 3$ **39.** $d^3 + 3$

Lesson 6.6 Skill C (pages 250–251)

Try This Example 1

a. $4x^2 - 20x + 25$ **b.** $25d^2 - 20d + 4$

Try This Example 2

$25c^2 - 36d^2$

Exercises

1. $a = 2d$ and $b = 4$ **3.** $a = 3k$ and $b = 5$

5. $k^2 + 2k + 1$ **7.** $x^2 - 4x + 4$ **9.** $4h^2 + 12h + 9$

11. $4t^2 + 12t + 9$ **13.** $t^2 - 1$ **15.** $4z^2 - 1$

17. $9 - d^2$ **19.** $4d^2 - 9$ **21.** $4u^2 + 20uv + 25v^2$

23. $25r^2 - 16s^2$ **25.** $4r^2s^2 + 12rs + 9$ **27.** $16p^2q^2 - 1$

29. $(a - b)(a - b) = a(a - b) - b(a - b)$
$\qquad = a^2 - ab - ba + b^2$
$\qquad = a^2 - 2ba + b^2$
$\qquad = a^2 - 2ab + b^2$

31. $n = 0$ **33.** $a^2 + 2ab + b^2 - c^2$

35. $38 \cdot 42 = (40 - 2)$
$(40 + 2) = 40^2 - 2^2 = 1600 - 4 = 1596$

37. $7y^3$ **39.** $-\dfrac{1}{2ab}$ **41.** $\dfrac{n}{m}$ **43.** $\dfrac{1}{2b^3}$

Lesson 6.7 Skill A (pages 252–253)

Try This Example 1

a. $2n^2 + \dfrac{1}{2}$ **b.** $uv - 1$

Try This Example 2

$6m^2 - \dfrac{3}{m}$

Exercises

1. $5n + 2$ **3.** $y^2z^2 - z$ **5.** $4a + 5$; yes

7. $-5x + 2$; yes   **9.** $4c - 4$; yes   **11.** $\dfrac{5}{2}s^2 - 5s$; yes

13. $a^2b^2 + ab$; yes   **15.** $k^2 - 7kp^2$; yes   **17.** $s - \dfrac{1}{s}$; no

19. $-4d + \dfrac{2}{d}$; no   **21.** $2z^2 - \dfrac{4}{z}$; no   **23.** $x + \dfrac{1}{x}$; no

25. $2z^2 - 6$; yes

27. $\dfrac{a - b}{c} = \dfrac{a + (-b)}{c}$ ← definition of subtraction

$\qquad = \dfrac{1}{c}[a + (-b)]$ ← definition of division

$\qquad = \dfrac{1}{c} \cdot a + \dfrac{1}{c} \cdot (-b)$ ← Distributive Property

$\qquad = \dfrac{a}{c} + \dfrac{-b}{c}$ ← definition of division

$\qquad = \dfrac{a}{c} - \dfrac{b}{c}$ ← definition of subtraction

29. $-27u^5$ **31.** $6d^4 + 4d^3 + 2d^2$ **33.** $a^5 + a^4 + a^3 + a^2$

Lesson 6.7 Skill B (pages 254–255)

Try This Example 1

$3y + 2$

Try This Example 2

$m + 7$

Exercises

1. $(x + 2)(x + 3) = x(x + 3) + 2(x + 3)$
$= x^2 + 3x + 2x + 6$
$= x^2 + 5x + 6$

3. $(x - 2)(x - 7) = x(x - 7) - 2(x - 7)$
$= x^2 - 7x - 2x + 14$
$= x^2 - 9x + 14$

5. $b + 4$ **7.** $u + 3$ **9.** $w - 7$ **11.** $3h + 5$

13. $4k + 1$ **15.** $5p - 2$ **17.** $x - 8$ **19.** $10z + 7$

21. $3a - b$

23. $a \leq 0$ or $a \geq 4$

25. no solution

27. $-3 \leq x \leq 3$

CHAPTER 7

Lesson 7.1 Skill A (pages 260–261)

Try This Example 1

Answers should include four of the following:
$1 \times 36, 2 \times 18, 3 \times 12, 4 \times 9, 6 \times 6$

Try This Example 2

$2^2 \times 3^2 \times 5^1$

Exercises

1. neither   **3.** prime

5. Answers should include three of the following:
$1 \times 24, 2 \times 12, 3 \times 8, 4 \times 6$

7. Answers should include three of the following:
$1 \times 40, 2 \times 20, 4 \times 10, 5 \times 8$

9. $2^3 \times 3^1$ **11.** prime   **13.** 5^2 **15.** $2^3 \times 3^2$

17. $2^2 \times 5^2$ **19.** $3^1 \times 17^1$ **21.** 5^3 **23.** 2^6

25. $243 = 3^5$ **27.** $625 = 5^4$

29. $12 = 2^2 \times 3^1$; $13 = 13^1$; the only common factor is 1.

31. $18 = 2^1 \times 3^2$; $25 = 5^2$; the only common factor is 1.

33. 10   **35.** $x = -3$ **37.** $x = -4.1$ **39.** $a = -4.9$

41. $z = 4.5$ **43.** $z = -4$

Lesson 7.1 Skill B (pages 262–263)

Try This Example 1

20

Try This Example 2

a. $18d$ **b.** 1

Try This Example 3

$3m^2n^2$

Try This Example 4

$4mn^2$

Exercises

1. z **3.** a and n **5.** 3   **7.** 2   **9.** 16   **11.** 1   **13.** 2

15. 9   **17.** $3x^2$ **19.** $3c$ **21.** 2   **23.** 7   **25.** $2x^2$

27. dz **29.** 2   **31.** $3a^3b^3$ **33.** $3ab$ **35.** 1   **37.** xy

39. p^2q **41.** 4   **43.** 1   **45.** $y - 4 = -\dfrac{1}{3}(x + 2)$

47. $y = -3x + 13$ **49.** $y = x + 2$

Lesson 7.1 Skill C (pages 264–265)

Try This Example 1

$3v(-v + 5)$

Try This Example 2

$13t^2(t^2 + 2t + 1)$

Try This Example 3

$3c^2d(-cd + 2d + 3)$

Exercises

1. $2x(x + 3) = (2x)(x) + (2x)(3)$
$= 2x^2 + 6x$

3. $3x^2(x + 1) = (3x^2)(x) + (3x^2)(1)$
$= 3x^3 + 3x^2$

5. $7(x + 2)$ **7.** $2b(5b + 6)$ **9.** $3z(2z - 3)$

11. $4n(3n^2 + n - 1)$ **13.** $5k(k^2 - 5k - 15)$

15. $2w^2(w^2 + 9w - 9)$ **17.** $3z^3(-4z^2 + 5z - 6)$

19. $11v^3(v^2 - v + 2)$ **21.** $rs(rs - 8s + 8r)$

23. $n(3mn + 10m^2n - 7m - 2)$

25. $7uv(u + 4v - v^3 + 1)$ **27.** $ab^2(1 + a - a^2b + ab^3)$

29. $3(a + b)^2(2a + b)$ **31.** $(m + n)^2(-2m + 5n)$

33. $x^4y^6(1 + x^5)$ **35.** $4x^2 - 8x - 5$

37. $x^2 + 9x + 14$ **39.** $-6y^2 + 13y - 6$ **41.** $4x^2 - 9$

Lesson 7.2 Skill A (pages 266–267)

Try This Example 1

  a. $(d - 9)^2$ **b.** $(3z + 1)^2$

Try This Example 2

  a. $(h - 10)(h + 10)$ **b.** $(4x - 7)(4x + 7)$

Exercises

  1. $a = x, b = 3$ **3.** $a = k, b = 3$ **5.** $(d + 1)^2$

  7. $(u - 4)^2$ **9.** $(h - 12)(h + 12)$

  11. $(k - 15)(k + 15)$ **13.** $(2y + 5)^2$ **15.** $(7m + 1)^2$

  17. $(4x - 11)(4x + 11)$ **19.** $(uv + 1)^2$ **21.** $(4mn - 2)^2$

  23. $(8cd - 5)(8cd + 5)$ **25.** $(11st - 9)(11st + 9)$

  27. $(5mn - 7xy)(5mn + 7xy)$

  29. $m = 5$ and $n = -5$ **31.** $6a^2 - 11a - 30$

  33. $-4n^2 - 4n + 15$ **35.** $9t^2 - 24tc + 16c^2$

  37. $2 \times 7 \times 11$ **39.** $3 \times 7 \times 11$

Lesson 7.2 Skill B (pages 268–269)

Try This Example 1

$(3 - y)(2a - b)$

Try This Example 2

$(3 + y)(3m + 5n)$

Try This Example 3

$(d + b)(d - 8)$

Exercises

  1. $3x - 7$ **3.** $x - y$ **5.** $(2 + h)(2p - 5)$

  7. $(-2 + t)(m + 7)$ **9.** $(-1 - w)(v - 3t)$

  11. $(c + 3)(x + y)$ **13.** $(x + y)(x - 10)$

  15. $(y + d)(y - 3)$ **17.** $(3c + 5d)(3n + 4m)$

  19. $(5 + 7t)(3a - 2b)$ **21.** $(2x + y)(x - 2)$

  23. $(4a + c)(b - 2)$ **25.** $(3 - 2a)(3x + 2)$

  27. $a = -5; (3 + n)(2y + 5)$ **29.** $a = 3$ and $b = -7$

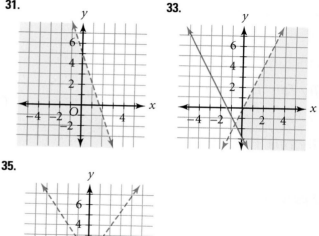

31.   **33.**

35.

Lesson 7.3 Skill A (pages 270–271)

Try This Example 1

$(a + 5)(a + 4)$

Try This Example 2

$(n - 9)(n - 4)$

Exercises

  1. 2 and 10; $(m + 2)(m + 10)$

  3. -1 and -20; $(k - 1)(k - 20)$

  5. -5 and -7; $(d - 5)(d - 7)$

  7. $(x + 5)(x + 7) = x(x + 7) + 5(x + 7) =$
  $x^2 + 7x + 5x + 35 = x^2 + 12x + 35$

  9. $(m + 1)(m + 4)$ **11.** $(z + 3)(z + 5)$

  13. $(k - 5)(k - 6)$ **15.** $(m - 1)(m - 7)$

  17. $(x + 5)(x + 8)$ **19.** $(y + 2)(y + 24)$

  21. $(t + 2)(t + 21)$ **23.** $(t - 2)(t - 21)$

  25. $(p - 10)(p - 40)$ **27.** $(ab + 3)(ab + 7)$

  29. Both factors are positive; both factors are negative.

  31. $8v^2$ **33.** $5n^2$ **35.** $4k^2$ **37.** $3h$ **39.** $3p(p^2 - 5p + 1)$

Lesson 7.3 Skill B (pages 272–273)

Try This Example 1

$(n + 8)(n - 5)$

Try This Example 2

$(n - 8)(n + 5)$

Try This Example 3

List all factor pairs for -8: -1×8, 1×-8, -2×4, 2×-4. None of these pairs has a sum of 5. Thus, $q^2 + 5q - 8$ cannot be factored.

Exercises

1. $b = -7, c = -18$ **3.** $b = -2, c = 35$

5. $(x - 7)(x + 1)$ **7.** $(v + 5)(v - 1)$

9. $(k - 2)(k + 1)$ **11.** $(w - 7)(w + 6)$

13. $(d + 9)(d - 7)$ **15.** $(h - 19)(h + 2)$

17. $(z + 16)(z - 6)$ **19.** $(t + 9)(t - 5)$

21. $(m - 43)(m + 1)$

23. List all factor pairs for 7: $1 \times 7, -1 \times -7$. None of these pairs has a sum of -9. Thus, $x^2 - 9x + 7$ cannot be factored.

25. List all factor pairs for 1: $1 \times 1, -1 \times -1$. None of these pairs has a sum of 3. Thus, $b^2 + 3b + 1$ cannot be factored.

27. List all factor pairs for -3: $1 \times -3, -1 \times 3$. None of these pairs has a sum of -5. Thus, $k^2 - 5k - 3$ cannot be factored.

29. The factor with the larger absolute value is positive and the factor with the smaller absolute value is negative.

31. $(n + 11)^2$ **33.** $(2a + 5)^2$ **35.** $(3y - b)(3y + b)$

37. $(2 + 3z)(z + 1)$ **39.** $(-1 + 3n)(n - 1)$

41. $(x - y)(x - 20)$

Lesson 7.4 Skill A (pages 274–275)

Try This Example 1

$(n + 5)(2n + 1)$

Try This Example 2

$(4a - 1)(2a - 3)$

Exercises

1. $a = 4, b = 9, c = 2, ac = 8$
List all factor pairs for 8: $1 \times 8, 2 \times 4, -1 \times -8, -2 \times -4$. The factor pair whose sum is 9 is 1 and 8.

3. $a = 2, b = 8, c = 8, ac = 16$
List all factor pairs for 16: $1 \times 16, 2 \times 8, 4 \times 4, -1 \times -16, -2 \times -8, -4 \times -4$. The factor pair whose sum is 8 is 4 and 4.

5. $20x + 6x$ **7.** $40x + 3x$ **9.** $-8y - 15y$

11. $-5v - 24v$ **13.** $(a + 3)(2a + 1)$

15. $(k + 1)(3k + 2)$ **17.** $(h + 2)(3h + 2)$

19. $(z + 3)(5z + 2)$ **21.** $(v + 1)(4v + 3)$

23. $(7x - 1)(x - 2)$ **25.** $(8c - 3)(c - 1)$

27. $-(3m - 1)(3m + 2)$ **29.** $-(2x - 2)(x + 3)$

31. $(2xy + 3)(2xy + 5)$ **33.** $\left(\dfrac{3}{7}, \dfrac{15}{14}\right)$

35. $\left(-\dfrac{2}{11}, -\dfrac{7}{22}\right)$ **37.** $\left(\dfrac{6}{13}, -\dfrac{30}{13}\right)$

Lesson 7.4 Skill B (pages 276–277)

Try This Example 1

$(3k - 2)(k + 2)$

Try This Example 2

$(3m + 7)(2m - 7)$

Try This Example 3

$a = -3, b = 15, c = -2, ac = 6$; there are no factors of 6 that add to 15, so $-3x^2 + 15x - 2$ cannot be factored.

Exercises

1. $-$ **3.** $+$ **5.** $(3v - 1)(v + 7)$ **7.** $(3d - 1)(d + 1)$

9. $(11x - 2)(x + 1)$ **11.** $(3h - 2)(h + 1)$

13. $(3c + 2)(c - 3)$ **15.** $(2m - 7)(m + 1)$

17. $(2x - 9)(x + 10)$ **19.** $(6x + 1)(x - 3)$

21. No factors of 60 add to 21.

23. No factors of -12 add to 3.

25. $a = 19, a = 11$, or $a = 9$. If $a < 0$, $a = -19, a = -11$, or $a = -9$.

27. inconsistent **29.** inconsistent **31.** inconsistent

Lesson 7.4 Skill C (pages 278–279)

Try This Example 1

a. $-(a - 3)(a - 4)$ **b.** $3(2n + 3)(n + 1)$

Try This Example 2

a. $3g(3g + 4)(3g - 4)$ **b.** $4d(d + 2)(d + 2)$

Try This Example 3

$3r(3r - 2)(r - 2)$

Exercises

1. No; x can be factored from $x^2 - x$.

3. Yes **5.** $-(x - 3)(x + 2)$ **7.** $-(u - 4)(u - 3)$

9. $-(z - 3)(z + 7)$ **11.** $3(v + 1)(v + 5)$

13. $7(x + 1)(x + 5)$ **15.** $-3(t - 6)(t - 5)$

17. $-5(2b - 1)(3b + 5)$

19. $7(3m - 1)(3m - 1)$, or $7(3m - 1)^2$

21. $7(2ab + b + 3)$ **23.** $5(5q + 3)(5q - 3)$

25. $3a(a + 11)(a - 1)$ **27.** $2x(x + 11)(x - 5)$

29. $n = 4$ **31.** $5xv^2(3x^2 + 4)$ **33.** $(5xy - 7)(5xy + 7)$

35. $(ab + xy)(ab - xy)$ **37.** $(a - 15)(a + 1)$

39. $(3x + 2)(x + 1)$ **41.** $(5k - 2)(k + 3)$

Lesson 7.5 Skill A (pages 280–281)

Try This Example 1

$d = -\dfrac{5}{3}$ or $d = 2$

Try This Example 2

$t = -\dfrac{5}{3}$ or $t = \dfrac{3}{2}$

Exercises

1. $x = -3$ or $x = 5$ 3. $n = -\dfrac{3}{5}$ or $n = -\dfrac{7}{2}$

5. $x = -2$ or $x = -5$ 7. $x = 5$ or $x = 7$

9. $n = -\dfrac{1}{5}$ or $n = 3$ 11. $t = -5$ or $t = 3$

13. $y = -8$ or $y = -5$ 15. $s = -3$ or $s = \dfrac{5}{2}$

17. $x = 4$ or $x = -11$ 19. $y = \dfrac{1}{5}$ or $y = 9$

21. $x = \dfrac{2}{3}$ 23. $x = \dfrac{3}{2}$ or $x = 1$ 25. $a = -\dfrac{1}{2}$ or $a = 4$

27. $x = \dfrac{m}{3}$ or $x = \dfrac{n}{2}$ 29. $x = -4$, $x = 4$, or $x = -5$

31. $6a^4 - 6a^3 - 6a^2$ 33. $-5m^6 - 5m^5 - 5m^4$
35. $y^6 - 2y^5 + 2y^4 - y^3$ 37. $8d^8 + 72d^6 + 8d^4$

Lesson 7.5 Skill B (pages 282–283)

Try This Example 1

$t = -\dfrac{5}{3}$ or $t = \dfrac{3}{2}$

Try This Example 2

$b = -4$ or $b = 9$

Try This Example 3

a. $c = -2$ or $c = 2$ b. $s = -1$

Exercises

1. $6x^2 - 7x - 5 = 0$ 3. $4n^2 - 12n + 9 = 0$

5. $15t^2 - 2t - 1 = 0$ 7. $x = -7$ or $x = 3$

9. $q = -11$ or $q = -3$ 11. $w = -7$ or $w = 6$

13. $b = -\dfrac{3}{5}$ or $b = \dfrac{1}{3}$ 15. $y = -\dfrac{3}{2}$ or $y = \dfrac{3}{7}$

17. $z = -1$ or $z = 11$ 19. $x = 1$ or $x = 9$

21. $x = 2$ or $x = -\dfrac{8}{3}$ 23. $x = -1$ or $x = -\dfrac{11}{14}$

25. $x = -\dfrac{1}{2}$ 27. $x = -\dfrac{1}{2}$ or $x = \dfrac{1}{2}$ 29. $v = -\dfrac{5}{2}$

31. $h = \dfrac{4}{5}$ 33. $x = \dfrac{2}{3}$ 35. $x = -1$ or $x = 11$

37. sometimes true, when $r = 0$ or $r = 7$

39. If $2(r - 5) + 3(r - 5) = 5(r - 5) + 1$, then $5r - 25 = 5r - 24$. This implies that $-25 = -24$. Since this is impossible, the equation is never true.

Lesson 7.5 Skill C (pages 284–285)

Try This Example 1

$k = 0$, $k = \dfrac{1}{2}$, or $k = \dfrac{4}{5}$

Try This Example 2

$z = 0$ or $z = 7$

Exercises

1. $x = 1$, $x = 2$, or $x = 3$ 3. $n = 0$, $n = -\dfrac{1}{2}$, or $n = \dfrac{2}{3}$

5. $y = 0$ or $y = 3$ 7. $x = 0$ or $x = \dfrac{3}{4}$

9. $v = 0$ or $v = 2$ 11. $k = 0$ or $k = 4$

13. $y = 0$ or $y = -3$ 15. $s = 0$ or $s = \dfrac{5}{7}$

17. $t = -9$, $t = -2$, or $t = 0$ 19. $x = 0$ or $x = 1$

21. $c = 0$ or $c = 3$ 23. $m = -\dfrac{5}{2}$, $m = 0$, or $m = \dfrac{7}{2}$

25. $x = -2$, $x = 0$, or $x = 2$ 27. $n = 0$ or $n = \dfrac{4}{5}$

29. $d = -7$, $d = 0$, or $d = 7$ 31. $b = -4, -3, -1, 1, 3,$ or 4

33. $z = \pm 1, \pm 2, \pm 3, \pm 4$

35. 1. Associative Property of Multiplication
    2. Substitution Principle
    3. Zero-Product Property
    4. Zero-Product Property

37. Because $2(r - 3) - (r - 3) = r - 3$, $2(r - 3) - (r - 3) \geq r - 3$ is always true.

39. sometimes true; if $r = -1$, the inequality is false. If $r = 2$, the inequality is true.

Lesson 7.6 Skill A (pages 286–287)

Try This Example

a. vertex: $(-2, -1)$; axis of symmetry: $x = -2$

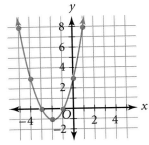

b. vertex: $(0, 2)$; axis of symmetry: $x = 0$

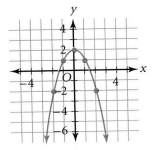

Exercises

1. vertex: $(-3, 2)$; axis of symmetry: $x = -3$

3. vertex: $(4, 0)$; axis of symmetry: $x = 4$

5. $a = -1, b = 2, c = 8$

7. $y = x^2 - 3x$; $a = 1, b = -3, c = 0$

9. $y = x^2 - 9$; $a = 1, b = 0, c = -9$

11. $y = x^2 + 4x + 4$; $a = 1, b = 4, c = 4$

13. upward   **15.** upward

17. a. vertex: $(0, -9)$; axis of symmetry: $x = 0$
  b. minimum
  c.

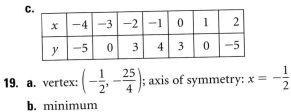

x	-4	-3	-2	-1	0	1	2
y	-5	0	3	4	3	0	-5

19. a. vertex: $\left(-\dfrac{1}{2}, -\dfrac{25}{4}\right)$; axis of symmetry: $x = -\dfrac{1}{2}$
  b. minimum
  c.

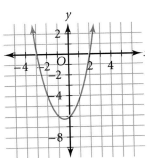

21. a. vertex: $(0, -1)$; axis of symmetry: $x = 0$
  b. minimum
  c.

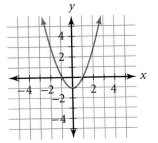

23. a. vertex: $(2, 1)$; axis of symmetry: $x = 2$
  b. maximum
  c.

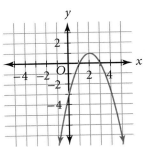

25. no x-intercept;
  y-intercept: -3.5

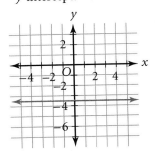

27. x-intercept: 6;
  y-intercept: 6

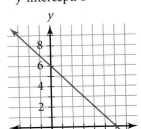

29. x-intercept: 4.5;
  y-intercept: 3

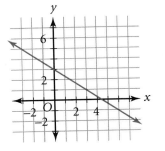

31. x-intercept: $\dfrac{7}{2}$;
  y-intercept: 1

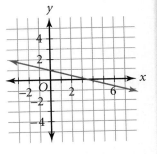

Lesson 7.6 Skill B (pages 288–289)

Try This Example 1

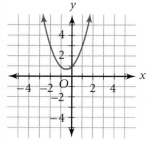

There are no x-intercepts

Try This Example 2

$5x^2 - 26x + 5 = (5x - 1)(x - 5)$; the equation has two real solutions. Therefore, there are two x-intercepts, 5 and $\frac{1}{5}$.

Exercises

1. two   **3.** one

5. two

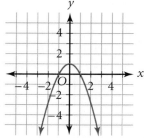

7. -1 and -6; two   **9.** -1 and -2; two

11. 3 and -3; two   **13.** $r > 0$ **15.** $r \neq 0$

17. $2v^3 - 3v^2 - 1$ **19.** $2t^3 + 3t^2 - 2t - 8$

21. $2a^2 + 2a - 40$ **23.** $-15d^2 - 13d - 2$

25. $49y^2 - 28y + 4$

Lesson 7.7 Skill A (pages 290–291)

Try This Example 1

$y \geq -\frac{25}{8}$

Try This Example 2

$y \leq \frac{45}{4}$

Exercises

1. $y \geq 2$ **3.** $y \geq -2$ **5.** $y \leq 11$ **7.** $y \geq -\frac{5}{4}$

9. $y \geq -\frac{25}{8}$ **11.** $y \geq -1$ **13.** $y \geq -13.5$ **15.** $y \leq 3$

17. $y \geq -1$ **19.** $y \leq 1$ **21.** $y \geq -\frac{961}{24}$ **23.** down

25. $0 \leq y \leq 4$ **27.** $-12n^7$ **29.** k^6 **31.** $\frac{1}{9}$ **33.** $\frac{1}{4g^2}$

Lesson 7.7 Skill B (pages 292–293)

Try This Example 1

$y = x^2 - 7x + 12$

Try This Example 2

$y = 2x^2 - 32$

Exercises

1. $y = a(x - 2)(x + 2)$ **3.** $y = a(x + 1)(x - 3)$

5. $y = a(x - 1)(x - 1)$ **7.** $y = -2x^2 + 12x + 32$

9. $y = -\frac{5}{9}x^2 + 5$ **11.** $y = 2x^2 + 14x - 16$

13. $y = x^2 - 2x - 3$ **15.** $y = x^2 - 6x + 8$

17. If r is the only x-intercept of the graph, then for some nonzero real number a, $y = a(x - r)(x - r) = a(x - r)^2$.

19. $a = 5$, or $a = -5$ **21.** $x = \frac{1}{2}$ or $x = \frac{3}{2}$

23. $c = -12$ or $c = \frac{11}{6}$ **25.** $n = 4$ or $n = -6$

27. $a = -\frac{65}{28}$ or $a = 0$

Lesson 7.7 Skill C (pages 294–295)

Try This Example 1

a. 64 feet
b. 2 seconds
c. 4 seconds

Try This Example 2

750 calendars; $22,500

Exercises

1. maximum   **3.** minimum   **5.** $x = 1$

7. The x-intercepts represent the times when the ball's height is 0.

9. 4000 units; $1,280,000   **11.** 350 units; $55,125

13. 400 units; $76,800   **15.** 450 units; $48,600

17. $(36)(2.5a)(2a)(4a)$ **19.** $(36)(1.5z)(1.5z)(2.5z)$

CHAPTER 8

Lesson 8.1 Skill A (pages 300–301)

Try This Example 1

a. ± 9 **b.** $\pm \dfrac{4}{5}$

Try This Example 2

-8 and -9; -8.77

Try This Example 3

a. ± 9 **b.** ± 7.94

Exercises

1. ± 11 **3.** ± 30 **5.** ± 12 **7.** ± 1 **9.** ± 0.5

11. $\pm \dfrac{11}{8}$ **13.** 8 and 9; 8.25   **15.** 5 and 6; 5.74

17. 1, 4, 9   **19.** 49, 64   **21.** 225, 256, 289, 324, 361, 400

23. $x = \pm 12$ **25.** $x = \pm 6.48$

27. Since $31^2 = 961$ and $32^2 = 1024$, there are no perfect squares between 961 and 1024.

29. If $x^2 - 5x + 4 = 0$, then $(x - 1)(x - 4) = 0$. By the Zero-Product Property, the solutions are 1 and 4. Since $1 \times 1 = 1$ and $2 \times 2 = 4$, the solutions are perfect squares.

31. $5^2 \cdot 29$ **33.** $2^4 \cdot 5^3$ **35.** 2401   **37.** 9

Lesson 8.1 Skill A (pages 302–303)

Try This Example 1

$6\sqrt{2}$

Try This Example 2

$\dfrac{\sqrt{10}}{7}$

Exercises

1. $12\sqrt{2}$ **3.** $7\sqrt{2}$ **5.** $\dfrac{\sqrt{7}}{8}$ **7.** $6\sqrt{3}$ **9.** $6\sqrt{7}$

11. $\dfrac{\sqrt{3}}{5}$ **13.** $\dfrac{\sqrt{2}}{7}$ **15.** $9\sqrt{5}$ **17.** $20\sqrt{3}$ **19.** $\dfrac{6\sqrt{2}}{7}$

21. $6\sqrt{13}$ **23.** $\dfrac{7\sqrt{7}}{5}$ **25.** $18\sqrt{11}$ **27.** $\dfrac{\sqrt{30}}{5}$ **29.** $\dfrac{1}{30}$

31. $\sqrt{\dfrac{a}{b}} \cdot \sqrt{\dfrac{a}{b}} = \sqrt{\dfrac{a}{b} \cdot \dfrac{a}{b}} = \sqrt{\dfrac{a^2}{b^2}} = \dfrac{\sqrt{a^2}}{\sqrt{b^2}} = \dfrac{a}{b}$

33. $-5s^6$ **35.** h^6 **37.** $-8a^6 b^{12}$ **39.** $a^4 z^6$ **41.** $\dfrac{2z^3}{3}$

43. $\dfrac{1}{125b^3}$

Lesson 8.1 Skill A (pages 304–305)

Try This Example

$$\sqrt{9a} = (9a)^{\frac{1}{2}} = (9)^{\frac{1}{2}} a^{\frac{1}{2}}$$
$$= (3^2)^{\frac{1}{2}} a^{\frac{1}{2}}$$
$$= 3^{\left(2 \times \frac{1}{2}\right)} a^{\frac{1}{2}} = 3a^{\frac{1}{2}} = 3\sqrt{a}$$

Exercises

1. 7   **3.** 12   **5.** definition of the exponent $\dfrac{1}{2}$

7. Multiplicative Inverse   **9.** definition of the exponent $\dfrac{1}{2}$

11. Commutative Property of Multiplication

13. definition of the exponent $\dfrac{1}{2}$

15. $\sqrt{\dfrac{a}{b}} = \sqrt{a \cdot \dfrac{1}{b}} = \sqrt{a} \cdot \sqrt{\dfrac{1}{b}} = \sqrt{a} \cdot \dfrac{1}{\sqrt{b}} = \dfrac{\sqrt{a}}{\sqrt{b}}$

17. $\dfrac{\sqrt{a^4 b^4}}{\sqrt{a^2 b^2}} = \sqrt{\dfrac{a^4 b^4}{a^2 b^2}} = \sqrt{a^2 b^2} = \sqrt{a^2}\sqrt{b^2} = a \cdot b = ab$

19. $\left(\dfrac{1}{2}, 4\right)$ **21.** $\left(3, \dfrac{9}{11}\right)$ **23.** $(-128, 147)$

Lesson 8.2 Skill A (pages 306–307)

Try This Example 1

$d = \pm 10$

Try This Example 2

$t = \pm 2\sqrt{6}$

Try This Example 3

$z = \pm 3\sqrt{3}$

Exercises

1. $t^2 = 4$ **3.** $z^2 = 4$ **5.** $x = \pm 1$ **7.** $n = \pm 4$

9. $s = \pm 2$ **11.** $z = \pm 10$ **13.** $d = \pm \sqrt{3}$

15. $a = \pm \sqrt{7}$ **17.** $t = \pm 2\sqrt{3}$ **19.** $v = \pm 2\sqrt{3}$

21. $u = \pm \sqrt{2}$ **23.** $u = \pm 2\sqrt{2}$ **25.** $n = \pm 4\sqrt{2}$

27. $r = \pm 2\sqrt{2}$ **29.** $9 < a < 16$ **31.** $-48 < c < -27$

33. $m = \dfrac{n(y - a)}{x}$ **35.** $z = \dfrac{-8}{y - x}$ **37.** $x = \dfrac{3a + 2b}{b - a}$

Lesson 8.2 Skill B (pages 308–309)

Try This Example 1
8.46 meters

Try This Example 2
27.1 feet

Exercises

1. $\pi r^2 = 1200$; r represents the radius of the circle.

3. 12.6 feet   5. 10.0 inches   7. 15.9 units

9. 433.0 inches

11. **a.** $AD = \sqrt{7.16^2 - 6.32^2}$
    b. $AD \approx 3.36$

13. $32.8x$

Lesson 8.3 Skill C (pages 310–311)

Try This Example 1
a. $7i$ **b.** $2i\sqrt{2}$

Try This Example 2
$a = \pm 5i$

Try This Example 3
$z = \pm 5i\sqrt{5}$

Exercises

1. real   3. pure imaginary   5. $4i$ 7. $10i$

9. $4i\sqrt{2}$ 11. $10i\sqrt{2}$ 13. $x = \pm 3i$ 15. $n = \pm 7i$

17. $m = \pm 2i$ 19. $p = \pm 4i$ 21. $m = \pm 2i\sqrt{3}$

23. $d = \pm 3i\sqrt{5}$ 25. $y = \pm 6i\sqrt{2}$ 27. $h = \pm 12i\sqrt{6}$

29. $x = \pm i\sqrt{2}$ 31. $z = \pm i\sqrt{5}$ 33. $x = \pm i\sqrt{3}$

35. $x = \pm 2i\sqrt{3}$

37. $i^3 = -i$, $i^4 = 1$, $i^5 = i$, and $i^6 = -1$.

$$i^n = \begin{cases} 1 \text{ if } n = 4k \\ i \text{ if } n = 4k + 1 \\ -1 \text{ if } n = 4k + 2 \\ -i \text{ if } n = 4k + 3 \end{cases}$$

where k is an integer

39. $5\sqrt{6}$ 41. $6\sqrt{10}$ 43. $\dfrac{20}{7}$ 45. $\dfrac{6\sqrt{7}}{5}$

Lesson 8.3 Skill A (pages 312–313)

Try This Example 1
$z = -13$ or $z = 3$

Try This Example 2
$n = -9 - 5\sqrt{3}$ or $n = -9 + 5\sqrt{3}$

Exercises

1. Take the square root of each side of the equation. Then add 4 to each side of the equation.

3. Multiply each side of the equation by 3. Take the square root of each side of the equation. Then add 3 to each side of the equation.

5. $x = -5$ or $x = 1$ 7. $x = -5$ or $x = 15$

9. $x = -7$ or $x = 11$ 11. $x = -10$ or $x = -4$

13. $m = -5 \pm 5\sqrt{3}$ 15. $t = -4 \pm 3\sqrt{3}$

17. $n = -2 \pm 2\sqrt{5}$ 19. $p = -10 \pm 2\sqrt{7}$

21. $n = -2$ or $n = \dfrac{8}{3}$ 23. $c = 1$ or $c = \dfrac{3}{5}$

25. $x = -8$ or $x = 4$ 27. $z = -\dfrac{7}{3}$ or $z = 1$

29. $k = -\dfrac{3}{5}$ or $k = \dfrac{1}{5}$

31. **a.** Division Property of Equality
    b. Definition of square root
    c. Addition Property of Equality

33. $n > 5$ 35. $243a^{10}b^{18}$

Lesson 8.3 Skill C (pages 314–315)

Try This Example 1
$z = -3$ or $z = 5$

Try This Example 2
$y = -1 \pm \dfrac{\sqrt{5}}{3}$

Exercises

1. 1   3. 36

5. $(s + 1)^2 = 4$
$s + 1 = \pm 2$
$s = -1 \pm 2$
$s = -3$ or $s = 1$

7. $(n - 6)^2 = 49$
$n - 6 = \pm 7$
$n = 6 \pm 7$
$n = -1$ or $n = 13$

9. $m = -5$ or $m = 1$ 11. $q = -1$ or $q = 7$

13. $h = -9$ or $h = 1$ 15. $r = 1$ and $r = 17$

17. $f = -1$ or $f = 11$ 19. $m = -2$ or $m = 18$

21. $z = -1 \pm \dfrac{2\sqrt{5}}{3}$ 23. $t = 3 \pm \sqrt{2}$

25. $m = -\dfrac{3}{2} \pm \sqrt{5}$ 27. $q = 2 \pm \dfrac{\sqrt{35}}{3}$ 29. $m < -9$

31. $y = x$ or $y = -x - 2$ 33. $\dfrac{3917}{2}$ 35. 1936   37. -92

Lesson 8.3 Skill C (pages 316–317)

Try This Example

Boat A: about 6.6 miles per hour

Boat B: about 7.6 miles per hour

Exercises

1.

	speed (miles per hour)	distance (miles)
Boat A	r	$2r$
Boat B	$r + 5$	$2(r + 5)$

3. Boat A: ≈ 13.8 miles per hour
Boat B: ≈ 20.8 miles per hour

5. Boat A: ≈ 11.4 miles per hour
Boat B: ≈ 18.4 miles per hour

7. Boat A: 8.4 miles per hour
Boat B: 5.4 miles per hour

9. $\begin{cases} x \geq -2 \\ y \geq \frac{1}{2}x + 3 \\ y \leq -\frac{5}{6}x + 7 \end{cases}$

Lesson 8.4 Skill A (pages 318–319)

Try This Example 1

$w = -1$ or $w = 4$

Try This Example 2

$a = \frac{3}{2} \pm \frac{\sqrt{29}}{2}$

Exercises

1. $x^2 - 5x - 6 = 0$;
$a = 1, b = -5, c = -6$

3. $2x^2 + 3x - 20 = 0$;
$a = 2, b = 3, c = -20$

5. $x = -6$ or $x = -5$ **7.** $w = -7$ or $w = -8$

9. $z = 1$ or $z = 7$ **11.** $x = -3$ or $x = 11$

13. $k = \frac{5}{2}$ or $k = 3$ **15.** $x = \frac{1}{2}$ or $x = \frac{5}{3}$

17. $y = 3 \pm \sqrt{7}$ **19.** $p = 4 \pm \sqrt{5}$ **21.** $x = 2 \pm \sqrt{2}$

23. $n = 1 \pm \frac{\sqrt{5}}{2}$ **25.** $y = \frac{4}{3} \pm \frac{\sqrt{22}}{3}$

27. $x = 0$ or $x = -\frac{b}{a}$ **29.** $b = \pm 12$

31. $n = -\frac{\sqrt{15}}{3}, \frac{\sqrt{15}}{3}, -\frac{\sqrt{2}}{2}, \text{ or } \frac{\sqrt{2}}{2}$

33. vertex: $(-5, 0)$;
axis of symmetry:
$x = -5$; minimum

35. vertex: $(0, 4)$;
axis of symmetry:
$x = 0$; maximum

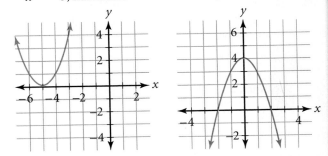

37. vertex: $(2, -4)$; axis of symmetry: $x = 2$; minimum

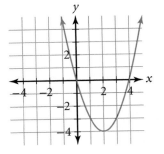

Lesson 8.4 Skill A (pages 320–321)

Try This Example 1

Try This Example 2

one

Exercises

1. none **3.** one **5.** two **7.** none **9.** one

11. two **13.** none **15.** one **17.** one **19.** none

21. two **23.** none **25.** two **27.** two

29. Answers may vary. Sample answer: $x^2 + r = 0$, where r is any positive real number

31. Answers may vary. Sample answer: $x^2 + r = 0$, where r is any negative real number

33. $c > \frac{49}{20}$

35. If $\frac{1}{2}(a + 3) = 5$, then by the Multiplication Property of Equality, $a + 3 = 10$. If $a + 3 = 10$, then by the Subtraction Property of Equality, $a = 7$. Therefore, if $\frac{1}{2}(a + 3) = 5$, then $a = 7$.

37. If $-4t + 5 + 5t = 9$, then by the Commutative Property of Addition, $-4t + 5t + 5 = 9$. If $-4t + 5t + 5 = 9$, then by the Distributive Property, $t + 5 = 9$. If $t + 5 = 9$, then by the Subtraction Property of Equality, $t = 4$.

39. If $4m + 5m = m + 2$, then by the Distributive Property, $9m = m + 2$. If $9m = m + 2$, then by the Subtraction Property of Equality, $8m = 2$. If $8m = 2$, then by the Division Property of Equality, $m = \frac{1}{4}$.

Therefore, if $4m + 5m = m + 2$, then $m = \frac{1}{4}$.

Lesson 8.4 Skill C (pages 322–323)

1. Subtraction Property of Equality

3. Subtraction Property of Equality

5. Addition Property of Equality

7. Take the square root of each side of the equation.

9. Multiplicative Identity

11. Quotient Property of Square Roots

13. Subtraction Property of Equality

15. Distributive Property (factoring)

17. Take the square root of each side of the equation.

19. $x = \pm 5$ **21.** $z = \pm 2i\sqrt{6}$

23. $v = -1 \pm \frac{3\sqrt{5}}{2}$ **25.** 31.0 feet

Lesson 8.5 Skill A (pages 324–325)

Try This Example 1

Answers may vary. Sample answer: The equation $-3x^2 = -48$ has exactly the same form as $7x^2 = 175$. The steps in the solution would be the same but the numbers involved would be different.

Try This Example 2

Answers may vary. Sample answer: The Zero-Product Property requires that 0 be one of the sides of the given equation. That is not the case for $(2x + 1)(3x - 7) = 3$.

Try This Example 3

Answers may vary. Sample answer: Application of the quadratic formula would involve more steps than the solution given in Example 1. For example, it is not necessary to subtract 175 from each side of the equation in order to put it in standard form.

484   Selected Answers

Exercises

In Exercises 1–22, answers may vary.

1. factoring and the Zero-Product Property

3. Yes; before taking square roots, divide each side of the equation by 2.

5. Yes; the given equation is ready for the application of the Zero-Product Property.

7. quadratic formula because the coefficients involve decimals.

9. taking a square root after applying the Addition Property of Equality followed by the Division Property of Equality

11. completing the square because $n^2 + 2n + 1 = 4$ is equivalent to $(n + 1)^2 = 4$

13. Zero-Product Property because the given equation is given in the form $ab = 0$

15. quadratic formula because the coefficients involve decimals

17. Observe that the Zero-Product Property will give two distinct solutions. Two distinct solutions indicate two distinct x-intercepts.

19. Graph $y = x^2 - 2$. Observe that the graph crosses the x-axis in two distinct points.

21. Calculate the discriminant and observe that it is a negative number. A negative discriminant indicates that there are no x-intercepts.

23. Write the discriminant $n^2 - 4(n + 1)$, or $n^2 - 4n - 4$. Solve $n^2 - 4n - 4 = 0$. You will get $2 \pm \sqrt{2}$ as solutions. The graph of $y = x^2 + nx + n + 1$ will have one x-intercept when $n = 2 - 2\sqrt{2}$ or when $n = 2 + 2\sqrt{2}$.

25. $y = -3x + 7$; -173

Lesson 8.5 Skill B (pages 326–327)

Try This Example 1

1 $x^2 + 10x + 21 = 0$
given

2 $x^2 + 10x + 21 = (x + 7)(x + 3)$
Distributive Property (factoring)

3 $(x + 7)(x + 3) = 0$
Substitution Principle with **1** and **2**

4 $x + 7 = 0$ or $x + 3 = 0$
Zero-Product Property

5 $x = -7$ or $x = -3$
Addition Property of Equality
Therefore, if $x^2 + 10x + 21 = 0$, then $x = -7$ or $x = -3$.

Try This Example 2

1 $x^2 + 8x = 20$
given

2 $x^2 + 8x + 16 = 20 + 16$
Addition Property of Equality

3 $(x + 4)^2 = 36$
Distributive Property (factoring)

4 $x + 4 = \pm 6$
Take the square root of each side.

5 $x = -10$ or $x = 2$
Subtraction Property of Equality
Therefore, if $x^2 + 8x = 20$, then $x = -10$ or $x = 2$.

Exercises

1. a. Distributive Property (factoring)
   b. Zero-Product Property
   c. Addition Property of Equality

3. $x^2 - 6x + 5 = 32$
given
$x^2 - 6x + 5 + 4 = 32 + 4$
Addition Property of Equality
$x^2 - 6x + 9 = 36$
addition of numbers
$(x - 3)^2 = 36$
Distributive Property (factoring)
$x - 3 = \pm 6$
Take the square root of each side.
$x = 9$ or $x = -3$
Addition Property of Equality

**In Exercises 5–10, arguments may vary.
Sample arguments are given.**

5. $n^2 - 5n = 0$
given
$n(n - 5) = 0$
Distributive Property (factoring)
$n = 0$ or $n - 5 = 0$
Zero-Product Property
$n = 0$ or $n = 5$
Addition Property of Equality

7. $8b^2 + 2b - 3 = 0$
given

$(4b + 3)(2b - 1) = 0$
Distributive Property (factoring)
$4b + 3 = 0$ or $2b - 1 = 0$
Zero Product Property
$4b = -3$ or $2b = 1$
Subtraction and Addition Properties of Equality
$b = -\dfrac{3}{4}$ or $b = \dfrac{1}{2}$

Division Property of Equality

9. $4r^2 - 9^2 = 0$
given
$(2r - 9)(2r + 9) = 0$
Distributive Property (factoring)
$2r - 9 = 0$ or $2r + 9 = 0$
Zero-Product Property
$2r = 9$ or $2r = -9$
Addition and Subtraction Properties of Equality
$r = \pm\dfrac{9}{2}$

Division Property of Equality

11. By definition, the x-intercepts of the graph of
$y = x^2 - 7x + 12$ are the real solutions to
$x^2 - 7x + 12 = 0$. By the Distributive Property, the
x-intercepts are the real solutions to $(x - 4)(x - 3) = 0$.
By the Zero-Product Property, the solutions are 4
and 3. Since 4 and 3 are positive, the x-intercepts
are positive.

13. $\left(\dfrac{7}{2}, \dfrac{19}{4}\right)$ **15.** $\left(\dfrac{3}{7}, \dfrac{5}{14}\right)$ **17.** $(-5, 10)$

Lesson 8.6 Skill A (pages 328–329)

Try This Example 1

Yes, the pattern is quadratic.

4	5	6
-14	-23	-36

Exercises

1. 5, 5, 5, 5, 5, 5; yes   **3.** quadratic   **5.** quadratic

7. linear   **9.** quadratic; $(6, 36), (7, 49), (8, 64)$

11. quadratic; $(3, -9), (4, -20), (5, -35)$

13. a. $y = an^2 + bn + c$
   b. $a(n + 1)^2 + b(n + 1) + c$
   $= an^2 + 2an + a + bn + b + c$
   c. $an^2 + 2an + a + bn + b + c - (an^2 + bn + c)$
   $= an^2 + 2an + a + bn + b + c - an^2 - bn - c$
   $= 2an + a + b = (2a)n + (a + b)$

Lesson 8.6 Skill B (pages 330–331)

Try This Example 1

$y = n^2 + 1; 2501$

Exercises

1.

n	1	2	3	4
y	3	5	11	21

3. $y = (n + 1)^2 + 1; 3722$ 5. $y = 2(n - 2)^2 + 3; 6731$

7. $a = -4$ or $a = -7$ 9. $z = -\dfrac{1}{7}$ or $z = \dfrac{4}{3}$

11. $y = -\dfrac{1}{2}$ 13. $r = -\dfrac{7}{3}$ or $r = \dfrac{4}{3}$ 15. $c = -\dfrac{7}{4}$

17. $m = 10$ or $m = -\dfrac{11}{3}$

Lesson 8.6 Skill C (pages 332–333)

Try This Example 1

$y = 2x^2 + 3x + 1$

Try This Example 2

$y = 2x^2 - 3$

Exercises

1. Because the second differences are constant (-10), the table represents a quadratic pattern.

3. Because the second differences are constant (4), the table represents a quadratic pattern.

5. $y = 5x^2$ 7. $y = -2x^2 + 2$ 9. $y = x^2 - 2x + 4$

11. $y = -2x^2 + x - 6$

13. $y = -\dfrac{1}{2}x^2 - 50, 60.5, -72, -84.5, -98, -112.5, -128$

15. $v = \pm\dfrac{2}{9}$ 17. $u = \pm 2\sqrt{6}$ 19. $m = \pm 2\sqrt{5}$

21. $p = \pm 4\sqrt{14}$

Lesson 8.7 Skill A (pages 334–335)

Try This Example 1

2 seconds: 64 feet and 5336 feet
4 seconds: 256 feet and 5144 feet
6 seconds: 576 feet and 4824 feet

Try This Example 2

11.1 seconds

Exercises

1. $t = \dfrac{5\sqrt{47}}{4}$ 3. $t = \sqrt{165}$ 5. 256 feet and 5344 feet

7. 400 feet and 4400 feet   9. 484 feet and 5816 feet

11. 9.0 seconds   13. 16.0 seconds   15. 21.7 seconds

17. $\dfrac{\sqrt{2h}}{8}$ seconds   19. 84.9 feet   21. 19.2 yards

23. 259.8 meters

Lesson 8.7 Skill B (pages 336–337)

Try This Example 1

after 2 seconds: 242 feet; after 3 seconds: 290 feet

Try This Example 2

after 1.8 seconds and after 4.2 seconds

Exercises

1. $-16t^2 + 96t - 120 = 0$

$t = \dfrac{-96 \pm \sqrt{96^2 - 4(-16)(-120)}}{2(-16)}$

3. $-16t^2 + 96t - 180 = 0$

$t = \dfrac{-96 \pm \sqrt{96^2 - 4(-16)(-180)}}{2(-16)}$

5. after 3 seconds: 194 feet
after 5 seconds: 130 feet

7. after 3 seconds: 336 feet
after 5 seconds: 336 feet

9. after 3 seconds: 160 feet
after 5 seconds: 0 feet

11. 1.5 seconds and 4.5 seconds

13. 0 seconds and 6 seconds   15. 6.6 seconds

17. a. 3 seconds   b. 192 feet

19. 1.25×10^6 21. 8.1×10^6

23. 1.5×10^{-1} 25. 5.0×10^{-3}

CHAPTER 9

Lesson 9.1 Skill A (pages 342–343)

Try This Example 1

all real numbers except $-\dfrac{4}{5}$

Try This Example 2

a. all real numbers except $-\frac{1}{2}$ and $\frac{1}{3}$ **b.** all real numbers

Exercises

1. $y - 2 = 0$ **3.** $6n^2 - 11n - 35 = 0$

5. all real numbers except -3

7. all real numbers except 0

9. all real numbers except $\frac{3}{2}$ and $-\frac{2}{3}$

11. all real numbers except 1 and 4

13. all real numbers except $-\frac{5}{4}$ and $-\frac{5}{2}$

15. all real numbers except $-\frac{1}{4}$ and $\frac{1}{4}$

17. all real numbers

19. all real numbers except $-\frac{5}{7}$ and $-\frac{1}{2}$

21. all real numbers except $\frac{7}{3}$ and $\frac{7}{2}$

23. a. $\frac{10}{3}$ units   **b.** $\frac{2b}{a}$ units

25. $(c + 7)(c + 9)$ **27.** $(a - 1)(a + 13)$

29. $(p - 25)(p + 4)$

Lesson 9.2 Skill A (pages 344–345)

Try This Example 1
$\frac{1}{3r - 7}$; $r \neq \frac{7}{3}$

Try This Example 2
a. $\frac{3c + 1}{c + 3}$; $c \neq -3$

b. $\frac{4n - 4}{n + 3}$; $n \neq -3$

Try This Example 3
a. -3; $m \neq 12$ **b.** $4n + 1$; $n \neq 0$

Exercises

1. $\frac{x + 4}{x - 4}$; $x \neq 4$ **3.** $\frac{2}{x - 1}$; $x \neq 1$ **5.** $\frac{-4}{b + 5}$; $b \neq -5$

7. 3; $h \neq 5$ **9.** $\frac{2a + 4}{a + 3}$; $a \neq -3$ **11.** $\frac{c - 3}{c + 2}$; $c \neq -2$

13. $-\frac{7}{3}$; $v \neq 3$ **15.** $\frac{1}{3}$; $z \neq -8$ **17.** $a = -15$ **19.** $x - 3$

21. a. $a = c$ and $b = d$
b. $a = kc$ and $b = kd$ for some nonzero real number k

23. $4(a + 1)^2$ **25.** $(6 - x)(x + 5)$ **27.** $(x - 3)^2$

Lesson 9.2 Skill B (pages 346–347)

Try This Example 1
$\frac{t + 3}{t + 2}$; $t \neq 3, -2$

Try This Example 2
$\frac{p + 1}{p + 7}$; $p \neq 0, -1, -7$

Exercises

1. $\frac{(z + 1)(z + 1)}{(z + 1)(z - 1)}$ **3.** $\frac{(n + 5)(n + 3)}{(n + 1)(n + 5)}$

5. $\frac{1}{z - 4}$; $z \neq 4, -4$ **7.** $3u + 7$; $u \neq \frac{7}{3}$

9. $\frac{c + 10}{c - 10}$; $c \neq 10, -10$ **11.** $\frac{m + 6}{m - 8}$; $m \neq 8, 3$

13. s; $s \neq 1, -1$ **15.** $\frac{k + 1}{k^2}$; $k \neq 0, 1$

17. $\frac{y^2 - 2y + 1}{y^2 - 2y + 3}$; $y \neq 0$ **19.** $\frac{h - 7}{h + 1}$; $h \neq 0, -1, -5$

21. $\frac{a - 16}{a + 1}$; $a \neq 0, -1, 4$ **23.** $\frac{a^2 b^2}{4}$ **25.** 2

27. xy^2 **29.** $4m^2 n^3$

Lesson 9.2 Skill C (pages 348–349)

Try This Example 1
$\frac{1}{b + 2}$

Try This Example 2
$\frac{c + d}{c - d}$

Try This Example 3
$-\frac{1}{5}$

Exercises

1. $a^2 b$ **3.** $4q^2 - 81$ **5.** $\frac{1}{s + c}$ **7.** $\frac{wz}{wz + 1}$ **9.** 1

11. $\frac{n - 1}{b - 1}$ **13.** p **15.** $\frac{1 - r}{s - 1}$ **17.** $\frac{1}{2}$ **19.** $\frac{3y + 5z}{3y - 5z}$

21. $\frac{2a - 5c}{a + 3c}$ **23.** $a \neq 0, -3b$ and $b \neq 0$ **25.** $\frac{21}{5}$

27. $\frac{1}{8}$ **29.** $\frac{8}{27}$ **31.** $\frac{7}{32}$

Lesson 9.3 Skill A (pages 350–351)

Try This Example 1
$\frac{7z}{18a}$

Try This Example 2
$\frac{-12}{y^2 - 12y + 35}$

Exercises

1. $\dfrac{2a^3}{3b^2}$ **3.** $\dfrac{5m^2}{3}$ **5.** $\dfrac{2}{3x}$ **7.** $\dfrac{11n}{3m}$ **9.** $\dfrac{3k}{2a}$ **11.** $\dfrac{1}{10y}$

13. $\dfrac{6}{x-3}$ **15.** $\dfrac{1}{t^2+10t+25}$ **17.** $\dfrac{x^2-1}{x^2-25}$

19. $\dfrac{m-1}{m+8}$ **21.** $\dfrac{s^2+7s+12}{s^2+4s+4}$ **23.** $\dfrac{v+6}{v}$ **25.** $\dfrac{14}{9}$

27. $\dfrac{1}{12}$ **29.** $-2x+3$ **31.** $7t+6$

Lesson 9.3 Skill B (pages 352–353)

Try This Example 1

$\dfrac{9}{4st^2}$

Try This Example 2

$\dfrac{m^2-5m-6}{m^2+5m-6}$

Exercises

1. $\dfrac{3x+1}{2x+3}$ **3.** $\dfrac{x^2-5x+6}{x^2+7x+6}$ **5.** $\dfrac{4}{3}$ **7.** $\dfrac{3}{x}$ **9.** $\dfrac{10b}{3a}$

11. $\dfrac{14}{9p^4y}$ **13.** $\dfrac{10}{3}$ **15.** $\dfrac{10}{3}$ **17.** $2k$ **19.** $\dfrac{45d}{2}$ **21.** $\dfrac{1}{c}$

23. $\dfrac{10g}{3}$ **25.** r **27.** $\dfrac{x^2-25}{x^2-8x+7}$ **29.** 1 **31.** $\dfrac{n-1}{n+1}$

33. $7a+\dfrac{9}{4}$ **35.** $6t-25$ **37.** $\dfrac{5a}{6}+\dfrac{5}{3}$

Lesson 9.3 Skill C (pages 354–355)

Try This Example 1

$\dfrac{1}{18r^4s^5}$

Try This Example 2

$\dfrac{1}{x^2-16}$

Exercises

1. $\dfrac{2m^2n}{5mn^2}\cdot\dfrac{15mn^2}{12m^2n^3}\cdot\dfrac{2mn}{5m^3n}$

3. $\dfrac{x^2-5x+6}{x^2-9}\cdot\dfrac{x^2-6x+9}{x^2-4x+4}\cdot\dfrac{1}{x-2}$

5. a^2 **7.** $\dfrac{2}{3w}$ **9.** $\dfrac{1}{x^2y^4}$ **11.** $\dfrac{20d^2c^2}{49}$ **13.** $\dfrac{1}{6xy}$ **15.** $\dfrac{1}{a}$

17. $\dfrac{1}{p^4q^6}$ **19.** $\dfrac{1}{x-5}$ **21.** 1 **23.** $\dfrac{a+5}{a-5}$ **25.** $\dfrac{(x+y)^2}{(m+n)^2}$

27. $\dfrac{31}{35}$ **29.** $\dfrac{47}{98}$ **31.** $\dfrac{37}{24}$ **33.** $\dfrac{31}{20}$

Lesson 9.4 Skill A (pages 356–357)

Try This Example 1

a. $-\dfrac{1}{3n}$ **b.** $-n+5$

Try This Example 2

$a-9$

Exercises

1. $5n-6$ **3.** $6a^2$ **5.** $\dfrac{1}{3n}$ **7.** $-\dfrac{1}{3k}$ **9.** $\dfrac{2}{c^2}$

11. $-2n+3$ **13.** 2 **15.** $\dfrac{-4}{u}$ **17.** 9 **19.** 3

21. 1 **23.** $\dfrac{1}{2y+6}$ **25.** $\dfrac{1}{y-3}$ **27.** $x+6$

29. $3a+4$ **31.** $m=\dfrac{1}{2}$ and $n=\dfrac{3}{2}$ **33.** $2a^2+a-1$

35. $-4b^3+b^2-6b$ **37.** $n^4-n^3+n^2+n-1$

Lesson 9.4 Skill B (pages 358–359)

Try This Example 1

a. $\dfrac{2b-5}{b^3}$

b. $\dfrac{4n^2-2n+9}{6n}$

Try This Example 2

$\dfrac{10p^2+14p+3}{6p^2}$

Exercises

1. $20z^2$ **3.** $30n^2$ **5.** $\dfrac{13}{8z}$ **7.** $\dfrac{-23}{12b}$ **9.** $\dfrac{79}{42w^2}$

11. $\dfrac{12t+5}{5t}$ **13.** $\dfrac{2n+7}{6n}$ **15.** $\dfrac{5b^2+3b+3}{15b}$

17. $\dfrac{5m^2+8m-3}{15m}$ **19.** $\dfrac{3v^2-3v+2}{3v^2}$

21. $\dfrac{10r^2-9r+2}{5r^2}$ **23.** $\dfrac{2a^3+3a^2+3a+2}{a^3}$

25. $\dfrac{11z^2-z-3}{z^2}$ **27.** $\dfrac{u^3+2u+3}{u^3}$ **29.** $\dfrac{19s+1}{9s^3}$

31. $\dfrac{x^{n-1}+x^{n-2}+\cdots+x^1+x^0}{x^n}$ **33.** $s\le 1$

35. $x\ge 0$ **37.** $y<-10$

Lesson 9.4 Skill C (pages 360–361)

Try This Example 1

$\dfrac{6b-4}{(b-1)(b+1)},\text{ or }\dfrac{6b-4}{b^2-1}$

Try This Example 2

$\dfrac{8}{(2c+1)(2c-1)},\text{ or }\dfrac{8}{4c^2-1}$

Exercises

1. $z^2+3z=z(z+3)$; LCD $=z(z+3)$

3. $k^2-1=(k+1)(k-1)$; $k^2-k=k(k-1)$;
LCD $=k(k+1)(k-1)$

5. $\dfrac{3z-2}{z(z+1)}$, or $\dfrac{3z-2}{z^2-z}$ **7.** $\dfrac{14y-15}{y(3y-5)}$, or $\dfrac{14y-15}{3y^2-5y}$

9. $\dfrac{5r^2+5r-25}{r^2(r-5)}$, or $\dfrac{5r^2+5r-25}{r^3-5r^2}$

11. $\dfrac{5d+7}{(d+1)(d+2)}$, or $\dfrac{5d+7}{d^2+3d+2}$

13. $\dfrac{7q+29}{(q+2)(q+5)}$, or $\dfrac{7q+29}{q^2+7q+10}$

15. $\dfrac{4t^2+10t+1}{(t+3)(t+2)}$, or $\dfrac{4t^2+10t+1}{t^2+5t+6}$

17. $\dfrac{4u^2+2u}{(u-1)(u+1)}$, or $\dfrac{4u^2+2u}{u^2-1}$

19. $\dfrac{g^2+5g}{(g-2)(g+2)}$, or $\dfrac{g^2+5g}{g^2-4}$

21. $\dfrac{6z+8}{(z+2)(z-2)}$, or $\dfrac{6z+8}{z^2-4}$

23. $\dfrac{5b-2}{(b-2)(b+2)(b+2)}$, or $\dfrac{5b-2}{b^3+2b^2-4b-8}$

25. $\dfrac{a^2+3a+3}{(a+1)^3}$, or $\dfrac{a^2+3a+3}{a^3+3a^2+3a+1}$

27. $\dfrac{5m^2+9m+3}{m(m+1)^2}$, or $\dfrac{5m^2+9m+3}{m^3+2m^2+m}$

29. $(0,0)$ **31.** $\left(-\dfrac{1}{3},\dfrac{5}{2}\right)$ **33.** $(3,-2)$

Lesson 9.5 Skill A (pages 362–363)

Try This Example 1   **Try This Example 2**

a. $\dfrac{5}{n}$ **b.** 2 3

Try This Example 3

$3t-7$

Exercises

1. $\dfrac{3t-1}{t}+\dfrac{-2t+2}{t}$ **3.** $\dfrac{3v^2-3}{v-7}+\dfrac{-2v^2-3}{v-7}$

5. $\dfrac{1}{v}$ **7.** $\dfrac{1}{7}$ **9.** 3 **11.** 5 **13.** 1 **15.** $\dfrac{4s-1}{3s-2}$

17. $\dfrac{12p^2-294}{2p+7}$ **19.** $\dfrac{4}{2z+1}$ **21.** $\dfrac{2n+1}{2n-3}$

23. $\dfrac{1}{2}$ **25.** $\dfrac{4}{35}$ **27.** $-\dfrac{5}{18}$ **29.** $-\dfrac{2}{11}$

Lesson 9.5 Skill B (pages 364–365)

Try This Example 1   **Try This Example 2**

a. $\dfrac{6-3m}{m^2}$ **b.** $\dfrac{-3p+7}{(p-1)}$ $\dfrac{5y-2}{5y-5}$

Try This Example 3

$$\dfrac{-4}{a^2+8a+16}$$

Exercises

1. like   **3.** unlike; $(x-3)(x+3)$, or x^2-9

5. $-\dfrac{1}{3x}$ **7.** $\dfrac{14}{15z}$ **9.** $\dfrac{-1}{x(x-1)}$ **11.** $\dfrac{6z^2-5z+10}{2z(z-2)}$

13. $\dfrac{2(x-4)}{5(x-3)}$ **15.** $\dfrac{5z-17}{5(z+2)}$ **17.** $\dfrac{7u^2+23u-6}{6(u^2-9)}$

19. $\dfrac{-(3x+1)}{x(x+1)}$ **21.** $\dfrac{-4a}{a^2-1}$ **23.** $\dfrac{1}{b+1}$

25. $\dfrac{-4y}{(y-1)(y+1)^2}$ **27.** $\dfrac{-2g^2+g+2}{(g+2)^2(g-2)}$

29. $\dfrac{a^2+b^2}{a^2-b^2}$ **31.** $-\dfrac{2a}{3b^2}$ **33.** -1 **35.** $-\dfrac{3yz^2}{7}$ **37.** $\dfrac{5d}{c}$

Lesson 9.5 Skill C (pages 366–367)

Try This Example

$A=\dfrac{x^2}{3}$

Exercises

1. large square: x^2; small square: $\dfrac{x^2}{4}$

3. $\dfrac{13x^2}{9}$ **5.** $\dfrac{3x^2(n^2-1)}{8n^2}$ **7.** $x\neq-3,-7$

9. $\dfrac{2u+3}{u-3}$ **11.** $\dfrac{1}{(x+2)(x+1)}$, or $\dfrac{1}{x^2+3x+2}$

13. $\dfrac{2}{v^5}$ **15.** $\dfrac{z+10}{6z^2}$ **17.** $k+1$ **19.** $\dfrac{5a^2}{12}$

Lesson 9.6 Skill A (pages 368–369)

Try This Example 1   **Try This Example 2**

a. $\dfrac{8}{3}$ **b.** $\dfrac{5}{9}$ **c.** $15k$ $\dfrac{a^2b^2}{b^2-a^2}$

Exercises

1. $\dfrac{2}{15}\cdot\dfrac{35}{12}=\dfrac{2}{3\cdot5}\cdot\dfrac{7\cdot5}{2\cdot2\cdot3}$

3. $\dfrac{2(a-1)}{12}\cdot\dfrac{4}{3a-3}=\dfrac{2(a-1)}{2\cdot2\cdot3}\cdot\dfrac{2\cdot2}{3(a-1)}$

5. $\dfrac{5}{3}$ **7.** $\dfrac{12n}{5}$ **9.** $\dfrac{23}{7}$ **11.** $\dfrac{10}{3}$ **13.** $\dfrac{3}{7}$ **15.** $\dfrac{a-b}{a+b}$

17. $\dfrac{s+r}{s-r}$ **19.** $x=3$ **21.** $d=\dfrac{9}{5}$ **23.** $x=28$

25. $w = -9$ or $w = -\frac{11}{3}$ **27.** $h = \frac{1}{4}$ **29.** $a = -\frac{10}{7}$

Lesson 9.6 Skill B (pages 370–371)

Try This Example 1

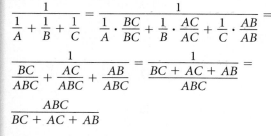

$$\frac{1}{\frac{1}{A} + \frac{1}{B} + \frac{1}{C}} = \frac{1}{\frac{1}{A} \cdot \frac{BC}{BC} + \frac{1}{B} \cdot \frac{AC}{AC} + \frac{1}{C} \cdot \frac{AB}{AB}} =$$

$$\frac{1}{\frac{BC}{ABC} + \frac{AC}{ABC} + \frac{AB}{ABC}} = \frac{1}{\frac{BC + AC + AB}{ABC}} =$$

$$\frac{ABC}{BC + AC + AB}$$

Try This Example 2

about 46.8 miles per hour

Exercises

1. Division Property of Equality and Reflexive Property of Equality

3. $\frac{60}{9}$, or $6\frac{2}{3}$ ohms **5.** 5.5 Ω **7.** 5 Ω

9. about 42.4 miles per hour

11. $\dfrac{2d}{\frac{d}{r} + \frac{d}{r}} = \dfrac{2d}{\frac{2d}{r}} = \dfrac{2d}{1} \cdot \dfrac{r}{2d} = r$

13. 0.5 second and 5.5 seconds

15. 1.7 seconds and 4.3 seconds

Lesson 9.7 Skill A (pages 372–373)

Try This Example 1

a. $y = 5$ **b.** $n = \frac{8}{7}$

Try This Example 2

a. $n = 2$ or $n = 4$
b. $r = 2$

Exercises

1. $(5)(7) = 2z$ **3.** $5a = 2(a - 1)$ **5.** $w = 15$

7. $t = \frac{10}{3}$ **9.** $b = \frac{12}{11}$ **11.** $y = \frac{35}{12}$ **13.** $n = -\frac{1}{8}$

15. $k = -\frac{6}{11}$ **17.** $y = 2$ **19.** $x = 2$ or $x = 4$

21. $x = \frac{bc}{ad}$ **23.** $n = \frac{sb}{rt - as}$ **25.** $10\sqrt{2}$ **27.** $7\sqrt{10}$

29. a. If $\frac{a}{b} = \frac{c}{d}$, then $ad = bc$ by the Cross Product Property. If $ad = bc$, then $a = \frac{bc}{d}$ by the Division Property of Equality. Thus, if $\frac{a}{b} = \frac{c}{d}$, then $a = \frac{bc}{d}$.

b. If $\frac{a}{b} = \frac{c}{d}$, then $ad = bc$ by the Cross Product Property. If $ad = bc$, then $d = \frac{bc}{a}$ by the Division Property of Equality. If $d = \frac{bc}{a}$, then $\frac{d}{c} = \frac{b}{a}$ by the Division Property of Equality. If $\frac{d}{c} = \frac{b}{a}$, then $\frac{b}{a} = \frac{d}{c}$ by the Symmetric Property of Equality.

Thus, if $\frac{a}{b} = \frac{c}{d}$, then $\frac{b}{a} = \frac{d}{c}$.

c. The information in parts a and b allows you to solve a proportion in one step, rather than the two steps required by the Cross Product Property.

31. $y \le 6$ **33.** $y \le -\frac{41}{4}$ **35.** $y \le \frac{9}{4}$

Lesson 9.7 Skill B (pages 374–375)

Try This Example 1

a. $d = 2$ or $d = -\frac{3}{2}$

b. $s = 3$ or $s = \frac{6}{5}$

Try This Example 2

$a = -2$

Exercises

1. -1 and $\frac{5}{2}$ are both solutions.

3. 1 is extraneous. **5.** $x = 5$ or $x = -2$

7. $a = \frac{2}{3}$ or $a = -\frac{1}{3}$ **9.** $y = -1$ or $y = -\frac{1}{2}$

11. $t = \frac{1}{3}$ or $t = \frac{3}{2}$ **13.** no solution

15. $x = -2$ **17.** $x = \frac{a + 1}{a}$; $a \ne 0, -1$ **19.** two

21. $\sqrt{81a} = (81a)^{\frac{1}{2}}$

$= 81^{\frac{1}{2}} a^{\frac{1}{2}}$

$= (9^2)^{\frac{1}{2}} a^{\frac{1}{2}}$

$= 9^{\left(2 \cdot \frac{1}{2}\right)} a^{\frac{1}{2}}$

$= 9a^{\frac{1}{2}} = 9\sqrt{a}$

23. $\sqrt{\dfrac{a}{25}} = \left(\dfrac{a}{25}\right)^{\frac{1}{2}} = \dfrac{a^{\frac{1}{2}}}{25^{\frac{1}{2}}} = \dfrac{a^{\frac{1}{2}}}{(5^2)^{\frac{1}{2}}}$

$= \dfrac{a^{\frac{1}{2}}}{5^{2 \cdot \frac{1}{2}}} = \dfrac{a^{\frac{1}{2}}}{5} = \dfrac{\sqrt{a}}{5}$

25. $\sqrt{\dfrac{64a}{121}} = \left(\dfrac{64a}{121}\right)^{\frac{1}{2}} = \dfrac{(64a)^{\frac{1}{2}}}{121^{\frac{1}{2}}} = \dfrac{(8^2 \cdot a)^{\frac{1}{2}}}{(11^2)^{\frac{1}{2}}}$

$= \dfrac{(8^{2 \cdot \frac{1}{2}})a^{\frac{1}{2}}}{11^{2 \cdot \frac{1}{2}}} = \dfrac{8a^{\frac{1}{2}}}{11} = \dfrac{8\sqrt{a}}{11}$

Lesson 9.7 Skill C (pages 376–377)

Try This Example 1

8 days

Try This Example 2

Alan: 20 hours
Louis: 60 hours

Exercises

1. $\dfrac{1}{12}$; $w = \dfrac{t}{12}$

3. Let t represent the amount of time Juan needs to complete the project working alone; $\dfrac{18}{t} + \dfrac{18}{1.5t} = 1$

5. 11.25 days   **7.** Lee: 36 hours; Kim: 36 hours

9. $a = \dfrac{bt}{b - t}$

11. 3 hours at 50 miles per hour and 5 hours at 54 miles per hour

Lesson 9.8 Skill A (pages 378–379)

Try This Example 1

$y = 12.5$

Try This Example 2

$42\dfrac{3}{16}$, or 42.1875 pounds per square inch

Exercises

1. 32 miles per hour, 16 miles per hour, 8 miles per hour, 4 miles per hour, 2 miles per hour, 2 miles per hour

3. $y = 4$ **5.** $x = 15$

7. If (x_1, y_1) and (x_2, y_2) satisfy $xy = k$, then $x_1 y_1 = k$ and $x_2 y_2 = k$. By the Transitive Property of Equality, $x_1 y_1 = x_2 y_2$. Divide each side of the equation by $y_1 x_2$. The result is $\dfrac{x_1}{x_2} = \dfrac{y_2}{y_1}$.

9. 84 pounds per square inch   **11.** $y = 7.5$

Lesson 9.8 Skill B (pages 380–381)

Try This Example 1

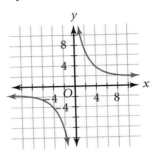

Try This Example 2

a. yes
b. domain: all real numbers except 0; range: all real numbers except 0
c. $y = 1.2$
d. If $xy = 144$ and $x = 80$, then $y = 1.8 \neq 0.25$. The point $(80, 0.25)$ is not on the graph.

Exercises

1. $k = 49$

3. If $xy = 49$, then $y = \dfrac{49}{x}$. Since $\dfrac{49}{x}$ is not defined for $x = 0$, there is no point on the graph corresponding to $x = 0$.

5.

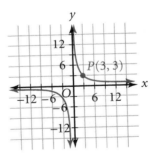

7. a. yes
b. domain: all real numbers except 0; range: all real numbers except 0
c. $y = \dfrac{32}{75}$
d. No; if $x = 100$, then $y = 0.64 \neq 0.25$.

9. a. yes
b. domain: all real numbers except 0; range: all real numbers except 0
c. $y = \dfrac{32}{75}$
d. No; if $x = 100$, then $y = 0.48 \neq 0.25$.

11. Sample counterexample: If $n = 0$, then $(3n - 1)(2n + 3) = (-1)(3) = -3$, and $-3 > -6$.

13. Sample counterexample: If $n = 0$, then
$(-3n)(n + 1) = (0)(1) = 0$, which is not positive.

Lesson 9.8 Skill C (pages 382–383)

Try This Example 1

$y = \dfrac{1}{64}$

Try This Example 2

$\dfrac{I_B}{I_A} = \dfrac{1}{9}$

Exercises

1.

x	1	2	3	4	5
y	1	$\dfrac{1}{4}$	$\dfrac{1}{9}$	$\dfrac{1}{16}$	$\dfrac{1}{25}$

3. $y = \dfrac{1}{45}$ **5.** $x = \pm 10$

7.
$$\frac{I_B}{I_A} = \frac{\dfrac{k}{(2d)^2}}{\dfrac{k}{d^2}}$$
$$= \frac{k}{(2d)^2} \cdot \frac{d^2}{k}$$
$$= \frac{k}{4d^2} \cdot \frac{d^2}{k}$$
$$= \frac{k}{k} \cdot \frac{d^2}{d^2} \cdot \frac{1}{4}$$
$$= \frac{1}{4}$$

9. $\dfrac{1}{r^2}$ times illumination at distance d

11. 30 feet   **13.** 333.33 feet

Lesson 9.9 Skill A (pages 384–385)

Try This Example 1

$x = \dfrac{a}{b - 1}$; $x \neq 0$ and $b \neq 1$

Try This Example 2

If $\dfrac{a}{b} = \dfrac{c}{d}$, then by the Cross Product Property of Proportions, $ad = bc$. By the Multiplication Property of Equality, $\dfrac{bd}{cd} \cdot \dfrac{a}{b} = \dfrac{bd}{cd} \cdot \dfrac{c}{d}$. By definition of multiplication of fractions, $\dfrac{bda}{cdb} = \dfrac{bdc}{cdd}$. By the Commutative and Associative Properties of Multiplication, $\dfrac{bda}{bdc} = \dfrac{bdc}{ddc}$. By definition of multiplication of fractions, $\dfrac{bd}{bd} \cdot \dfrac{a}{c} = \dfrac{dc}{dc} \cdot \dfrac{b}{d}$. By the Inverse Property of

Multiplication, $1 \cdot \dfrac{a}{c} = 1 \cdot \dfrac{b}{d}$. By the Identity Property of Multiplication, $\dfrac{a}{c} = \dfrac{b}{d}$.

Exercises

1. $b \neq 0$ **3.** $x = \dfrac{bc}{ad}$; $b \neq 0$, $d \neq 0$, $a \neq 0$

5. $x = \dfrac{a - c}{b}$; $x \neq -\dfrac{c}{b}$ and $b \neq 0$

7. $x = \dfrac{a - bc}{c - 1}$; $x \neq -b$ and $c \neq 1$

9. $x = \dfrac{b(1 + c)}{a(1 - c)}$; $x \neq -\dfrac{b}{a}$, $a \neq 0$, $c \neq 1$

11. $x = \dfrac{-b \pm 1}{a}$; $a \neq 0$, $x \neq -\dfrac{b}{a}$

13. definition of multiplication of fractions

15. If $\dfrac{1}{b} = \dfrac{c}{d}$, then by the Cross Product Property of Proportions, $d = bc$. By the Division Property of Equality, $\dfrac{d}{dc} = \dfrac{bc}{dc}$. By the definition of multiplication of fractions, $\dfrac{d}{d} \cdot \dfrac{1}{c} = \dfrac{b}{d} \cdot \dfrac{c}{c}$. By the Inverse Property of Multiplication, $1 \cdot \dfrac{1}{c} = \dfrac{b}{d} \cdot 1$. By the Identity Property of Multiplication, $\dfrac{1}{c} = \dfrac{b}{d}$. By the Symmetric Property of Equality, $\dfrac{b}{d} = \dfrac{1}{c}$.

17. If $\dfrac{a}{b} = \dfrac{c}{d}$, then $\dfrac{a}{b} - 1 = \dfrac{c}{d} - 1$ by the Subtraction Property of Equality. By the Substitution Principle, $\dfrac{a}{b} - \dfrac{b}{b} = \dfrac{c}{d} - \dfrac{d}{d}$. Thus, by the definition of subtraction of fractions, $\dfrac{a - b}{b} = \dfrac{c - d}{d}$.

19. Suppose that $\dfrac{a}{b} = \dfrac{c}{d}$ and $\dfrac{a}{d} = \dfrac{c}{b}$. Then $ad = bc$ and $ab = dc$. Thus, $ad - ab = bc - dc$.
So, $a(d - b) = -c(d - b)$. If $b \neq d$, then $a = -c$.
The statement is true when $b = d$ or when $a = -c$.

21. $a \neq \pm\dfrac{1}{2}$ **23.** $x \neq -\dfrac{11}{12}$ **25.** $x \neq 0, 4$

CHAPTER 10

Lesson 10.1 Skill A (pages 390–391)

Try This Example

a. $x \geq -\dfrac{5}{3}$ **b.** domain: $x \geq -\dfrac{1}{2}$; range: $y \geq 0$

Exercises

1. $3m \geq 0$ **3.** $3(x - 4) - 12 \geq 0$ **5.** $n \geq 0$

7. $a \leq 0$ **9.** $a \geq -\dfrac{9}{2}$ **11.** $m \geq -5$

13. domain: $t \geq 0$; range: $y \geq 0$

15. domain: $g \geq -3$; range: $y \geq 0$

17. domain: $s \geq \dfrac{9}{7}$; range: $y \geq 0$

19. domain: $x \geq 1$; range: $y \geq 0$

21. domain: $r \leq 1$; range: $y \geq 0$

23. $a = \dfrac{3}{5}$

25. The domain of $y = \sqrt{2x - 7}$ is all real numbers such that $2x - 7 \geq 0$, that is, all real numbers such that $x \geq \dfrac{7}{2}$. The domain of $y = \sqrt{2x - 5}$ is all real numbers such that $2x - 5 \geq 0$, that is, all real numbers such that $x \geq \dfrac{5}{2}$. Thus, the domain of $y = \sqrt{2x - 7}$ is contained in the domain of $y = \sqrt{2x - 5}$. The number 3 is in the domain of $y = \sqrt{2x - 5}$ but not in the domain of $y = \sqrt{2x - 7}$.

27. $n = \pm 5$ **29.** $x = \pm\sqrt{3}$ **31.** $v = \pm\sqrt{7}$ **33.** $u = \pm 3$

Lesson 10.2 Skill A (pages 392–393)

Try This Example 1
a. $v = \dfrac{1}{4}$ **b.** $m = 6.25$

Try This Example 2
$x = 22$

Try This Example 3
$x = 1$ or $x = 10$

Exercises

1. yes **3.** no **5.** $c = 81$ **7.** $d = 225$ **9.** $d = 4$

11. $n = \dfrac{11}{5}$ **13.** $p = -\dfrac{4}{3}$ **15.** $z = 23$ **17.** $n = 3$

19. $b = \dfrac{25}{2}$ **21.** $x = -5$ or $x = 5$ **23.** $u = -9$ or $u = 3$

25. $z = -4$ or $z = 7$ **27.** $q = 1$ or $q = 3$

29. $x = \pm\sqrt{1 - b^2}$ **31.** $2x^2 + 3x - 5$

33. $4n^2 - 4n - 8$ **35.** $-abx^2 - ax + bx + 1$

37. $b^2x^2 + 2abx + a^2$ **39.** $a^2 - b^2x^2$

Lesson 10.2 Skill B (pages 394–395)

Try This Example 1
about 1.76 kilometers

Try This Example 2
about 9.2 miles north or 9.2 miles south

Exercises

1. $PQ = \sqrt{(1 - 0)^2 + (7 - 4)^2}$

3. $PQ = \sqrt{[3 - (-1)]^2 + [5 - (-1)]^2}$

5. about 0.07 kilometer **7.** about 0.10 kilometer

9. about 11.3 units above or below $P(4, 0)$; $Q(4, 11.3)$ and $Q'(4, -11.3)$

11. about 6.2 units above or below $Z(7, 2)$; $Q(7, 8.2)$ and $Q'(7, -4.2)$

13. $h = \left(\dfrac{d}{113.14}\right)^2$ **15.** 5 days **17.** 5 days

Lesson 10.2 Skill C (pages 396–397)

Try This Example 1
$t = 8$

Try This Example 2
$x = 1$ or $x = 4$

Exercises

1. $3(x - 2) + 5 = 3x - 1$; linear

3. $r^2 - 2r = 3r$; quadratic **5.** no real solution

7. $b = 0$ **9.** no real solution **11.** $g = \dfrac{9}{5}$ **13.** $n = 2$

15. $z = -4$ **17.** $a = 2$ or $a = 6$ **19.** no real solution

21. If $\sqrt{3t + 12} = \sqrt{3(4 + t) - 1}$, then $3t + 12 = 3(4 + t) - 1$. This implies that $3t + 12 = 3t + 11$, and that $12 = 11$. Since this is impossible, $\sqrt{3t + 12} = \sqrt{3(4 + t) - 1}$ has no solution.

23. $x = \dfrac{d - b}{a - c}$; if $a = c$, then $b = d$; otherwise $a \neq c$

25. $OA = \sqrt{(6 - 0)^2 + (r - 0)^2} = \sqrt{36 + r^2}$; $OB = \sqrt{(r - 0)^2 + (6 - 0)^2} = \sqrt{r^2 + 36}$; Since $36 + r^2 = r^2 + 36$ for all real numbers r, then $\sqrt{36 + r^2} = \sqrt{r^2 + 36}$ for all real numbers r. Thus, $OA = OB$ for all nonzero values of r. This implies that two sides of triangle OAB are equal in length.

27. $7\sqrt{3}$ **29.** $9\sqrt{11}$ **31.** $\dfrac{\sqrt{3}}{10}$ **33.** $\dfrac{7\sqrt{2}}{11}$

Lesson 10.3 Skill A (pages 398–399)

Try This Example 1
a. $t = 2\sqrt{2}$ **b.** $7 - 2\sqrt{2}$

Try This Example 2
a. $-4 + 8\sqrt{2}$ **b.** 5

Exercises

1. $\dfrac{3}{4}$ **3.** $\sqrt{2}$ **5.** $5\sqrt{2}$ **7.** $5\sqrt{2}$ **9.** $-20\sqrt{2}$

11. $12 + 4\sqrt{2}$ **13.** 8 **15.** 0 **17.** $-14 - 2\sqrt{2}$

19. $10 - 10\sqrt{2}$ **21.** $r = -4, s = -2$ **23.** $r = 4, s = 2$

25. $(a + b\sqrt{2}) + [(-a) + (-b)\sqrt{2}]$
$= [a + b\sqrt{2} + (-a)] + (-b)\sqrt{2})$
$= a + (-a) + b\sqrt{2} + =(-b)\sqrt{2}$
$= 0 + [b + (-b)]\sqrt{2}$
$= 0 + 0$
$= 0$

27. $6a^2 + 16a + 8$ **29.** $-4a^2 + 2a + 6$ **31.** $b^2 - b - 2$

Lesson 10.3 Skill B (pages 400–401)

Try This Example 1
a. $-36 + 30\sqrt{2}$ **b.** $-57 - 40\sqrt{2}$

Try This Example 2
a. $\dfrac{5\sqrt{2}}{6}$ **b.** $\dfrac{14 + 5\sqrt{2}}{6}$ **c.** $\dfrac{28 + \sqrt{2}}{17}$

Exercises

1. $-34 - 2\sqrt{2}$ **3.** $6 + 15\sqrt{2}$ **5.** $-2 - 14\sqrt{2}$

7. $4 + 2\sqrt{2}$ **9.** $30 - 5\sqrt{2}$ **11.** $\left(\dfrac{1}{2}\right)\sqrt{2}$

13. $\left(-\dfrac{2}{5}\right)\sqrt{2}$ **15.** $-1 + 12\sqrt{2}$ **17.** $-38 + 34\sqrt{2}$

19. $54 - 20\sqrt{2}$ **21.** $\dfrac{8}{17} + \left(\dfrac{3}{34}\right)\sqrt{2}$

23. $-\dfrac{5}{7} + \left(\dfrac{3}{7}\right)\sqrt{2}$ **25.** $\dfrac{3}{73} - \left(\dfrac{25}{73}\right)\sqrt{2}$

27. $7 + 5\sqrt{2}$ **29.** $116 + 90\sqrt{2}$

31. $(a + b\sqrt{2})(a - b\sqrt{2})$
$= a(a - b\sqrt{2}) + b\sqrt{2}(a - b\sqrt{2})$
$= a^2 - ab\sqrt{2} + ab\sqrt{2} - (b\sqrt{2})(b\sqrt{2})$
$= a^2 - (b\sqrt{2})(b\sqrt{2})$
$= a^2 - 2b^2$

33. $9\dfrac{5}{21}$ **35.** $14\dfrac{4}{5}$ **37.** $1\dfrac{8}{9}$ **39.** $\dfrac{17}{99}$

Lesson 10.3 Skill C (pages 402–403)

Try This Example 1

By the Commutative and Associative Properties of Addition,
$(-7 + 3\sqrt{2}) + [1 + (-5)\sqrt{2}] = (-7 + 1) + 3\sqrt{2} + (-5)\sqrt{2}$.
By the Distributive Property, $(-7 + 1) + 3\sqrt{2} + (-5)\sqrt{2} = (-7 + 1) + [3 + (-5)]\sqrt{2}$.
By addition, $(-7 + 1) + [3 + (-5)]\sqrt{2} = -6 + (-2)\sqrt{2}$.
The sum has the form $a + b\sqrt{2}$, where $a = -6$ and $b = -2$, both rational numbers.

Try This Example 2
$x = -\dfrac{1}{7} + \dfrac{2}{7}\sqrt{2}$

Exercises

1. $2.5 - 5\sqrt{2}$ **3.** $-4 + 1.4\sqrt{2}$

5. $(2 - \sqrt{2}) + (5 + 4\sqrt{2}) = (2 + 5) + (-1)\sqrt{2} + 4\sqrt{2} = (2 + 5) + [(-1) + 4]\sqrt{2} = 7 + 3\sqrt{2}$
The sum has the form $a + b\sqrt{2}$, where $a = 7$ and $b = 3$, both rational numbers.

7. $n = -1 + 1\sqrt{2}$ **9.** $k = \dfrac{8}{7} + \left(\dfrac{5}{7}\right)\sqrt{2}$

11. $n = 2 - \sqrt{2}$ **13.** $n = -2 + \dfrac{7}{2}\sqrt{2}$ **15.** $x \geq -4$

17. $v = \dfrac{3}{2}$ **19.** $r = $ no real solution **21.** $-7 - 17\sqrt{2}$

23. $-\dfrac{1}{3} + \dfrac{5}{6}\sqrt{2}$ **25.** $v = \dfrac{10}{41} - \dfrac{3}{41}\sqrt{2}$

Lesson 10.4 Skill A (pages 404–405)

Try This Example 1
18

Try This Example 2
a. 5 **b.** $2\sqrt{3}$

Try This Example 3
a. $\dfrac{6\sqrt{7}}{7}$ **b.** $\dfrac{5\sqrt{2}}{4}$

Exercises

1. $5\sqrt{7}$ **3.** 15 **5.** $7\sqrt{2}$ **7.** $11\sqrt{3}$ **9.** 12

11. $7\sqrt{2}$ **13.** 10 **15.** 7 **17.** $\sqrt{3}$

19. $\sqrt{3}$ **21.** $\dfrac{\sqrt{15}}{3}$ **23.** $\dfrac{\sqrt{10}}{4}$

25. $\dfrac{\sqrt{6}}{20}$ **27.** $\dfrac{7\sqrt{21}}{15}$ **29.** $\sqrt{\dfrac{x+2}{x+3}}$; $x < -3$ or $x \geq -2$

31. $\sqrt{\dfrac{2z-3}{z+11}}$; $z < -11$ or $z \geq \dfrac{3}{2}$ **33.** 1 **35.** 15

37. 10 **39.** $\dfrac{9}{5}$

Lesson 10.4 Skill B (pages 406–407)

Try This Example 1
a. $-5\sqrt{5}$ **b.** -5

Try This Example 2
a. $16\sqrt{3}$ **b.** 0

Try This Example 3
a. $5 + 34\sqrt{3}$ **b.** $-7 + 14\sqrt{3}$

Exercises

1. $6 - 4\sqrt{11}$ **3.** $6\sqrt{3}$ **5.** $4 + 5\sqrt{5}$ **7.** -7
9. -15 **11.** $6 + 5\sqrt{5}$ **13.** 6 **15.** $6\sqrt{3}$
17. $4\sqrt{5}$ **19.** $7 - 19\sqrt{3}$ **21.** $-6 - 84\sqrt{2}$
23. $3 + 2\sqrt{2}$ **25.** $4a + 14b\sqrt{2}$
27. $(1 + x + x^2 + x^3 + x^4 + x^5 + x^6 + x^7 + x^8)\sqrt{x}$
29. -7 **31.** $-21 - 50\sqrt{2}$ **33.** $-4.5 + 4\sqrt{2}$
35. $-3 - 2\sqrt{2}$

Lesson 10.4 Skill C (pages 408–409)

Try This Example 1
a. $70 + 14\sqrt{14}$
b. $20 - 5\sqrt{2}$

Try This Example 2
a. $3 + 13\sqrt{3}$
b. -23

Try This Example 3
a. $\dfrac{5}{22} + \dfrac{7}{22}\sqrt{5}$ **b.** $-\dfrac{19}{3} - \dfrac{7}{3}\sqrt{7}$

Exercises

1. $\sqrt{5}(5) - 3\sqrt{5}\sqrt{5}$
3. $(2)(7) - (2)3\sqrt{7} - \sqrt{7}(7) + 3\sqrt{7}\sqrt{7}$
5. $\dfrac{2 - 5\sqrt{3}}{2 - 5\sqrt{3}}$ **7.** $10 + 3\sqrt{2}$ **9.** $7 - 4\sqrt{2}$
11. -29 **13.** $-12 + 6\sqrt{2}$ **15.** $-35 + 15\sqrt{5}$
17. $-\dfrac{3}{10} - \dfrac{3}{10}\sqrt{11}$ **19.** $\dfrac{5}{4} + \dfrac{3}{4}\sqrt{5}$ **21.** $\dfrac{5}{2} - \dfrac{1}{2}\sqrt{7}$

23.a. No; for example, if $a = -1$ and $b = -1$, then
$\sqrt{(-1)^2 + (-1)^2} \leq -1 + (-1)$ is false.

b. Squaring both sides of $\sqrt{a^2 + b^2} \leq a + b$ gives $a^2 + b^2 \leq a^2 + 2ab + b^2$. This inequality is true when $a = -1$ and $b = -1$. However, from part **a**, the original inequality is not true when $a = -1$ and $b = -1$.

So, squaring both sides of an inequality does not always give an equivalent inequality. Therefore, squaring both sides is not a valid method for solving an inequality.

25. n^5 **27.** z^7 **29.** $25p^7$ **31.** $\dfrac{7}{22}b^9$

Lesson 10.5 Skill A (pages 410–411)

Try This Example 1
a. $3^{\frac{1}{4}}$ **b.** $30^{\frac{1}{3}}$ **c.** $\sqrt[4]{190}$ **d.** $\sqrt[3]{75}$

Try This Example 2
a. 3 **b.** 4 **c.** 2 **d.** 5

Exercises

1. $7 = \sqrt{49}$ and $7 = 49^{\frac{1}{2}}$

3. $10 = \sqrt[3]{1000}$ and $10 = 1000^{\frac{1}{3}}$

5. $13^{\frac{1}{3}}$ **7.** $\sqrt[3]{100}$ **9.** $43^{\frac{1}{4}}$ **11.** $\sqrt[3]{90}$ **13.** $55^{\frac{1}{5}}$
15. $\sqrt[3]{33}$ **17.** 2 **19.** 7 **21.** 1 **23.** 8 **25.** a
27. n **29.** ab **31.** rs
33. r and s are nonzero real numbers such that $r = \dfrac{1}{s}$.

35. $5\sqrt{11}$ **37.** 22 **39.** $\dfrac{1}{10}$ **41.** $\dfrac{2\sqrt{6}}{7}$

Lesson 10.5 Skill B (pages 412–413)

Try This Example 1
a. 16 **b.** $5\sqrt{5}$

Try This Example 2
$a^{\frac{11}{2}}$

Exercises

1. $5^{\frac{2}{3}}$ **3.** $3^{\frac{3}{4}}$ **5.** $\sqrt[3]{11^2}$ **7.** $\sqrt{3^5}$ **9.** 625 **11.** 2187
13. $3\sqrt{3}$ **15.** $2\sqrt[3]{4}$ **17.** $x^{\frac{3}{2}}$ **19.** $t^{\frac{7}{3}}$ **21.** $n^{\frac{5}{3}}$
23. $b^{\frac{7}{6}}$ **25.** $t^{\frac{37}{12}}$

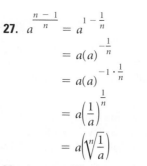

27. $a^{\frac{n-1}{n}} = a^{1-\frac{1}{n}}$

$= a(a)^{-\frac{1}{n}}$

$= a(a)^{-1 \cdot \frac{1}{n}}$

$= a\left(\frac{1}{a}\right)^{\frac{1}{n}}$

$= a\left(\sqrt[n]{\frac{1}{a}}\right)$

29. $2300   **31.** $4679.43   **33.** $10,123.64

Lesson 10.5 Skill C (pages 414–415)

Try This Example

$6768.14

Exercises

1. two   **3.** six   **5.** $1694.59   **7.** $1793.85   **9.** no

11. 0.41 second and 7.59 seconds

13. 1.68 seconds and 6.32 seconds

15. 0 seconds and 8 seconds

Lesson 10.6 Skill A (pages 416–417)

Try This Example 1

The statement is false. The expression $\sqrt{x} - \sqrt{x-1}$ is only defined when $x \geq 1$. For these real numbers, the statement is true.

Try This Example 2

Writing $\sqrt{5^2} - \sqrt{4^2} = \sqrt{5^2 - 4^2}$ in Step **1** is incorrect. It assumes that $\sqrt{a - b} = \sqrt{a} - \sqrt{b}$ for all real numbers a and b for which the expressions are defined. However, $\sqrt{a - b} = \sqrt{a} - \sqrt{b}$ is not always true.
$\sqrt{5^2} - \sqrt{4^2} = \sqrt{25 - 16} = \sqrt{9} = 3$

Try This Example 3

If $x = \frac{1}{4}$, then $\frac{1}{4} > \sqrt{\frac{1}{4}}$, or $\frac{1}{2}$. This inequality is false. If $x = 4$, then $4 > \sqrt{4}$, or 2. This inequality is true. Therefore, the statement is sometimes true.

Exercises

1. Answers may vary, but all values of x given should be negative.

3. $x = 0$

5. $\sqrt{pq} = \sqrt{a^2 b^2} = \sqrt{a^2}\sqrt{b^2} = ab$
Since a and b are positive integers, ab is a positive integer. Thus, $\sqrt{pq}$ is a positive integer.

7. The order of operations was not followed. Operations under the square-root symbol take precedence over division by 5.

$$\frac{\sqrt{10(3 + 7)}}{5} = \frac{\sqrt{10(10)}}{5} = \frac{\sqrt{100}}{5} = \frac{10}{5} = 2$$

9. If $\sqrt{\frac{t}{2}} = \frac{\sqrt{t}}{2}$, then, by squaring each side of the equation, $\frac{t}{2} = \frac{t}{4}$. This equation is true only when $t = 0$. Since t is a positive real number, the statement is never true.

11. $x = 4$ **13.** $x = 6$ **15.** $y = \frac{4}{25}$

Lesson 10.6 Skill B (pages 418–419)

Try This Example 1

69.28 miles per hour

Try This Example 2

On dry asphalt, the speed of the vehicle before braking would be $\sqrt{2}$, or about 1.4, times the speed on wet asphalt.

Exercises

1. An increase in speed will cause an increase in stopping distance.

3. about 27.4 miles per hour

5. about 49.0 miles per hour

7. The second vehicle was traveling at about 110% of the speed of the first vehicle.

9. If $0 < r < 1$, the second vehicle will have the lesser speed. If $r = 1$, both vehicles will have the same speed. If $r > 1$, the first vehicle will have the lesser speed.

11. 24 cubic inches

Lesson 10.6 Skill C (pages 420–421)

Try This Example

1.73

Exercises

1. between 2.2 and 2.3   **3.** between 4 and 4.5

5. 3.32   **7.** 6.71

9. between 14 and 15   **11.** between 15 and 16

13. $\frac{1}{\sqrt{73}}, \frac{1}{\sqrt{37}}, \frac{1}{\sqrt{30.7}}, \frac{1}{\sqrt{7.3}}, \frac{1}{\sqrt{3.7}}, \frac{1}{\sqrt{3.07}}$

15. $x = \pm 5i$ **17.** $n = \pm 2i$ **19.** $a = \pm i\sqrt{2}$

21. $y = \pm i\sqrt{13}$

Index

Definitions of boldface entries can be found in the glossary.

A

Absolute value, 44

Absolute-value equations, 176

Absolute-value inequalities, 178

Adding fractions with like denominators, 356

Addition
  Distributive Property, 66
  fractions, 356–361
  horizontal, 234
  inverse property, 8, 56
  polynomials, 234, 238
  properties of, 8
  rational expressions, 356–361
  of real numbers, 46, 48
  solving inequalities with, 146, 148, 152
  square-root expressions, 398, 402, 406
  vertical, 234

Addition Property of Equality, 20, 62, 72, 76

Addition Property of Inequality, 148

Additive Identity Property, 8, 20

Additive inverse, 50, 56, 402

Algebraic expressions, 12–13

"And," 170

Applications
  academics, 6, 162, 376
  agriculture, 214
  banking and finance, 34, 54, 72, 80, 168, 414
  chemistry, 168, 200
  electricity, 370
  engineering, 100, 104, 116
  entertainment, 34, 124
  food preparation, 174
  gardening, 74, 174, 208
  geometry, 6, 130, 214, 232, 238, 244, 308, 366
  inventory, 152
  landscaping, 194
  manufacturing, 120, 180
  personal finance, 13, 100, 116, 152, 158
  physics, 228, 294, 334, 336, 378, 382, 418–419
  retail/sales, 6, 102, 104, 124, 158, 238, 294
  sports/fitness, 152
  temperature, 26, 54, 72
  travel/transportation, 110, 120, 228, 316, 370, 394, 418–419
  veterinary science, 194
  voting, 162

Approximations of functions, 420

Assessments, 36–37, 88–89, 138–139, 184–185, 216–217, 256–257, 296–297, 338–339, 386–387, 422–423. *See also* Mid-chapter reviews

Associative Property of Addition, 8, 18, 48, 54, 402

Associative Property of Multiplication, 8, 402

Axis of symmetry, 286

B

Base (of an exponential expression), 220

Basic Properties of Equality, 18–19

Basic Properties of Numbers, 8

Binomials, 230, 254, 268

Boundary lines, 206–207, 210

Braces, 10, 14

Brackets, 10

Braking distance, 228, 418–419

C

Canceling, 60

Circles, 308

Closed sets, 234, 252, 398, 400, 402

Closure
  Property of Addition, 8, 398
  Property of Multiplication, 8
  of numbers of the form $a + b\sqrt{2}$, 398, 400, 402

Coefficient of friction, 418

Coefficients, 66

Collinear points, 108

Combining like terms, 66

Common binomial factor, 268

Commutative Property of Addition, 8, 14, 22, 54, 66, 402

Commutative Property of Multiplication, 8, 402

Completing the square, 312, 314, 316

Complex fractions, 368–371

Composite numbers, 260

Compound inequalities, 146, 170–175

Conclusions, 84

Conditional statements, 84

Congruent rectangles, 238

Conjectures, 16

Conjugates, 400

Conjunctions, 146, 170

Consistent systems, 202

Constant of variation, 102, 378, 382

Constant slope, 110

Converse (of a statement), 384

Coordinate axes, 92

Coordinate plane, 92–95, 126

Counterexamples, 16

Critical Thinking, 5, 9, 15, 19, 21, 23, 25, 29, 31, 33, 35, 41, 43, 47, 55, 57, 61, 65, 71, 73, 77, 79, 81, 85, 87, 93, 97, 99, 107, 109, 111, 113, 117, 119, 123, 127, 129, 131, 133, 135, 147, 149, 151, 155, 157, 159, 161, 165, 169, 171, 173, 175, 177, 179, 183, 191, 197, 199, 203, 205, 211, 213, 221, 223, 225, 227, 229, 231, 235, 241, 245, 247, 249, 251, 253, 261, 267, 271, 273, 277, 281, 283, 285, 289, 293, 295, 300, 303, 307, 311, 313, 315, 317, 325, 329, 333, 335, 343, 345, 347, 349, 353, 357, 359, 361, 365, 369, 371, 373, 375, 381, 385, 393, 397, 405, 407, 409, 411, 419, 421

Cross-Product Property of Proportions, 372, 384

Cube-root radicals, 410

Cubic equations, 230, 284

Cubic polynomials, 346

Cylinders, 232, 256

D

Data, 124

Decimals, 40, 42, 420

Deductive reasoning, 84, 134, 182, 326, 384

Density Property of Real Numbers, 420

Dependent systems, 204

Dependent variables, 100

Difference of two squares, 250, 266

Difference, 4. *See also* Subtraction

constant, 132, 328

first, 328

of two squares, 250, 266

second, 328

subtraction, 4

Direct variation, 102, 226

alternative form of, 104

Discriminant, 320

Disjunctions, 172

Distance Formula, 394

Distributive Property

application, 56, 70, 72, 78, 82

combining like terms, 66

definition, 8

multiplication and division, 68, 252

polynomials, 236, 246, 248, 252

quadratic equations, 326

Division

Distributive Property, 68, 252

inequalities, 182, 206, 390

monomials, 242, 244

negative numbers, 206

polynomials, 252–255, 354

rational expressions, 352, 354

of real numbers, 58

simplifying expressions, 60

solving inequalities, 156, 158

square-root expressions, 400, 404, 408

Division Property of Equality, 30, 32, 34, 64, 70, 312

Division Property of Inequality, 156

Domain

functions, 96, 102

graphs, 98, 100

quadratic functions, 290

rational expressions, 342

square-root expressions, 390

Dot patterns, 330

Double root, 288, 316

E

Electrical resistance, 370–371

Elimination method, 196, 198, 200, 202

Ellipsis, 42

Equality of Squares Property, 392, 396

Equations. *See also* Quadratic equations; Systems of linear equations

absolute-value, 176

cubic, 230, 284

equivalent, 70

graphing, 94

linear, 112, 114, 188

and logical reasoning, 14–19

multistep, 76–81

parametric, 136

point-slope form of, 118, 120, 122

rational, 372–377

simplification of, 10, 40, 52

slope-intercept form of, 114–115, 118, 122, 124, 126, 202

square-root, 392–397, 416, 418

standard linear form, 122

systems of, 202–205

from two points, 122–125

two-step, 70–75

Equilateral triangles, 87

Equivalent equations, 70

Evaluating an expression, 12, 232

Excluded values, 342, 344, 346

Exponential expressions, 220, 222, 410, 412

Exponents

division, 242

fractional, 304, 410, 412, 414

integer, 220–225

multiplication, 240

negative, 222

properties of, 220, 222

rational, 350, 352, 410, 412, 414

scientific notation, 224

zero, 222

Expressions, 4–7

Extraneous solution, 374, 396

F

Factoring

common binomial, 268

complete, 278

cubic equations, 284

greatest common, 262

greatest common monomial, 264

grouping, 268, 274

integers, 260

monomial, 344

no solution, 272, 276

prime, 260–262

quadratic trinomials, 270, 272, 274, 276, 280

rational expressions, 344, 346, 350

simplifying before, 278

solving by, 280–285

special polynomials, 266–269

splitting the middle term, 274

summary of steps, 284

Fields, 402

Finite number of digits, 40

FOIL method, 248

Formula

definition, 12, 72

quadratic, 318, 322–323

Fourth-root radicals, 410

Fractional exponents, 304, 410, 412, 414

Fractions. *See also* Rational expressions

adding, 356–361

complex, 368–371

multiplying and dividing, 350–355

simplifying, 40

Free-falling objects, 229, 256, 294, 334, 336

Functions

definition, 96

direct variation, 102

linear, 132

power, 226–229

rational, 342

reciprocal, 380

square-root, 390, 418, 420

variables, 100

vertical-line test, 96, 380

G

GCF (greatest common factor), 262, 264, 278, 348

Geometric mean, 373

Geometry. *See also* Triangles

area, 250, 366

coordinate plane, 92–95, 126

cylinders, 232

perimeter, 6, 238

polygons, 130–131, 214, 366

Pythagorean Theorem, 308, 316, 394

surface area, 12, 232, 244–245, 256

volume, 244

Graphing

absolute value, 176, 178

continuous and discrete graphs, 116

domain and range, 98

equations in two variables, 94, 112

functions, 96, 98, 100

inequalities, 142, 144, 146, 148, 150, 154

intercepts, 112, 114, 188

inverse variation, 380

linear equations, 112, 114

linear inequalities, 206, 208

on the number line, 94

ordered pairs, 92

points on a line, 108

quadratic functions, 286, 288, 290

slope and orientation, 106, 114–115

systems of linear equations, 188

systems of linear inequalities, 210–215

Gravity, acceleration of, 229, 256, 294, 334, 336

Greatest common factor (GCF), 262, 264, 278, 348

Grouping symbols, 10

H

Horizontal-addition format, 234

Horizontal lines, 112, 122

Horizontal-multiplication format, 248

Hypothesis, 84

I

i, definition of, 310

Identity, 14

Identity Property of Addition, 8, 402

Identity Property of Multiplication, 8, 18, 28, 402

If-then statements, 84

Imaginary roots, 310, 396

Imaginary units, 310

INDEX

Inconsistent systems, 202

Independent systems, 204

Independent variables, 100

Index of the radical, 410

Indirect reasoning, 86

Inductive reasoning, 134, 136

Inequalities, 142–185. *See also* Solving inequalities

  absolute-value, 178

  approximations, 420

  compound, 146, 170–175

  deductive reasoning and, 182

  definition, 142

  division, 182, 390

  graphing, 144, 146, 148–150, 154

  multistep, 164, 166, 168

  with no solution, 166, 172, 212

  one-step, 160

  ranges of quadratic functions, 290

  restrictions, 208

  standard form, 212

  symbols, 142, 164

  systems of linear, 210–215 in two variables, 206–209

Infinite number of digits, 40

Integer exponents, 220–225

Integers, 40

Intercepts, 112, 114, 188, 292

Interest rate calculations, 13, 80, 168, 414

Inverse operations, 24, 28, 32

Inverse Property of Addition, 8, 56

Inverse Property of Multiplication, 8

Inverse-square variation, 382

Inverse variation, 378–383

  alternative form of, 378

Irrational numbers, 42–43, 420

Isolated variables, 192

Isosceles triangles, 87

Key words, 26

L

Leading coefficient, 230

Leading term, 230

Least common denominator (LCD), 358, 360, 364, 368

Least common multiple (LCM), 76, 358

Like terms, 66

Linear equations, 112, 114–116. *See also* Equations; Systems of linear equations

Linear functions, 132

Linear inequalities

  in **standard form**, 212

    systems of, 210–215

    in two variables, 206, 210–215

Linear patterns, 132, 328

Lines

  boundary, 206–207, 210

  horizontal, 112, 122

  number, 44, 46, 94

  parallel, 109, 126, 130

  perpendicular, 128, 130

  in a plane, 126

  slopes of, 106, 108–110, 114–115

  vertical, 112, 122

Logic, 18

Logical arguments, 84

Logical reasoning

  and/or, 146

  correctness, 82

  deductive, 84, 134, 182, 326, 384

  indirect, 86

  inductive, 134, 136

  in solving quadratic equations, 324

M

Mathematical expressions, 4–7

Mathematical models, 124

Mathematical proofs

  closed sets, 402

  counterexamples and, 16

  definition, 16

  direct variation, 104

  dividing a negative number by a positive number, 58

  multiplying two real numbers, 56

  point-slope form, 118

  quadratic formula, 322–323

  square-root statements, 416

Mathematical statements, 4–7, 16, 42, 86, 182

Mathematical symbols

  inequalities, 142, 164, 170

  radical signs, 300, 390

  words and, 4–7

Maximum, 158, 174, 286

Mid-chapter reviews, 19, 61, 109, 163, 201, 233, 279, 323, 367, 403. *See also* Assessments

Minimum, 158, 174, 286

Monomials, 66, 230, 240–246, 252, 264, 358

Multiplication. *See also* Factoring

  Distributive Property, 68

  elimination method, 198, 200

  monomials, 240, 244

  polynomials, 246–251

  properties of, 8

  rational expressions, 350, 354

  of real numbers, 56

same/opposite signs, 56, 58

simplifying expressions, 60

solving inequalities, 154, 156, 158, 162

square-root expressions, 400, 402, 404, 408

Multiplication Property of Equality, 28, 32, 64, 72, 76, 312

Multiplication Property of Inequality, 154, 160

Multiplication Property of Zero, 280

Multiplicative identity, 8

Multiplicative inverse, 58, 402

Multiplicative Property of −1, 52, 54

N

Natural numbers, 40

Negations, 86

Negative exponents, 222

Negative numbers, 52

Net change, 54

Number lines

absolute value, 176

equations in two variables, 94

inequalities, 142, 144, 146, 148, 154

real numbers, 44, 46

Numbers. *See also* Real numbers

basic properties, 8

classification, 42

composite, 260

imaginary, 310

integers, 40

irrational, 42–43, 420

natural, 40

negative, 52

prime, 260–261, 406

rational, 40, 42

signs, 56, 58

whole, 40

Numerical expressions, 10–11, 13

O

Open sentences, 14

Opposite of a Difference, 52, 236, 348

Opposite of a Sum, 52, 236

Opposites, 8, 44, 52, 236, 348

"Or," 172

Ordered pairs, 92, 132, 328, 332

Order of operations, 10

Orientation of a line, 106

Origin, 44, 92

P

Parabolas, 286, 288, 292, 294, 332

Parallel lines, 109, 126, 130

Parallelograms, 131

Parametric equations, 136

Parentheses, 4, 10, 52

Patterns

algebraic, 134, 136

graphical, 136

linear, 132, 134

Percents, 18

Perfect squares, 300

Perfect-square trinomials, 250, 266, 278

Perpendicular lines, 128, 130

Pi, 43

Points

collinear, 108

finding equations from, 122–125

on a line, 108

ordered pairs, 92

Point-slope form, 118, 120, 122

Polygons. *See also* Triangles

parallelograms, 131

quadrilaterals, 130

similar, 366

trapezoids, 130

Polynomials. *See also* Factoring

addition and subtraction, 234–239

cubic, 346

degree of, 230

descending order of, 230

division, 252–255

FOIL method, 248

in geometry formulas, 232

leading coefficient of, 230

leading term of, 230

multiplication, 246–251

quotients of, 342, 354

solving by factoring, 280–285

special products, 250, 266–269

types of, 230

Power functions, 226–229

Power-of-a-Fraction Property of Exponents, 222

Power-of-a-Power Property of Exponents, 220, 304, 412

Power-of-a-Product Property of Exponents, 220, 242, 266, 304

Prime factorization, 260–262

Prime numbers, 260–261, 406

Principal, 13

Principal square root, 300

Problem-solving plans, 74, 308. *See also* Solving equations

Problem-Solving Strategies

choose a strategy, 74

guess and check, 16, 86

look for a pattern, 330

look for key words, 26, 30, 34, 54, 142, 152, 158, 214

make a diagram, 74, 180, 260, 316

make a graph, 100, 394

make an organized list, 16, 208, 270, 272

make a table, 14, 94, 116, 134, 136, 168, 200, 260, 286, 288, 316, 334, 336

problem-solving plans, 74, 308

understand the problem, 74

use a formula, 168, 194, 208, 336, 376, 414

write an equation, 30, 34, 74, 80, 136, 194, 226, 228, 316, 376, 394

write an inequality, 152, 158, 162, 174, 214

Product Property of Exponents, 220, 246, 304, 412

Product Property of Square Roots, 302, 304

Products, 4

Projectile paths, 229, 256, 294, 334, 336

Proofs. *See* Mathematical proofs

Proportions, 104, 372, 384

Pure imaginary numbers, 310

Pythagorean Theorem, 308, 316, 394

Quadrants, 92

Quadratic equations

applications of, 294, 308

completing the square, 312, 314, 316

degree of, 230

factoring, 270, 272, 274

of the form $a(x - r)^2 = s$, 312

free-falling objects, 334, 336

graphing, 286, 288, 290, 292

imaginary roots, 310

logical reasoning and, 324, 326

with no real roots, 320

in rational expressions, 342

real roots, 320

taking the square root, 300, 302, 324, 326

trinomials, 270, 272, 274, 276, 280

using the quadratic formula, 318, 322–324

using the Zero-Product Property, 280, 282, 284, 324, 326

Quadratic formula, 318, 320, 322, 324

Quadratic patterns, 328–333

Quadratic trinomials, 230, 270, 272, 274, 276, 280

Quadrilaterals, 130–131

Quotient Property of Exponents, 222, 242

Quotient Property of Square Roots, 302, 404

Quotients, 4, 222, 368. *See also* Division

Radical expressions, 410, 412

Radical signs, 300, 390, 410

Radicand, 390

Radius of a circle, 308

Range, 96, 98, 100, 102, 290, 390

Rate of change, 110

Rate problems, 102, 110, 376, 418

Ratio, 102, 106

Rational equations, 372–377

Rational exponents, 304, 410, 412, 414

Rational expressions

addition, 356–361

definition, 342

domains, 342

in geometry, 366

monomial denominators, 358

with more than one variable, 348

multiplication and division, 350–355

polynomial factoring, 346

quotients, 368

simplification, 344–349

subtraction, 362–367

Rational functions, 342

Rationalizing the denominator, 404, 408

Rational numbers, 40, 42

Real numbers

adding, 46, 48, 52

multiplying, 56

number lines, 44

properties of, 398–403

subtracting, 50, 52

types of, 40, 42

Reciprocal functions, 380

Reciprocal, 8, 58

Reflexive Property of Equality, 18

Relations, 96, 98

Relatively prime numbers, 261

Repeating decimals, 40

Replacement sets, 14

Right triangle, 130

Rise, 106, 110

Roots, 280, 288, 306, 318, 320

Run, 106, 110

Satisfying equations, 14

Scalene triangles, 87

Scientific notation, 224

Second differences, 328

Sequences, 132

Simplest radical form, 302

Simplification

complex fractions, 368

correctness of, 82

using the Distributive Property, 66, 68

fractions, 40

inequalities, 166, 182

multistep equations, 78

using the order of operations, 10

numerical expressions, 10

polynomials, 246, 278

rational expression, 344–350, 354, 356, 358, 362, 364

real numbers, 52

square roots, 302, 306, 310, 404

Slope-intercept form, 114–115, 118, 122, 124, 126, 202

Slope of a line, 106, 108–110, 114–115, 126, 128

Solution region, 206

Solutions, 14

Solution to a system, 188. *See also* Systems of linear equations

Solving equations

absolute-value, 176

combining like terms, 66

cubic, 284

by elimination, 196, 198, 200, 202

by factoring, 280–285

inverse-variation, 382

isolating the variable, 24

judging correctness, 82

logical arguments, 84

multistep, 76–81

problem-solving plans, 74

quadratic, 280–285, 310

rational, 372–377

involving real numbers, 62

replacement sets, 14

for a specified variable, 72

square-root, 396

by substitution, 190, 192

systems, 188, 190, 192, 194, 196

in two steps, 70

in two variables, 94

using addition, 20, 24

using division, 30, 32

using multiplication, 28, 32

using subtraction, 20, 22, 24

Solving inequalities

absolute-value, 178

addition and subtraction, 144, 148, 152, 162, 164

compound, 170–175

multiplication and division, 154, 156, 162

with no solution, 166, 172

number lines, 144

one-step, 160

systems, 210–215

Special quadrilaterals, 130–131

Splitting the middle term, 274

Square roots, 300–305, 324, 326, 416–421

Square-root equations, 392–397, 416, 418

Square-root expressions, 390, 396, 400, 402, 404, 406, 408, 416

Square-root functions, 390, 418, 420

Standard form

of a cubic equation, 284

of a linear equation, 122

of a linear inequality, 212

of a quadratic equation, 280, 282

Statements, mathematical, 4–7, 16

Stopping distance, 228, 418–419

Subset, 42, 398

Substitution method, 190, 192

Substitution Principle, 12–13

Subtraction

Distributive Property, 66

polynomials, 236, 238, 268, 274

rational expressions, 362–367

real numbers, 50

solving inequalities, 146, 150, 152, 162, 164

square-root expressions, 398, 406

Subtraction Property of Equality, 22–24, 62, 70, 76

Subtraction Property of Inequality, 150, 160

Sums, 4. *See also* Addition

Surface areas

cube, 244

cylinder, 232, 256

rectangular box, 12, 245

Symbols. *See* Mathematical symbols

Symmetric Property of Equality, 18, 20, 24, 34, 76, 84

Systems of linear equations

classification of, 202–205

consistent/inconsistent, 202–205

dependent/independent, 204

solving by elimination, 196, 198, 200, 202

solving by graphing, 188

quadratic patterns, 332

solving by substitution, 190, 192, 194

Systems of linear inequalities, 210–215

T

Terminating decimal, 40

Term of a polynomial, 230

Term of a sequence, 132

Tolerance, 180

Transitive Property of Deductive Reasoning, 84

Transitive Property of Equality, 18

Transitive Property of Inequality, 146

Trapezoids, 130

Triangles

  Pythagorean theorem, 308, 316, 394

  right, 130, 308

  similar, 366

  types of, 87

Trinomials

  perfect-square, 250, 266, 278

  quadratic, 230, 270, 272, 274, 276, 280

Try This, 4, 6, 8, 10, 12, 14, 16, 18, 20, 22, 24, 26, 28, 30, 32, 34, 40, 42, 44, 46, 48, 50, 52, 54, 56, 58, 60, 62, 64, 66, 68, 70, 72, 74, 76, 78, 80, 82, 84, 86, 92, 94, 96, 98, 100, 102, 104, 106, 108, 110, 112, 114, 116, 118, 120, 122, 124, 126, 128, 130, 132, 134, 136, 142, 144, 146, 148, 150, 152, 154, 156, 158, 160, 162, 164, 166, 168, 170, 172, 174, 176, 178, 180, 182, 188, 190, 192, 194, 196, 198, 200, 202, 204, 206, 208, 210, 212, 214, 220, 222, 224, 226, 228, 230, 232, 234, 236, 238, 240, 242, 244, 246, 248, 250, 252, 254, 262, 264, 266, 268, 270, 272, 274, 276, 278, 280, 282, 284, 286, 288, 290, 292, 294, 300, 302, 304, 306, 308, 310, 312, 314, 316, 318, 320, 324, 326, 330, 332, 334, 336, 342, 344, 346, 348, 350, 352, 354, 356, 358, 360, 362, 364, 366, 368, 370, 372, 374, 376, 380, 382, 384, 390, 392, 394, 396, 398, 400, 402, 404, 406, 410, 412, 414, 416, 418, 420

U

Unit rates, 110

Unlike terms, 66

V

Variables, 12, 24, 72, 94, 100, 192

Variation

  constant of, 102, 378, 382

  as the cube, 226

  direct, 102, 104, 226

  inverse, 378–383

  inverse-square, 382

  as the square, 226

Vertex, 286, 290, 294

Vertical-addition format, 234

Vertical lines, 112, 122

Vertical-line test, 96, 380

Vertical-multiplication format, 248, 250

Volume, 244

W

Whole numbers, 40

Work rate, 376

X

x-axis, 92

x-coordinate, 92

x-intercept, 112, 288, 292

Y

y-axis, 92

y-coordinate, 92

y-intercept, 112, 114

Z

Zero, 44

Zero denominator, 374

Zero exponent, 222

Zero-Product Property, 280, 282, 284, 324, 326